Tolerance to Environmental Contaminants

Environmental and Ecological Risk Assessment

Series Editor
Michael C. Newman
College of William and Mary
Virginia Institute of Marine Science
Gloucester Point, Virginia

Published Titles

Coastal and Estuarine Risk Assessment
Edited by Michael C. Newman, Morris H. Roberts, Jr., and Robert C. Hale

Risk Assessment with Time to Event Models
Edited by Mark Crane, Michael C. Newman, Peter F. Chapman,
and John Fenlon

Species Sensitivity Distributions in Ecotoxicology
Edited by Leo Posthuma, Glenn W. Suter II, and Theo P. Traas

**Regional Scale Ecological Risk Assessment:
Using the Relative Risk Method**
Edited by Wayne G. Landis

**Economics and Ecological Risk Assessment:
Applications to Watershed Management**
Edited by Randall J.F. Bruins

**Environmental Assessment of Estuarine Ecosystems:
A Case Study**
Edited by Claude Amiard-Triquet and Philip S. Rainbow

Environmental and Economic Sustainability
Edited by Paul E. Hardisty

Tolerance to Environmental Contaminants
Edited by Claude Amiard-Triquet, Philip S. Rainbow, and Michèle Roméo

Tolerance to Environmental Contaminants

Edited by
Claude Amiard-Triquet
Philip S. Rainbow • Michèle Roméo

CRC Press
Taylor & Francis Group
Boca Raton London New York

CRC Press is an imprint of the
Taylor & Francis Group, an **informa** business

CRC Press
Taylor & Francis Group
6000 Broken Sound Parkway NW, Suite 300
Boca Raton, FL 33487-2742

First issued in paperback 2019

ISBN-13: 978-1-4398-1770-4 (hbk)
ISBN-13: 978-0-367-38311-4 (pbk)

Library of Congress Cataloging-in-Publication Data

Tolerance to environmental contaminants / edited by Michèle Roméo, Philip S. Rainbow, Claude Amiard-Triquet.
    p. cm. -- (Environmental and ecological risk assessment)
    Includes bibliographical references and index.
    ISBN 978-1-4398-1770-4 (hardcover : alk. paper)
    1. Pollution--Environmental aspects. 2. Nature--Effect of human beings on. 3. Environmental toxicology. 4. Threshold limit values (Industrial toxicology) I. Roméo, Michèle. II. Rainbow, P. S. III. Amiard-Triquet, C.

QH545.A1T65 2011
571.9'5--dc22

                                  2010044044

Visit the Taylor & Francis Web site at
http://www.taylorandfrancis.com

and the CRC Press Web site at
http://www.crcpress.com

Contents

SECTION   *Ecological and Ecophysiological Aspects of Tolerance*

SECTION   *Case Studies*

Preface

Tolerance may be defined as the ability of organisms to cope with stress, particularly the chemical stress resulting from the anthropogenic input of one or more of many different toxic contaminants into the environment. Tolerance has been described in many organisms from bacteria to fungi, from phytoplankton to terrestrial flowering plants, and from invertebrates like worms to vertebrates like fish and amphibians. There are two generally agreed methods by which organisms can become tolerant to a toxic contaminant. First, tolerance may be gained by physiological acclimation during the exposure of an individual organism to a sublethal bioavailability of the toxicant; this tolerance is not transferable to future generations. Second, tolerance may also be acquired as a consequence of genetic adaptation in populations exposed over generations to the toxic contaminant, through the action of natural selection on genetically based individual variation in resistance; this tolerance is transferable to future generations. This latter genetic adaptation may be lost in the absence of continuing exposure to the contaminant, again by natural selection, if, as appears to be usual, the genetically based tolerance has a metabolic cost that brings a selective disadvantage in the absence of contaminant. Indeed, the presence of a genetically tolerant population is direct evidence that the bioavailability of the toxic contaminant in the local environment is sufficient to be ecotoxicologically relevant.

So what? In fact, the gaining of tolerance, be it by physiological acclimation or genetic adaptation, can have great consequences for the local biodiversity, and hence the ecology and ecosystem functioning of many of the world's habitats. Contamination by toxicants can lead to decreased production of biological resources, including agricultural or fishery products, and the interruption of key ecological processes, such as decomposition and nutrient cycling. Tolerant species, particularly bacteria in sediments or primary producers like phytoplankton, may play key functional roles in ecosystems. Understanding the frequency of the occurrence of tolerance therefore has tremendous implications for the sustainability of biodiversity and ecosystem functioning. Metabolic processes involved in tolerance are energetically expensive, and thus may interfere with the allocation of energy in an organism, thereby governing the success of reproduction and growth. Reduction of the overall amount of genetic variation in populations exposed to a strong selective toxic pressure can result in increased sensitivity to new stresses in organisms otherwise tolerant to one source of stress. Thus, the adaptive benefit of being tolerant may have negative knock-on effects in the long term. Beyond effects on the crucial ecosystem functioning of the habitats around us, delivering the vital ecosystem services on which we depend (food, clean water, etc.), the acquisition of chemical tolerance may be a more direct source of concern to humans in that it allows the survival of harmful species (insecticide-resistant mosquitoes, antibiotic-resistant pathogenic bacteria) and the presence of highly contaminated links in food webs, including those leading to humans.

The book is an up-to-date compilation of the views of international experts on the phenomenon of tolerance of living organisms to toxic contaminants, usually of anthropogenic origin. The general principles governing the acquisition and biological consequences of tolerance, genetically or physiologically based, are examined at different levels of biological organisation, taxonomically from bacteria and archaea to flowering plants and vertebrates, and within organisms from molecular biology and biochemistry through physiology to whole organism, community, and ecosystem levels of organisation. Thus, part of the book is specifically devoted to mechanisms of defence involved in the acquisition of tolerance to different classes of environmental contaminants, taking into account the limits above which such mechanisms are overwhelmed. Another part of the book examines the ecological consequences of tolerance in terms of both positive (conservation of biodiversity in contaminated environments) and negative (physiological costs of tolerance with consequences on growth and reproduction, transfer of contaminants in the food webs) aspects. The final section of the book considers specific aspects of tolerance that can have major impacts for the environment and for society (tolerance in bacteria, plants, and insects).

Thus, this volume presents a state-of-the-art synthesis of the many aspects of the phenomenon of tolerance to environmental contaminants. Ecotoxicologists have made good progress in the understanding of the mechanisms that allow organisms to cope with pollutants in their environment, but the links with potential effects at higher levels of organisation need to be more strongly established. While the positive effects of tolerance at supra-organismal levels (population, community, ecosystem) for environment and health protection are often considered, the relative importance of any negative effects of tolerance are not typically fully assessed. The reviews offered in each chapter of this book contribute to the provision of tools to carry out relevant risk-benefit analyses in a more informed fashion. From an operational point of view, tolerance must be taken into account when biological responses (biochemical, behavioural, genetic biomarkers) are applied for environmental biomonitoring. Mechanisms of defence may be profitably used as biomarkers, revealing the exposure of organisms to contaminants but within limits that this book helps to define. The problem of over- versus underestimation of risk is also a core question for the development of toxicity reference values. The contaminant exposure history of populations, and whether the local biota have acquired tolerance or not, are clear confounding factors in the interpretation of bioassays that must be understood and taken into account. The reviews presented here can only assist ecotoxicologists to produce more informed and therefore more reliable risk assessments when assessing the ecotoxicological risks to life in any of the contaminated habitats that now surround us in our industrialised society.

We have deliberately sought to put together a synthesis that takes a multidisciplinary approach across contaminant types, habitats, organisms, biological levels of organisation, scientific disciplines, and approaches. The volume presents science at the frontier of research in the subject compiled by international experts from across the world. It is our aim that the book has relevance to environmental scientists and other stakeholders from government to the public. It should also prove invaluable to final-year undergraduate and master's students across the world, and contribute to graduate students in PhD programs, under a wide range of subject heads that include ecotoxicology, ecology, marine and freshwater biology, microbiology, environmental management, and environmental regulation. The book has great relevance, both to readers in developed countries seriously addressing problems of environmental contamination, including North America, Europe, Asia, Australia, and New Zealand, and to those in developing countries with industrial expansion and associated real and potential problems of environmental contamination (Central and South America, Eastern Europe, Africa, India, China, East and Southeast Asia). It is our hope that we have succeeded in our objectives, and that this book serves as an important taking-off point for further understanding of the very wide significance of the phenomenon of tolerance to environmental contaminants.

Claude Amiard-Triquet

Philip S. Rainbow

Michèle Roméo

About the Editors

Dr. Claude Amiard-Triquet is a Research Director in the CNRS (French National Research Center), based at the University of Nantes, France. She was awarded the degree of DSc in 1975, for her research in radioecology at the French Atomic Energy Commission. Dr. Amiard-Triquet's topics of research interest include metal ecotoxicology, biomarkers and, more recently, emerging contaminants (endocrine disruptors, nanoparticles). As the head of multi-disciplinary research programmes, she has managed research collaborations between specialists in organic and inorganic contaminants, and chemists and biologists involved in studies from the molecular to ecosystem levels, with a constant concern for complementarity between fundamental and applied research. Dr. Amiard-Triquet regularly acts as an expert for the assessment of scientific proposals (e.g., the European Framework Program for Research and Development, the International Foundation for Science, and the Sea Grant Administration, Oregon State), and is also in demand as a referee for a dozen or so international journals. She has authored or co-authored more than 170 research papers, and has authored 7 chapters in books. Dr. Amiard-Triquet has also co-authored one book, *La Radioécologie des Milieux Aquatiques* with J.C. Amiard, and co-edited two books, *L'Évaluation du Risque Écologique à l'Aide de Biomarqueurs* with J.C. Amiard and *Environmental Assessment of Estuarine Ecosystems: A Case Study* with P.S. Rainbow. She has given or contributed to more than 90 presentations at international meetings.

Professor Philip Rainbow is the Head of the Department of Zoology at the Natural History Museum, London, leading a staff of more than 100 working scientists in one of the premier museums of the world. He holds the degrees of PhD (1975) and DSc (1994) from the University of Wales. Philip Rainbow was appointed (1994) to a personal chair in the University of London, where he was Head of the School of Biological Sciences at Queen Mary (1995–1997) and is now a Visiting Professor. Professor Rainbow has been an editor of the *Journal of Zoology* and is on the editorial boards of *Environmental Pollution, Marine Environmental Research* and the Journal of the Marine Biological Association, UK. In 2002 Philip Rainbow was invited to give the Kenneth Mellanby Review Lecture by the journal *Environmental Pollution* at the Society of Environmental Toxicology and Chemistry annual meeting at Salt Lake City, Utah. He has more than 200 peer-reviewed publications including 5 co-edited books and two co-authored books. The first (*Biomonitoring of Trace Aquatic Contaminants*, with DJH Phillips) went to two editions. The second has recently been published (2008) by Cambridge University Press, co-authored with Professor Sam Luoma — *Metal Contamination in Aquatic Environments: Science and Lateral Management.* Philip Rainbow's recent research has focused on the factors affecting the bioavailability of trace metals to aquatic invertebrates from both solution and the diet, and the biodynamic modelling of trace metal bioaccumulation.

Michèle Roméo earned her PhD at the University of Nice–Sophia Antipolis in 1975 and has belonged to the French National Institute for Health and Medical Research (INSERM) since that time. She has been working as a researcher in the Marine Biological Station of Villefranche-sur-Mer and, since 1992, in the University of Nice–Sophia Antipolis and in the laboratory ROSE (Responses of Organisms to Environmental Stress), which became ECOMERS (Ecology of Marine Ecosystems and Response to Stress), where she is presently the head of the ecotoxicology team. Her general field of research concerns the response of aquatic organisms to chemical stress (metals and persistent organic pollutants) considered in terms of biomarkers (biomarkers of oxidative stress and of general damage leading to behavioural alterations). The chosen models are bivalve molluscs: mussels and clams and their larvae. Since 2004, research has evolved to the cloning of some genes coding for proteins used as biomarkers and to the use of a DNA microarray technique (genomics) and proteomics. These last two techniques allow measuring of the simultaneous expression of hundreds of genes (genomics) or proteins (proteomics) with very high sensitivity.

Roméo has been the head of the research project PNETOX with the French Ministry of the Ecology and Sustainable Development, and participated in another directed by Dr. Amiard-Triquet (University of Nantes, France). She has authored and coauthored around one hundred papers. She has participated in writing chapters for several books (for Elsevier, Lavoisier Tec and Doc, Taylor and Francis, and Humana Press). She is referee for more than ten journals. Her teaching activities concern ecotoxicology lectures in the master of applied biology of the University of Corsica and the master of environment management and sustainable development of the University of Nice–Sophia Antipolis.

Contributors

Claude Amiard-Triquet
CNRS
Université de Nantes
Nantes, France
Claude.Amiard-Triquet@univ-nantes.fr

Maura Benedetti
Dipartimento di Biochimica, Biologia e
    Genetica
Università Politecnica delle Marche
Ancona, Italy
m.benedetti@univpm.it

Brigitte Berthet
ICES
Université de Nantes
Nantes, France
Brigitte.Berthet@univ-nantes.fr

Françoise Bringel
Génétique Moléculaire, Génomique,
    Microbiologic
CNRS
Université de Strasbourg
Strasbourg, France
francoise.bringel@unistra.fr

Anne Créach
Laboratoire de Génétique et Evolution
    des Populations Végétales
CNRS
Université des Sciences et Technologies
    de Lille
Villeneuve d'Ascq, France
anne.creach@univ-lille1.fr

Gautier Damiens
Laboratory of Ecotoxicology
University of Le Havre
Le Havre, France
gautierdams@gmail.com

Hélène Frérot
Laboratoire de Génétique et Evolution
    des Populations Végétales
CNRS
Université des Sciences et Technologies
    de Lille
Villeneuve d'Ascq, France
Helene.Frerot@univ-lille1.fr

Murad Ghanim
Department of Entomology
Agricultural Research Organization
The Volcani Center
Bet Dagan, Israel
ghanim@volcani.agri.gov.il

Maria Elisa Giuliani
Dipartimento di Biochimica, Biologia e
    Genetica
Università Politecnica delle Marche
Ancona, Italy
m.e.giuliani@univpm.it

Christiaan Hummel
Vrije Universiteit
Amsterdam, The Netherlands
cahummel@home.nl

Herman Hummel
Netherlands Institute of Ecology
Centre for Estuarine and Marine
    Ecology
Yerseke, The Netherlands
H.Hummel@nioo.knaw.nl

Gwenaël Imfeld
Laboratory of Hydrology and
    Geochemistry of Strasbourg
CNRS
Université de Strasbourg
Strasbourg, France
gwenael.imfeld@engees.u-strasbg.fr

Isaac Ishaaya
Department of Entomology
Agricultural Research Organization
The Volcani Center
Bet Dagan, Israel
vpisha@volcani.agri.gov.il

Emma L. Johnston
Evolution and Ecology Research Centre
School of Biological, Earth and
    Environmental Sciences
University of New South Wales
Sydney, Australia
e.johnston@unsw.edu.au

Patrick de Laguérie
Laboratoire de Génétique et Evolution
    des Populations Végétales
CNRS
Université des Sciences et Technologies
    de Lille
Villeneuve d'Ascq, France
Patrick.Laguerie@univ-lille1.fr

Kenneth M. Y. Leung
The Swire Institute of Marine Science
    and School of Biological Sciences
The University of Hong Kong
Hong Kong, China
kmyleung@hkucc.hku.hk

Priscilla T. Y. Leung
The Swire Institute of Marine Science
    and School of Biological Sciences
The University of Hong Kong
Hong Kong, China
ptyleung@hkucc.hku.hk

Samuel N. Luoma
John Muir Institute of the Environment
University of California at Davis
Davis, California
and
The Natural History Museum
Department of Zoology
London, United Kingdom
snluoma@ucdavis.edu

Claire-Lise Meyer
Laboratoire de Génétique et Evolution
    des Populations Végétales
CNRS
Université des Sciences et Technologies
    de Lille
Villeneuve d'Ascq, France
claire-lise.meyer@ed.univ-lille1.fr

Christophe Minier
Laboratory of Ecotoxicology
University of Le Havre
Le Havre, France
minier@univ-lehavre.fr

Bernard Montuelle
Cemagref
Unité Milieux Aquatiques, Ecologie et
    Pollution
Lyon, France
bernard.montuelle@cemagref.fr

Catherine Mouneyrac
CEREA
Université Catholique de l'Ouest
Nantes, France
catherine.mouneyrac@uco.fr

Maxime Pauwels
Laboratoire de Génétique et Evolution
    des Populations Végétales
CNRS
Université des Sciences et Technologies
    de Lille
Villeneuve d'Ascq, France
Maxime.Pauwels@univ-lille1.fr

Philip S. Rainbow
The Natural History Museum
Department of Zoology
London, United Kingdom
p.rainbow@nhm.ac.uk

Francesco Regoli
Dipartimento di Biochimica, Biologia e
  Genetica
Università Politecnica delle Marche
Ancona, Italy
f.regoli@univpm.it

Michèle Roméo
Faculté des Sciences
Université de Nice–Sophia Antipolis
Nice, France
romeo@unice.fr

Pierre Saumitou-Laprade
Laboratoire de Génétique et Evolution
  des Populations Végétales
CNRS
Université des Sciences et Technologies
  de Lille
Villeneuve d'Ascq, France
pierre.saumitou-laprade@univ-lille1.fr

Adam Sokolowski
University of Gdansk
Institute of Oceanography
Gdynia, Poland
adams@ocean.univ.gda.pl

Ahmed Tlili
Cemagref
Unité Milieux Aquatiques, Ecologie et
  Pollution
Lyon, France
ahmed.tlili@cemagref.fr

Stéphane Vuilleumier
Génétique Moléculaire, Génomique,
  Microbiologie
CNRS
Université de Strasbourg
Strasbourg, France
vuilleumier@unistra.fr

Judith S. Weis
Department of Biological Sciences
Rutgers University
Newark, New Jersey
jweis@andromeda.rutgers.edu

Sander Wijnhoven
Netherlands Institute of Ecology
Centre for Estuarine and Marine
  Ecology
Yerseke, The Netherlands
s.wijnhoven@nioo.knaw.nl

Isaac Wirgin
Nelson Institute of Environmental
  Medicine
New York University School of
  Medicine
Tuxedo, New York
Isaac.Wirgin@nyumc.org

1 Pollution Tolerance

From Fundamental Biological Mechanisms to Ecological Consequences

Claude Amiard-Triquet

CONTENTS

1.1 INTRODUCTION

Tolerance may be defined as the ability of organisms to cope with stress, either natural, such as temperature changes, salinity variations, oxygen level fluctuations, and plant toxins, or chemical, depending on anthropogenic inputs of many different classes of contaminants into the environment. Resistance is frequently used in the scientific literature as a synonym for tolerance. Several authors have tried to clarify these terms (Lotts and Stewart 1995; Morgan et al. 2007). However, the definitions they proposed were strongly different, and none of them is currently generally adopted. In this book, most of the authors use the term *tolerance* in acceptance of the

1

general definition above. However, the use of the term *resistance* has been preferred here by some authors, particularly those interested in the genetic basis of an organism's ability to survive in a contaminated environment. In these cases, the authors will clearly specify their choice of terminology in their chapters.

Tolerance has been described in many taxonomic groups exposed to toxicants at sublethal levels. Tolerance may be achieved by many biological processes responsible for physiological acclimation or genetic adaptation. Among tolerant species, some have a key role in ecosystems. Understanding the frequency of occurrence of tolerance has tremendous implications for the sustainability of biodiversity. Processes involved in tolerance are energetically expensive, and thus may interfere with the allocation of energy, thereby governing the success of reproduction and growth. Reduction of the overall amount of genetic variation in populations exposed to a strong selective pressure can result in increased sensitivity to new stresses in tolerant organisms. Thus, the adaptive benefit of being tolerant may have negative counterparts in the long term. On the other hand, tolerance may be a source of concern in that it allows the survival of harmful species (mosquitoes, pathogenic bacteria) and the presence of highly contaminated links in food webs. From an operational point of view, mechanisms involved in tolerance may be a source of biomarkers of exposure. On the other hand, the history of experimental populations, either tolerant or sensitive, may be a confounding factor in the interpretation of bioassays. This volume brings together reviews on these several aspects of the tolerance of organisms to pollutants, with the ultimate aim of understanding the ecological consequences of such tolerance.

1.2   HOW MAY TOLERANCE BE ASSESSED?

The existence of tolerance in a given species or in one or more of its constituent populations may be revealed in many different biological responses to stress, the common feature being that higher levels of stress are necessary to induce an impairment of response in tolerant organisms than in their nontolerant (more sensitive) counterparts. Comparative survival to acute toxicity doses has been used in a number of studies with various species and contaminants, such as in the case of metal exposure of the worm *Nereis diversicolor* (Ait Alla et al. 2006 and literature cited therein), and different fish species (Lotts and Stewart 1995; Hollis et al. 1999; Chowdhury et al. 2004). Differential survival has also been observed in organisms exposed to organic compounds, for instance, in the decapod crustacean *Palaemonetes pugio* exposed to fluoranthene (Harper-Arabie et al. 2004) or the European eel *Anguilla anguilla* exposed to pesticides (Peña-Llopis et al. 2001, 2003).

Less harsh and simplistic experimental approaches to detect the presence of tolerance are frequently preferred, based on sublethal doses of exposure. Under these conditions, the toxicological parameters of interest may be measured in the medium or long term and include longevity or functional impairments. For instance, in the crustacean *Daphnia magna* exposed to the herbicide molinate for two generations, Sánchez et al. (2004) have observed increased longevity and reproduction in specimens belonging to the second generation compared to their parents. In the fish species *Catostomus commersoni*, the fertilisation rate and the quality of gametes were

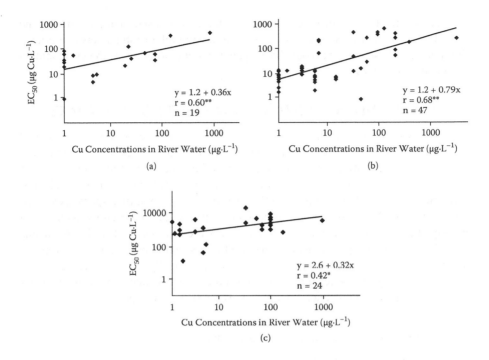

FIGURE 1.1 Cu concentrations in riverine waters giving 50% inhibition of photosynthesis in microalgae belonging to three different taxa: (a) Cyanophyceae, (b) Bacillariophyceae, and (c) Chlorophyceae (Chlorococcales). *, significant at the 95% level; **, significant at the 99% level. (After Takamura, N., et al., *J. Appl. Phycol.*, 1, 39–52, 1989.)

used to reveal tolerance to metals in field populations living in an area impacted by mining activities (Munkittrick and Dixon 1988). Burlinson and Lawrence (2007) based a comparison of tolerant and nontolerant populations of the worm *N. diversicolor* on behavioural disturbances.

In primary producers, growth is commonly used as a toxicological parameter to determine the noxious effects of contaminants. In addition, photosynthesis inhibition is considered a valuable measure to compare tolerance between taxa or populations within the same species, as exemplified in Figure 1.1. Takamura et al. (1989) have determined the effective concentrations of copper that reduce photosynthesis (EC_{50}) in many different strains of microalgae originating from rivers that are relatively clean or impacted by mining activities.

In any case, when the genes for resistance are well known, tolerance may be assessed by quantifying these genes. In populations or communities exposed to pollution in their environment, an increase of the tolerant genes is expected. However, high gene flow from neighbouring populations less exposed to contamination may be responsible for a moderate increase of the suspected tolerant genes, as suggested, for instance, in the European flounder, *Platichthys flesus*, along the French Atlantic Coast (Marchand et al. 2004). In some cases, for instance, in bacteria, the genes involved in resistance are well known, such as the merA gene, which encodes for a protein involved in the reduction of the toxic mercuric ion Hg^{2+} into the volatile and less toxic elemental mercury Hg⁰. Thus, in water, soil, and sediment, the merA gene is a relevant model in ecological studies for assessing relationships between microbial mercury resistance and bioavailable mercury contamination of the environment (Ramond et al. 2008).

1.3   INTER- AND INTRASPECIFIC VARIABILITY OF TOLERANCE

The existence of tolerant versus sensitive species may be exemplified in many taxa from microorganisms to invertebrates and vertebrates. It is necessary to keep this in mind when extrapolating from ecotoxicological data determined in a small number of standard species in the laboratory to the huge number of animals in the real environment. In addition to this inherent tolerance, populations within a species, exposed to environmental contaminants, can develop tolerance, based on either acclimation or adaptation. Among specialists, *acclimation* is the term used when the organisms are able to cope with pollutants as a result of nonheritable physiological processes. When the mechanisms involved in tolerance are genetically based and can be transferred to the progeny, the term *adaptation* is preferred. The latest thinking on physiological acclimation and genetic adaptation is reported in Chapter 2.

1.3.1   INTERSPECIFIC VARIABILITY OF TOLERANCE

In microalgae, the effects of metals have been studied considering many species belonging to different taxa, some of them represented by a consistent number of species (Figure 1.1). Among Chlorophyceae, the EC_{50} for copper was generally between 100 and 10,000 $\mu g \cdot L^{-1}$ in river waters with low levels of copper, whereas it was between 1 and 100 $\mu g \cdot L^{-1}$ for species belonging to Cyanophyceae and Bacillariophyceae. For cadmium and zinc, the differences of sensitivity between species belonging to these taxa also differed by orders of magnitude (Takamura et al. 1989).

In invertebrates, it has been demonstrated that important differences occur even between species belonging to a restricted taxonomic group—the bivalves—which in addition share the same mode of feeding (filter feeders). The oyster *Crassostrea gigas* and the scallop *Chlamys varia* are strong accumulators of silver, whereas mussels from both freshwater (*Dreissena polymorpha*) and seawater (*Mytilus edulis*) are weak accumulators. Among the strong bioaccumulators, one of the species is tolerant (*C. gigas*), whereas the other is sensitive (*C. varia*). Similarly, among weak accumulators, the freshwater mussel is tolerant, whereas the marine mussel is sensitive (Berthet et al. 1992). In addition, the impairments are very different for each species, being limited to metabolic effects in oysters, whereas in marine mussels, gill cells were damaged, and in scallops, byssus secretion was inhibited with consequences on behaviour, leading eventually to death.

Biomonitoring programmes implemented after the numerous oil spills that have impacted coastal areas have revealed consistent features concerning the relative sensitivity of invertebrate taxa to petroleum hydrocarbons (Gómez Gesteira and Dauvin 2000). Because there has been a very low impact of the spills on polychaetes, but a high one on amphipod crustaceans, these authors have proposed the use of a polychaete/amphipod ratio to monitor temporal changes of macrofauna in soft-bottom communities. Similarly among species from the meiobenthos, nematodes show a greater tolerance to petroleum hydrocarbons (and also to hypoxia) than copepods.

In invertebrates, interspecific differences in tolerance are generally attributed to the variety of the processes developed to cope with stress. In vertebrates, regulation physiology is better developed and less diversified, with a greater homogeneity of physiological strategies allowing survival in a contaminated environment. However, even in fish, interspecific differences are well documented, such as differences in the behavioural responses of marine teleosts to the presence of copper (Scarfe et al. 1982), or differences in the toxicity of organophosphorus pesticides to freshwater fish, which can reach one to two orders of magnitude (Keizer et al. 1995). Oliveira et al. (2007) have carried out an extensive comparison of the responses of more than twenty species of marine fish to methyl paraoxon with a view to the use of acetylcholinesterase activity (an enzyme activity that is affected by exposure to organophosphorus and carbamate pesticides) as a pesticide biomarker. Their results suggest a possible evolutionary linkage of AChE sensitivity to methyl paraoxon.

Interspecific variability of tolerance is at the basis of the pollution-induced community tolerance (PICT) concept proposed by Blanck et al. (1988) and revisited in this book (Chapters 4 and 14). A biological community is composed of different species, the inherent sensitivity of which toward a given toxicant is highly variable, as documented above. Thus, in a contaminated environment, the most sensitive organisms are lost as a consequence of pollutant pressure, whereas tolerant organisms are maintained. Consequently, the new community as a whole is more tolerant to the toxicant responsible for selection than another community, initially identical, but which has never been exposed to this toxicant. Such a PICT has been demonstrated in many studies on periphyton (e.g., Blanck et al. 1988, 2003) and nematodes (Millward and Grant 1995, 2000), and the same philosophy is behind the use of lichen communities in the monitoring of air pollution (Hawksworth and Rose 1976, quoted in Millward and Grant 1995).

1.3.2 Tolerance Acquired in Populations Previously Exposed to Pollutants

In addition to the tolerance characteristics of different species, it is well established that, within the same species, populations previously exposed to chemicals in their environment are able to cope more efficiently than "naive" individuals. Carbamate resistance in mosquitoes was described as early as 1966 (Georghiou et al. 1966), and the importance of insecticide resistance has been recognised for ecological and evolutionary aspects as well as for management (McKenzie 1996; Denholm et al. 1999; Ishaaya 2001; Hemingway et al. 2004; Coleman and Hemingway 2007; Labbé et al. 2007). For decades, plant tolerance to herbicides (LeBaron and Gressel 1982) and metals (Shaw 1989) has been recognised, and because tolerant plants are often strong accumulators, their role in remediation of a metal-contaminated environment has given rise to many studies (Li 2006). Among the best known examples, the resistance of bacteria to antibiotics (and other chemicals) (Ramos et al. 2002; De 2004) has been a topic of major interest because of the consequences for human health (limiting the efficiency of many drugs) and ecosystem functioning (allowing subnormal biogeochemical cycles and also remediation) (De 2004). Individual chapters in this book are devoted to the responses of bacteria (Chapter 14), metal-tolerant plants (Chapter 15), and resistant insects (Chapter 16). Metal tolerance in aquatic organisms is well established after a review by Klerks and Weis (1987). More recently, Amiard-Triquet et al. (2008) drew attention to acquired tolerance in microalgae and Cyanobacteria after exposure to metals, polychlorinated biphenyls (PCBs), and different classes of pesticides; in different taxa of annelids exposed to several metals; in crustaceans exposed to metals and pesticides; and in fish exposed to metals, PCBs, polychlorinated dibenzo-p-dioxins (PCDDs), and polyaromatic hydrocarbons (PAHs).

Many studies on contaminant effects focus on acute exposures and short-term effects. In these studies, the detoxificatory processes, which will be reviewed below, are immediately overwhelmed (see Figure 1.3) and do not allow organisms to cope with contaminants as they do in the field. However, at sublethal doses, these mechanisms are functional, and many of them respond on the scale of days or weeks. For instance, in worms *Nereis diversicolor* exposed for ninety-six hours to sediments from reference and contaminated sites in the Mira and Sado estuaries (Portugal), Moreira et al. (2006) examined a number of biochemical parameters involved in detoxification processes, antioxidant defence, and an indicator of oxidative stress. Even after this relatively short duration of exposure, compensation mechanisms were already activated, enzyme activities showing generally higher values for worms from the contaminated sites. Preexposure to chemicals is responsible for the induction of detoxification processes, subsequently allowing a different response between preexposed and naive individuals, as shown, for instance, in the mussel *Perna viridis* exposed to cadmium for only one week (Ng et al. 2007). Lotts and Stewart (1995) have demonstrated that minnows were able to acclimate to total residual chlorine (TCR) within a short period of exposure (seven to twenty-one days). Their study may explain why several species of fish were observed in aquatic systems where TCR concentrations were large enough that fish kills would have been expected. Previous papers reported by these authors indicated similar fast acclimation to TCR in crayfish or copepods. However, the protection provided by acclimation is limited since de-acclimation of minnows can also occur over a short period (seven days). In the fish *Catostomus commersoni* living in copper- and zinc-contaminated lakes, the tolerance of larvae at the yolk-supported stage is lost when larvae begin feeding, twenty-four days after hatching (Munkittrick and Dixon 1988).

When animals are chronically exposed to contaminants in the field or over several generations in the laboratory (particularly in this case, when many generations are produced over a short duration, as known for microorganisms or small invertebrates like *Daphnia*, mosquitoes, etc.), there is a possibility that they acquire tolerance as a result of genetic adaptation. This phenomenon represents a protection both for individuals living in an impacted environment and for their progeny, allowing the durability of local populations in contaminated environments.

1.3.3 CHOICE OF TESTS ORGANISMS AND SENTINEL SPECIES

Sentinel species may be used for different objectives, including (1) risk assessment based on bioassays developed in agreement with national and international regulatory bodies; (2) monitoring of biodiversity, including the use of indicator species, the presence or absence of which reveals environmental changes at work; and (3) the use of sensitive and early biological responses (biomarkers) able to reveal noxious effects in biota well before local extinction of species occurs (Berthet 2008). In all these three cases, intra- or interspecific tolerance must be considered for a relevant choice of sentinel species (Chapter 3).

To assess environmental risk in habitats exposed to chemicals, the following procedure is often recommended. After bioassays on a number of species (for instance, the European Community Water Framework Directive recommends the use of (1) algae or macrophytes, (2) *Daphnia* or marine organisms, (3) fish) have been carried out, security factors are applied and guidelines are established, indicating the maximum concentrations (in water, sediments or soils, food species) below which the environmental quality is considered good. These guidelines are published for individual contaminants, neglecting the fact that in natural environments, living organisms are generally exposed to complex mixtures of contaminants that can interact with each other, but also that changes in natural conditions can interfere deeply with the ability of biota to cope with chemical stress. For decades, it has been hypothesised that those species that are able to cope with natural stress in their environment (e.g., estuarine species exposed to large and fast fluctuations of salinity, oxygen, and temperature) are able to tolerate additional stress due to chemical inputs linked to anthropogenic activities. This assumption has been revisited, for instance, by Hummel et al. (1997) and Heugens et al. (2001), and their conclusions are not so optimistic. Thus, it is indispensable to discuss the state of the art in this field (Chapter 5).

1.4 MECHANISMS OF DEFENCE INVOLVED IN TOLERANCE TO CHEMICAL STRESS

Biochemical mechanisms allowing aquatic biota to cope with the presence of chemicals in their environment have been recently reviewed (Amiard-Triquet et al. 2008). Many of them are based on processes involved in defence against natural substances. It is particularly evident in the case of metals (Chapter 6) that are normal constituents of the earth's crust and are present everywhere in our environment as traces, some of them (essential metals such as copper or zinc) even being indispensable for a number of vital functions. However, at very high doses, even essential metals can turn toxic, and anyway, this class of contaminants includes nonessential metals (e.g., mercury, lead) that are known only for their toxicity.

Organisms are also exposed to natural organic compounds, such as plant toxins, and a number of species are well equipped to face the challenge of deriving energy from food containing molecules that are highly toxic for others. Some examples include coniferyl benzoate, a secondary metabolite of aspen buds, which are commonly used as food by grouse *Bonasa umbellus* (Guglielmo et al. 1996); terpenoids, alkaloids, and ranunculin present in the plants preferentially consumed by the tortoise *Testudo horsfieldi* (Lagarde et al. 2003); terpenes in eucalyptus, an important food items for possums (Sorensen et al. 2007); etc.

In addition to these biochemical mechanisms, behavioural responses can contribute to the defence of organisms in environments submitted to toxic chemical inputs. Detection of a toxicant (in water, soil or sediment, food) can induce an avoidance response. Avoidance may be a strategy to escape exposure, sometimes in the short term only (e.g., valve closure in bivalves), but in certain cases also in the long term (e.g., food selection).

1.4.1 BIOCHEMICAL MECHANISMS

The cytochrome P450s (CYPs) are a group of enzymes responsible for the oxidative metabolism of a wide range of organic compounds. The CYPs are a well-supported counterdefence mechanism employed by herbivores to metabolise, and subsequently eliminate, ingested plant secondary metabolites (PSMs). Specifically, CYP3A is an important metaboliser of PSMs in a variety of herbivores (Sorensen et al. 2007). Biotransformation of organic xenobiotics such as hydrocarbons, organochlorine insecticides, and polychlorinated biphenyls (PCBs) also involves the cytochrome P450 monooxygenase system (Newman and Unger 2003). Metabolites resulting from this phase I reaction can be acted on in phase II reactions that involve conjugation with endogenous compounds (carbohydrate derivatives, amino acids, glutathione, or sulphate). The biotransformation of a highly lipophilic compound into a more water-soluble metabolite, more prone to elimination, is very often termed detoxification (Newman and Unger 2003). However, it must be kept in mind that biotransformation can also result in the production of reactive compounds that can be responsible for toxicity to cellular macromolecules, leading to toxification instead of detoxification, particularly by producing oxidative stress (Chapter 8).

Oxidative stress, i.e., damage to biomolecules from free oxyradicals (oxygen peroxide (H_2O_2), superoxide radical ($O_2^{\cdot-}$), and hydroxyradical ($^{\cdot}OH$)), is experienced potentially by all aerobic life. Reactive oxygen species (ROS) are produced in organisms as a consequence of processes involving both endogenous and xenobiotic compounds. Because oxyradical-generating compounds are normally produced by aerobic metabolism, organisms are well equipped to face oxidative stress (Chapter 7), being able to produce antioxidants that react with oxyradicals (vitamins C and E, β-carotene, glutathione, etc.) and enzymes that reduce the amount of oxyradicals (superoxide dismutase (SOD), catalase (CAT), glutathione peroxidase) (Van der Oost et al. 2005). However, when antioxidant defences are overcome by prooxidant forces, increased concentrations of free radicals can cause a number of dysfunctions, including lipid peroxidation and changes in the structure and function of biomolecules, including DNA, with potential consequences such as genotoxic effects and increased risk of cancer (Newman and Unger 2003).

Lesions of biomolecules induce stress proteins able to recognise denatured or aggregated proteins that are then unfolded and refolded properly in order to restore their functions. When the damage is too great to be repaired, stress proteins ensure the breakdown and elimination of nonfunctional proteins. Because of their role, stress proteins are also termed *chaperones* (Frydman 2001). They were first recognised in organisms submitted to drastic temperature changes, the reason why they were initially known as heat shock proteins. In fact, such cellular stress responses may be elicited by both physical (temperature, ultraviolet radiation) and chemical (metals, organic xenobiotics) agents. Stress proteins have evolved very conservatively, and their induction has been recognised in many bacterial, plant, and animal species. They appear as parts of a universal process, able to contribute to tolerance in biota exposed to environmental contaminants (Chapter 9).

Because the study of the ecotoxicology of metals has taken advantage of efficient analytical procedures long before they were reliable for organic xenobiotics, the knowledge of adaptive strategies adopted by organisms exposed to metals in their environment is particularly well developed. Briefly, they include the control of metal uptake (for instance, by binding metals at the surface of the cell as a result of the secretion of mucus), the control of intracellular metal speciation (mainly by biomineralisation or binding to detoxification proteins such as metallothioneins in animals and phytochelatins in plants), and the elimination of excess metals (Chapter 6). In the case of organic xenobiotics, little is known about the first process. Contrary to what happens with incorporated metals, detoxification of organic compounds is based mainly on biotransformation, even if certain molecules such as organochlorines are not metabolically active when stored in reserve lipids.

A transmembrane P-glycoprotein (P-gp) has been recognised as responsible for the resistance of some tumour cells to anticancer drugs (multidrug resistance (MDR)). It prevents the accumulation of cytotoxic drugs, by acting as an energy-dependent efflux pump. P-gp-like proteins have been described in many nonmammalian organisms, including even invertebrates (sponges, mussels, oysters, clams, worms). In addition to natural products, environmental contaminants may be translocated, thus preventing cellular accumulation in exposed biota. With reference to MDR, this phenomenon has been termed *multixenobiotic resistance* (MXR). As early as 2000, Bard suggested that the induction of a multixenobiotic defence mechanism may explain why some species are able to face the challenge of surviving in polluted environments. However, it is necessary to examine if it is a general phenomenon and what are the limits of MXR as a protective system against environmental contaminants (Chapter 10). For instance, in the case of metals, the detoxificatory role of MXR is not clearly established. In the Gironde estuary strongly contaminated by metals (Cd, Cu, Pb, Zn), no influence on the expression of a MXR-type system was observed in oysters (Minier et al. 1993). In addition, the protective role of MXR may be counteracted in the presence of inhibitors (emerging contaminants, natural substances produced by certain invasive species) at doses that may be encountered in the natural medium (Smital et al. 2004; Luckenbach et al. 2004). In the Seine estuary, despite enhanced levels of MXR proteins being determined in the freshwater mussel *Dreissena polymorpha*, impaired condition index and decreased lysosomal stability were also observed (Minier et al. 2006).

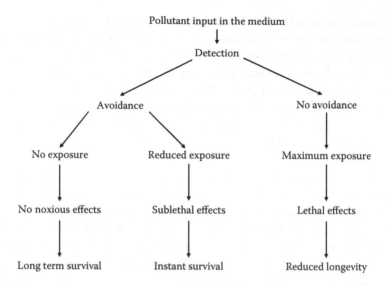

FIGURE 1.2 Different types of animal responses to the presence of contaminants and consequences for the fate of individuals. (After Amiard, J. C., *Océanis Doc. Océanogr.*, 9, 465–80, 1983.)

1.4.2 BEHAVIOURAL RESPONSES

Facing a chemical input in its environment, an animal will react differently, depending on its ability to detect or not a pollutant (Figure 1.2). If the toxic chemical is not detected, no avoidance will take place and the organism will be submitted to maximum exposure with drastic potential effects. If the pollutant is readily detected, the animal will be able to avoid at least partly the contaminated water mass, sediment/soil, or food, thus leading to a reduced exposure with limited effect on long-term survival when the animal leaves the contaminated area.

Avoidance reactions have long been studied, and the capability of avoiding metals and organic compounds is well known in both invertebrates and vertebrates (for a review, see Amiard-Triquet and Amiard 2008). Biological early warning systems have taken advantage of avoidance responses for monitoring water quality, particularly in freshwaters, but some of them are also applicable in estuaries and coastal waters. Fish monitors have been developed for the protection of potable water intakes and are sensitive devices particularly for the detection of a wide range of metals and some polyelectrolytes. In bivalves, the closure of the shell is a typical example of an escape behaviour response under stress. Warning systems have been developed using the valve movement response of both marine mussels and the freshwater mussel *Dreissena polymorpha* (see review by Baldwin and Kramer 1994).

Food selection is an important component of defence against toxic molecules. Plant secondary metabolites (PSMs) are a major constraint to the ingestion of food by herbivores. Recent reviews indicate that herbivores can use diet choice and the rate and amount of PSM consumption to prevent the concentration of PSM in blood from reaching levels able to produce significant adverse effects (Marsh et al. 2006; McLean and Duncan 2006). Avoidance of metal-enriched food has been recognised in the freshwater crustacean *Gammarus pulex* exposed to zinc (Wilding and Maltby 2006). In the terrestrial isopod crustacean *Oniscus asellus*, individuals from a control population ingested artificially metal-enriched leaf litter material at the same rate as unpolluted food. In contrast, a population from a site in the vicinity of a mine was able to distinguish between metal-enriched and clean food, reducing food intake drastically during exposure when only contaminated food was available. In the field, this selective feeding enables *O. asellus* to survive under high metal concentrations (Köhler et al. 2000).

In sediment-dwelling species, different types of environmental contaminants can disturb burrowing behaviour (Amiard-Triquet 2009; Bonnard et al. 2009). Such behavioural impairments have been observed even in organisms (the ragworm *Nereis diversicolor* and the clam *Scrobicularia plana*) exposed to naturally contaminated sediments. Cross experiments between specimens and sediments originating from contaminated versus reference sites have allowed us to distinguish between physiological disturbance and avoidance. In both species, animals are able to recognise contaminated sediments, and in this case, decreased burrowing speeds were registered (Mouneyrac et al. 2010; Boldina-Cosqueric et al. 2010). However, this avoidance behaviour has a poor adaptive value since disturbed burrowing facilitates predation, as demonstrated for the littleneck clam *Protothaca staminea*, allowed to burrow in clean sand or in sand mixed with Prudhoe Bay crude oil. Shallow burial and slow reburrowing in oiled sand were responsible for increased predation as a consequence of increased accessibility of clams to the Dungeness crab *Cancer magister* (Pearson et al. 1981). Nevertheless, when the conditions get particularly harsh, even species considered sedentary can get involved in migrations, such as *N. diversicolor* observed by Essink (1978, reported in Essink 1985) as they escape a confined area after the onset of organic waste discharge.

Anyway, it would be wrong to consider that all living organisms are able to escape polluted environments. The ecological context is indeed very important in determining whether or not pollutant-induced avoidance will occur in the wild. For instance, freshwater fish *Coregonus clupeaformis* avoid metals at low concentrations under standardised conditions of light, but in the case of competing gradients of light and metals, the fish prefer the contaminated shade to the uncontaminated high light intensity, except at the highest concentration tested (Scherer and McNicol 1998).

1.4.3 LIMITS OF DEFENCE MECHANISMS

Mechanisms of defence provide useful biomarkers that have been widely studied in the laboratory and in the field (Van der Oost et al. 2005; Amiard-Triquet et al. 2008). However, low concentrations of metallothioneins or ethoxyresorufin O-deethylase (EROD; a phase I biotransformation enzyme) in contaminant-exposed organisms do not necessarily indicate stress insensitivity. Although many heavy metals have been shown usually to induce metallothioneins in many organisms, including both vertebrates and invertebrates (Amiard et al. 2006), the concentrations of these proteins

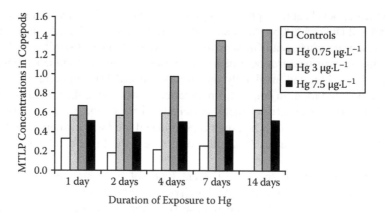

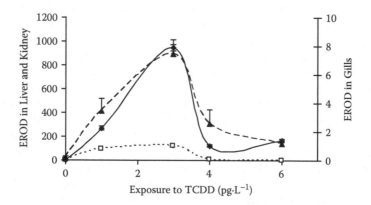

FIGURE 1.3 Different responses of biomarkers along a pollution gradient. Top: Metallothionein-like protein content (MTLP in $\mu g \cdot g^{-1}$ wet weight) in the copepod *Tigriopus brevicornis* exposed for 1 to 14 days to mercury; EROD (ethoxyresorufin O-deethylase in pmol res.$^{-1}$ min.$^{-1}$ mg protein) in kidney (dotted line), gills (dashed line), and liver (solid line) of the fish *Sparus aurata* exposed to TCDD for 20 days. (Data kindly provided by Barka, S., et al., *Comp. Biochem. Physiol.*, 128C, 479–93, 2001; Ortiz-Delgado, J. B., et al., *Ecotoxicol. Environ. Saf.*, 69, 80–88, 2008.)

follow a curve with a maximum at rather high but not extreme concentrations (Figure 1.3). Despite it being considered a reliable marker of exposure to dioxin, PCBs, and PAHs in fish, a similar bell-shaped relationship may be observed for EROD (Figure 1.3). Similar patterns have been mentioned by Dagnino et al. (2007) for GST (a phase II biotransformation enzyme) and catalase (an enzyme involved in antioxidant defence). In response to very high toxicant concentrations, the protein levels or the enzymatic activities decrease, most probably due to severe pathological impact upon target organs as explained by Köhler et al. (2000) in the case of hsp 70. Consequently, interpreting these biochemical indices in terms of organism health is not a simple task.

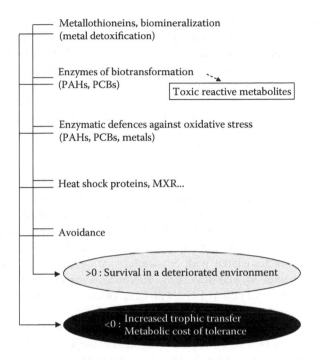

Metallothioneins, biomineralization (metal detoxification)

Enzymes of biotransformation (PAHs, PCBs)

Toxic reactive metabolites

Enzymatic defences against oxidative stress (PAHs, PCBs, metals)

Heat shock proteins, MXR...

Avoidance

>0 : Survival in a deteriorated environment

<0 : Increased trophic transfer Metabolic cost of tolerance

FIGURE 1.4   Some major causes and consequences of tolerance.

1.5   ECOLOGICAL AND ECOPHYSIOLOGICAL ASPECTS OF TOLERANCE

All the mechanisms of defence described above allow the survival of the most toler-ant species and, within each species, the populations that have been able to acclimate or adapt to the presence of environmental contaminants (Figure 1.4). Thus, tolerance has an obvious positive aspect by contributing to the (partial) conservation of biodi-versity—at least species diversity (Chapter 11)—since adaptations are suspected of being a source of reduction of genetic diversity (Chapter 2). On the other hand, the constitutive formation of mucus, enzymes, and other proteins at a high rate would be an energy-intense strategy to counteract the symptoms of toxicity, and would be likely to alter processes of energy allocation with potential consequences on fitness (Chapter 12). Significant modification of a population's genetic composition, includ-ing the selection of resistant genotypes, can reduce mean fitness, erode evolutionary potential, and contribute to the likelihood of local or even global extinction (cited opinion, Laroche et al. 2002). However, because a number of resistant species are able to cope with environmental pollution, they can constitute highly contaminated links in the food web, thus inducing a potential risk of trophic transfer (Chapter 13).

1.5.1   CONSERVATION OF BIODIVERSITY

In some cases, tolerance to chemical stress is efficient only for a short period, for instance, when an animal is able to isolate itself from its medium when it has detected the presence of a toxic chemical. In bivalves, valve closure is an efficient behavioural response at least for short-term exposure, but the bivalves suffer respiration disturbances that can become lethal within days, whereas in the long term, partial closure results in a reduction of food intake and absorbed energy, with potential effects on fitness. In mobile species, avoidance can result in animals' flight out of polluted media, a situation that is beneficial at the individual and species level, but from an ecosystem point of view, it plays the same role as a local extinction, decreasing local biodiversity. Other examples have been cited (see Section 1.2.2) showing a temporary acclimation that permits survival in the short or medium term, for instance, in the cases of a pollution accident or pulse inputs of effluents.

On the other hand, when the acquisition of tolerance results from genetic adaptation in populations that have been exposed for generations to selection pressure due to the presence of toxicants in their environment, it represents an efficient protection for the local population of a species, allowing survival and reproduction, even if tolerance is not an all-purpose remedy (see Section 1.5.2 about the cost of tolerance and major problems associated with cross-tolerance of bacteria to antibiotics and contaminants (Chapter 14) and insecticide tolerance in insects (Chapter 16)). Many, but not all, populations in polluted areas do have an increased tolerance, and in particular, there is more evidence for the evolution of tolerance in microorganisms and small invertebrates than in macrofauna. In addition, Klerks and Weis (1987) have identified a potential bias in the reporting of field examples of tolerance since it is questionable whether negative results have an equal chance of being published in the relevant scientific literature.

Ascending from population to species level, it is clear that acute pollution resulting from accidents will be responsible for the local loss of the most sensitive species. Even in the case of chronic contamination, when chemical stress increases, sensitive species will disappear first, followed by less sensitive species. For instance, in a river impacted by historical mining activities, lotic insect species that store metals in nondetoxified form were rare or absent in the most contaminated sections, whereas tolerant species, which have efficient detoxification strategies, remained present (Cain et al. 2004). As mentioned above (Section 1.2.1), interspecific variability of sensitivity is at the origin of communities showing pollution-induced community tolerance (PICT). PICT is an important phenomenon in the conservation of ecosystem functioning even in areas submitted to toxicant pressure high enough to provoke local species extinction. For instance, in the Seine estuary, one of the largest estuaries in France, diatom assemblages—which are the main constituents of the microphytobenthos on the surface of the mudflats—were consistently different from those observed in a small reference estuary (Sylvestre 2009). Nevertheless, the photosynthetic performances of the two microphytobenthic communities, measured under the same environmental conditions (light and temperature), were quite similar, as was also the gross community primary production in $mgC.m^{-2}.h^{-1}$ (Migné et al. 2007; Amiard-Triquet et al. 2009).

Depending on the functional role of different species in the community, indirect effects of tolerance will appear following different patterns (Fleeger et al. 2003). If a sensitive species is a prey or a host species, its extinction will lead to a depletion of its predator or symbiont populations (Figure 1.5). On the contrary, if a sensitive species

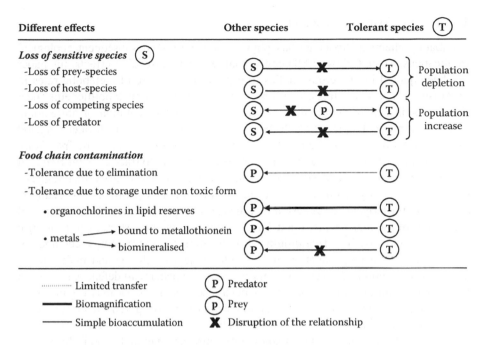

| Different effects | Other species | Tolerant species (T) |

Loss of sensitive species (S)
  -Loss of prey-species
  -Loss of host-species
  -Loss of competing species
  -Loss of predator

} Population depletion

} Population increase

Food chain contamination
  -Tolerance due to elimination
  -Tolerance due to storage under non toxic form
      • organochlorines in lipid reserves
      • metals → bound to metallothionein
                → biomineralised

———— Limited transfer   (P) Predator
———— Biomagnification   (p) Prey
———— Simple bioaccumulation   **X** Disruption of the relationship

FIGURE 1.5 Indirect ecological effects of tolerance. (After Moore, N. W., in *Advances in Ecological Research*, ed. J. B. Cragg, Academic Press, New York, 1967, Vol. 4, pp. 75–129; Amiard-Triquet, C., et al., in *Les biomarqueurs dans l'évaluation de l'état écologique des milieux aquatiques*, ed. J. C. Amiard and C. Amiard-Triquet, Lavoisier, Paris, 2008, pp. 55–94.)

is a competitor or a predator of a tolerant species, the extinction of the former will favour the latter, the populations of which will increase.

Among organisms that have strategies (at the individual, population, or species level) allowing them to cope with chemical stress, some of them are keystone species, with important roles in ecosystem functioning, on which numerous species will depend. Thus, resistant bacteria will be able to intervene in biogeochemical cycling of nutrients. Primary producers using these nutrients will contribute to the normal functioning of food webs, the basis of which they constitute, and so on to successive trophic levels. Earthworms in terrestrial environments or endobenthic organisms in aquatic environments will remain able to rework soil or sediment, an important process for the oxygenation of porewater and fate of organic matter and nutrients. Within a given type of ecosystem, several species able to fulfil the same functional role are often present concomitantly. However, in other media, the natural conditions are harsh enough to limit the number of such equivalent species even in the absence of any anthropogenic stress. For instance, it is well established that estuarine species are much less numerous than freshwater or marine species (McLusky 1989). Thus, in estuaries that are among the most polluted areas across the world, it is evident that the extinction of even a small number of species is sufficient to affect ecosystem function.

1.5.2 THE COST OF TOLERANCE

Combating chemical toxicants has physiological costs and subsequent ecological implications. Since Calow (1991) conceptualised these cascading effects, his paper has been quoted in numerous studies to explain the increase of metabolic rates in organisms exposed to stress, resulting in the induction of metallothioneins, heat shock proteins, biotransformation enzymes, or the implementation of antioxidant defences (Amiard-Triquet et al. 2008). In several species (the freshwater decapod *Palaemonetes paludosus*, the banded water snake *Nerodia fasciata*, the frog *Rana catesbeiana*) differing deeply in their phylogeny, exposure to trace element-rich coal combustion waste induced elevated maintenance costs, a similarity that suggests that this may be a general response to metal contaminants (Rowe 1998). Leung et al. (2000), similarly, have suggested that metabolic depression in cadmium-exposed specimens of the marine gastropod *Nucella lapillus* may be a strategy to minimise the uptake and toxicity of cadmium while meeting the extra energy demand for detoxification (mucus production and metallothionein synthesis) and maintenance (e.g., repair of cellular damage). The energy cost of biochemical defences is an even more general phenomenon since it has been also shown for various organic substances in different taxa (biotransformation of plant toxins or pyrene, respectively, by the grouse *Bonasa umbellus*, studied by Guglielmo et al. (1996), and in hepatocytes of the trout *Oncorhynchus mykiss* studied by Bains and Kennedy 2004). In the fish *Leporinus elongatus*, it appears that a significant part of the energy available under severe O_2 restriction may be directed preferentially to the synthesis of antioxidants at the expense of food consumption and weight gain, which consistently decreased (Wilhelm Filho et al. 2005). Even if the cost of tolerance is well documented, only a few studies have allowed a precise quantification of the fraction of total absorbed energy, which is devoted to defence mechanisms. In addition, when impairments (changes in oxygen consumption, levels of energy reserves, condition indices, growth, reproduction) occur in organisms exposed to chemical stress, separating the relative contribution of the cost of tolerance and direct toxic effects of contaminants is far from clear. Thus, it is crucial to determine the ramifications of pollution-induced modifications in energetics on individual- and population-level parameters before importance can be ascribed to the cost of tolerance (Chapter 12).

Genetic adaptation evolved under the pressure of environmental contaminants can be assimilated to a maladaptation in another context. For instance, recent studies with laboratory lines of fish have shown that increased resistance to chemicals has a cost, consisting of a higher susceptibility to other stress factors, such as UV, hypoxia, or increasing temperatures (Meyer and Di Giulo 2003; Xie and Klerks 2004). This may be a major concern in the context of climate change in the near future.

1.5.3 TROPHIC TRANSFER OF ENVIRONMENTAL CONTAMINANTS

The trophic transfer of environmental contaminants is partly linked to mechanisms involved in the tolerance of prey species (Figure 1.5). For instance, organochlorines stored in lipids have no toxic interference with animal metabolism until reserves are remobilised (for instance, during sexual maturation or migrations over long distances), but during digestion by a predator or human consumer, they can be transferred to the next trophic level. Biotransformation efficiency and the chemical characteristics of the resulting metabolites are also crucial factors controlling the bioaccumulation and biomagnification of lipophilic environmental contaminants along a food chain. When metabolicity is low, or when it is high but leads to the formation of hydrophobic metabolites, high food chain bioaccumulation is likely to occur. When phase I metabolism is important but results in highly polar metabolites, the latter are easily excreted, thus limiting bioaccumulation in the food web (Chapter 13). Metals detoxified in the cytosol by binding to metallothioneins are more prone to assimilation by a predator than biomineralisation products such as crystals of silver sulphide or mercury selenide, which are very stable compounds resistant to degradation during digestion, whereas the fate of metals associated with other inorganic granules, for instance, may be more questionable (Chapter 13).

Fish are considered a very healthy food because they are rich in proteins, poor in saturated fats, and can be protective against coronary heart disease (CHD), since marine fish oils are rich in omega-3, an essential fatty acid, known to reduce CHD risk. However, lipophilic contaminants (such as mercury as methyl mercury, dioxins, PCBs, and brominated flame retardants) with high $K_{ow}s$ are easily bioaccumulated (Chapter 13) and are present at the highest concentrations in the most lipid-rich fish. Intake of these environmental contaminants from fish may counterbalance beneficial effects. Virtually all humans who consume fish and shellfish have at least trace quantities of methyl mercury present in their tissues. Health effects at current levels of exposure in populations showed various effects with most evidence for neurodevelopmental and cardiovascular symptoms. A major outbreak of severe methyl mercury poisoning occurred in Minamata, Japan, caused by the presence of mercury in effluent from a chemical factory that contaminated the surrounding bay and its living resources; led to very high human exposure to methyl mercury via fish consumption; and resulted in severe injuries to people, of whom hundreds died (Harada 1995). Within the past decade there have been clinically obvious cases of mercury poisoning in the Songhua River region of China (Chun et al. 2001, cited by Mahaffey 2004). Artisanal gold mining has been a source of mercury exposure in the Brazilian Amazon and other countries in South America (Venezuela, Colombia, Bolivia, French Guyana, Guyana, Ecuador, and Peru) since the 1980s. Inorganic mercury, subsequently methylated and incorporated into fish, has produced severe exposures and neurological symptoms among people routinely consuming fish from these waters. Many reviews have described the nature and extent of this environmental pollution problem (e.g., Malm 1998; Maurice-Bourgoin et al. 2000). Other important human exposures are well documented among the Faroese population and the Inuit of Nunavik, as a consequence of the consumption of whale meat in addition to fish (Dewailly 2004).

Still in the aquatic environment, the presence of dioxins and other persistent organic pollutants (POPs) in farmed salmon received great media coverage (Hites 2004). The contamination of prey species, particularly seafood, able to tolerate high levels of POPs is also at the origin of increasing levels of these pollutants in the polar bear, a talisman species for conservation, with possible effects at the population level (Derocher et al. 2003).

Itai-itai disease (IID) was officially recognised in 1968 as the first disease induced by environmental pollution in Japan. IID was found initially in the cadmium-polluted Jinzu River basin in Toyama Prefecture. The patients of IID suffered from renal anemia, tubular nephropathy, and osteopenic osteomalacia. The degree of pollution in different parts of the endemic area, determined by analysing Cd concentrations in the soils of the paddy fields, was correlated with the prevalence of IID, whereas no phytotoxicity occurred; thus, rice was able to play the role of a pollution vector.

Another talisman species is the bald eagle, officially adopted as the U.S. national emblem on June 20, 1782. Use of the pesticide DDT after World War II poisoned eagles' foods and weakened eggshells, making them too thin to support the weight of brooding parents. A 1972 ban on DDT led to gradual improvements in the bald eagle population. Bald eagles eat fish, waterfowl, and small to medium-sized mammals. As supercarnivores, they are at the top of the food chain and particularly exposed to lipophilic POPs such as DDT, bioaccumulated to a lesser degree, and therefore tolerable in their prey species. According to the U.S. National Parks Conservation Association (http://www.npca.org), PCBs are emerging as another chemical threat due to food chain contamination, but nonchemical threats such as habitat loss resulting from development in coastal areas must not be ignored.

1.6 CONCLUSIONS

It is important to establish the state of the art in the field of tolerance to environmental contaminants because of many important aspects for science, management, and society. In 1987, Klerks and Weis wrote: "It seems dangerous to relax water quality criteria on the assumption that all populations in polluted environments will evolve an increased resistance" (p. 173). Is it now possible to go further in risk assessment and the implementation of toxicity reference values by regulation bodies, taking into account our knowledge about tolerance? Ecotoxicologists have made good progress in the knowledge of mechanisms, allowing organisms to cope with pollutants in their environment, but the links with potential effects at higher levels of organisation need to be more strongly established (Amiard and Amiard-Triquet 2008). While positive effects of tolerance at supra-organismal levels (population, community, ecosystem) for environment and health protection are often considered, the relative importance of any negative effects of tolerance is not typically fully assessed. The reviews offered in each chapter of this book will contribute to the provision of tools to carry out the appropriate risk-benefit analysis.

REFERENCES

Ait Alla, A., et al. 2006. Tolerance and biomarkers as useful tools for assessing environmental quality in the Oued Souss estuary (Bay of Agadir, Morocco). *Comp. Biochem. Physiol.* 143C:23–29.

Amiard, J. C. 1983. Les tests de toxicité subléthale: Approches biochimiques et éthologiques. *Océanis Doc. Océanogr.* 9:465–80.

Amiard, J. C., and C. Amiard-Triquet. 2008. *Les biomarqueurs dans l'évaluation de l'état écologique des milieux aquatiques*. Paris: Lavoisier.

Amiard, J. C., et al. 2006. Metallothioneins in aquatic invertebrates: Their role in metal detoxification and their use as biomarkers. *Aquat. Toxicol.* 76:160–202.

Amiard-Triquet, C. 2009. Behavioral disturbances: The missing link between sub-organismal and supra-organismal responses to stress? Prospects based on aquatic research. *Hum. Ecol. Risk Assess.* 15:87–110.

Amiard-Triquet, C., and J. C. Amiard. 2008. L'écotoxicologie du comportement. In *Les biomarqueurs dans l'évaluation de l'état écologique des milieux aquatiques*, ed. J. C. Amiard and C. Amiard-Triquet, 211–40. Paris: Lavoisier.

Amiard-Triquet, C., et al. 2009. Tolerance in organisms chronically exposed to estuarine pollution. In *Environmental assessment of estuarine ecosystems. A case study*, ed. C. Amiard-Triquet and P. S. Rainbow, 131–53. Boca Raton, FL: CRC Press.

Amiard-Triquet, C., C. Cossu-Leguille, and C. Mouneyrac. 2008. Les biomarqueurs de défense, la tolérance et ses conséquences écologiques. In *Les biomarqueurs dans l'évaluation de l'état écologique des milieux aquatiques*, ed. J. C. Amiard and C. Amiard-Triquet, 55–94. Paris: Lavoisier.

Bains, O. S., and C. J. Kennedy. 2004. Energetic costs of pyrene metabolism in isolated hepatocytes of rainbow trout, *Oncorhynchus mykiss*. *Aquat. Toxicol.* 67:217–26.

Baldwin, I. G., and K. J. R. Kramer. 1994. Biological early warning systems (BEWS). In *Biomonitoring of coastal waters and estuaries*, ed. K. J. R. Kramer, 1–28. Boca Raton, FL: CRC Press.

Bard, S. M. 2000. Multixenobiotic resistance as a cellular defense mechanism in aquatic organisms. *Aquat. Toxicol.* 48:357–89.

Barka, S., J. F. Pavillon, and J. C. Amiard. 2001. Influence of different essential and nonessential metals on MTLP levels in the copepod *Tigriopus brevicornis*. *Comp. Biochem. Physiol.* 128C:479–93.

Berthet, B. 2008. Les espèces sentinelles. In *Les biomarqueurs dans l'évaluation de l'état écologique des milieux aquatiques*, ed. J. C. Amiard and C. Amiard-Triquet, 121–48, Paris: Lavoisier.

Berthet, B., et al. 1992. Bioaccumulation, toxicity and physico-chemical speciation of silver in bivalve molluscs: Ecotoxicological and health consequences. *Sci. Total Environ.* 125:97–122.

Blanck, H., S. A. Wängberg, and S. Molander. 1988. Pollution-induced community tolerance— A new tool. In *Function testing of aquatic biota for estimating hazards of chemicals*, ed. J. J. Cairns and J. R. Pratt, 219–30. STP 998. Philadelphia: American Society for Testing and Materials.

Blanck, H., et al. 2003. Variability in zinc tolerance, measured as incorporation of radio-labeled carbon dioxide and thymidine, in periphyton communities sampled from 15 European river stretches. *Arch. Environ. Contam. Toxicol.* 44:17–29.

Boldina-Cosqueric, I., et al. 2010. Biochemical, physiological and behavioural markers in the endobenthic bivalve *Scrobicularia plana* as tools for the assessment of estuarine sediment quality. *Ecotoxicol. Environ. Saf.*, submitted.

Bonnard, M., M. Roméo, and C. Amiard-Triquet. 2009. Effects of copper on the burrowing behavior of estuarine and coastal invertebrates, the polychaete *Nereis diversicolor* and the bivalve *Scrobicularia plana*. *Hum. Ecol. Risk Assess.* 15:11–26.

Burlinson, F. C., and A. J. Lawrence. 2007. Development and validation of a behavioural assay to measure the tolerance of *Hediste diversicolor* to copper. *Environ. Pollut.* 145:274–278.

Cain, D. J., S. N. Luoma, and W. G. Wallace. 2004. Linking metal bioaccumulation of aquatic insects to their distribution patterns in a mining-impacted river. *Environ. Toxicol. Chem.* 23:1463–73.

Calow, P. 1991. Physiological costs of combating chemical toxicants: Ecological implications. *Comp. Biochem. Physiol.* 100C:3–6.

Chowdhury, M. J., E. F. Pan, and C. M. Wood. 2004. Physiological effects of dietary cadmium acclimation and waterborne cadmium challenge in rainbow trout: Respiratory, ionoregulatory, and stress parameters. *Comp. Biochem. Physiol.* 139C:163–73.

Coleman, M., and J. Hemingway. 2007. Insecticide resistance monitoring and evaluation in disease transmitting mosquitoes. *J. Pest. Sci.* 32:69–76.

Dagnino, A., et al. 2007. Development of an expert system for the integration of biomarker responses in mussels into an animal health index. *Biomarkers* 12:155–72.

De, J. 2004. Mercury-resistant marine bacteria and their role in bioremediation of certain toxicants. PhD thesis, Goa University.

Denholm, I., J. A. Pickett, and A. L. Devonshire. 1999. *Insecticide resistance from mechanisms to management.* Wallingford, UK: CAB International.

Derocher, A. E., et al. 2003. Contaminants in Svalbard polar bear samples archived since 1967 and possible population level effects. *Sci. Total Environ.* 301:163–74.

Dewailly, E. 2004. Le cas du mercure. In *Les valeurs toxicologiques de référence en santé environnementale et en santé au travail: les comprendre, les appliquer.* http://www.inspq.qc.ca/jasp/archives.

Essink, K., et al. 1985. Population dynamics of the ragworm *Nereis diversicolor* in the Dollard (Ems estuary) under changing conditions of stress by organic pollution. In *Marine biology of polar regions and effects of stress on marine organisms*, ed. J. S. Gray and M. E. Christiansen, 585–600. New York: John Wiley & Sons Ltd.

Fleeger, J. W., K. R. Carman, and R. M. Nisbet. 2003. Indirect effects of contaminants in aquatic ecosystems. *Sci. Total Environ.* 317:207–33.

Frydman, J. 2001. Folding of newly translated proteins *in vivo*: The role of molecular chaperones. *Annu. Rev. Biochem.* 70:603–47.

Georghiou, G. P., R. L. Metcalf, and F. E. Gidden. 1966. Carbamate-resistance in mosquitoes: Selection of *Culex pipiens fatigans* Wield (= *Culex quinquefasciatus*) for resistance to Baygon. *B. World Health Organ.* 35:691–708.

Gómez Gesteira, J. L., and J. C. Dauvin. 2000. Amphipods are good bioindicators of the impact of oil spills on soft-bottom macrobenthic communities. *Mar. Pollut. Bull.* 40:1017–27.

Guglielmo, C. G., W. H. Karasov, and W. J. Jakubas. 1996. Nutritional costs of a plant secondary metabolite explain selective foraging by Ruffed Grouse. *Ecology* 77:1103–15.

Harada, M. 1995. Minamata disease: Methylmercury poisoning in Japan caused by environmental pollution. *Crit. Rev. Toxicol.* 25:1–24.

Harper-Arabie, R. M., et al. 2004. Protective effects of allozyme genotype during chemical exposure in the grass shrimp, *Palaemonetes pugio. Aquat. Toxicol.* 70:41–54.

Hemingway, J., et al. 2004. The molecular basis of insecticide resistance in mosquitoes. *Insect Biochem. Mol.* 34:653–65.

Heugens, E. H. W., et al. 2001. A review of the effects of multiple stressors on aquatic organisms and analysis of uncertainty factors for use in risk assessment. *Crit. Rev. Toxicol.* 31:247–84.

Hites, R. A. 2004. Polybrominated diphenyl ethers in the environment and in people: A meta-analysis of concentrations. *Environ. Sci. Technol.* 38:945–56.

Hollis, L., et al. 1999. Cadmium accumulation, gill Cd binding, acclimation, and physiological effects during long-term sublethal Cd exposure in rainbow trout. *Aquat. Toxicol.* 46:101–19.

Hummel, H., et al. 1997. Sensitivity to stress in the bivalve *Macoma balthica* from the most northern (Arctic) to the most southern (French) populations: Low sensitivity in Arctic populations because of genetic adaptations? *Hydrobiologia* 355:127–38.

Ishaaya, I. 2001. *Biochemical sites of insecticide action and resistance.* Heidelberg: Springer.

Keizer, J., et al. 1995. Enzymological differences of AChE and diazinon hepatic metabolism: Correlation of *in vitro* data with the selective toxicity of diazinon of fish species. *Sci. Total Environ.* 171:213–20.

Klerks, P. L., and J. S. Weis. 1987. Genetic adaptation to heavy metals in aquatic organisms: A review. *Environ. Pollut.* 45:173–205.

Köhler, H. R., et al. 2000. Selection favours low hsp70 levels in chronically metal-stressed soil arthropods. *J. Evol. Biol.* 13:569–82.

Labbé, P., et al. 2007. Forty years of erratic insecticide resistance evolution in the mosquito *Culex pipiens*. *PLoS Genetics* 3:e205.

Lagarde, F., et al. 2003. Foraging behaviour and diet of an ectothermic herbivore: *Testudo horsfieldi*. *Ecography* 26:236–42.

Laroche, J., et al. 2002. Genetic and physiological responses of flounder (*Platichthys flesus*) populations to chemical contamination in estuaries. *Environ. Toxicol. Chem.* 21:2705–12.

LeBaron, H., and J. Gressel, 1982. *Herbicide resistance in plants.* Hoboken, NJ: John Wiley & Sons.

Leung, K. M. Y., A. C. Taylor, and R. W. Furness. 2000. Temperature-dependent physiological responses of the dogwhelk *Nucella lapillus* to cadmium exposure. *J. Mar. Biol. Ass. U.K.* 80:647–60.

Li, M. S. 2006. Ecological restoration of mineland with particular reference to the metalliferous mine wasteland in China: A review of research and practice. *Sci. Total Environ.* 357:38–53.

Lotts, J. W., and A. J. Stewart. 1995. Minnows can acclimate to total residual chlorine. *Environ. Toxicol. Chem.* 14:1365–74.

Luckenbach, T., I. Corsi, and D. Epel. 2004. Fatal attraction: Synthetic musk fragrances compromise multixenobiotic defense systems in mussels. *Mar. Environ. Res.* 58:215–19.

Mahaffey, K. R. 2004. Fish and shellfish as dietary sources of methylmercury and the [omega]-3 fatty acids, eicosahexacnoic acid and docosahexaenoic acid: Risks and benefits. *Environ. Res.* 95:414–28.

Malm, O. 1998. Gold mining as a source of mercury exposure in the Brazilian Amazon. *Environ. Res.* 77:73–78.

Marchand, J., et al. 2004. Physiological cost of tolerance to toxicants in the European flounder *Platichthys flesus*, along the French Atlantic Coast. *Aquat. Toxicol.* 70:327–43.

Marsh, K., et al. 2006. The detoxification limitation hypothesis: Where did it come from and where is it going? *J. Chem. Ecol.* 32:1247–66.

Maurice-Bourgoin, L., et al. 2000. Mercury distribution in waters and fishes of the upper Madeira rivers and mercury exposure in riparian Amazonian populations. *Sci. Total Environ.* 260:73–86.

McKenzie, J. A. 1996. *Ecological and evolutionary aspects of insecticides resistance.* Austin, TX: Academic Press.

McLean, S., and A. Duncan. 2006. Pharmacological perspectives on the detoxification of plant secondary metabolites: Implications for ingestive behavior of herbivores. *J. Chem. Ecol.* 32:1213–28.

McLusky, D. S. 1989. *The estuarine ecosystem.* Glasgow: Blackie.

Meyer, J. N., and R. T. Di Giulo. 2003. Heritable adaptation and fitness costs in killifish (*Fundulus heteroclitus*) inhabiting a polluted estuary. *Ecol. Appl.* 13:490–503.

Migné, A., et al. 2007. Photosynthetic activity of intertidal microphytobenthic communities during emersion: *In situ* measurements of chlorophyll fluorescence (PAM) and CO_2 flux (IRGA). *J. Phycol.* 43:864–73.

Millward, R. N., and A. Grant. 1995. Assessing the impact of copper on nematode communities from a chronically metal-enriched estuary using pollution-induced community tolerance. *Mar. Pollut. Bull.* 30:701–6.

Millward, R. N., and A. Grant. 2000. Pollution-induced tolerance to copper of nematode communities in the severely contaminated Restronguet Creek and adjacent estuaries, Cornwall, United Kingdom. *Environ. Toxicol. Chem.* 19:454–61.

Minier, C., F. Akcha, and F. Galgani. 1993. P-Glycoprotein expression in *Crassostrea gigas* and *Mytilus edulis* in polluted seawater. *Comp. Biochem. Physiol.* 106B:1029–36.

Minier, C., et al. 2006. A pollution-monitoring pilot study involving contaminant and bio-marker measurements in the Seine estuary, France, using zebra mussels (*Dreissena polymorpha*). *Environ. Toxicol. Chem.* 25:112–19.

Moore, N. W. 1967. A synopsis of the pesticide problem. In *Advances in ecological research*, ed. J. B. Cragg, 75–129. Vol. 4. New York: Academic Press.

Moreira, S. M., et al. 2006. Effects of estuarine sediment contamination on feeding and on key physiological functions of the polychaete *Hediste diversicolor*: Laboratory and *in situ* assays. *Aquat. Toxicol.* 78:186–201.

Morgan, A. J., P. Kille, and S. R. Stürzenbaum. 2007. Microevolution and ecotoxicology of metals in invertebrates. *Environ. Sci. Technol.* 41:1085–96.

Mouneyrac, C., H. Perrein-Ettajani, and C. Amiard-Triquet. 2010. Influence of anthropogenic stress on fitness and behaviour of a key-species of estuarine ecosystems, the ragworm *Nereis diversicolor*. *Environ. Pollut.* 158:121–28.

Munkittrick, K. R., and D.G. Dixon. 1988. Evidence for a maternal yolk factor associated with increased tolerance and resistance of feral white sucker (*Catostomus commersoni*) to waterborne copper. *Ecotoxicol. Environ. Saf.* 15:7–20.

Newman, M. C., and M. A. Unger. 2003. *Fundamentals of ecotoxicology*. Boca Raton, FL: Lewis Publishers.

Ng, T. Y. T., et al. 2007. Metallothionein turnover, cytosolic distribution and the uptake of Cd by the green mussel *Perna viridis*. *Aquat. Toxicol.* 84:153–61.

Oliveira, M. M., et al. 2007. Brain acetylcholinesterase as a marine pesticide biomarker using Brazilian fishes. *Mar. Environ. Res.* 63:303–12.

Ortiz-Delgado, J. B., et al. 2008. Tissue-specific induction of EROD activity and CYP1A protein in *Sparus aurata* exposed to B(*a*)P and TCDD. *Ecotoxicol. Environ. Saf.* 69:80–88.

Pearson, W. H., et al. 1981. Effects of oiled sediment on predation on the Littleneck Clam, *Protothaca staminea*, by the Dungeness Crab, *Cancer magister*. *Estuar. Coast. Shelf Sci.* 13:445–54.

Peña-Llopis, S., M. D. Ferrando, and J.B. Peña. 2003. Fish tolerance to organophosphate-induced oxidative stress is dependent on the glutathione metabolism and enhanced by N-acetylcysteine. *Aquat. Toxicol.* 65:337–60.

Peña-Llopis, S., et al. 2001. Glutathione-dependent resistance of the European eel *Anguilla anguilla* to the herbicide molinate. *Chemosphere* 45:671–81.

Ramond, J. B., et al. 2008. Relationships between hydrosedimentary processes and occurrence of mercury-resistant bacteria (*merA*) in estuary mudflats (Seine, France). *Mar. Pollut. Bull.* 56:1168–76.

Ramos, J. L., et al. 2002. Mechanisms of solvent tolerance in gram-negative bacteria. *Annu. Rev. Microbiol.* 56:743–68.

Rowe, C. L. 1998. Elevated standard metabolic rate in a freshwater shrimp (*Palaemonetes paludosus*) exposed to trace element-rich coal combustion waste. *Comp. Biochem. Physiol.* 121A:299–304.

Sánchez, M., E. Andreu-Moliner, and M. D. Ferrando. 2004. Laboratory investigation into the development of resistance of *Daphnia magna* to the herbicide molinate. *Ecotoxicol. Environ. Saf.* 59:316–23.

Scarfe, A. D., et al. 1982. Locomotor behavior of four marine teleosts in response to sublethal copper exposure. *Aquat. Toxicol.* 2:335–53.

Scherer, E., and R. E. McNicol. 1998. Preference-avoidance responses of lake whitefish (*Coregonus clupeaformis*) to competing gradients of light and copper, lead, and zinc. *Water Res.* 32:924–29.

Shaw, J. 1989. *Heavy metal tolerance in plants: Evolutionary aspects*. Boca Raton, FL: CRC Press.

Smital, T., et al. 2004. Emerging contaminants—Pesticides, PPCPs, microbial degradation products and natural substances as inhibitors of multixenobiotic defense in aquatic organisms. *Mutat. Res.-Fund. Mol. M.* 552:101–17.

Sorensen, J. S., et al. 2007. Tissue distribution of cytochrome P450 3A (CYP3A) in brushtail possums (*Trichosurus vulpecula*) exposed to *Eucalyptus* terpenes. *Comp. Biochem. Physiol.* 145C:194–201.

Sylvestre, F. 2009. Modern diatom distribution in the Seine and Authie estuaries. In *Environmental assessment of estuarine ecosystems: A case study*, ed. C. Amiard-Triquet and P. S. Rainbow, 241–54. Boca Raton, FL: CRC Press.

Takamura, N., F. Kasai, and M.M. Watanabe. 1989. Effects of Cu, Cd and Zn on photosynthesis of freshwater benthic algae. *J. Appl. Phycol.* 1:39–52.

Van der Oost, R., C. Porte-Visa, and N. W. Van den Brink. 2005. Biomarkers in environmental assessment. In *Ecological testing of marine and freshwater ecosystems*, ed. P. J. Den Besten and I. F. Munawar, 87–152. Boca Raton, FL: Taylor & Francis.

Wilding, J., and L. Maltby. 2006. Relative toxicological importance of aqueous and dietary metal exposure to a freshwater crustacean: Implications for risk assessment. *Environ. Toxicol. Chem.* 25:1795–801.

Wilhelm Filho, D., et al. 2005. Effect of different oxygen tensions on weight gain, feed conversion, and antioxidant status in piapara, *Leporinus elongatus* (Valenciennes, 1847). *Aquaculture* 244:349–57.

Xie, L., and P. L. Klerks. 2004. Fitness cost of resistance to cadmium in the least killifish (*Heterandria formosa*). *Environ. Toxicol. Chem.* 23:1499–503.

2 Tolerance to Contaminants

Evidence from Chronically Exposed Populations of Aquatic Organisms

Emma L. Johnston

CONTENTS

2.1 INTRODUCTION

Contaminants may reduce the survival, recruitment, growth, and reproductive success of an organism, and hence they are potentially powerful agents of selection. Ecotoxicologists have long been interested in the rapid evolution of tolerance to contaminants (Luoma 1977), and it is generally accepted that this may occur after only a few generations (Klerks and Levinton 1989). There are substantial social and environmental consequences of the evolution of contaminant tolerance. The evolution of pesticide and herbicide resistance is a costly concern for agriculture, while pollution-tolerant bacteria and plants are increasingly used in the remediation of contaminated waste. Moreover, if guidelines for the protection of the environment are based on sensitivity data from adapted populations, they may not be adequately protective. It is therefore crucial that we understand the frequency at which contaminant tolerance occurs and the extent to which this phenomenon modifies the response of plants and animals to contaminant exposure.

A number of excellent reviews have recently been published regarding various aspects of evolutionary ecotoxicology (e.g., Grant 2002; Klerks 2002; Nacci et al. 2002b; Hoffmann and Daborn 2007; Medina et al. 2007; Morgan et al. 2007; Janssens et al. 2009), and additional theses are published in this book. The purview of this chapter is to critically review and summarise the last decade of research on tolerance (resistance) to contaminants in chronically exposed field populations of aquatic invertebrates and fish. A systematic approach to literature selection was taken in order to describe the range of approaches and organisms tested, and the frequency at which differential tolerance has been observed. This review is deliberately constrained to studies that assess an endpoint of direct relevance to an organism's fitness (e.g., survival, recruitment, growth, or reproductive success). It does not review direct genotoxic effects of contaminants on the molecular structure of genetic material, which tend to result in detrimental effects on fitness. Nor does it describe the impacts of contaminants on biomarkers, bioaccumulation, or the genetic variability of populations *unless* such differences were also associated with a direct change in the fitness of the exposed populations.

2.1.1 PHYSIOLOGICAL ACCLIMATION AND GENETIC ADAPTATION

Differential fitness in response to contaminant exposure may result from physiological acclimation or genetic adaptation, and the mechanism will vary with the organism and toxicant in question. This chapter will use the term *tolerance* synonymously with the term *resistance*, to include two subcategories: (1) heritable adaptive responses (genetic adaptation) that affect the mean tolerance to contaminants in a population, and (2) physiological acclimation of individuals. Tolerance will thus also include instances where the mechanism is unknown (Morgan et al. 2007). Physiological acclimation and genetic adaptation are not mutually exclusive, as the ability to acclimate is conferred through the genetics of a species. Some species are inherently capable of tolerating a contaminant through existing physiological acclimation mechanisms. In this case, initial exposure to a contaminant may reduce the effective toxicity of future exposures of that individual to the contaminant. The response is usually a physiological one that upregulates an existing detoxification mechanism, but it may also be a behavioural response that effectively reduces exposure. In this case, an observation of differential tolerance between populations may or may not be the result of anthropogenic contamination acting as a selection agent. Where the ambient contamination levels have been sufficiently different to trigger physiological acclimation at one site, differential tolerance to acute toxicant exposure should be observed in field-collected

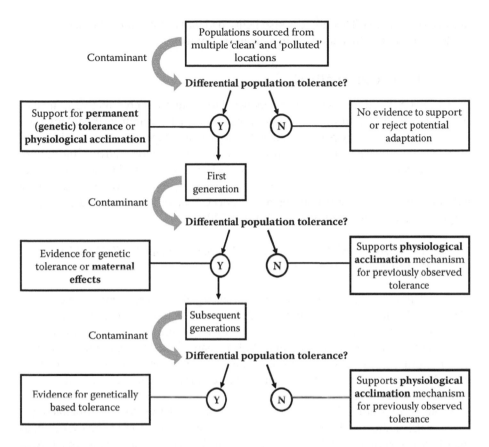

FIGURE 2.1 Process for assessing evidence for contaminant tolerance in chronically exposed field populations.

organisms (Figure 2.1). Tolerance should be rapidly lost if exposures are not maintained.

If the selective force applied by a toxicant is maintained across multiple generations, it may result in differences in fitness that are associated with changes to population genetics. Heritable adaptive responses of a population to a contaminant may manifest in different ways. There is potential for selection to increase the frequency of phenotypically plastic organisms that can rapidly upregulate a toxicant defence mechanism. This is the same effect as described above (physiological acclimation), but the frequency of its occurrence will be higher in adapted populations. Alternatively, selection may favour individuals with permanently increased tolerance to a contaminant. Distinguishing between selection for phenotypic plasticity in tolerance and selection for elevated mean tolerance requires bioassays on the offspring of individuals collected from the field and raised in a controlled environment. Preferably, testing should be done on the first and second generations so that maternal effects derived from the mother's response to a contaminant (essentially a form of physiological acclimation transferred to the first generation) may be explicitly defined (Figure 2.1). This process also helps to identify any potential fitness costs to a tolerant organism that is raised in a clean environment.

2.2 LITERATURE SEARCH

A systematic literature review was conducted to capture a representative sample of the aquatic tolerance (resistance) literature. A systematic search methodology does not aim to capture every paper published on a particular topic. It is a repeatable sampling approach designed to capture a selection of studies demonstrative of the current trends and focus in a research field. I used specific search terms in the Scopus database, which covers >4,300 titles in the life sciences (peer-reviewed journals, trade publications, conference papers, or book series). Searches were limited to English language studies published between 1999 and 2009:

> Search: ALL ("contam*") AND ALL ("genetic adaptation") OR ALL ("physiological acclimation")

All of the abstracts of the papers that emerged in these searches were read (332), and I selected for review those that (1) had an aquatic invertebrate or fish focus and (2) tested the effects of a contaminant on the tolerance of at least two populations collected from the field. At least one of the test endpoints had to be a test relevant to the fitness of an individual (e.g., relative survival, reproduction, growth, or recruitment). In total, thirty-four research articles satisfied the criteria for inclusion in the review.

From these studies I extracted qualitative data relating to contaminant type, organism(s) studied, and endpoints assessed. I then collated data on the overall finding of the research (reduced, increased, or no difference in contaminant tolerance or resistance between the populations tested) as concluded by the authors and the direction and magnitude of the change. Many of the studies performed no formal tests of hypotheses, and I considered descriptive data from these studies.

2.3 RESULTS: SUMMARY OF LITERATURE SEARCH

Of the 332 abstracts examined, 106 studies dealt explicitly with tolerance (or resistance) to contaminants in aquatic vertebrates or invertebrates. Of these, thirty-four included a test directly related to fitness, i.e., relative survival, reproduction, growth, or recruitment, and hence were included in this review (summarised in Table 2.1). A further twenty-nine examined the relative response of a biomarker, bioaccumulation, or genetic variability between field-collected populations; thirty-two studies examined differential tolerance through laboratory breeding studies; and fourteen reviewed the literature regarding the evolution of contaminant tolerance (Figure 2.2a).

The average number of sites from which animals were collected and tested in the review studies was 3.4 (range 2–14; Figure 2.2b). Eight studies only tested populations from two sites, making it difficult to attribute differential tolerance to the contaminant exposure at the site. Contaminated habitats may differ from other locations in many environmental features; e.g., contaminated sites in estuaries tend to

TABLE 2.1
Summary of the Focal Species, Contaminant, Test Endpoints, and Evidence for Differential Tolerance for All of the Papers Selected for This Review (bold denotes significant difference between populations for that endpoint)

Taxonomic Group	Species	Contaminant	Differential Tolerance in Field-Collected Populations	Differential Tolerance in Subsequent Generations	Endpoints Directly Relevant to Fitness (Assayed on Field-Collected Populations)	Other Study Endpoints	Reference
Copepod	*Tisbe holothuriae*	Co, Cr	Y	Y	**48 h LC50,** reproductive output	NA	Miliou et al. 2000
Copepod	*Microarthridion littorale*	Contaminated sediment	N	NA	Mortality, reproductive output	Genetic relatedness greater between polluted and reference populations than between polluted populations	Kovatch et al. 2000
Cladoceran	*Daphnia magna*	Model pesticide	Y	Y	**48 h EC50**	Measurement of neutral genetic marker indicates genetic diversity positively correlated with land use intensity	Coors et al. 2009
Cladoceran	*Daphnia magna*	Cd	Y	Y	**48 h EC50**	Lower expression of stress protein hsp70 in more sensitive clones	Haap and Köhler 2009
Cladoceran	*Ceriodaphnia dubia*	Zn	Y	N	**48 h EC50, carapace length, reproductive output**	NA	Muyssen and Janssen 2001

(continued)

TABLE 2.1

Summary of the Focal Species, Contaminant, Test Endpoints, and Evidence for Differential Tolerance for All of the Papers Selected for This Review (bold denotes significant difference between populations for that endpoint) (Continued)

Taxonomic Group	Species	Contaminant	Differential Tolerance in Field-Collected Populations	Differential Tolerance in Subsequent Generations	Endpoints Directly Relevant to Fitness (Assayed on Field-Collected Populations)	Other Study Endpoints	Reference
Cladoceran	*Moinodaphnia macleayi*	Uranium	N	N	48 h EC50, LOEC, NOEC	NA	Semaan et al. 2001
Cladoceran	*Daphnia longispina*	Cu	Y	Y	**Mortality**, feeding inhibition	NA	Lopes et al. 2004
Cladoceran	*Ceriodaphnia pulchella*	Contaminated water	Y	Y	**Mortality, reproductive output**	Support for genetic erosion due to higher frequency of sensitive individuals in reference population	Lopes et al. 2005
Cladoceran	*Daphnia longispina*	Contaminated water, Cu	Y	Y	**Mortality, fecundity**	NA	Lopes et al. 2006
Amphipod	*Orchestia gammarellus*	Cd, Cu, Zn	N	NA	17 d LC50	Significant differences in MTLP levels in field populations, but response could not be elicited in lab	Mouneyrac et al. 2002
Caddisfly larva	*Hydropsyche betteni*	Zn	Y	NA	Mortality, **feeding, weight**	NA	Balch et al. 2000
Chironomid larva	*Chironomus riparius*	Cd	Y	Y	**4 d EC50**	Reciprocal crosses suggesting absence of maternal effects	Groenendijk et al. 2002

Flatworm	*Polycelis tenuis*	Cd	N	N	28 d LC50 and EC50	NA	Indeherberg et al. 1999
Oligochaete	*Tubifex tubifex*	Hg	Y	NA	**10 d LC50**	Elevated levels of accumulated tissue Hg levels in metal-exposed populations	Vidal and Horne 2003
Polychaete	*Nereis diversicolor*	Ag, Cd, Cu, Zn	Y	NA	**21 d LC50**	Lower levels of detoxificatory enzymes and MTLP in metal-exposed field populations	Mouneyrac et al. 2003
Polychaete	*Nereis diversicolor*	Cd, Cu, Zn	Y	NA	LT50	Stored energy reserves lower in metal-exposed populations	Durou et al. 2005
Polychaete	*Nereis diversicolor*	Cu, Zn	Y	NA	**Mortality**	Higher rate of Cu accumulation in heavily contaminated population	Zhou et al. 2003
Bryozoa	*Bugula neritina*	Cu	Y	Y	Mortality, **growth**, settlement	NA	Piola and Johnston 2006b
Snail	*Physella columbiana*	Cd, Pb, Zn	Y	Y	LC50, **avoidance behaviour**	NA	Lefcort et al. 2004
Benthic invertebrates		Cd, Cu, Zn	Y	NA	Presence/absence	NA	Courtney and Clements 2000
Sessile invertebrates		Cu, Zn, TBT	Y	NA	Presence/absence	NA	Dafforn et al. 2009
Sediment infauna		Contaminated soil	Y	NA	**Emergence**	NA	Bahrndorff et al. 2006

(continued)

TABLE 2.1

Summary of the Focal Species, Contaminant, Test Endpoints, and Evidence for Differential Tolerance for All of the Papers Selected for This Review (bold denotes significant difference between populations for that endpoint) (Continued)

Taxonomic Group	Species	Contaminant	Differential Tolerance in Field-Collected Populations	Differential Tolerance in Subsequent Generations	Endpoints Directly Relevant to Fitness (Assayed on Field-Collected Populations)	Other Study Endpoints	Reference
Fish	Melanotaenia nigrans	Cu	Y	NA	**96 h EC50**	Lower rate of metal uptake in populations from contaminated sites; further, lower heterozygosity at loci associated with copper sensitivity, suggesting selection of copper-tolerant allozymes	Gale et al. 2003
Fish	Fundulus heteroclitus	PCB	Y	N	**7 d LC20**, bacterial tolerance	NA	Nacci et al. 2009
Fish	Fundulus heteroclitus	PCB	Y	Y	**28 d LC20 and LC50**	Higher frequency of sensitive individuals at cleaner sites, suggesting inheritance of tolerance rather than local selection	Nacci et al. 2002a
Fish	Gambusia affinis	Cd	Y	Y	**48 h and 96 h LC50**	NA	Annabi et al. 2009
Fish	Fundulus heteroclitus	Contaminated sediment	Y	Y	**Mortality**	NA	Meyer and Di Giulio 2003

Fish	*Gobio gobio*	Cd	Y	NA	**Mortality**	Higher rate of MTLP induction in populations with metal exposure history	Knapen et al. 2004
Fish	*Fundulus heteroclitus*	Model PAH	N	NA	Survival	EROD activity induced with effect lower in metal-exposed populations	Nacci et al. 2002c
Fish	*Gillichthys mirabilis*	Mixture	N	NA	Growth	NA	Forrester et al. 2003
Fish	*Menidia menidia*	Model PCB	Y	Y	**28 d LC20**	PGM genotypes potentially associated with PCB tolerance	Roark et al. 2005
Fish	*Fundulus heteroclitus*	Model pro-oxidant	Y	Y	4.4 mM LT50	Upregulation of antioxidant defences in populations with previous metal exposure; demonstrated potential heritable transmission	Meyer et al. 2003
Amphibian	*Rana temporaria*	Nitrates	Y	NA	**Mortality, growth,** metamorphosis	NA	Johansson et al. 2001

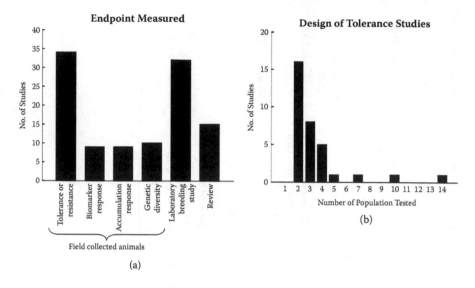

(a)

FIGURE 2.2 (a) Summary of Scopus literature search for all studies published between 1999 and 2009 on aquatic vertebrates and invertebrates associated with the search terms: "contam*" and "genetic adaptation" or "physiological acclimation." Studies have been categorised by the endpoints measured. "Tolerance" or "resistance" includes all studies that compared mortality, reproduction, or growth as an endpoint in a toxicological assay of field-collected populations. "Biomarker response," "accumulation response," and "genetic diversity" include all studies that exposed organisms to a toxicant and measured a response of field-collected animals that suggests differential fitness but did not directly record mortality, growth, or reproduction. "Laboratory breeding study" includes all studies that examined the tolerance of laboratory-bred individuals only. (b) Number of populations compared in studies of tolerance of field-collected populations (from Figure 2.3a).

have a higher proportion of fine sediment, which dramatically affects the ecology of infauna. Differential tolerance may then covary with any number of other biological and environmental differences that may affect the underlying tolerance (e.g., hypoxic stress or population demography). Researchers would do well to test organisms from at least two suitable reference sites when examining the relative tolerance of an impacted population.

While evolution of tolerance to contaminants is regularly observed in selection studies conducted under controlled laboratory environments with strong selection pressure, some authors have concluded that genetically based field tolerance will occur infrequently (Woods and Hoffmann 2000; Klerks 2002). Of the thirty-four studies selected for this review, 85% (twenty-nine) concluded that there were differences in population tolerance that correlated with predicted differences in chronic field exposures to a contaminant (see Figure 2.3a for vertebrate studies and Figure 2.3b for invertebrate studies). Seventy-four percent (twenty-five) of the studies reported actual measures of the contaminant from the field or referred to studies that have made such measurements. Where possible (impacted populations: eleven studies), I calculated the ratio of the test endpoint value between the most tolerant and the most sensitive population. The ratio of differences in tolerance

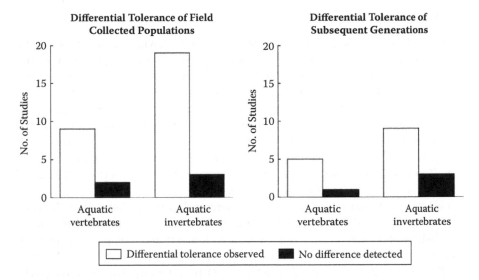

FIGURE 2.3 Studies uncovered by the literature search and published between 1999 and 2009 that tested for: (a) differential tolerance to a contaminant between populations of organisms collected from the field or (b) differential tolerance to a contaminant in subsequent generations. Studies in (b) are a subset of the studies in (a). White bars represent studies that found a difference between populations in tolerance to a contaminant. Black bars represent studies that found no difference.

between populations was between 1.4 and 8, with the notable exception that one study reported a population of *Fundulus heteroclitus* (killifish) that was 347,000 times more tolerant to PCB 126 than a reference population (LC50 concentration ratio; Nacci et al. 2002a).

Of the twenty-nine studies that did detect differential tolerance to a contaminant among different field-collected populations, just half (14/29) went on to assess the genetic basis of the tolerance traits (Figure 2.3). These studies tested for tolerance by running bioassays on subsequent generations raised under clean conditions (using either sexual or clonal lineages). All but three of these confirmed contaminant tolerance in subsequent generations (Figure 2.3b), although maternal effects were not explicitly ruled out in many of these studies because there was no test of whether tolerance in the first generation was different from tolerance in later generations. The frequency of tolerance (11/14 studies = 79%) does not necessarily imply that most contaminant tolerance is heritable, but it is likely to reflect the greater ease of publishing statistically significant results. I would therefore urge editors, reviewers, and authors to publish brief, well-designed studies that find no tolerance to contaminants (Klerks 2002).

A brief summary of the thirty-four studies is provided below and in Table 2.1. I begin with the aquatic vertebrates and move on to the invertebrate studies. Supplementary texts are referred to where they assist in the interpretation of the current studies.

2.3.1 Aquatic Vertebrates

Various fish species have been key model organisms with which to investigate evolutionary ecotoxicology. In particular, fish used for regulatory testing of contaminants such as rainbow trout (*Oncorhynchus mykiss*) and mosquitofish (*Gambusia affinis*) have been popular in laboratory breeding experiments. However, a common estuarine fish, *Fundulus heteroclitus* (the killifish or mummichog), from the U.S. coast has been key to the study of contaminant tolerance in field populations. A recent review of evolved tolerance discusses key insights from this species (Burnett et al. 2007), as does a review of resistance in North American fish populations (Wirgin and Waldman 2004). I will highlight some of the most recent work on differential tolerance in field populations of this species.

Genetically based tolerance to methylmercury has been observed in the early life stages of *Fundulus heteroclitus* (Weis 2002), while reproductively active populations of *F. heteroclitus* inhabit sites with dioxin-like compound concentrations exceeding those that are lethal to fish from clean reference sites (Nacci et al. 2002a). Nacci et al. (2002c) used a model polyaromatic hydrocarbon (PAH) (benzo[a]pyrene (B[a]P)) to test the tolerance of two populations of killifish that differ in their tolerance to a dioxin-like compound. They found no difference in survival between fish from the two populations, but some biomarker responses were different. In response to B[a]P exposure, reference fish had elevated EROD activity (hepatic microsomal activity for ethoxyresorufin-O-deethylase) relative to fish from the polychlorinated biphenyl (PCB)-contaminated harbor. B[a]P exposure also caused higher DNA adduct levels in the reference fish livers. DNA adducts are pieces of DNA bonded to cancer-causing chemicals and are considered to be an origin of cancerous cells. They are used as biomarkers of cancer in animals, and the results suggest that fish from the contaminated site will be less likely to suffer from PAH-mediated cancers. Fish from the contaminated site carried higher contaminant body burdens, and it is not yet known whether the observed differences are genetic or environmentally based.

A study that examined the response of two subsequent laboratory generations of killifish from a highly PAH-contaminated site found that they had greater tolerance to the toxicity of sediments from that site than did offspring of killifish from a reference site (Meyer and Di Giulio 2003). This tolerance was greater in the first than in the second generation, which suggests that the tolerant phenotype is based on both maternal provisioning and heritable adaptive responses (Meyer and Di Giulio 2003). Larval first- and second-generation offspring from killifish collected from a highly contaminated site were also more tolerant to the toxicity of *t*-butyl hydroperoxide (a model pro-oxidant) than reference offspring (Meyer et al. 2003). In laboratory experiments the first and second generations were exposed to contaminated sediments or a reference oxidative stressor and several antioxidant parameters were measured. Upregulated antioxidant defences were demonstrated in both the immediate physiological and heritable multigenerational (Meyer et al. 2003) response of the offspring. Some fitness disadvantages were observed in *F. heteroclitus* resident in contaminated sites, including increased sensitivity to low oxygen and phototoxicity (Meyer and Di Giulio 2003).

It is more likely that differential tolerance will be observed in species that are consistently exposed to a contaminant through all phases of the life cycle and where dispersal rates are relatively low. The killifish is highly resident and nonmigratory, whereas an interesting study recently published by Roark et al. (2005) presents tests of genetic adaptation to local contaminants in populations of a migratory marine fish, *Menidia menidia*. One population of this fish only resides seasonally in an industrial harbor with extreme polychlorinated biphenyl (PCB) contamination. In response to the seasonal exposure regime, the embryonic offspring of *M. menidia* collected from the impacted harbor were significantly less sensitive to a dioxin-like PCB congener than offspring of *M. menidia* from a reference site.

Differential tolerance to metals has recently been reported in field-collected individuals of black-banded rainbow fish (*Melanotaenia nigrans*) (Gale et al. 2003) and a goby (*Gobio gobio*) (Knapen et al. 2004). Gale et al. (2003) investigated the copper tolerance of a population of black-banded rainbow fish with a forty-year history of copper exposure due to leachate from a uranium-copper mine. The 96 h EC50 of field-collected fish was 8.3 times higher than that of reference fish, and reduced copper accumulation was observed in the tolerant fish (see Chapter 6). The copper tolerance of subsequent generations was not tested, but reduced allozyme variation observed at the impacted location suggested that selection had occurred to reduce the genetic diversity of this population. Gobies from a metal-impacted site survived acute exposures better than fish from a reference site (Knapen et al. 2004). This effect was more pronounced after an extra acclimation period to a sublethal Cd concentration. In this case, neither the average Cd accumulation rate nor the lethal body concentrations differed between fishes from each site. Tolerance to cadmium is also reported for mosquitofish (*Gambusia affinis*) from a polluted site in Tunisia (Annabi et al. 2009). A genetic basis to the tolerance was inferred from consistent results of bioassays conducted on the F1 generation.

No between-population difference in tolerance was observed in a goby species (*Gallichthys mirabilis*) exposed to a mixture of contaminants in marine sediments (Forrester et al. 2003). In this unique study, gobies from three estuaries subject to different contamination levels were reciprocally transplanted to each estuary and deployed in cages. Fish showed no difference in growth rate relative to their source estuary, but reduced growth was correlated with increasing sediment metal and organic contamination. This suggests that there had been no adaptation or acclimation to past contaminant exposure.

The literature search uncovered one study of eutrophication-associated impacts on frog larvae (Johansson et al. 2001). In Scandinavia, a chronic experiment was used to test whether tadpoles of the common frog (*Rana temporaria*) from the north were less well adapted to cope with high nitrate concentrations than those from the south. The bioassays accounted for two major potential confounding environmental latitudinal differences (temperature and photoperiod), and the study did find some differential effects of nitrates on the first-generation frog larvae. High concentrations of nitrate reduced the growth rates and metamorphic size of larvae from the north, but not the south. However, the differences were slight and current nitrate inputs are not expected to cause major problems for these frog populations.

2.3.2 Aquatic Invertebrates

All but one of the twenty-three aquatic invertebrate studies captured by the literature search examined tolerance to metals. This likely reflects the persistent and ubiquitous nature of metal contamination in aquatic systems as well as the relative focus of contaminant expertise in aquatic systems. In the case of some trace metal contaminants that are essential elements but become toxic at higher concentrations, we might expect that a few organisms will have endogenous biochemical systems that engender the ability to acclimate following contaminant exposure. It may be that the evolution of tolerance to metals is more likely than the evolution of tolerance to other contaminants, particularly if multiple molecular mechanisms exist across phylogenetically distinct groups (Janssens et al. 2009). Certain crustaceans, for example, decapods, are known for their general ability to regulate the accumulation of some essential metals, and twelve of the twenty-three studies in this review use crustaceans as the focal species. Klerks and Weis (1987) reviewed genetic adaptation to heavy metals in aquatic organisms, and there is an excellent review of microevolution and ecotoxicology of metals in invertebrates by Morgan and colleagues (2007). Here, I summarise the most recent studies of tolerance in chronically exposed populations of aquatic invertebrates.

2.3.2.1 Cladocerans

Cladocerans are small freshwater crustaceans sometimes called water fleas. They are commonly used in ecotoxicological testing and are generally easy to culture in the laboratory. Five studies of metal tolerance in cladocerans have recently been published, and all but one demonstrated differential tolerance to the metal contaminant involved. I begin by describing an interesting series of studies on cladocerans by Lopes and coworkers, who aimed to distinguish between immediate environmental acclimation and genetically based adaptation to contaminants. Lopes et al. (2006) established cloned lineages from four field populations of *Daphnia longispina*: two from sites historically impacted by acid mine drainage (AMD) and two from reference sites. Both acclimated and nonacclimated individuals from the impacted populations were significantly less sensitive to mine contaminated waters than those from the reference populations. Interestingly, differences in tolerance were more obvious in the lethal bioassays than the sublethal reproductive output assays. This confirmed the convergence hypothesis that differences in population lethal responses at higher concentrations are more likely to occur because they will involve fewer and more specific mechanisms than sublethal responses to low concentrations (Lopes et al. 2005).

In order to study the effects of environmental pollution on the genetic diversity of this species, Lopes et al. (2004) collected *Daphnia longispina* and established clonal lineages from one reference and one mine-impacted site. They created over 125 acclimated cloned lineages from each population and exposed them to copper concentrations for twenty-four hours. The impacted population did not include the most sensitive lineages; instead, the most tolerant ones were present in the reference population. This suggests a reduction in genetic diversity (genetic erosion) at the impacted location. Using a different cladoceran species (*Ceriodaphnia pulchella*), Lopes et al. (2005) tested whether the tolerance that they observed in field populations living at the site of an abandoned cupric-pyrite mine was due to the disappearance of sensitive individuals or the appearance of highly tolerant ones. They concluded that the most sensitive individuals only occurred in the reference populations, while the most tolerant were only present in the stressed population. This was based on laboratory toxicity studies in which 15% of the reference population individuals had died before any of the individuals from the impacted population had. Moreover, when all of the reference populations were dead, there were still 50% of the individuals present from the impacted population.

Haap and Köhler (2009) were interested in variation in Cd sensitivity of seven clones of *Daphnia magna* from different geographical regions. In this study, Cd tolerance correlated well with gene expression of the stress protein hsp70 (the most tolerant population had the lowest expression), but had no correlation with Cd bioaccumulation. Haap and Köhler (2009) found substantial variation in relative Cd sensitivity as measured by an immobility bioassay. Their least sensitive clone was 2.2 times more tolerant to Cd than their most sensitive. While greater differences in sensitivity have been observed in other species, this result still has serious consequences for the interpretation of many toxicological assays that test the response of only a single clonal lineage. This point is further emphasised in a study by Muyssen and Janssen (2001), who tested subsequent generations of two field-collected natural clonal lineages versus a clone that had been raised for many generations in the absence of natural levels of zinc. The field-collected *D. magna* clones were initially more tolerant to acute zinc toxicity (up to a factor of 4), but gradually lost their Zn tolerance and became more similar to the laboratory clone. They question the relevance of toxicity data obtained with organisms that have adapted or acclimated to laboratory conditions containing unnaturally low levels of trace elements.

A substantial study investigated differential tolerance of *Daphnia magna* to two model toxicants, the pesticide carbaryl and the metal potassium dichromate (Coors et al. 2009). In this study, clonal lineages of cladoceran populations were established from dormant egg banks collected from ten different ponds. The study precluded acclimation and provides evidence of local adaptation of particular pond populations to both the pesticide and the metal. Evidence for genetic erosion by contamination was also observed. Polymorphic allozyme diversity (a neutral marker of genetic diversity) was negatively correlated with increasing agricultural land use intensity.

One study of a tropical cladoceran species (*Moinodaphnia macleayi*) failed to observe differential tolerance to uranium (Semaan et al. 2001). A laboratory stock population that had been maintained for ten years was compared to a reference population and an impacted population located on a uranium mine site in Australia. Although the species was more sensitive to uranium than any other cladoceran tested, there was little difference in the response of the three populations to acute or chronic exposures to uranium. This justified the use of laboratory stock populations in further regulatory testing.

2.3.2.2 Other Crustaceans

Occasionally a study is published in which organisms are chronically exposed to very high contaminant loads but do not develop enhanced tolerance. Mouneyrac et al. (2002) found no difference in the Cd, Cu, or Zn sensitivity of three populations of an amphipod from more or less polluted areas. They also write a cautionary tale (Mouneyrac et al. 2002) regarding the use of metallothionein (MTLP) as a biomarker of metal exposure. While the concentrations of MTLP were higher in amphipods from the contaminated site (measured directly after collection), the same effect could not be simulated in laboratory exposures. Exposure to dissolved Cd, Cu, or Zn did not induce different levels of MTLP between amphipods from the clean or contaminated sites (but see also Chapter 6). A second study using whole-sediment reproduction bioassays found no difference in the population sensitivity of meiobenthic copepods (*Microarthridion littorale*) exposed to a highly contaminated sediment mixture (Kovatch et al. 2000). This was despite the fact that significant genetic differentiation was found between copepod populations from the control and the two contaminated sites (measured as DNA sequences of the mitochondrial gene, cytochrome apoenzyme b). The authors suggested that the genetic relationship between the impacted copepod populations and the reference populations may be more related to geography than contamination history. The two polluted sites in this study were geographically closer to each other (8.2 km) than they were to the reference population (92 km).

Some studies have observed differential tolerance in wild populations that persists for many generations in clean laboratory conditions. A metal-tolerant copepod (*Tisbe holothuriae*) was collected from a contaminated site and maintained in laboratory culture for more than forty generations (Miliou et al. 2000). The animal was still more tolerant to Co^{2+} and Cr^{6+} than those collected from a nonpolluted area and maintained under identical conditions. This may suggest that under benign laboratory conditions (i.e., in the absence of biotic or abiotic stress) there may be little cost to maintaining metal tolerance. Alternatively, the tolerant phenotype may have become fixed.

2.3.2.3 Annelids

Annelids have also been interesting objects of study for evolutionary ecotoxicologists. Sediments are renowned as being sinks for contaminants in aquatic systems, and many aquatic annelids (polychaetes and oligochaetes) spend substantial time associated with sediments (even ingesting them). Hence, the exposure of reproductively mature annelids is likely to be consistent and high. However, polychaetes and oligochaetes have a mobile larval stage that may be highly dispersive, and this represents a substantial opportunity for genetic mixing, which may reduce the opportunity for strong selection by sediment-bound contaminants.

This review uncovered one study of oligochaete tolerance to mercury published in the last decade and three studies of polychaete tolerance to metals. Vidal and Horne (2003) investigated variable tolerance to mercury among five populations of the sediment-dwelling oligochaete *Sparganophilus pearsei*. Worms collected from contaminated sediments were eight to ten times more tolerant to mercury exposure in the laboratory than reference site worms. They also had two to eight times higher levels of mercury in their tissues. Vidal and Horne (2003) did not go on to test offspring of these worms, but they suggest that genetic adaptation is possible based on earlier work with another oligochaete, *Limnodrilus hoffmeisteri*. For this species, genetic adaptation to Cd-contaminated sediments was shown to occur after only a few generations (Klerks and Levinton 1989) and is thought to be controlled by a single gene (Martinez and Levinton 1996). However, until a greater understanding of the mechanism of genetically based adaptive tolerance is understood, it is not wise to suggest that this process is consistent between species, even when they are within the same order (see Chapter 6).

The polychaete worm *Nereis diversicolor* has been used as a biomonitor of metal (particularly copper) contamination in sediments (Bryan 1971). Species that are good biomonitors of contamination are probably suitable for evolutionary studies because, by definition, some proportion of them survive in locations that are heavily contaminated. Durou et al. (2005) observed differential tolerance to Zn between two populations of the polychaete *N. diversicolor.* However, tolerance to Cd and Cu did not differ between these same populations. Worms from the impacted site (the Seine estuary) were generally smaller and had lower energy reserves than those from the reference site. The authors suggest that this may indicate a cost of Zn tolerance. However, many other factors may differ between the two sites, including food availability, and further experimental work would be necessary to confirm the link between tolerance and energy consumption. *N. diversicolor* populations from three sites in the UK were tested for differential tolerance to Cu and Zn (Zhou et al. 2003). No difference in Zn or Cd tolerance was observed. There were only slight differences in copper tolerance, but the most tolerant worms were not from the most contaminated site, despite it being one of the most metal-contaminated sites in northwestern Europe (Dulas Bay, north Wales). Mouneyrac et al. (2003) tested *Nereis (Hediste) diversicolor* originating from a strongly metal-contaminated area and a relatively clean site against a range of metals. They observed increased tolerance to Cd, Cu, and Zn of the impacted worms, whereas the opposite was shown for Ag.

2.3.2.4   Other Aquatic Invertebrates

One of the most interesting studies of contaminant tolerance to be published in recent years is work on the dynamic nature of contaminant tolerance in riverine chironomid (midge) larvae in Belgium (Groenendijk et al. 2002). Previous studies had identified temporally variable life history strategies and tolerance to metals in the midge. They had hypothesised that this variation was related to seasonally variable mixing rates (gene flow) between populations of chironomids (Groenendijk et al. 1999). High levels of gene flow should reduce the speed of adaptation to any selective force. They simulated this seasonally based mixing by running cross-breeding studies in the laboratory. Populations that originally displayed tolerance lost this tolerance when cross-bred with midges collected from uncontaminated sites. This rapid loss of metal adaptation was observed in the very first generation of hybrid offspring. Tolerance will therefore be a result of selection within a contaminated site and the rate of recruitment or mixing with populations in clean sites. If breeding is seasonal, then tolerance may be more frequently observed during the nonbreeding season. High gene flow may explain why genetically adapted populations are not often observed in the field in this species and possibly many others.

An interesting study monitored a behavioural response as well as weight loss in caddisfly larvae (*Hydropsyche betteni*) (Balch et al. 2000). Larvae were collected from near a zinc mining operation and from an unpolluted site and were exposed in the laboratory to elevated concentrations of zinc. Abnormalities in the construction of the nets that the larvae use to capture food and body mass loss were both monitored. Slightly greater weight loss and slightly more abnormalities in net architecture were observed in reference larvae when exposed to zinc. This suggested that chronic preexposures may be triggering differential tolerance. Further generations were not tested.

A second behavioural study observed the evolution of contaminant detection responses in a freshwater snail *Physella columbiana* (Lefcort et al. 2004). Snails from a site that had been contaminated with metals for about 120 years were better able to detect and avoid metals in a laboratory choice study than snails from a reference site. Offspring of snails from the polluted site were somewhat less tolerant than their parents, but more tolerant than the reference snails. Lefcort et al. (2004) suggest that the snails can exist in the polluted site, not because they are more tolerant to contaminant exposure, but because they are better at detecting and avoiding hot spots of contamination within the site. This situation would only occur when the distribution of contaminants is patchy and the species are sufficiently mobile.

Interspecies differences in contaminant sensitivity are the general rule, and it might be predicted that simple organisms with large surface areas and limited waste excretion mechanisms might be more susceptible to contaminant exposure than others. As such, it is surprising that so few toxicity studies are conducted on one of the simplest metazoan groups, the platyhelminths or flatworms (Lee and Johnston 2006). This review uncovered a single study on platyhelminths using five populations of the planarian *Polycelis tenuis*, collected from four locations in a metal-contaminated stream and one location in an unpolluted tributary (Indeherberg et al. 1999). The study compared the worms' chronic (twenty-eight-day) sensitivity to cadmium using survival, reproduction, and body size changes. One of the populations that occurred downstream from an old lead-zinc mine displayed an increased rate of elimination and a greater ultimate LC50 value than the other four populations.

2.3.3 COMMUNITY-LEVEL STUDIES

In a series of field-based microcosm studies, Bahrndorf et al. (2006) found conflicting evidence for the adaptation of chironomids to metals. They used microcosms containing different levels of contaminants placed in polluted and reference locations. Chironomids recruited to the sediment boxes in the field over a five-week period. The boxes then returned to the laboratory and the emergence of chironomids was monitored over a two-month period. Emergence is representative of both fecundity and survival of the chironomids from immature stages to the adult. The response of one chironomid species, *Chironomus februarius,* suggested local adaptation, while another species, *Kiefferulus intertinctus,* did not. In a second experiment, Bahrndorf et al. (2006) chose two polluted and two unpolluted sites. In this experiment, species responded differently to contamination and there was no evidence for adaptation. Moreover, species diversity (a community-level measure) did not respond to pollution—a result consistent with many previous studies of chironomid assemblages. The authors concluded that adaptation to heavy metals may be uncommon in the field and also species specific.

In another community-level field study, this time in a marine system, Dafforn et al. (2009) detected several species whose presence was positively correlated with ambient copper contamination levels. Two serpulid polychaetes were positively related to background levels of Cu across the sixteen sites in four estuaries, and these species also showed a positive response to experimentally applied copper-based paints. Tolerance to experimentally applied copper was only observed at the sites where background Cu levels were already elevated (recreational boating areas). This indicates a series of locally adapted populations, but would require further testing of subsequent generations in the laboratory. In a series of field and laboratory studies, Piola and Johnston (2006a,b, 2009) show that nonindigenous fouling marine invertebrates are generally more tolerant to copper than native ones, and that this facilitates the invasion of marine communities. This series of assays did not test beyond the first-generation offspring from field-collected parents, and hence it is possible that they are observing the ecological effects of physiological acclimation to contaminants mediated through changes to maternal provisioning.

Studies such as those outlined above give us the opportunity to test for population tolerance in entire communities. Pollution-induced community tolerance (PICT) (see Chapter 4) is observed when communities from a polluted site are generally less sensitive to contaminant exposure in the laboratory (Grant 2002). This may be observed as smaller changes in the diversity or abundance of a community when exposed to a contaminant. Courtney and Clements (2000) conducted microcosm studies on intact benthic invertebrate assemblages collected from two sites in a metal-polluted river and a reference site in a clean river in Colorado. They exposed the assemblages to either metals (Zn, Cu, Cd) or pH 4.5 in stream microcosms, and their responses corresponded to the exposure history of the assemblage: reference communities were more sensitive to metals, while metal-impacted assemblages were more sensitive to acidic pH. The authors suggest that the enhanced metal tolerance results from a combination of assemblage-level differences in community structure and population-level differences in sensitivity among sites.

2.4   DISCUSSION

This review uncovers plenty of recently published evidence for differential contaminant tolerance in chronically exposed field populations of aquatic animals. Aquatic vertebrates exhibit tolerance to complex organic contaminants that represent a combination of short-term physiological responses (including ones that can be transmitted to the first generation only) and longer-term structural shifts in population genetics. Invertebrates display tolerance to a wide range of metals, and where tests have been done, there is substantial evidence that tolerance is a result of selection for tolerant genotypes. There is a great deal more research required to elucidate the underlying mechanisms of tolerance, and this should be based on well-designed studies on the offspring of chronically exposed populations compared with multiple reference populations.

In general, differential tolerance involved populations that were more than twice as tolerant to contaminant exposure than unimpacted populations. From an environmental perspective, the major concern regarding these observations is that guidelines for the protection of the environment may not be adequately protective if they are based on sensitivity data from adapted populations. In order to deal with this concern, the variability in tolerance traits should be included in ecological risk assessment safety margins. If differences in tolerance ratios are relatively small, then they may be encompassed by conventional safety factors used to establish protective guidelines. This is consistent with the results of Medina et al. (2007), who argued that existing safety margins are likely to be protective of regulations based on adapted populations. However, given the large range in tolerance ratios that we observed, safety margins may not be adequate for all contaminant classes, and it is crucial that we continue to investigate the extent to which tolerance can modify an organism's response. Moreover, we must pay attention to the conditions under which laboratory test organisms are reared. There is evidence that test populations may evolve to be hypersensitive to trace elements if they are reared in unnaturally depleted waters. Alternatively, test populations may become unusually hardy if they are maintained in contaminated environments.

2.4.1 ECOLOGICAL CONSEQUENCES OF DIFFERENTIAL TOLERANCE

Contamination is strongly associated with substantial reductions in the biodiversity of aquatic communities (Johnston and Roberts 2009), but some organisms survive. It is likely that the evolution of tolerance to a contaminant will be species specific and may result in the ecological dominance of pollution-tolerant species (Rygg 1985). The consequences of differential tolerance for wild populations is rarely examined, and studies of the ecosystem cost of tolerance are equally rare (Wirgin and Waldman 2004). Tolerance has been shown to facilitate marine invasions in the field resulting in the biotic homogenisation of contaminated habitats (Piola and Johnston 2008). Nonindigenous marine invertebrates introduced on ships' hulls are generally more tolerant to copper than natives in Australian waters (Piola and Johnston 2008; Dafforn et al. 2009; Piola and Johnston 2009), and this may lend them a competitive advantage in disturbed locations.

Other evidence suggests that the evolution of tolerance may result in small observable ecological differences in communities at impacted and control locations (Kelaher et al. 2003). Indeed, it was originally suggested that the presence of increased tolerance might be a useful indicator of the presence of ecotoxicologically significant bioavailabilities of a toxic contaminant (Luoma 1977). However, it is considered more likely that ecological impacts will occur and be detected before adaptation is observed (Klerks 2002). Tolerance does not always develop, and while it may provide strong evidence of impact by a particular contaminant, it will be of limited use as a monitoring or early warning tool (Grant 2002).

2.4.2 WHEN TOLERANCE DOES NOT OCCUR

There are several cases where one might expect to see the evolution of tolerance in response to a strong selective agent of toxicity, but none is observed. Kovatch et al. (2000) found no difference in the toxicant sensitivity of copepod populations, although they did find differences in population genetics. Rainbow et al. (2003) present an example in which accumulation and detoxification mechanisms appear to be sufficient to cope with massively elevated metal exposures without selection for metal-tolerant populations. In many cases it may be that dispersal between contaminated and reference populations is sufficient to obscure and prevent the observation of differential tolerance, particularly if contamination is inconsistent.

Fitness costs of adaptation may also counterbalance the selective advantage of increased tolerance to contamination (Medina et al. 2007). The frequency and extent of trade-offs between tolerance and other fitness traits are reviewed in Chapter 12. In this chapter we see that tolerance may result in reduced fitness in other contexts. Killifish tolerant to contaminated sediments were less tolerant to photoenhanced toxicity and natural hypoxia (Meyer and Di Giulio 2003). Piola et al. (2006a) detected differential tolerance to copper in laboratory-grown populations that resulted in reduced growth rates of a colonial bryozoan. Differential growth rates in the laboratory, however, were not relevant to success in the field, where the major selective pressure was competition for space rather than metal toxicity. Further studies are needed to examine tolerance costs and benefits under realistic environmental and ecological conditions. The ramifications of adaptation are not always costly (Harper et al. 1997), particularly where the genetic mechanism of adaptation is known to be associated with few loci (e.g., in plants).

2.5 CONCLUSIONS

While we continue to observe cases of tolerance to contaminants in field populations, there are few generalities that can currently be made for aquatic organisms. Studies comparing only the tolerance of field-collected populations, while interesting, are really assessing the impact of a particular contaminant. These effects may be accompanied by adaptive change if elevated contaminants alter patterns of natural selection. This review did not describe many studies with statistically powerful genetic breeding experiments on populations from multiple field sites. Breeding studies do exist but are generally conducted on individuals from a single population. They allow us to calculate how heritable a trait such as pollution tolerance is and the additive genetic variance for that population. This heritability influences the rate of response to selection, and breeding studies allow us to observe potential constraints on responses to selection (e.g., genetic covariances between traits) (Lau et al. 2007). The next major advances will probably result from crossbreeding studies between conspecific populations with different tolerances. These studies will utilise the new wave of genomic technologies that will aid our search for toxicant action pathways and the genes or gene expression active in adapted populations. Until a greater understanding of the mechanism of genetically based adaptive tolerance is achieved, we cannot assume that the process will be consistent even between closely related species. Molecular and quantitative genetics studies of the heritability of contaminant tolerance will substantially improve our understanding of pollution tolerance. Well-designed field studies will enable us to assess the costs of pollution tolerance to populations, communities, and ecosystems.

ACKNOWLEDGMENTS

The author was supported by the Australian Research Council while preparing this review through its Australian Research Fellowship and a linkage grant. I thank

Graeme Clark, Louise McKenzie, William Sherwin, and the book editors for comments that improved earlier drafts of this chapter. Mailie Gall assisted with literature search and production of figures.

REFERENCES

Annabi, A., et al. 2009. Comparative study of the sensitivity to cadmium of two populations of *Gambusia affinis* from two different sites. *Environ. Mon. Assess.* 155:459–65.

Bahrndorff, S., et al. 2006. A microcosm test of adaptation and species specific responses to polluted sediments applicable to indigenous chironomids (Diptera). *Environ. Pollut.* 139:550–60.

Balch, G. C., et al. 2000. Weight loss and net abnormalities of *Hydropsyche betteni* (caddisfly) larvae exposed to aqueous zinc. *Environ. Toxicol. Chem.* 19:3036–43.

Bryan, G. W. 1971. The effects of heavy metals (other than mercury) on marine and estuarine organisms. *Proc. R. Soc. Lond.* 177:389–410.

Burnett, K. G., et al. 2007. *Fundulus* as the premier teleost model in environmental biology: Opportunities for new insights using genomics. *Comp. Biochem. Physiol. D* 2:257–86.

Coors, A., et al. 2009. Land use, genetic diversity and toxicant tolerance in natural populations of *Daphnia magna*. *Aquat. Toxicol.* 95:71–79.

Courtney, L. A., and W. H. Clements. 2000. Sensitivity to acidic pH in benthic invertebrate assemblages with different histories of exposure to metals. *J. N. Am. Benth. Soc.* 19:112–27.

Dafforn, K. A., T. M. Glasby, and E. L. Johnston. 2009. Links between estuarine condition and spatial distributions of marine invaders. *Diversity Distributions* 15:807–21.

Durou, C., C. Mouneyrac, and C. Amiard-Triquet. 2005. Tolerance to metals and assessment of energy reserves in the polychaete *Nereis diversicolor* in clean and contaminated estuaries. *Environ. Toxicol.* 20:23–31.

Forrester, G. E., et al. 2003. Growth of estuarine fish is associated with the combined concentration of sediment contaminants and shows no adaptation or acclimation to past conditions. *Mar. Environ. Res.* 56:423–42.

Gale, S. A., et al. 2003. Insights into the mechanisms of copper tolerance of a population of black-banded rainbow fish (*Melanotaenia nigrans*) (Richardson) exposed to mine leachate, using 64/67Cu. *Aquat. Toxicol.* 62:135–53.

Grant, A. 2002. Pollution-tolerant species and communities: Intriguing toys or invaluable monitoring tools? *Human Ecol. Risk Assess.* 8:955–70.

Groenendijk, D., et al. 1999. Fluctuating life-history parameters indicating temporal variability in metal adaptation in riverine chironomids. *Arch. Environ. Contam. Toxicol.* 37:175–81.

Groenendijk, D., et al. 2002. Dynamics of metal adaptation in riverine chironomids. *Environ. Pollut.* 117:101–9.

Haap, T., and H. R. Köhler. 2009. Cadmium tolerance in seven *Daphnia magna* clones is associated with reduced hsp70 baseline levels and induction. *Aquat. Toxicol.* 94:131–37.

Harper, F. A., S. E. Smith, and M. R. Macnair. 1997. Where is the cost in copper tolerance in *Mimulus guttatus*? Testing the trade-off hypothesis. *Function. Ecol.* 11:764–74.

Hoffmann, A. A., and P. J. Daborn. 2007. Towards genetic markers in animal populations as biomonitors for human-induced environmental change. *Ecol. Lett.* 10:63–76.

Indeherberg, M. B. M., N. M. Van Straalen, and E. R. Schockaert. 1999. Combining life-history and toxicokinetic parameters to interpret differences in sensitivity to cadmium between populations of *Polycelis tenuis* (Platyhelminthes). *Ecotoxicol. Environ. Saf.* 44:1–11.

Janssens, T. K. S., D. Roelofs, and N. M. Van Straalen. 2009. Molecular mechanisms of heavy metal tolerance and evolution in invertebrates. *Insect Sci.* 16:3–18.

Johansson, M., K. Räsänen, and J. Merilä. 2001. Comparison of nitrate tolerance between different populations of the common frog, *Rana temporaria*. *Aquat. Toxicol.* 54:1–14.

Johnston, E. L., and D. Roberts. 2009. Contaminants reduce the richness and evenness of marine communities: A review and meta-analysis. *Environ. Pollut.* 157:1745–52.

Kelaher, B. P., et al. 2003. Changes in benthos following the clean-up of a severely metal-polluted cove in the Hudson River Estuary: Environmental restoration or ecological disturbance? *Estuaries* 26:1505–16.

Klerks, P. L. 2002. Adaptation, ecological impacts, and risk assessment: Insights from research at Foundry Cove, Bayou Trepagnier, and Pass Fourchon. *Human Ecol. Risk Assess.* 8:971–82.

Klerks, P. L., and J. S. Levinton. 1989. Rapid evolution of metal resistance in a benthic oligochaete inhabiting a metal-polluted site. *Biol. Bull.* 176:135–41.

Klerks, P. L., and J. S. Weis. 1987. Genetic adaptation to metals in aquatic organisms: A review. *Environ. Pollut.* 45:173–205.

Knapen, D., et al. 2004. Resistance to water pollution in natural gudgeon (*Gobio gobio*) populations may be due to genetic adaptation. *Aquat. Toxicol.* 67:155–65.

Kovatch, C. E., et al. 2000. Tolerance and genetic relatedness of three meiobenthic copepod populations exposed to sediment-associated contaminant mixtures: Role of environmental history. *Environ. Toxicol. Chem.* 19:912–19.

Lau, J. A., et al. 2007. Strong ecological but weak evolutionary effects of elevated CO_2 on a recombinant inbred population of *Arabidopsis thaliana*. *New Phytol.* 175:351–62.

Lee, K. M., and E. L. Johnston. 2006. Low levels of copper reduce the reproductive success of a mobile invertebrate predator. *Mar. Environ. Res.* 64:336–46.

Lefcort, H., et al. 2004. Aquatic snails from mining sites have evolved to detect and avoid heavy metals. *Arch. Environ. Contam. Toxicol.* 46:478–84.

Lopes, I., D. J. Baird, and R. Ribeiro. 2004. Genetic determination of tolerance to lethal and sublethal copper concentrations in field populations of *Daphnia longispina*. *Arch. Environ. Contam. Toxicol.* 46:43–51.

Lopes, I., D. J. Baird, and R. Ribeiro. 2005. Resistance to metal contamination by historically-stressed populations of *Ceriodaphnia pulchella*: Environmental influence versus genetic determination. *Chemosphere* 61:1189–97.

Lopes, I., D. J. Baird, and R. Ribeiro. 2006. Genetic adaptation to metal stress by natural populations of *Daphnia longispina*. *Ecotoxicol. Environ. Saf.* 63:275–85.

Luoma, S. N. 1977. Detection of trace contaminant effects in aquatic ecosystems. *J. Fish. Res. Board Can.* 34:436–39.

Martinez, D., and J. S. Levinton. 1996. Adaptation to heavy metals in the aquatic oligochaete *Limnodrilus hoffmeisteri*: Evidence for control by one gene. *Evolution* 50:1339–43.

Medina, M. H., J. Correa, and C. Barata. 2007. Micro-evolution due to pollution: Possible consequences for ecosystem responses to toxic stress. *Chemosphere* 67:2105–14.

Meyer, J. N., and R. T. Di Giulio. 2003. Heritable adaptation and fitness costs in killifish (*Fundulus heteroclitus*) inhabiting a polluted estuary. *Ecol. Appl.* 13:490–503.

Meyer, J. N., et al. 2003. Antioxidant defenses in killifish (*Fundulus heteroclitus*) exposed to contaminated sediments and model prooxidants: Short-term and heritable responses. *Aquat. Toxicol.* 65:377–95.

Miliou, H., et al. 2000. Influence of life-history adaptations on the fidelity of laboratory bioassays for the impact of heavy metals (Co^{2+} and Cr^{6+}) on tolerance and population dynamics of *Tisbe holothuriae*. *Mar. Pollut. Bull.* 40:352–59.

Morgan, A. J., P. Kille, and S. R. Sturzenbaum. 2007. Microevolution and ecotoxicology of metals in invertebrates. *Environ. Sci. Technol.* 41:1085–96.

Mouneyrac, C., et al. 2002. Partitioning of accumulated trace metals in the talitrid amphipod crustacean *Orchestia gammarellus*: A cautionary tale on the use of metallothionein-like proteins as biomarkers. *Aquat. Toxicol.* 57:225–42.

Mouneyrac, C., et al. 2003. Trace-metal detoxification and tolerance of the estuarine worm *Hediste diversicolor* chronically exposed in their environment. *Mar. Biol.* 143:731–44.

Muyssen, B. T. A., and C. R. Janssen. 2001. Multigeneration zinc acclimation and tolerance in *Daphnia magna*: Implications for water-quality guidelines and ecological risk assessment. *Environ. Toxicol. Chem.* 20:2053–60.

Nacci, D. E., T. R. Gleason, and W. R. Munns Jr. 2002b. Evolutionary and ecological effects of multi-generational exposures to anthropogenic stressors. *Human Ecol. Risk Assess.* 8:91–97.

Nacci, D. E., et al. 2002a. Predicting the occurrence of genetic adaptation to dioxin-like compounds in populations of the estuarine fish *Fundulus heteroclitus*. *Environ. Toxicol. Chem.* 21:1525–32.

Nacci, D. E., et al. 2002c. Effects of benzo[a]pyrene exposure on a fish population resistant to the toxic effects of dioxin-like compounds. *Aquat. Toxicol.* 57:203–15.

Nacci, D. E., et al. 2009. Evolution of tolerance to PCBs and susceptibility to a bacterial pathogen (*Vibrio harveyi*) in Atlantic killifish (*Fundulus heteroclitus*) from New Bedford (MA, USA) harbor. *Environ. Pollut.* 157:857–864.

Piola, R., and E. L. Johnston. 2008. Pollution reduces endemic diversity and increases invader dominance in hard-substrate communities. *Diversity Distributions* 14:329–42.

Piola, R., and E. L. Johnston. 2009. Comparing differential tolerance of native and non-indigenous marine species to metal pollution using novel assay techniques. *Environ. Pollut.* 157:2853–64.

Piola, R. F., and E. L. Johnston. 2006a. Differential resistance to extended copper exposure in four introduced bryozoans. *Mar. Ecol. Progr. Ser.* 311:103–14.

Piola, R. F., and E. L. Johnston. 2006b. Differential tolerance to metals among populations of the introduced bryozoan *Bugula neritina*. *Mar. Biol.* 148:997–1010.

Rainbow, P. S., G. Blackmore, and W. X. Wang. 2003. Effects of previous field-exposure history on the uptake of trace metals from water and food by the barnacle *Balanus amphitrite*. *Mar. Ecol. Progr. Ser.* 259:201–13.

Roark, S. A., et al. 2005. Population genetic structure and tolerance to dioxin-like compounds of a migratory marine fish (*Menidia menidia*) at polychlorinated biphenyl-contaminated and reference sites. *Environ. Toxicol. Chem.* 24:726–32.

Rygg, B. 1985. Distribution of species along pollution induced diversity gradients in benthic communities in Norwegian fjords. *Mar. Pollut. Bull.* 16:469–74.

Semaan, M., D. A. Holdway, and R. A. Van Dam. 2001. Comparative sensitivity of three populations of the cladoceran *Moinodaphnia macleayi* to acute and chronic uranium exposure. *Environ. Toxicol.* 16:365–76.

Vidal, D. E., and A. J. Horne. 2003. Mercury toxicity in the aquatic oligochaete *Sparganophilus pearsei*. I. Variation in resistance among populations. *Arch. Environ. Contam. Toxicol.* 45:184–89.

Weis, J. S. 2002. Tolerance to environmental contaminants in the mummichog, *Fundulus heteroclitus*. *Human Ecol. Risk Assess.* 8:933–53.

Wirgin, I., and J. R. Waldman. 2004. Resistance to contaminants in North American fish populations. *Mutation Res. Fund. Mol. Mechanisms Mutagenesis* 552:73–100.

Woods, R., and A. A. Hoffmann. 2000. Evolution in toxic environments: Quantitative versus major gene approaches. In *Demography in ecotoxicology*, ed. E. Kammenga and R. Laskowski, 129–46. Chichester, UK: Wiley.

Zhou, Q., P. S. Rainbow, and B. D. Smith. 2003. Tolerance and accumulation of the trace metals zinc, copper and cadmium in three populations of the polychaete *Nereis diversicolor*. *J. Mar. Biol. Ass. U.K.* 83:65–72.

3 Inter- and Intraspecific Variability of Tolerance
Implications for Bioassays and Biomonitoring

Brigitte Berthet, Kenneth M. Y. Leung,
and Claude Amiard-Triquet

CONTENTS

3.1 INTRODUCTION

Although the increasingly precise techniques of analytical chemistry allow us to identify and to quantify, in all physical or biological matrices, an increasing number of xenobiotics and their metabolites, they do not provide access to all the toxic molecules of interest. Moreover, these measures supply no information about toxic or ecotoxicological risks, for environmental conditions may modify the chemical characteristics of the compounds, and thus their ecotoxicity. Laboratory bioassays on a few sentinel species are used to determine the nature of the risk, as well as the dose-effect relationship. Nevertheless, these tests do not take into account the influence of the multiple parameters of the environment that intervene in natural conditions: abiotic factors (such as temperature, salinity, and oxygenation), specific diversity, interactions between species, and interactions between pollutants or between pollutants and environmental factors. The objective of biomonitoring is to integrate these parameters using *in situ* sentinel species and biotic indices based on relative tolerance/sensitivity of individual species or higher taxa. The use of sentinel species, in the perspective of biodiversity conservation, allows an evaluation of the damage to organisms, which has the potential, within the short, medium, or long term, to have repercussions on populations, communities, and ecosystems. Bioassays and biomonitoring at different levels of biological organisation (biomarkers, bioindicators) have to be used in complementarity. For instance, Smolders et al. (2004) have compared *in situ* and laboratory bioassays using two test species to evaluate the impact of effluent discharges, reflecting the views of Chapman (2002), who indicates that, often, toxicological studies neglect ecological considerations.

Sentinel species have been looked for in all ecosystems. Recently the prerequisite characteristics of these species have been reviewed by Berthet (2008 and references therein), but differences may be encountered according to the authors and their priorities, notably according to the type of exposure, in the laboratory or *in situ*. In order to include more "eco" in ecotoxicology, Chapman (2002) recommended the choice of dominant or keystone species in the area being assessed, whereas in the ECOMAN project, Galloway et al. (2004) recommended a multispecies approach where species are chosen in consideration of the diversity of their feeding types (filter feeding, grazing, or predation) and habitat requirements (sands, rocky shores, muddy sand). However, while this ecological approach is indispensable, it is not necessarily sufficient in itself, for the extent to which tolerance has evolved in local populations of a chosen sentinel species must be taken into account. Indeed, as shown in several chapters of this book, tolerant species have been identified in many taxa, from microorganisms to vertebrates or plants, while populations within a single species, exposed to environmental pollution, can often develop tolerance to pollutants. These intra- and interspecific variabilities between tested organisms in response to the same toxicants cause problems for interpretation, and can lead to an underestimation of risk. On the other hand, since many bioassays are carried out over short time periods and under conditions of acute toxicity, mechanisms of defence do not have time to come into effect as they would in cases of chronic toxicity, leading to a possibility of overestimation of risk. According to the three objectives described in Figure 3.1, this chapter will develop the implications of tolerance in sentinel species used (1) in risk assessment based on bioassays, (2) as models for monitoring based on biomarkers and/or chemical quantification, and (3) in the monitoring of biodiversity.

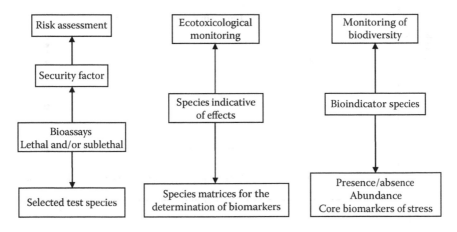

FIGURE 3.1 Different uses of sentinel species according to the objectives of the study.

3.2 TOLERANCE AND CHOICE OF BIOLOGICAL MODELS FOR BIOASSAYS

Bioassays are procedures that involve exposure of organisms to various levels of contaminants to determine if adverse effects occur. These procedures are also called toxicity tests, although McDonald et al. (1992) reserve the first term for determination of the toxicity threshold of a specific substance, whereas the latter is used to determine the toxicity of a whole sample, and not its components. Bioassays may be performed in laboratories, too often in acute toxicity conditions, involving exposure to relatively high concentrations of one or few pollutants, in order to compress the test period into a short and easily manageable duration. Field exposures across a pollution gradient are a good complementary alternative. Certainly, it is not often possible to establish a link between a precise substance and observed effects on organisms, but interactions between the different compounds in mixtures, between this mixture and environmental factors, and between species are taken into account. Whatever the type of test used, the choice of the species tested must be made by respecting numerous criteria such as immobility, sufficient population size, measurable response to the pollutant, function in the ecosystem, etc. (Berthet 2008 and references therein; Niemeijer and de Groot 2008 and references therein). Nevertheless, few studies have focussed on the implications of tolerance on this choice of species.

3.2.1 INTERSPECIFIC VARIABILITY OF TOLERANCE

The validation of a taxon as a sentinel organism is fundamentally based on the possibilities of extrapolating the results obtained to all the species of the ecosystem. However, even between very closely related taxa, sensitivity to xenobiotics can vary by several orders of magnitude. Takamura et al. (1989) have illustrated this variability in testing the photosynthetic activity of seven classes, comprising 118 species, of freshwater microalgae exposed to Cu, Cd, or Zn. Their work has shown differences

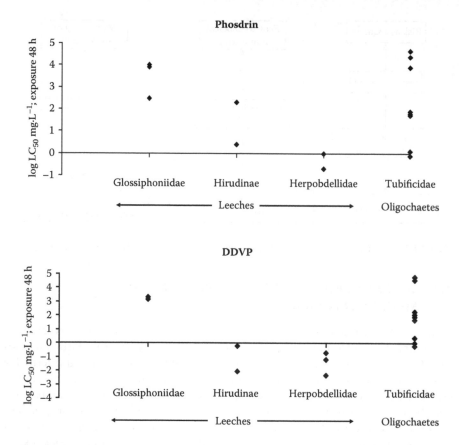

FIGURE 3.2 Comparison of tolerance of various species of leeches and oligochaetes to organophosphate pesticides (DDVP = dimethyldichlorovinyl phosphate). (After Lapkina, L. N., and N. R. Arkhipova, *Russ. J. Ecol.*, 5, 336–40, 2000.)

of several orders of magnitude in the inhibition of 50% of the photosynthetic activity of microalgae belonging to the same taxon (Figure 1.1). Lapkina and Arkhipova (2000) also illustrated this variability among the phylum Annelida, exposing eleven oligochaete and thirteen leech species to several pesticides. As shown by some of their results presented in Figure 3.2, they have observed important differences in lethal concentrations between the classes of annelids, and even between species in the same family.

In fact, relatively tolerant and sensitive species are numerous and known in many taxa, from microorganisms to vertebrates or plants; examples are illustrated throughout the different chapters of this book.

Furthermore, differences are also observed according to the ecotoxicological endpoint measured (physiological, cellular or biochemical biomarkers, survival, growth inhibition) and the type of test (acute or more or less prolonged). Brown et al. (2004) have illustrated the differential sensitivity to copper of three marine invertebrates from different functional feeding groups. They have exposed blue mussels (filter

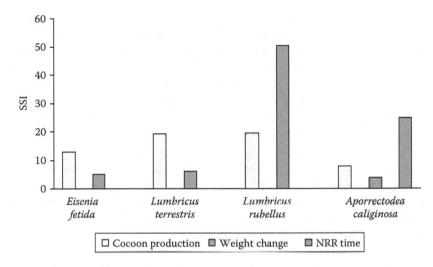

FIGURE 3.3 Comparison of sublethal sensitivity indices (SSIs) for cocoon production, weight change, and NRR time for four species of earthworms exposed to zinc. (After Spurgeon, D. J., et al., *Environ. Toxicol. Chem.*, 19, 1800–8, 2000.)

feeders), common limpets (grazers), and shore crabs (omnivores) to environmentally realistic concentrations of copper for seven days, and first observed that the order of relative sensitivity differed from that derived from acute toxicity tests (limpets > crabs > mussels in their experiments, limpets > mussels > crabs according to the 48 h LC50). They also observed that all of the five sublethal biomarkers measured in the limpets exhibited significant responses above the weakest tested concentrations (6.1 µg.L^{-1}). Mussels, in contrast to the acute toxicity data, were relatively insensitive compared with crabs, presenting only effects for one of the sublethal biomarkers (lysosomal activity), and only at 68.1 µg Cu·L^{-1}, whereas crabs demonstrated modification of three biomarkers (lysosomal activity, MT concentration, and AChE activity) at this concentration of copper.

Kammenga and Riksen (1996) have tested the effect of copper on reproduction and population growth rate in two species of terrestrial nematodes. Reproduction (daily production of eggs, juvenile survival, survival during the reproductive period, length of the juvenile period, and length of the reproductive period) indicated differences in tolerance to copper between the two species, but the impact on the population growth rate was similar for both species. In the same context, Spurgeon et al. (2000) compared the responses of four earthworm species to zinc, through two series of endpoints: life cyle traits (survival, weight changes, and cocoon production rate) and a biomarker (neutral red retention (NRR) time by coelomocyte lysosomes). To compare the relative sensitivities of life cycle and biomarker responses, they applied a sublethal sensitivity index (SSI = LC50/EC50) (Figure 3.3). Their results indicate that according to the species, maximum sensitivity is observed for different parameters (cocoon production for *E. fetida* and *L. terrestris*, NRR time for *L. rubellus* and *A. caliginosa*). The threshold concentration for the effects of zinc on weight was the least sensitive.

Improving risk assessments needs to put more "eco" into ecotoxicological tests, using keystone species characteristic of the studied area. For example, estuarine ecosystems are submitted to physicochemical changes (temperature, salinity, turbidity, oxygen concentration, organic matter, etc.) over tidal cycles with additional seasonal fluctuations. However, Chapman and Wang (2001) underline that as far as estuarine ecosystems are concerned, in too many cases, sediment toxicity tests are carried out on freshwater, or on a small number of marine species, but few on characteristic species of estuaries. Moreover, they regret that the bioassays do not try to reproduce the interstitial conditions of salinity, and those of overlying water as found *in situ*. Salinity variations, but also concentrations of organic matter, dissolved oxygen, are well known to change the bioavailability and toxicity of pollutants. Thus, the results obtained in laboratory, under rather unrealistic conditions, must be considered very carefully before extrapolation to the real ecosystem to avoid erroneous conclusions.

On the other hand, organisms able to cope with fluctuating parameters, such as observed in estuaries, have often been considered able to tolerate additional stress due to xenobiotics. However, this assumption is not confirmed by several authors, such as Hummel et al. (1997) and Heugens et al. (2001). This aspect will be developed in Chapter 5.

Because the sensitivity of a given species depends on the pollutant involved, the ecotoxicological endpoint measured, and the type of test used, several species and endpoints have to be investigated in risk assessment. The variability of interspecific responses constitutes a major problem, thus limiting the possibilities of extrapolating bioassays to the whole ecosystem. Nethertheless, this variability is used in the concept of the pollution-induced community tolerance (PICT) developed by Blanck et al. (1988) and revisited in Chapters 4 and 14.

3.2.2 INTRASPECIFIC VARIABILITY OF TOLERANCE

In order to reduce the problems of interpretation of results due to the interspecific variability of tolerance, it is recommended to use standardised tests with selected species. However, intraspecific variabilities are also a core issue for the development of toxicity reference values. This variability can have its sources in the geographic origin of the organisms, but also in some intrinsic parameters.

3.2.2.1 Origins of the Tested Organisms

3.2.2.1.1   Wild Populations versus "Laboratory Rats"

From an operational point of view, organisms used in bioassays can be reared/cultivated in laboratory or taken from the natural environment. Among the evoked advantages of the first ones are their availability and their genetic stability. Thus, species with parthenogenetic reproduction, such as daphnid crustaceans, which produce genetically identical neonates, are usually used in lethal or sublethal toxicity tests. Several authors (Baird 1992; Lam and Gray 2001) have suggested that, because the selection of test organisms has often been established on their tolerance to laboratory conditions, they are fairly hardy and could exhibit a significantly increased tolerance to various stresses. On the other hand, studies carried out on various species seem to prove the opposite. Barata et al. (2002) have tested the tolerance to cadmium of *Daphnia magna* clones from four geographically separate European populations. Their results suggest that optimal laboratory conditions will favour individuals with high fitness or reproductive performance, resulting in reared populations with similar or lower tolerance to toxic stress than the original field population. They conclude that minimising genetic diversity in toxicity tests will increase the uncertainty attendant in extrapolating from the laboratory to the field. Comparing the responses (biological fitness, biochemical parameters) of a field and a laboratory-derived population of a dipteran, the midge *Chironomus riparius*, to four insecticides (malathion, parathion, propxur, and DDT), Hoffman and Fisher (1994) suggested that the field population, as a result of selection pressures, had developed resistance against insecticides. On the other hand, no such resistance mechanisms were observed in the laboratory-derived population, which was less tolerant to these molecules. Nowak et al. (2007), studying the effects of cadmium on *C. riparius*, think that, for populations well represented in the natural environment, and characterised by significant mobility, high gene flow exists between adjacent habitats, in contrast to the laboratory, where genetic impoverishment is observed. The same observations are true for other taxa, such as the zebrafish *Danio rerio*, commonly used in freshwater toxicological studies (Coe et al. 2009), the amphipod *Hyalella azteca* (Duan et al. 1996), or the parasitic trematode flatworm *Schistosoma mansoni* (Stohler et al. 2004). This genetic impoverishment of laboratory populations has to be regarded as an artificial bias for ecotoxicological tests, and regular refreshments with individuals obtained from the field will restore genetic variation. Bossuyt and Janssen (2003) have tested the influence of copper acclimation of laboratory *D. magna* cultures over three generations on the results of acute and chronic toxicity tests. As the internationally recommended culture and test media for daphnids may be deficient in essential trace elements like copper, they hypothesise that this may affect the health of the daphnids, and thus the interpretation of toxicity tests may be difficult.

Thus, laboratory strains established from different geographical regions or impacted habitats might show different levels of stress susceptibility, which will lead to variability in their sensitivity or their tolerance according to the applied pollutant. Then, there exist no generally sensitive or tolerant strains, and if using clones in ecotoxicological bioassays can reduce the genetic variability, we have to recognise that this strain is possibly not as sensitive as another one of this species in the field, leading to under- or overestimation of the risk for natural populations. That is taken into account by the application of safety factors (Figure 3.1). Nevertheless, keeping up the genetic variability by collecting organisms from natural environments would lead to other biases, in particular from an ignorance of their life history exposure to contaminants.

Moreover, in most cases, toxicity tests are conducted on a single species. Nevertheless, Chapman et al. (1982), using mixed populations of freshwater oligochaetes (*Limnodrilus hoffmeisteri* and *Tubifex tubifex*), concluded that combinations of species may supply data different from those obtained on isolated organisms: higher tolerance to Cd, Hg, or sodium pentachlorophenol and, on the contrary, lower tolerance to anoxia. Similarly, Keilty et al. (1988) showed that when exposed to sediment contaminated by endrine, *L. hoffmeisteri* benefits from the presence of the oligochaete *Stylodrilus heringianus*, with a decrease of its mortality, an increase of its mass, and a stimulation of its bioturbation.

3.2.2.1.2   Naive Populations versus Exposed Ones

Within a same population, all the individuals do not present the same sensitivity to contaminants. When the population is exposed at sublethal doses, the sensitive individuals disappear, and only the most resistant remain. The tolerance can result from the implementation of various mechanisms of defence, such as involvement of metallothioneins, biomineralisation in the case of metal pollution, enzymes of biotransformation, or enzymatic defences against oxidative stress in the case of exposure to polyaromatic hydrocarbons (PAHs), polychlorinated biphenyls (PCBs), etc. (Chapters 6 to 10). Moreover, some organisms also have the ability to repair damage induced by toxic insults, as shown by Ching et al. (2001), who conclude that when green-lipped mussels are exposed to benzo[a]pyrene, observed increases in DNA adducts and DNA strand breaks responses probably indicate exposure, but an apparent lack of response could be due to either an absence of inducers (e.g., sufficiently high levels of genotoxic agents) or the action of an efficient repair system.

These mechanisms require from a few days to some weeks to come into effect, and thus have no time to be effective in acute toxicity tests, leading to an overvaluation of mortalilty, compared to the chronic exposures that often occur in the field. On other hand, the use of preexposed individuals will increase the chance of a false negative, and thus will underestimate the risk in relatively clean habitats. However, sometimes, the protection may be limited in time, since de-acclimation can also occur over a short period if the concentration of the pollutant decreases enough (Lotts and Stewart 1995).

A knowledge of previous exposure is thus essential before choosing test organisms, not only for the pollutant of concern, but possibly also for others, as shown by McGeer et al. (2007), studying cross-acclimation to metals on acute Cd toxicity in rainbow trout *Oncorhynchus mykiss*. Their results demonstrate that chronic sublethal exposure to Cu produces enhanced resistance to Cd. Norwood et al. (2007) expressed regret that there is no consistent method of quantifying the effects of metal mixtures, for their work on the amphipod *Hyalella azteca* showed that Co, Cd, and Ni bioaccumulation was significantly inhibited, and As bioaccumulation enhanced, with an increase in the number of metals in the test medium during a one-week exposure. Such cross-acclimation also exists between organic compounds like PCBs and DDT, as demonstrated for marine diatoms by Cosper et al. (1987). Since numerous sites suffer from multipollution, this phenomenon of cross-resistance must be carefully taken into account to avoid false positives or negatives in the conclusions of toxicological tests.

Tolerance may be genetically based (Chapter 2), and the selection of tolerant individuals in a contaminated area may lead to the establishment of a tolerant population. Today, numerous cases of tolerance are known for the problems that they cause in agriculture or for the health of man and the domesticated species. For example, in a recent book on pesticide resistance in arthropods, Whalon (2008) reported more than 7,747 cases of pesticide resistance, with more than 331 insecticide compounds involved. Whalon (2008) also counted 553 arthropod species with resistance to insecticides. We could also quote the resistance of bacteria against antibiotics, those of plants to herbicides, and others (Chapters 14 to 16). If the examples of transmission of tolerance to offspring are many (Chapter 2), there are also counterexamples: Sánchez et al. (1999) have exposed daphnids, *D. magna*, to the insecticide diazinon and observed an increase in susceptibility to the pesticide over generations.

3.2.2.1.3 Stable Environments versus Fluctuating Ones

In laboratory studies, test organisms are almost always kept in optimal conditions of temperature, oxygen, salinity, food, hygrometry, etc., in order to optimise performance in the control, and to isolate the effects of the tested chemical. In their natural environments, organisms are rarely in optimal conditions, but on the contrary have to cope with environmental stresses. Euryecic species have been considered for a long time to be more tolerant to additional stress, such as chemicals, than stenoecic ones (Slobodkin and Sanders 1969; Fisher et al. 1973; Fisher 1977). Estuarine organisms show good patterns of strong ecological valency, and several works have shown that this assertion is still not verified. Heugens et al. (2001) have shown that metal toxicity is smaller for the various individuals living near their iso-osmotic salinity area, because of the lack of the physiological cost of osmoregulation. In a Cd-polluted environment, the cytoprotective upregulation of heat shock proteins is partially suppressed in the oyster *Crassostrea virginica* (Ivanina et al. 2009), which may contribute to a reduced thermotolerance shock that is observed at low tide in summer. In the same species, elevated temperature (28°C) and anoxia (as during emersion at low tide) increased sensitivity to Cd at the mitochondrial level, resulting in a fast decrease in the capacity to synthesise ATP (Lannig et al. 2006; Eilers et al. 2009). Moreover, Hummel et al. (1996; see also Chapter 5) have studied the sensitivity of the estuarine bivalve *Macoma balthica* to copper from its southern to its northern distribution limit. Their work showed that when aquatic organisms live at the limit of tolerance to natural changes, as may be the case in estuaries, they are generally more sensitive to an additional stress (exposure to xenobiotics, for example). Studying the same species in the Baltic Sea, at the edge of its salinity tolerance, Sokołowski et al. (2004) suggested that any environmental change (natural and/or anthropogenic) may disturb the physiological functioning of the clam (resulting in strong deformation of the shell, high metal accumulation).

In the blue crab *Callinectes sapidus*, a euryhaline crab, the response to lowering salinities is characterised in gills by an increased oxygen consumption rate and approximately double metallothionein-like protein (MTLP) concentrations (De Martinez Gaspar Martins and Bianchini 2009). Such variations in MTLP concentrations were also observed in other crabs, *Carcinus maenas* and *Pachygrapsus marmoratus* (Legras et al. 2000; Mouneyrac et al. 2001). As MTLPs play a primary role in the homeostasis of essential metals (Cu, Zn, etc.), as well as the detoxification of nonessential ones, particularly Cd, changes in MTLP concentrations may affect metal toxicity in euryhaline crustaceans.

In small freshwater ponds, temperature and food supply may be highly variable. Heugens et al. (2006) have tested the population growth of the common water flea *D. magna* under the joint effects of temperature (within the range of temperatures observed in the field), food level (ranging from concentrations used in the standard

Daphnia reproduction test to lower food levels that are common in aquatic eco-systems), and Cd concentration. They observed that negative effects of Cd were enhanced when temperature increases, but that high food levels protect the daphnids from the effects of Cd, and concluded that a substantial part of uncertainty in toxicity tests could be explained by natural factors.

Thus, spatiotemporal heterogeneity of ecosystems as regards the distributions of xenobiotics causes organisms to suffer different exposures, for different periods, and in different concentrations and bioavailabilities. The different histories of the experimental specimens offer so many possible confounding factors to be taken into account to correctly interpret the results of toxicity tests and to extrapolate to the field. But, intrinsic parameters also have to be taken into account.

3.2.2.2   Intrinsic Parameters

In nature, populations continously change quantitatively and qualitatively (ratio of adults to juveniles, of males to females), whereas seasonal external factors (temperature, oxygenation, nutritional status of the organisms) will modify physiological conditions. On the contrary, toxicity tests are carried out with homogeneous populations (sex, age, size, etc.), as, for example, standard tests with the water flea *D. magna*, which are performed with less than 24 h old organisms (OECD 1998). Thus, several authors have investigated the influence of biological parameters on the sensitivity of organisms to chemicals with respect to fluctuations in mechanisms of defences (induction of metallothioneins, antioxidant defences, biotransformation enzymes, etc.) or disturbance in their functioning (e.g., inhibition of cholinesterases, thiobarbituric acid reactive substance (TBARS) concentrations). These confounding factors are many; they can affect each other and can have antagonistic or synergic effects between them. As they are quite different depending on the species concerned, they cause problems for the interpretation of bioassays, and are considered a challenge to be taken into account in ecotoxicology.

3.3   CHOICE OF SENTINEL SPECIES FOR ECOTOXICOLOGICAL MONITORING

The interspecific variability of biological responses that may be used as biomarkers of exposure to, or effects of, chemicals represents a serious challenge for environmental risk assessment since different species are more or less tolerant to the presence of toxic compounds in their medium. This variability can be linked to differences in exposure (for instance, through feeding habits), in bioaccumulation pattern—some species being well recognised as strong accumulators (Rainbow 2002)—or in intrinsic biological characteristics, particularly in the case of defence mechanisms deployed by organisms when submitted to toxicants.

In this section, we will examine how interspecific and interpopulation differences interfere with biomarkers such as biochemical responses of defence that contribute to tolerance, but also biomarkers of damage that reveal the tolerance capacity has been overwhelmed. This discussion will be based on three major cases involving species of interest as sentinels: bivalves representative of the water column, namely, mussels and oysters commonly used in Mussel Watch-type programmes; endobenthic invertebrates, which are key species for the structure and functioning of estuarine and coastal areas; and among vertebrates, estuarine fish living in these areas, which are characterised by both their high biological richness and their high degree of pollution.

3.3.1 MUSSELS AND OYSTERS

It is a striking feature of the literature that body concentrations of accumulated trace metals, essential or nonessential, vary greatly between invertebrates, often between closely related species. Concentration data from biomonitoring programmes based on metal analyses in different mussel and oyster species reveal that the latter are much stronger accumulators, particularly in the case of Cu and Zn (Luoma and Rainbow 2008). In the framework of the French Mussel Watch (RNO 2006), contamination ratios calculated from metal concentrations determined in oysters (*Crassostrea gigas*) and mussels (*Mytilus* spp.) collected from the same site reached 2.5 for Cd, 10 for Cu, 15 for Zn, and 50 for Ag. A key factor in appreciating the significance of different metal concentrations in marine invertebrates is an understanding of the metal accumulation pattern adopted. On the other hand, interspecific differences were small and nonsignificant for organic contaminants.

It is unclear if these differences in bioaccumulation are associated or not with interspecific differences in tolerance. Because embryo-larval tests are widely used to assess the toxicity of marine pollutants as well as marine environmental quality, the relative sensitivity/tolerance of different species of oysters and mussels has received considerable attention. A review by His et al. (1999) concluded that various species used in embryo toxicity bioassays appear to have similar sensitivities to environmental contaminants. "Green" oysters (*Ostrea edulis, Crassostrea gigas*) containing concentrations of several thousands μg Cu.g^{-1} dry weight, representing a threat to human consumers, were observed in fisheries located in contaminated coastal environments such as Restronguet Creek, UK, submitted to long-term contamination by mining wastes (Bryan et al. 1987), or the coastal environment in Taiwan (Han et al. 1994). On the other hand, mussels *Mytilus edulis*, which might be expected in Restronguet Creek based on evidence from other creeks in the same estuarine system, were absent (Bryan et al. 1987). In the metal-contaminated Gironde estuary, recruitment of mussel (*M. edulis*) and oyster (*C. gigas*) spat was observed, but adult mussels were absent, whereas oysters were covering all hard surfaces available in the polyhaline area (Geffard et al. 2004). However, in mussels translocated from a comparatively clean site to the Gironde estuary, the condition index was not affected and no cellular pathology was observed. Because globally mussels and oysters share the same detoxification processes, Geffard et al. (2004) hypothesised that the absence of adult mussels in the Gironde may be due to competition with oysters for free surfaces for settlement as well as predation by crabs. Despite higher metal accumulated concentrations, oysters *Crassostrea angulata* collected from the South Atlantic Spanish littoral along a metal contamination gradient were found in the more polluted environments, unlike mussels *Mytilus galloprovincialis* (Funes et al. 2006).

Anyway, it is questionable if the ability of oysters to store large quantities of metals interferes or not with the responses of biomarkers linked to metal contamination. Metals are known as inducers of metallothioneins, the use of which is recognised as biomarkers of metal exposure in a number of species (Amiard et al. 2006), and some metals, particularly Cu, are also known to be responsible for oxidative stress in organisms (Gaetke and Chow 2003).

In both mussels and oysters, tolerance to metals is based on the control of intracellular metal speciation: sequestration by cellular ligands such as in lysosomes, mineralised or organic-based concretions, and metallothionein (MT). In both taxa, insoluble storage is important, but class I MTs, exhibiting considerable similarity to mammalian MT, have also been reported (see review by Langston et al. 1998). Interspecific and interorgan differences have been systematically investigated in bivalves from the metal-rich Gironde estuary (Amiard et al. 2006). Based on the assumption that MT is induced by metals, it would be expected that tissues with the highest accumulated metal concentrations should have the highest MT concentrations. Despite huge interspecific differences for metal concentrations, MT concentrations were not so very different, reaching a few hundred mg.kg^{-1} (wet weight) in the gills of both species and a few thousand in their digestive glands. Again, these remarkable interorgan differences in MT concentrations cannot be explained by interorgan comparisons in accumulated metal concentrations. The accumulated concentrations of each of the three studied metals (Cd, Cu, Zn) were of the same order of magnitude in both gills and digestive gland of mussels, and the same pattern was shown in the case of oysters. From an operational point of view, the most important thing is to obtain a signal significantly higher than the background. When background levels are low, any change would be theoretically easy to assess, suggesting that gills would be the best matrix for MT determination in bivalves. In fact, because a number of confounding factors can also intervene (season, sexual maturity, and associated weight changes), it has been suggested that MT concentrations in the digestive gland of mussels represented the biomarker reflecting raised metal exposure with the best reliability (Geffard et al. 2005), even if oyster gills could also be a useful tool (Geffard et al. 2002a). In order to avoid seasonal effects as well as changes linked to sexual maturity, mussel and oyster larvae may be recommended since MT induction in these organisms constitutes a sensitive indicator of metal contamination (Geffard et al. 2002b, 2007).

At the moment, most published works only provide total MT concentrations, whereas isoforms coexist within organisms. Molecular biological advances using MT gene amplification or duplication have confirmed that the functions of different isoforms are different, some of the isoforms being involved in metal homeostasis and others in nonessential metal detoxification. These general findings are documented in mussels *Mytilus edulis* (Lemoine et al. 2000) and oysters *Crassostrea gigas* (Tanguy et al. 2002) and *C. virginica* (Jenny et al. 2006). However, in another oyster, *Ostrea edulis*, Tanguy et al. (2003) observed no significant induction of MTs, either in the gills or the digestive gland, nor MT-RNA expression. These results led to the conclusion that MTs are not involved in metal detoxification in this oyster, consistent with the lack of genetic diversity in field populations. The determination of MT concentrations as a whole is adapted to biomonitoring aiming at assessing the global degree of metal contamination. If the objective is to characterise undefined metal contamination in a given ecosystem, the use of MT isoforms that are responsive to particular metal(s) may be envisaged. In the present state of knowledge, it remains impossible to recommend the use of a peculiar species, organ (Lemoine et al. 2000; Géret and Cosson 2002), or life stage (with isoforms expressed differently in larvae or adults, as shown by Jenny et al. 2006 in *C. virginica*).

Biomarkers of oxidative stress were determined in oysters *Crassostrea angulata* and mussels *Mytilus galloprovincialis* collected from the South Atlantic Spanish littoral (Funes et al. 2006). Oysters, with the higher metal loads of the two species, showed increased antioxidant defences (superoxide dismutase, catalase, glutathione peroxidases, glutathione S-transferases) and less extensive oxidative damage (malonedialdehyde and 8-oxodG contents). In contrast, mussels, which accumulated much lower metal concentrations, showed clear increases in oxidised biomolecules, in agreement with their low increases in antioxidant defence mechanisms.

According to the recommendation of Funes et al. (2006), oysters can be chosen as model sentinel organisms for adaptation studies, while mussels, more sensitive to pollution, can be more useful as indicator organisms. In addition, the tolerance of oysters provides a biological matrix, even in highly contaminated environments, very useful for the determination of contaminant concentrations as well as biomarkers, provided that interspecific differences are well established (RNO 2006).

3.3.2 THE BIVALVE *SCROBICULARIA PLANA* AND THE POLYCHAETE *NEREIS DIVERSICOLOR*

Because of the lack of ecologically relevant sentinels for sediment toxicity evaluations in estuarine and coastal areas, recent research efforts have been directed to the development of biomarkers in sediment-dwelling invertebrates.

In addition to being good biomonitors of several metal and organic contaminant bioavailabilities (Bryan et al. 1985; Ruus et al. 2005), *S. plana* and *N. diversicolor* represent an important part of the biomass in European intertidal mudflats. Their role in the functioning of estuarine ecosystems is well established, as briefly reviewed by Moreira et al. 2006 for *N. diversicolor*. Recent papers have again highlighted their role in transferring contaminants from sediment to higher trophic levels (Coelho et al. 2008), and their influence on oxygen diffusion via their burrows with cascading effects on the distribution of other species (Bouchet et al. 2009).

Both classical biomarkers—such as those involved in metal detoxification (Chapter 6), antioxidative defence (Chapter 7), or organic contaminant biotransformation (Chapter 8)—and physiological or behavioural markers with added ecological value have been tested in the bivalve and/or the polychaete (Scaps and Borot 2000; Pérez et al. 2004; Moreira et al. 2006; Durou et al. 2007; Mouneyrac et al. 2008, 2009a,b; Sun and Zhou 2008; Bergayou et al. 2009; Kalman et al. 2009; Solé et al. 2009).

The results obtained by Solé et al. (2009) suggested a different response in the two sediment-dwelling organisms collected along a pollution gradient in southwest Spain. A number of biochemical and behavioural markers were also examined in both species originating from a multipolluted estuary (Loire estuary) and a reference site (Bay of Bourgneuf) in France (Figures 3.4 and 3.5). For glutathione S-transferase (GST), a phase II enzyme involved in the metabolism of lipophilic organic contaminants (van der Oost et al. 2005), the activity was consistently lower in the polychaete than in the bivalve, as shown also in the invertebrates originating from Spain. For both species in both studies, significant intersite differences were rare, but considering the whole sampling period, a clear trend was observed for a higher enzyme activity in bivalves from the Loire than those from the reference site (Figure 3.4).

Acetylcholinesterase (AChE) activity, inhibited by some neurotoxic pesticides, can also respond to the presence of metals, detergents, and algal toxins in marine invertebrates. AChE may thus prove to be a useful biomarker of general physiological stress in aquatic organisms (Roméo et al. 2009). AChE activity was consistently lower in the bivalve than in the polychaete. In the former, significant intersite differences were scarce, whereas the polychaete was much more responsive with significantly lower activities in the contaminated sites (Figure 3.4 and Solé et al. 2009).

Again, in the case of digestive enzymes, the responses of the bivalve and the polychaete were contrasted, with few intersite differences in *S. plana* and significant and consistent intersite differences in *N. diversicolor* (Figure 3.4). For amylase, one month after a small oil spill that occurred in the Loire estuary in March 2008, the activity was sixty-five times lower than in worms from the reference site. The ratio decreased to 13 and 20 for the two following months, and then stabilised at 3 to 4 after October 2008. Anyway, the conditions chronically prevailing in the Loire estuary are sufficient to induce an intersite difference. The pattern for carboxymethylcellulase (CMCase) is typically the same, even if intersite differences of activities in *N. diversicolor* are not so large as in the case of amylase (Figure 3.4).

For metallothioneins (MT or metallothionein-like proteins (MTLPs)), no relationship was found between any bioaccumulated metal and MT concentrations in the study by Solé et al. (2009). The authors proposed that this lack of response can be explained by the small metal gradient represented in their sites. In the case of *N. diversicolor*, even in highly contaminated sites or in experiments, no increased concentrations of MTLPs were observed despite these molecules representing a relatively important store for metals. In worms from metal-contaminated Restronguet Creek, the percentages of cytosolic metals bound to MTLPs were 47% for Cd, 35% for Cu, and 19% for Zn (Berthet et al. 2003). It remains likely that MTLPs intervene in detoxification via an increased turnover (Ng et al. 2008), since it has been proposed that after their degradation, the associated metals may become stored in lysosomal vesicles or granules that are abundant in this species (Mouneyrac et al. 2003). For *S. plana*, in the multipolluted Loire estuary, concentrations of MTLPs were generally significantly different from those determined in specimens from the reference site (not shown).

Burrowing behaviour is also a response that is different between these two biological model species. In parallel with biochemical measurements shown in Figure 3.4 in the multipolluted Loire estuary and the comparatively clean site of the Bay of Bourgneuf, the burrowing kinetics of worms and clams in their sediments of origin were examined (Figure 3.5). For worms of both origins, nearly all the specimens had burrowed within a quarter of an hour, whereas in the case of the bivalves, some

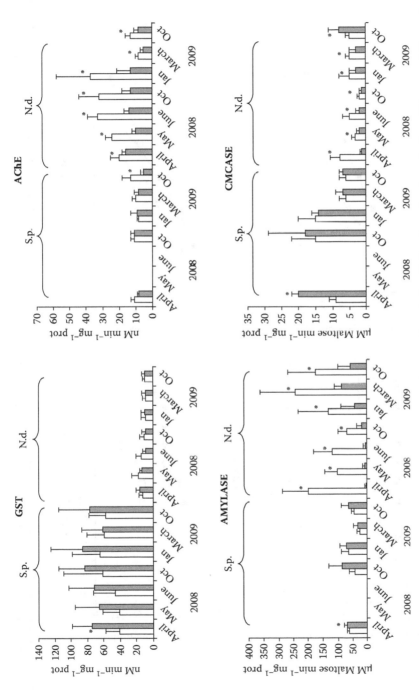

FIGURE 3.4 Biomarker responses in *Scrobicularia plana* (*S.p.*) and *Nereis diversicolor* (*N.d.*) originating from the Bay of Bourgneuf, France (reference site, white bars) and the multipolluted Loire estuary, France (grey bars). Asterisks indicate significant intersite differences (at least $p < 0.05$). (Data provided by I. Boldina-Cosqueric, P. E. Buffet, O. Fossi Tankoua, and J. Kalman, PhD students.)

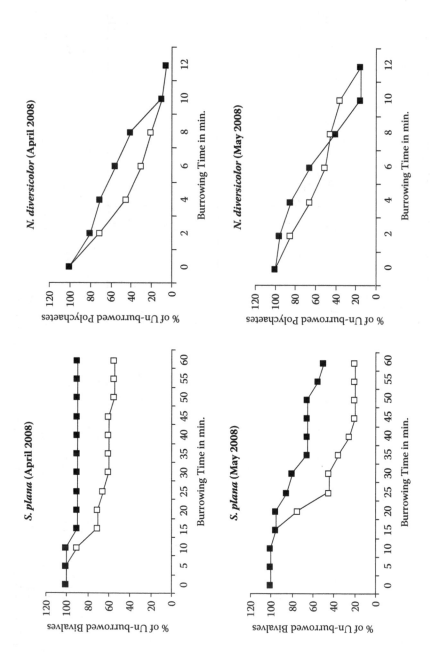

FIGURE 3.5 Burrowing kinetics of *Scrobicularia plana* and *Nereis diversicolor* originating from the Bay of Bourgneuf, France (reference site, open symbols) and the multipolluted Loire estuary, France (filled symbols). (Data provided by I. Boldina-Cosqueric and J. Kalman, PhD students.)

specimens are still at the surface of the sediment after one hour. Another difference is that in the reference site—independently of any pollution event—burrowing kinetics did not show any seasonal trend for *N. diversicolor*, whereas the number of unburrowed clams was lower in late spring and autumn than in early spring and winter (Figure 3.5 and five additional data—not shown—from June 2008 to October 2009). After the small oil spill mentioned above, the behaviour of *S. plana* was disturbed for two months (Figure 3.5), the burrowing kinetics being then restored, whereas no impairment occurred for *N. diversicolor*. This difference between bivalves (*Mercenaria mercenaria*, Olla et al. 1983; *Protothaca staminea*, Pearson et al. 1981) and polychaetes (*Nereis virens*, Olla et al. 1984) has also been observed in the presence of oiled sediment.

Solé et al. (2009) suggested that the different responses in the two sediment-dwelling organisms was attributable to different contamination exposures as they distinguished between the sediment eating of the polychaete and the proposed water filtering of the clam. However, the difference in their feeding habits is by no means so contrasted. On the contrary, Hughes (1969) describes *S. plana* primarily as a deposit feeder, which obtains only some of its food by filtering suspended matter from the sea. *N. diversicolor*, as an opportunistic species, is considered able to fulfil its energy needs using different kinds of diet (Olivier et al. 1995).

Anyway, concentrations of metals in both species were different (Solé et al. 2009). These differences were strongly enhanced in a highly contaminated area such as Restronguet Creek (Bryan et al. 1985). In general, concentrations were higher in the bivalve than in the polychaete, with the exception of Cu, which was about ten times higher in *N. diversicolor* from Restronguet Creek than in *S. plana*. The distribution of organisms in Restronguet Creek suggests a better tolerance of *N. diversicolor* than *S. plana* since the latter is absent from the farthest upstream sites (Bryan et al. 1987). The tolerance of *N. diversicolor* is well documented (Berthet et al. 2003; Mouneyrac et al. 2003 and literature quoted therein) and its genetic basis has been established (Grant et al. 1989; Hateley et al. 1989). Less is known about the bivalve's metal tolerance, but whereas Zn concentrations as high as 3,160 mg.kg^{-1} were determined in specimens living in Restronguet Creek (Bryan et al. 1985), concentrations nearly one order of magnitude lower (651 mg.kg^{-1}) were high enough to provoke lethality in specimens from a comparatively clean site (Amiard, personal communication). However, metals do not tell the whole story. In the multipolluted Seine estuary (France), historical records have shown a depletion of *N. diversicolor* populations (Bessineton, 2009) whereas populations of *S. plana* have not been affected (Bessineton, personal communication).

To summarise, when comparing the tolerance of the two sediment-dwelling invertebrates *S. plana* and *N. diversicolor*, it is necessary to consider both the levels of exposure and the nature of the pollutants involved. When metal contamination is the key factor as in Restronguet Creek, the spatial distribution suggests a better tolerance for *N. diversicolor* even if laboratory experiments have demonstrated a certain degree of tolerance in the bivalves from Restronguet Creek to at least one of the main contaminants at this site (Zn). *N. diversicolor* is very well equipped to cope with metals, being able to store metals under detoxified forms, namely, MTLPs and biomineralised structures. In the bivalve *S. plana*, metal handling strategies include induction of MTLPs and storage of metals mainly in hemocytes as sulphides for Ag and Cu and probably as phosphates for Pb and Zn (Truchet et al. 1990). Because for many metals, the bivalve is a strong accumulator compared to the polychaete (but the contrary was shown for Cu), it remains possible that defence mechanisms of *S. plana* are overwhelmed in highly contaminated sites.

In areas such as large European estuaries, mixed contaminations are present, and in this case, the conservation of populations is based upon a large array of defence mechanisms that in addition can be influenced by abiotic stress (salinity, temperature, hypoxia) as underlined in other chapters (see, for instance, Chapter 9). Under these conditions, *N. diversicolor* showed a number of impairments, including decreased feeding rate, enzyme activities (AChE, CMCase, amylase), and energy reserves (Pérez et al. 2004; Moreira et al. 2005, 2006; Durou et al. 2007; Kalman et al. 2009; Solé et al. 2009), and finally, effects at the population and ecosystem levels (Gillet et al. 2008; Mouneyrac et al. 2009a). Such situations can arise despite defences against metals, the induction of the biotransformation enzyme GST (not in the Loire estuary or the sites studied by Solé et al. 2009, in southwest Spain, but in the Seine, France, studied by Durou et al. 2007 and the Sado, Portugal, studied by Moreira et al. 2006), and the production of LDH (Moreira et al. 2006) to face costs of tolerance (see Chapter 12), because defence mechanisms become overwhelmed.

For behavioural tests, in the case of petroleum hydrocarbons, bivalves seem more responsive than polychaetes, but at a high degree of pollution, such as in the Seine, delayed burrowing was also observed in *N. diversicolor* (Mouneyrac et al. 2009b).

3.3.3 THE ESTUARINE FISH *FUNDULUS HETEROCLITUS* AND *MICROGADUS TOMCOD*

Fish species have often been used in studies assessing biological and biochemical responses to environmental contaminants because they can be found widely in the aquatic environment and play a major ecological role in aquatic food webs. Despite limitations, such as a relatively high mobility, fish are generally considered to be good organisms for pollution monitoring in aquatic ecosystems (van der Oost et al. 2003). They are known to bioaccumulate high levels of aromatic hydrocarbons (AHs) such as PCBs, PCDDs, and PCDFs, whereas PAHs, which are easily biodegradable compounds, do not tend to accumulate strongly in fish tissues (van der Oost et al. 2003). However, between different species, variations of certain biomarkers of environmental pollution may become apparent. In a review, Wirgin and Waldman (2004) studied the resistance of fish populations, in particular the Atlantic killifish (*Fundulus heteroclitus*) and the Atlantic tomcod (*Microgadus tomcod*) from North American estuaries known for the presence of a single class of AH pollutants (Table 3.1), but in fact contaminated by a variety of pollutants. In the contaminated estuaries, tomcod exhibited much higher levels of PCDDs, PCDFs, and PCBs than fish from cleaner rivers, whereas important bioaccumulation of PCBs was also indicated for killifish. This review indicated resistance to PCBs and TCDD in both tomcod and killifish populations, whereas only killifish population have developed resistance to PAHs (for details, see Chapter 8). Surprisingly, they reported a high prevalence of hepatic carcinomas in Hudson River tomcod that were nevertheless resistant to PCBs, whereas they indicated differences between killifish according to

TABLE 3.1
Comparison of Bioaccumulation, Molecular, Pathological, and Populational Effects of AH Exposure, among Species and Population from Different North American Estuaries

Species	Atlantic Tomcod *Microgadus tomcod*	Atlantic Killifish *Fundulus heteroclitus*			
Estuaries	Hudson River	Hudson River	New Bedford Harbor	Newark Bay	Elizabeth River
Major contaminants	PCBs	PCBs	PCBs	PCDDs	PAHs
Resistance to	TCDD and PCBs not to PAHs		TCDD, PCBs, and PAHs		
AH bioaccumulation	PCBs, PCDDs, PCDFs	PCBs	PCBs	PCBs	PCBs
Early-life-stage toxicity[a]	No with TCDD, PCBs	n.i.	No with PCBs, TCDD	No with PCBs, TCDD	Decrease in F_1 fitness (exposed to contaminants other than PAHs)
Pathological and population effects					
Hepatic carcinomas	40–90%	n.i.	No	No	Moderate to high frequencies
Abundance	Decrease			No dramatic fluctuations	
Age structure	Truncated			n.i.	
Molecular effects					
CYP1A expression	x 15 versus control	n.i.	Similar to control	n.i.	n.i.
CYP1A inducibility[a]	PAHs: similar to control ↘ with PCBs, TCDD	n.i.	↘ with PCB126, PAHs	↘ with PCBs, TCDD	↘ with PCB126, 3-methylcholanthrene (PAH)

Source: After Wirgin, I., and J. R. Waldman, *Mutation Res.*, 552, 73–100, 2004, and literature cited.

n.i. = no information.

[a] Determined after experimental exposure.

their origin. The fish from the Elizabeth River, although resistant to PAHs, presented neoplasms in moderate to high frequencies, whereas no carcinomas were encountered for the fish from New Bedford Harbor and the lower Hudson River Estuary. Wirgin and Waldman (2004 and literature cited) also reported differences in the damage caused to the two fish populations. They evoked a decrease in abundance of the Hudson River tomcod population, as well as a dramatic truncated age structure compared to populations from cleaner rivers. On the contrary, no important fluctuations were reported in killifish population size.

Resistance to AH contaminants in these two fish species was assessed considering the degree of inducibility of cytochrome P4501A (CYP1A) and sensitivities of early life stages when experimentally exposed to these contaminants. In tomcod from the Hudson River, CYP1A mRNA expression was about fifteen-fold those encountered in fish from cleaner rivers. On the other hand, in these fish allowed to depurate, then reexposed in the laboratory, reduced CYP1A mRNA inducibility was observed, compared with the reference population, when exposed to PCBs or TCDD, but not to PAHs. Moreover, early life stages of tomcod from the Hudson River were less sensitive to TCDD and PCBs than those from cleaner rivers (Table 3.1). Wirgin and Waldman (2004 and literature cited) found differences in resistance to AH contamination between tomcod and killifish, as well as intraspecific differences in killifish populations. No differences in larval mortality or growth were observed in killifish from Newark Bay or New Bedford Harbor and from a reference site experimentally treated with PCBs or TCDD. However, no CYP1A inducibility was observed in fish from Newark Bay, whereas it was highly inducible in fish from the reference site. In killifish from New Bedford Harbor, levels of CYP1A expression were similar to those of fish from cleaner sites, and consistent with this, inducibility by PCB126 or PAHs was reduced compared to fish from reference sites. In the same manner, evidence of resistance of the Elizabeth River killifish was shown by the absence of inducibility of CYP1A when exposed to 3-methylcholanthrene (a PAH) or PCB126. On the other hand, when reared in clean environments, or exposed to contaminants other than PAHs, F_1 Elizabeth River offspring presented reductions in fitness, which seems to indicate significant evolutionary costs.

According to the studies carried out on fish species from North American estuaries, it seems generally considered that resistance is associated to reduced inducibility of CYP1A. Enzymes of the cytochrome P4501A system, particularly EROD (ethoxyresorufin O-deethylase), have long been considered as contributing to organic compound detoxification (Newman and Unger 2003). Since deleterious effects may result from biotransformation of these compounds, it is necessary to understand what are the other processes involved in resistance (for more details and interpretation, see Chapter 8).

EROD and more recently, following the development of molecular biology, CYP genes are frequently used as core biomarkers of exposure to organic xenobiotics. However, the examples of tomcod and killifish underline that, depending on species and sites, responses may be different with the consequent possibility of false negatives or positives when interpreting the results of a biomonitoring programme. Moreover, observations on hepatic carcinomas reveal interspecific differences in sites polluted by the same major class of contaminants (PCBs in Hudson River or

New Bedford Harbor). Again, population effects (abundance) are much marked in the Atlantic tomcod *M. tomcod* compared to in the killifish *F. heteroclitus*. In terms of environmental quality assessment, the use of one species or the other can result in contrasted interpretations, leading to either over- or underevaluations.

3.4   TOLERANCE AND THE MONITORING OF COMMUNITIES

Hypothetically, a gradual or large-scale mortality within a particular species can take place as a result of the direct toxic effects of chemical exposure that may eventually result in the elimination of a sensitive species in a community. The replacement of the eliminated sensitive species by less sensitive (or more tolerant) ones due to the decrease of competition will lead to an alteration of the species composition of a community. The PICT (pollution induced community tolerance) concept has been well developed for microbial communities such as biofilms (Chapter 4) or soil bacteria (Chapter 14). Considering macroorganisms, the removal of sensitive grazers or predators can result in dramatic changes in primary production and in biological interactions within food webs.

The extents and magnitude of chemical-mediated ecological impacts are also dependent on the level and duration of the chemical exposure. Medina et al. (2007) suggested five main possibilities of pollutant responses at community levels:

1. Elimination of sensitive species due to direct toxic effects
2. Replacement of these species by less sensitive species due to the release from competition
3. Shifts in food web interactions as a consequence of decreased predation or grazing of toxicant susceptible species
4. Physiological acclimation
5. Selection of genetically inherited tolerance (i.e., microevolution due to pollution)

The first three consequences are probably associated with high levels of chemical contamination, while the last two processes are dependent on the level and duration of the chemical exposure as well as the genetic makeup of the affected populations (Rubal et al. 2009).

3.4.1   RELATIVE TOLERANCE OF CERTAIN TAXA TO GIVEN CLASSES OF CONTAMINANTS

Pollutants can cause cascade direct and indirect effects to marine plankton community structure in the marine water column. A field-based manipulative study demonstrated that exposure to a highly toxic antifouling biocide, zinc pyrithione, can significantly reduce phytoplankton abundance, leading to diminishment of essential algal food sources to zooplankton, and hence resulting in a collapse of the entire zooplankton community (Hjorth et al. 2006). An elevated amount of available substrate from dead algae further promoted excessive growth of bacteria, and in turn changed the bacterial community (Hjorth et al. 2006). Interestingly, the surviving phytoplankton could develop an increased tolerance to zinc pyrithione over the time of exposure (Hjorth et al. 2006).

Prolonged exposure to tributyltin (TBT) *in situ* for five months triggered significant changes in the structure of macro- and meiofauna communities as well as the functional diversity of the microbial community in a marine sediment (Dahllof et al. 2001). It was reported that TBT not only affected the structure and recruitment of the community, but also reduced the number of species, diversity, biomass, and community similarity in contrast to the control. Dahllof et al. (2001) further demonstrated that crustacean species were the most tolerant and predominant organisms, whereas echinoderm species were the most sensitive to the toxic effect of TBT exposure. In mesocosms, Gustafsson et al. (2000) exposed surface sediment to two concentrations of TBT for seven months. Compared to the control, increased foraminiferal abundance was observed in the lowest contaminated mesocosm (0.02 nmol of $TBT.g^{-1}$ dry sediment), suggesting a higher tolerance of foraminiferans than that of their predators or competitors, leading to a decreased predation. On the contrary, decreased global abundance was observed in the highest TBT-contaminated mesocosm (2.00 nmol), although a few species particularly tolerant to TBT were more abundant. These results suggested that, whatever its concentration in the sediments, TBT will affect the foraminiferal community.

Carman et al. (2000) identified that crustaceans such as copepods and ostracods were the most sensitive taxonomic groups to hydrocarbons, and that reductions in abundance and grazing activity of crustaceans could lead to enhanced algal biomass in the benthic salt marsh community. Furthermore, the reduced copepod diversity subsequently altered competitive interactions among meiofauna (Carman et al. 2000). These authors underline that their findings are consistent with several other studies indicating that copepods as a group tend to be more sensitive to pollution than nematodes, thus suggesting the nematode/copepod ratio at the taxon level to be a useful index of petroleum hydrocarbon exposure. Gómez Gesteira and Dauvin (2000) have studied the effects of oil spills on infralittoral muddy-sand macrobenthic communities. They observed the disappearance of the amphipods, with a very low, but progressive recovery rate during the four years after the spill. On the contrary, polychaetes appeared globally to be resistant to high levels of hydrocarbons, with few changes in the sites where they dominated. In parallel with the nematode/copepod ratio proposed to reflect the temporal change in meiobenthos communities, these authors suggested the use of a polychaete/amphipod ratio to identify oil spill impacts on soft-bottom communities.

Cain et al. (2004) have used subcellular metal partitioning into proposed metal-sensitive and detoxified compartments to interpret vulnerabilities of five taxa of aquatic insects in a mining-impacted river. As expected, high accumulation of Cd, Cu, and Zn in metal-sensitive fractions (heat-denatured cytosolic proteins and microsomal fractions) was highest in mayflies, which were rare or absent from the most contaminated sites, but occurred at less contaminated ones. On the contrary, metal accumulations in these fractions were low for caddisflies, which were widely distributed along the river, and dominant in the most contaminated sections. Thus, in metal-contaminated sites, changes in species assemblages appear to operate through the progressive selection against the least tolerant to the most tolerant taxa within the considered community.

Rubal et al. (2009) recently compared the community structure of meiobenthos between a clean site and a site heavily polluted with trace metals associated with mining activities in Portugal, and noted that there was no clear separation based on the biodiversity data between the two sites. However, Rubal et al. (2009) further discovered that both density and egg clutch volume of a copepod *Paronychocamptus nanus* were significantly lower in the polluted site than in the clean site, while the copepods in the polluted site displayed a much greater tolerance to copper (as reflected by consistently higher median lethal concentration (LC50) values) than the control population. It is clear that the copper-tolerant copepods had reduced population fitness. While sublethal levels of pollution may not cause significant changes in community structure, it is still possible that chemical exposure can lead to reduced fitness in some sensitive species, even though they are able to acquire some tolerance.

According to Dauvin (2008), despite high pollution levels of metals, the Seine estuary still continues to support a very high benthic biomass and remain quite productive. Salinity appears to be an important factor in controlling contaminant distribution in the sediment and the overlying or interstitial waters and affecting their bioavailability. Of course, the response to contaminants varies greatly among species and assemblages, and metals explain only a small part of the variation in benthic community structure. Some species, such as the shrimp *Crangon crangon*, appear vulnerable to metal pollution, while other species, such as the clam *Scrobicularia plana*, are able to tolerate quite high levels of cadmium in their tissue (Dauvin 2008). However, metals are not the only source of chemical stress in the Seine (Amiard et al. 2009), and in this situation of multipollution, the clam *S. plana* was also more tolerant than another endobenthic species, the ragworm *Nereis diversicolor* (see Section 3.3.2).

3.4.2 Monitoring of Coastal and Estuarine Biodiversity

In recent years, several sets of legislation worldwide (Oceans Act in United States, Australia, or Canada, Water Framework Directive or Marine Strategy in Europe, National Water Act in South Africa, etc.) have been developed in order to address ecological quality or integrity, within estuarine and coastal systems. Several tools have been developed to fulfil the needs of these regulations, among which biotic indices have received particular attention (Borja et al. 2000; Simboura and Zenetos 2002; Diaz et al. 2004; Borja et al. 2008; Dauvin et al. 2009; Teixeira et al. 2009). To assess the composition of the macrobenthos, the AZTI's Marine Biotic Index (AMBI) (software freely available at http://www.azti.es) is widely used. It is based on the model first used to assess the response of soft-bottom macrobenthic communities to the introduction of organic matter in the system. Sensitive and pollution-tolerant indicator species have been ordered into five ecological groups as summarised by Grall and Glémarec (1997). Borja et al. (2000) have used the AMBI to detect different natural and anthropogenic changes in soft-bottom communities, from 1995 to 1998, along thirty stations of the Basque coastline (north of Spain). They observed an important change in the ecological group composition during this period. The AMBI increased from 2 (slighty polluted) to 5 to 6 (heavily polluted) during the three years, corresponding to important discharges of domestic and industrial wastes, including high amounts of organic matter from paper manufacturers. Positive correlations were particularly indicated between AMBI and metals (Cd, Cr, Cu, Ni, Pb, and Zn), and AMBI and PCBs. The AMBI has also been verified successfully in many different regions in Europe in relation to other environmental impact sources (e.g., drilling cuts with ester-based mud, submarine outfalls, industrial and mining wastes, jetties and sewerage works) (Borja et al. 2003). Borja et al. (2000) concluded that, contrary to other indices, AMBI may be used even in slightly polluted areas because it has a high sensitivity at these levels. However, some discrepancies have been highlighted between the responses of AMBI and other indices (Albayrak et al. 2006), the robustness of the former being reduced when a very low number of taxa exists.

The ecological classification of key species in the community and the balance expected between ecological groups of estuarine communities had great influence in the final ecological assessment (Teixeira et al. 2009). An example is presented for the Sea of Marmara by Albayrak et al. (2006), who showed that the assessment produced when employing the Shannon-Wiener index and BENTIX was similar in most sampling sites, while a discrepancy occurred for a shallow area, attributable to the dominance of the whelk *Nassarius reticulatus* (42% of specimens).

Choueri et al. (2009), working on sediment quality from Paranagua Estuarine System (PES), Brazil, revealed "a gradient of increasing degradation of sediments (i.e., higher concentrations of trace metals, higher toxicity, and impoverishment of benthic community structure) towards inner PES" (Figure 3.6). However, pollution is not the whole story, and the importance of a comprehensive approach in order to avoid false interpretation due to confounding factors must be underlined (see Section 3.4.3).

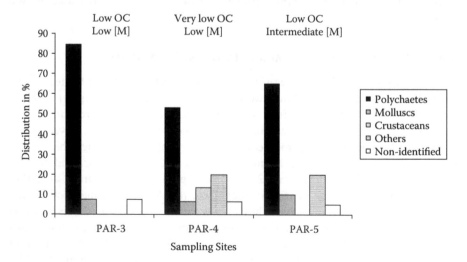

FIGURE 3.6 Dominance distribution of the main taxa in sediments from the Paranagua Estuarine System differing in their organic content (OC) and metal concentrations [M]. One hundred percent polychaetes in sediments with the highest OC and metal concentrations (not shown). (Adapted from Choueri, R. B., et al., *Ecotoxicol. Environ. Saf.*, 72, 1824–31, 2009.)

3.4.3 CONFOUNDING FACTORS

Although an assessment tool is required to accurately translate shifts in communities due to anthropogenic influence, this is often a hard goal to reach. Ideally, tools should be able to distinguish between natural and anthropogenic disturbances. This topic has been intensively discussed in the case of the European Water Framework Directive for coastal and estuarine waters (Dauvin 2007; Teixeira et al. 2009). The main problem appears to be that all the indices for determining anthropogenic stress examine the abundances of stress-tolerant species, which also are able to tolerate natural stressors such as those occurring in estuaries: salinity, nature of substrata, depth, level of fines in the sediment, and oxygen concentration (Dauvin et al. 2009). In northern Spanish estuaries, Puente et al. (2008) found an overall dominance of species tolerant to organic enrichment in all status categories and a low range of variation of the AMBI.

There are many confounding factors influencing the evolution of chemical tolerance under dynamic field settings. First, it has been shown that the fitness trade-off mediated by the chemical exposure is likely to be pronounced and stronger under limited food resource availability or high population density, and to be less apparent and weaker under abundant food or low-density conditions (Semlitsch et al. 2000). Second, the selective pressure (e.g., pollutant exposure) could interplay with other selective forces (e.g., UV radiation) and have implications on the costs of tolerance, leading to interacting effects on the impacted populations and communities. For example, Kashian et al. (2007) revealed that stream benthic communities from a metal-polluted site could be more tolerant to metal contamination but less tolerant to UV-B than reference communities. Thus, chemical exposure should not be treated in isolation from other key environmental factors when conducting field studies. Third, animals that have developed tolerance to one chemical may be able to tolerate other chemicals (i.e., the development of cross-tolerance; Cao and Han 2006); in this way, the overall cost of tolerance may be less than expected. Therefore, partitioning environmental and genetic variation in chemical tolerance is essential to understanding the mechanisms of genetic adaptation in the field, and to identify the most susceptible species, the amount of genetic variance present, the potential for adaptation to the contaminants, and the presence of fitness trade-offs (Semlitsch et al. 2000).

The importance of food resources as influencing toxic responses is not specific to aquatic environments. The effect of pesticides on population fitness is very complex in different field scenarios and likely confounded by the effect of density dependency (i.e., resource availability) and biological interaction in the communities. Hypothetically, insect populations with overcompensating density dependence will tend to evolve pesticide resistance faster than ones with "perfect" or undercompensating density dependence (May 1985). In other words, if an overcompensating population is perturbed by pesticides, leading to a reduced population size below its long-term average value in one generation, it will tend to bounce back above this value in the next generation. When the pesticide has killed off the majority of insects, there will be abundant resources available to the surviving insects, which can have accelerated population growth and thus be less affected by the migration of susceptible genes from regions untreated with the pesticide. In contrast, insects with undercompensating density dependence will be more vulnerable to the migration of susceptible genes from untreated regions. Interestingly, May (1985) noted that most field insect populations appeared to show undercompensating density dependence, but the majority of laboratory populations showed marked overcompensation. Apparently there are other natural mortality factors (e.g., biological interactions, natural enemies, seasonal migration, and diseases) shaping the field population that are seldom present in the laboratory experimental design (May 1985; Lenormand et al. 1999).

A careful assessment of pollution effects on communities in the field needs to take into account all the nonpollutant factors able to influence the fate and effects of toxic chemicals. The work by Choueri et al. (2009) quoted above is particularly demonstrative on this topic. A strong contrast was revealed in dominance distribution of the main taxa in sediments from the Paranagua Estuarine System, depending on metal contamination (Figure 3.6). However, at the sites at which the macrofauna was composed of 100% polychaetes, sediments had not only the highest organic content and metal concentrations, but also the highest percentages of fines (65% compared to 15 to 27% in the three other sites shown in Figure 3.6). Frequently, there is a positive correlation between particle size, organic carbon, and the contaminant level in sediments. It is generally expected that diversity and species variety decrease as sediments became muddier. However, according to Albayrak et al. (2006), studying ecological quality status in the Sea of Marmara, the diversity and number of species reflects not only the type of substratum, sandy (11 to 23 species per 0.1 m^2) to muddy (4 to 24 species per 0.1 m^2), but also the impact of anthropogenic stress received by the sites.

In aquatic environments, hydrodynamic characteristics are particularly important when considering the communities of the smaller organisms, such as bacteria, diatoms, foraminiferans, or nematodes (Amiard-Triquet and Rainbow 2009). These organisms are relatively abundant compared with macrofaunal taxa, so they can be investigated with small samples, relatively simple equipment, and minimised impact on the study site. Such taxa also tend to exhibit higher diversity than macrofauna, increasing the range of potential responses to pollutants and potential sensitivity for the detection of an impact. Data obtained have high information content, facilitating statistical analysis. The short generation times and low dispersal characteristics of most of these taxa are associated with rapid population response to environmental perturbations. Some meiofaunal taxa, particularly the nematodes, have also been found to contain species with very high tolerance to certain pollutants, such that nematodes can be the last metazoans to persist in grossly polluted systems, and thus provide the only model organisms to study in these conditions. The importance of hydrodynamic conditions on the fate of these organisms has been well illustrated in the Seine and Authie estuaries (Amiard-Triquet and Rainbow 2009).

To conclude, those confounding factors that cannot be controlled (salinity, granulometry, organic content, hydrodynamics, food availability, etc.) must at least be measured in parallel with ecological and ecotoxicological determinations.

3.5 CONCLUSIONS

The potential effects of chemicals are classically assessed by using single species biotests, allowing the determination of the toxicity of a given chemical to a given species. Many biotest methods are recommended by regulatory authorities (American Society for Testing and Materials (ASTM), International Council for the Exploration of the Sea (ICES), Convention for the Protection of the Marine Environment of the North-East Atlantic (OSPAR), Organisation for Economic Cooperation and Development (OECD), U.S. Environmental Protection Agency (USEPA)). For instance, those devoted to the evaluation of dredged material and sediments were recently reviewed (Nendza 2002). A large range of biological models is proposed, including many different taxa which have all shown a potential to acquire tolerance in contaminated environments (Tables 2.1, 13.2, 13.4). In addition, from a practical point of view, the species involved in biotests are generally tolerant in order to survive easily under laboratory conditions.

The determination from biotest results of the concentration of a chemical below which no deleterious effects will occur for many species in a field situation needs the use of uncertainty or security factors. This practice applied for the protection of species in the environment is derived from human toxicology (Vermeire et al. 1999). The extrapolation factors take into account inter- and intraspecific variability in order that the threshold thus obtained can protect efficiently both sensitive and tolerant organisms (Calabrese 1985; Burin and Saunders 1999). At each level, the use of a factor of 10 is generally recommended. The calculation of predicted no-effect concentrations (PNECs) for the protection of organisms in the environment is described in detail in the Technical Guidance Document on Risk Assessment in support of European Commission regulations (TGD 2003). In addition to security factors, the determination of the PNECs takes into account the number of bioassay results available for different species representative of different trophic levels and the availability of long-term toxicity data. However, the notion of long term in the framework of the methodology of bioassays (Nendza 2002) differs from the notion of long term when considering ecological consequences at higher levels of biological organisation.

Thus, according to Medina et al. (2007), despite the existence of empirical evidence showing that safety margins currently applied in the environmental risk assessment process may account for pollution-induced genetic changes in tolerance, information regarding long-term ecological consequences due to ecological costs (Chapter 12) is not explicitly considered in these procedures. In general, differential tolerance involves populations that are more than twice as tolerant to contaminant exposure as unimpacted populations (Chapter 2). However, in an expanded study, Nacci et al. (2009) reported that the LC20 (lethal concentration for 20% of the individuals tested) to PCB126 in killifish embryos from twenty-four estuarine sites in the United States, from Maine to Virginia, ranged over *four orders of magnitude*, and sensitivities of the response reflected sediment total PCB concentrations. As explained in Chapter 8, xenobiotic transformation by enzymes involves several forms of the aryl hydrocarbon receptor (AHR). In the fish *Microgadus tomcod*, both AHR2 alleles dose responsively drove gene expression, but the AHR2 allele common in all reference populations did so at *two orders of magnitude* lower TCDD or PCB126 doses than did the allele present in individuals from the PCB-polluted Hudson River. These results were consistent with the difference between tomcod from the Hudson River and reference populations in the inducibility of *in vivo* CYP1A expression. If differences in tolerance ratios are relatively small, then they may be encompassed by conventional safety factors used to establish protective guidelines. However, such examples suggest that safety margins may not be adequate for all contaminant classes.

Because so many contaminants—a number of one hundred thousand molecules is frequently put forward—are present in the environment, it is impossible to determine in a given site all the classes of chemicals able to impair biological and ecological systems. To overcome this difficulty, biomarkers that are direct manifestations of the exposure of an organism in the field to xenobiotics, toxins, or abiotic stress represent a relevant tool. The choice of biomarkers involves the utilisation of particular organisms in which the use of these biomarkers has been clearly established (Amiard-Triquet and Rainbow 2009). However, the differential responsiveness of populations depending on their exposure history traits can introduce a bias in the interpretation of biomarker data. To overcome this difficulty, the strategy of active biomonotoring proposed by de Kock and Kramer (1994) has been successfully adapted and used by many. It provides the potential of using individuals collected from the same site or even cultured, meaning that they share the same history or even the same genetic pattern.

At the interspecific level, depending on the tolerance of species to different kinds of contaminants and to different levels of exposure, it is indispensable to take into account a sufficient number of biological models. Because the determination of many different biomarkers in different species presents great resource implications, it is necessary to find relevant criteria for the selection of models. As recommended earlier in the case of estuaries—but the concepts are applicable to other environments—the choice of species to be studied must consider (1) their representativeness of their environment and (2) their role in the structure and functioning of ecosystems. By focussing on key species, it is assumed that impairments of their responses used as biomarkers will reveal a risk of cascading deleterious effects right through to the ecosystem level. Disturbances at the supra-individual levels affecting these organisms may result in ecological disorders such as changes in biogeochemical cycles and trophic relationships that govern nutrient/food availability (both as energy resource and risk of contaminant transfer) (Amiard-Triquet and Rainbow 2009). Depending on their sensitivity and thus their presence/absence at different sites, it is possible to concurrently use several species as currently done in Mussel Watch-type programmes (see Section 3.3.1), provided that differences in their level of responsiveness are clearly established.

The assessment of the ecological health status in our environment takes root in the fact that environmental stressors will be responsible for the loss of the more sensitive species/taxa, whereas tolerant species/taxa will be favoured. In this chapter, we have put the emphasis on aquatic environments, particularly transitional waters that pose complicated problems due to their high natural variability and the influence of many confounding factors. However, similar approaches have been developed for other environments, such as the use of lichens as bioindicators of air pollution (for a review, see Conti and Cecchetti 2001).

Based on interspecific tolerance, different strategies may be developed. They can take into account (1) the presence in a given ecosystem of bioindicator species, the tolerance of which is well documented (e.g., opportunistic species that proliferate under pronounced unbalanced situations); (2) the relative abundance of tolerant and sensitive species (e.g., monitoring of petroleum hydrocarbon pollution by using polychaete/

amphipod and nematode/copepod ratios); or (3) more comprehensive approaches (PICT analyses well developed for microorganisms and biotic indices). In order to obtain a more comprehensive view of the community, the use of biotic indices, together with other metrics, may be recommended. For instance, in sediments—a compartment that is the major sink for nearly all chemicals entering the aquatic environment—the use of combined approaches that incorporate the physicochemical condition of the substratum would be a straightforward approach to reduce the risk of failing in the assignment of ecological status category based upon biotic indices (Puente et al. 2008).

In response to abundance of tolerant species in transitional waters, Prato et al. (2009) consider that thresholds established in the biotic index scale values need to be modified with a view to a better ecological quality status assessment. Because it has been clearly established that organisms tolerant to a number of natural stressors are generally sensitive to any additional stressor (Chapter 5), it is questionable if such changes are justified.

Anyway, the presence or absence of species at time T is not sufficient to decide the environmental health status because tolerant species may be still alive but no longer healthy, being affected with carcinoma (Chapter 8) or reduced fitness and reproduction success (Chapter 12).

REFERENCES

Albayrak, S., et al. 2006. Ecological quality status of coastal benthic ecosystems in the Sea of Marmara. *Mar. Pollut. Bull.* 52:790–99.

Amiard, J. C., et al. 2006. Metallothioneins in aquatic invertebrates: Their role in metal detoxification and their use as biomarkers. *Aquat. Toxicol.* 76:160–202.

Amiard, J. C., et al. 2009. Quantification of contaminants. In *Environmental assessment of estuarine ecosystems. A case study*, ed. C. Amiard-Triquet and P. S. Rainbow, 31–57. Boca Raton, FL: CRC Press.

Amiard-Triquet, C., and P. S. Rainbow. 2009. Conclusions. In *Environmental assessment of estuarine ecosystems. A case study*, ed. C. Amiard-Triquet and P. S. Rainbow, 323–48. Boca Raton, FL: CRC Press.

Baird, D. J. 1992. Predicting population response to pollutants: In praise of clones. A comment on Forbes and Depledge. *Funct. Ecol.* 6:616–17.

Barata, C., et al. 2002. Among- and within-population variability in tolerance to cadmium stress in natural populations of *Daphnia magna*: Implications for ecological risk assessment. *Environ. Toxicol. Chem.* 5:1058–64.

Bergayou, H., et al. 2009. Oxidative stress responses in bivalves (*Scrobicularia plana, Cerastoderma edule*) from the Oued Souss estuary (Morocco). *Ecotoxicol. Environ. Saf.* 72:765–69.

Berthet, B. 2008. Les espèces sentinelles. In *Les biomarqueurs dans l'évaluation de l'état écologique des milieux aquatiques*, ed. J. C. Amiard and C. Amiard-Triquet, 121–48. Paris: Lavoisier.

Berthet, B., et al. 2003. Accumulation and soluble binding of Cd, Cu and Zn in the polychaete *Hediste diversicolor* from coastal sites with different trace metal bioavailabilities. *Arch. Environ. Contam. Toxicol.* 45:468–78.

Bessineton, C. 2009. Historical records of the *Nereis diversicolor* population in the Seine estuary. In *Environmental assessment of estuarine ecosystems. A case study*, ed. C. Amiard-Triquet and P. S. Rainbow, 183–97. Boca Raton, FL: CRC Press.

Blanck, H., S. A. Wängberg, and S. Molander. 1988. Pollution-induced community tolerance—A new tool. In *Function testing of aquatic biota for estimating hazards of chemicals*, ed. J. J. Cairns and J. R. Pratt, 219–30. Philadelphia: American Society for Testing and Materials.

Borja, A., J. Franco, and V. Pérez. 2000. A marine biotic index to establish the ecological quality of soft-bottom benthos within European estuarine and coastal environments. *Mar. Pollut. Bull.* 40:1100–14.

Borja, A., I. Muxika, and J. Franco. 2003. The application of a marine biotic index to different impact sources affecting soft-bottom benthic communities along European coasts. *Mar. Pollut. Bull.* 46:835–45.

Borja, A., et al. 2008. Overview of integrative tools and methods in assessing ecological integrity in estuarine and coastal systems worldwide. *Mar. Pollut. Bull.* 56:1519–73.

Bossuyt, B. T. A., and C. R. Janssen. 2003. Acclimation of *Daphnia magna* to environmentally realistic copper concentrations. *Comp. Biochem. Physiol.* C 136:253–64.

Bouchet, V. M. P., et al. 2009. Influence of the mode of macrofauna-mediated bioturbation on the vertical distribution of living benthic foraminifera: First insight from axial tomodensitometry. *J. Exp. Mar. Biol. Ecol.* 371:20–33.

Brown, R. J., et al. 2004. Differential sensitivity of three marine invertebrates to copper assessed using multiple biomarkers. *Aquat. Toxicol.* 66:267–78.

Bryan, G. W., et al. 1985. A guide to the assessment of heavy metal contamination in estuaries using biological indicators. *Occ. Publ. Mar. Biol. Ass. U.K.* 4.

Bryan, G. W., et al. 1987. Copper, zinc, and organotin as long-term factors governing the distribution of organisms in the Fal Estuary in Southwest England. *Estuaries* 10:208–19.

Burin, G. J., and D. R. Saunders. 1999. Addressing human variability in risk assessment—The robustness of the intraspecies uncertainty factor. *Regul. Toxicol. Pharm.* 30:209–16.

Cain, D. J., S. N. Luoma, and W. G. Wallace. 2004. Linking metal bioaccumulation of aquatic insects to their distribution patterns in a mining-impacted river. *Environ. Toxicol. Chem.* 23:1463–73.

Calabrese, E. J. 1985. Uncertainty factor and interindividual variation. *Regul. Toxicol. Pharm.* 5:190–96.

Cao, G. C., and Z. J. Han. 2006. Tebufenozide resistance selected in *Plutella xylostella* and its cross-resistance and fitness cost. *Pest Manage. Sci.* 62:746–51.

Carman, K. R., J. W. Fleeger, and S. M. Pomarico. 2000. Does historical exposure to hydrocarbon contamination alter the response of benthic communities to diesel contamination? *Mar. Environ. Res.* 49:255–78.

Chapman, P. M. 2002. Integrating toxicology and ecology: Putting the "eco" into ecotoxicology. *Mar. Pollut. Bull.* 44:7–15.

Chapman, P. M., M. A. Farrel, and R. O. Brinkhurst. 1982. Effects of species interactions on the survival and respiration of *Limnodrilus hoffmeisteri* and *Tubifex tubifex* (Oligochaeta, tubificidae) exposed to various pollutants and environmental factors. *Water Res.* 16:1405–08.

Chapman, P. M., and F. Wang. 2001. Assessing sediments contamination in estuaries. *Environ. Toxicol. Chem.* 20:3–22.

Ching, E. W. K., et al. 2001. DNA adduct formation and DNA strand breaks in green-lipped mussels (*Perna viridis*) exposed to benzo[a]pyrene: Dose and time dependent relationships. *Mar. Pollut. Bull.* 42:603–10.

Choueri, R. B., et al. 2009. Integrated sediment quality assessment in Paranagua Estuarine System, Southern Brazil. *Ecotoxicol. Environ. Saf.* 72:1824–31.

Coe, T. S., et al. 2009. Genetic variation in strains of zebrafish (*Danio rerio*) and implications for ecotoxicology studies. *Ecotoxicology* 18:144–50.

Coelho, J. P., et al. 2008. The role of two sediment-dwelling invertebrates on the mercury transfer from sediments to the estuarine trophic web. *Estuar. Coast. Shelf Sci.* 78:505–12.

Conti, M. E., and G. Cecchetti. 2001. Biological monitoring: Lichens as bioindicators of air pollution assessment—a review. *Environ. Pollut.* 114:471–92.

Cosper, E. M., et al. 1987. Induced resistance to polychlorinated biphenyls confers cross-resistance and altered environmental fitness in a marine diatom. *Mar. Environ. Res.* 23:207–22.

Dahllof, I., et al. 2001. The effect of TBT on the structure of a marine sediment community—A boxcosm study. *Mar. Pollut. Bull.* 42:689–95.

Dauvin, J. C. 2007. Paradox of estuarine quality: Benthic indicators and indices, consensus or debate for the future. *Mar. Pollut. Bull.* 55:271–81.

Dauvin, J. C. 2008. Effects of heavy metal contamination on the macrobenthic fauna in estuaries: The case of the Seine estuary. *Mar. Pollut. Bull.* 57:160–69.

Dauvin, J. C., et al. 2009. Benthic indicators and index approaches in the three main estuaries along the French Atlantic coast (Seine, Loire and Gironde). *Mar. Ecol. Prog. Ser.* 30:228–40.

de Kock, W. C., and K. J. M. Kramer. 1994. Active biomonitoring (ABM) by translocation of bivalve molluscs. In *Biomonitoring of coastal waters and estuaries,* ed. K. J. M. Kramer, 51–84. Boca Raton, FL: CRC Press.

De Martinez Gaspar Martins, C., and A. Bianchini. 2009. Metallothionein-like proteins in the blue crab *Callinectes sapidus*: Effect of water salinity and ions. *Comp. Biochem. Physiol. A* 152:366–71.

Diaz, R. J., M. Solan, and R. M. Valente. 2004. A review of approaches for classifying benthic habitats and evaluating habitat quality. *J. Environ. Manage.* 73:165–81.

Duan, Y., S. I. Guttman, and J. T. Oris. 1996. Genetic differentiation among laboratory populations of *Hyalella azteca*: Implications for toxicology. *Environ. Toxicol. Chem.* 16:691–95.

Durou, C., et al. 2007. Biomonitoring in a clean and a multi-contaminated estuary based on biomarkers and chemical analyses in the endobenthic worm *Nereis diversicolor*. *Environ. Pollut.* 148:445–58.

Eilers, S., et al. 2009. Effects of multiple stressors on energy metabolism in oysters. *Comp. Biochem. Physiol. A* 153:S172.

Fisher, N. S. 1977. On the differential sensitivity of estuarine and open-sea diatoms to exotic chemical stress. *Am. Nat.* 111:871–95.

Fisher, N. S., et al. 1973. Geographic differences in phytoplankton sensitivity to PCBs. *Nature* 241:548–49.

Funes, V., et al. 2006. Ecotoxicological effects of metal pollution in two mollusc species from the Spanish South Atlantic littoral. *Environ. Pollut.* 139:214–23.

Gaetke, L. M., and C. K. Chow. 2003. Copper toxicity, oxidative stress, and antioxidant nutrients. *Toxicology* 189:147–63.

Galloway, T. S., et al. 2004. Ecosystem management bioindicators: The ECOMAN project—A multi-biomarker approach to ecosystem management. *Mar. Environ. Res.* 58:233–37.

Geffard, A., et al. 2002b. Relationships between metal bioaccumulation and metallothionein levels in larvae of *Mytilus galloprovincialis* exposed to contaminated estuarine sediment. *Mar. Ecol. Prog. Ser.* 233:131–42.

Geffard, A., et al. 2004. Comparative study of metal handling strategies in bivalves *Mytilus edulis* and *Crassostrea gigas*: A multidisciplinary approach. *J. Mar. Biol. Ass. U.K.* 84:641–50.

Geffard, A., et al. 2007. Bioaccumulation of metals in sediment elutriates and their effects on growth, condition index, and metallothionein contents in oyster larvae. *Arch. Environ. Contam. Toxicol.* 53:57–65.

Geffard, A., J. C. Amiard, and C. Amiard-Triquet. 2002a. Use of metallothionein in gills of oysters *Crassostrea gigas* as a biomarker: Seasonal and intersite fluctuations. *Biomarkers* 7:123–37.

Geffard, A., C. Amiard-Triquet, and J. C. Amiard. 2005. Do seasonal changes interfere with metallothionein induction by metals in mussels *Mytilus edulis*? *Ecotoxicol. Environ. Saf.* 61:209–20.

Géret, F., and R. P. Cosson. 2002. Induction of specific isoforms of metallothionein in mussel tissues after exposure to cadmium or mercury. *Arch. Environ. Contam. Toxicol.* 42:36–42.

Gillet, P., et al. 2008. Response *of Nereis diversicolor* population (Polychaeta, Nereidida) to the pollution impact—Authie and Seine estuaries (France). *Estuar. Coast. Shelf Sci.* 76:201–10.

Gómez Gesteira, J. L., and J. C. Dauvin. 2000. Amphipods are good ioindicators of the impact of oil spills on soft-bottom macrobenthic communities. *Mar. Pollut. Bull.* 40:1017–27.

Grall, J., and M. Glémarec. 1997. Using biotic indices to estimate macrobenthic community perturbations in the Bay of Brest. *Estuar. Coast. Shelf Sci.* 44(Suppl. A):43–53.

Grant, A., J. G. Hateley, and N. V. Jones. 1989. Mapping the ecological impact of heavy metals in the estuarine polychaete *Nereis diversicolor* using inherited metal tolerance. *Mar. Pollut. Bull.* 20:235–38.

Gustafsson, M., et al. 2000. Benthic foraminiferal tolerance to tri-*n*-butyltin (TBT) pollution in an experimental mesocosm. *Mar. Pollut. Bull.* 40:1072–75.

Han, B. C., et al. 1994. Copper intake and health threat by consuming seafood from copper-contaminated coastal environments in Taiwan. *Environ. Toxicol. Chem.* 13:775–80.

Hateley, J. G., A. Grant, and N. V. Jones. 1989. Heavy metal tolerance in estuarine populations of *Nereis diversicolor*. In *Reproduction, genetics and distribution of marine organisms*, ed. J. S. Ryland and P. A. Tyler, 379–85. Proceedings of 23rd European Marine Biology Symposium. Olsen & Olsen, Fredensborg, Denmark.

Heugens, E. H. W., et al. 2001. A review of the effects of multiple stressors on aquatic organisms and analysis of uncertainty factors for use in risk assessment. *Crit. Rev. Toxicol.* 31:247–84.

Heugens, E. H. W., et al. 2006. Population growth of *Daphnia magna* under multiple stress conditions: Joint effects of temperature, food, and cadmium. *Environ. Toxicol. Chem.* 25:1399–407.

His, E., R. Beiras, and M. N. L. Seaman. 1999. The assessment of marine pollution—Bioassays with bivalve embryos and larvae. *Adv. Mar. Biol.* 37:1–178.

Hjorth, M., I. Dahllöf, and V. E. Forbes. 2006. Effects on the function of three trophic levels in marine plankton communities under stress from the antifouling compound zinc pyrithione. *Aquat. Toxicol.* 77:105–15.

Hoffman, E. R., and S. W. Fisher. 1994. Comparison of a field and laboratory-derived population of *Chironomus riparius* (Diptera: Chironomidae): Biochemical and fitness evidence for population divergence. *J. Econ. Entomol.* 87:318–25.

Hughes, R. N. 1969. A study of feeding in *Scrobicularia plana*. *J. Mar. Biol. Ass. U.K.* 49:805–23.

Hummel, H., et al.1996. Sensitivity to stress of the estuarine bivalve *Macoma balthica* from areas between the Netherlands and its southern limits (Gironde). *Neth. J. Sea Res.* 35: 315–321.

Ivanina, A. V., C. Taylor, and I. M. Sokolova. 2009. Effects of elevated temperature and cadmium exposure on stress protein response in eastern oysters *Crassostrea virginica* (Gmelin). *Aquat. Toxicol.* 91:245–54.

Jenny, M. J., et al. 2006. Regulation of metallothionein genes in the American oyster (*Crassostrea virginica*): Ontogeny and differential expression in response to different stressors. *Gene* 379:156–65.

Kalman, J., et al. 2009. Assessment of the health status of populations of the ragworm *Nereis diversicolor* using biomarkers at different levels of biological organisation. *Mar. Ecol. Prog. Ser.* 393:55–67.

Kammenga, J. E., and A. G. Riksen. 1996. Comparing differences in species sensitivity to toxicants: Phenotypic plasticity versus concentration-response relationships. *Environ. Toxicol. Chem.* 15:1649–53.

Kashian, D. R., et al. 2007. The cost of tolerance: Sensitivity of stream benthic communities to UV-B and metals. *Ecol. Appl.* 17:366–75.

Keilty, T. J., D. S. White, and P. F. Landrum. 1988. Sublethal responses to endrin in sediment by *Limnodrilus hoffmeisteri* (Tubificidae), and in mixed-culture with *Stylodrilus heringianus* (Lumbriculidae). *Aquat. Toxicol.* 13:227–49.

Lam, P. K. S., and J. S. Gray. 2001. Predicting effects of toxic chemicals in the marine environment. *Mar. Pollut. Bull.* 42:169–73.

Langston, W. J., M. J. Bebianno, and G. R. Burt. 1998. Metal handling strategies in molluscs. In *Metal metabolism in aquatic environments*, ed. W. J. Langston and M. J. Bebianno, 219–83. London: Chapman & Hall.

Lannig, G., A. S. Cherkasov, and I. M. Sokolova. 2006. Temperature-dependent effects of cadmium on mitochondrial and whole-organism bioenergetics of oysters (*Crassostrea virginica*). *Mar. Environ. Res.* 62:S79–82.

Lapkina, L. N., and N. R. Arkhipova. 2000. Comparative analysis of tolerance to pesticides in annelids. *Russ. J. Ecol.* 5:336–40.

Legras, S., et al. 2000. Changes in metallothionein concentrations in response to variation in natural factors (salinity, sex, weight) and metal concentration in crabs from a metal-rich estuary. *J. Exp. Mar. Biol. Ecol.* 246:259–79.

Lemoine, S., et al. 2000. Metallothionein isoforms in *Mytilus edulis* (Mollusca, Bivalvia): Complementary DNA characterization and quantification of expression in different organs after exposure to cadmium, zinc, and copper. *Mar. Biotechnol.* 2:195–203.

Lenormand, T., et al. 1999. Tracking the evolution of insecticide resistance in the mosquito *Culex pipiens*. *Nature* 400:861–64.

Lotts, J. W., and A. J. Stewart. 1995. Minnows can acclimate to total residual chlorine. *Environ. Toxicol. Chem.* 14:1365–74.

Luoma, S. N., and P. S. Rainbow. 2008. *Metal contamination in aquatic environments. Science and lateral management.* Cambridge: Cambridge University Press.

May, R. M. 1985. Evolution of pesticide resistance. *Nature* 315:12–13.

McDonald, D. A., et al. 1992. The coastal resource coordinator's bioassessment manual. Report HAZMAT 93-1. Seattle, WA: NOAA.

McGeer, J. C., et al. 2007. Influence of acclimation and cross-acclimation of metals on acute Cd toxicity and Cd uptake and distribution in rainbow trout (*Oncorhynchus mykiss*). *Aquat. Toxicol.* 84:190–97.

Medina, M. H., J. A. Correa, and C. Barata. 2007. Micro-evolution due to pollution: Possible consequences for ecosystem responses to toxic stress. *Chemosphere* 67:2105–14.

Moreira, S. M., et al. 2005. A short-term *in situ* toxicity assay with *Hediste diversicolor* (Polychaeta) for estuarine sediments based on post-exposure feeding. *Environ. Toxicol. Chem.* 24:2010–18.

Moreira, S. M., et al. 2006. Effects of estuarine sediment contamination on feeding and on key physiological functions of the polychaete *Hediste diversicolor*: Laboratory and *in situ* assays. *Aquat. Toxicol.* 78:186–201.

Mouneyrac, C., et al. 2001. Comparison of metallothionein concentration and tissue distribution of trace metals in crabs (*Pachygrapsus marmoratus*) from a metal-rich estuary, in and out of the reproductive season. *Comp. Biochem. Physiol. C* 129:193–209.

Mouneyrac, C., et al. 2003. Physico-chemical forms of storage and the tolerance of the estuarine worm *Nereis diversicolor* chronically exposed to trace metals in the environment. *Mar. Biol.* 143:731–44.

Mouneyrac, C., et al. 2008. Biological indices, energy reserves, steroid hormones and sexual maturity in the infaunal bivalve *Scrobicularia plana* from three sites differing by their level of contamination. *Gen. Comp. Endocr.* 157:133–41.

Mouneyrac, C., et al. 2009a. Linking energy metabolism, reproduction, abundance and structure of *Nereis diversicolor* populations. In *Environmental assessment of estuarine ecosystems. A case study*, ed. C. Amiard-Triquet and P. S. Rainbow, 159–81. Boca Raton, FL: CRC Press.

Mouneyrac, C., et al. 2009b. Influence of anthropogenic stress on fitness and behaviour of a key-species of estuarine ecosystems, the ragworm *Nereis diversicolor. Environ. Pollut.* 158:121–28.

Nacci, D., et al. 2009. Evolution of tolerance to PCBs and susceptibility to a bacterial pathogen (*Vibrio harveyi*) in Atlantic killifish (*Fundulus heteroclitus*) from New Bedford Harbor (MA, USA) *Environ. Pollut.* 157:857–64.

Niemeijer, D., and R. S. de Groot. 2008. A conceptual framework for selecting environmental indicator sets. *Ecol. Indic.* 8:14–25.

Nendza, M. 2002. Inventory of marine biotest methods for the evaluation of dredged material and sediments. *Chemosphere* 48:865–83.

Newman, M. C., and M. A. Unger. 2003. *Fundamentals of ecotoxicology*. Boca Raton, FL: Lewis Publishers.

Ng, T. Y. T., et al. 2008. Decoupling of cadmium biokinetics and metallothionein turnover in a marine polychaete after metal exposure. *Aquat. Toxicol.* 89:47–54.

Norwood, W. P., U. Borgmann, and D. G. Dixon. 2007. Interactive effects of metals in mixtures on bioaccumulation in the amphipod *Hyalella azteca. Aquat. Toxicol.* 84:255–67.

Nowak, C., et al. 2007. Consequences of inbreeding and reduced genetic variation on tolerance to cadmium stress in the midge *Chironomus riparius. Aquat. Toxicol.* 85:278–84.

OECD. 1998. *Daphnia magna* reproduction test. In *OECD guidelines for testing of chemicals.* Vol. 211. Paris: OECD.

Olivier, M., et al. 1995. Behavioural responses of *Nereis diversicolor* (O.F. Müller) and *Nereis virens* (Sars) (Polychaeta) to food stimuli—Use of specific organic matter (algae and halophytes). *Can. J. Zool.* 73:2307–17.

Olla, B. L., A. J. Bedja, and W. H. Pearson. 1983. Effects of oiled sediment on the burrowing behaviour of the hard clam, *Mercenaria mercenaria. Mar. Environ. Res.* 9:183–93.

Olla, B. L., et al. 1984. Sublethal effects of oiled sediment on the sand worm, *Nereis (Neanthes) virens*: Induced changes in burrowing and emergence. *Mar. Environ. Res.* 13:121–39.

Pearson, W. H., et al. 1981. Effects of oiled sediment on predation on the littleneck clam, *Protothaca staminea*, by the Dungeness crab, *Cancer magister. Estuar. Coast. Shelf Sci.* 13:445–54.

Pérez, E., J. Blasco, and M. Solé. 2004. Biomarker responses to pollution in two invertebrate species: *Scrobicularia plana* and *Nereis diversicolor* from the Cádiz bay (SW Spain). *Mar. Environ. Res.* 58:275–79.

Prato, S., et al. 2009. Application of biotic and taxonomic distinctness indices in assessing the ecological quality status of two coastal lakes: Caprolace and Fogliano lakes (Central Italy). *Ecol. Indic.* 9:568–83.

Puente, A., et al. 2008. Ecological assessment of soft bottom benthic communities in northern Spanish estuaries. *Ecol. Indic.* 8:373–88.

Rainbow, P. S. 2002. Trace metal concentrations in aquatic invertebrates: Why and so what? *Environ. Pollut.* 120:497–507.

RNO. 2006. *Surveillance de la qualité du milieu marin*. Paris: Ministère de l'Environnement and Institut français de recherche pour l'exploitation de la mer (Ifremer).

Roméo, M., L. Poirier, and B. Berthet. 2009. Biomarkers based upon biochemical responses. In *Environmental assessment of estuarine ecosystems. A case study*, ed. C. Amiard-Triquet and P. S. Rainbow, 59–81. Boca Raton, FL: CRC Press.

Rubal, M., L. M. Guilhermino, and M. H. Medina. 2009. Individual, population and community level effects of subtle anthropogenic contamination in estuarine meiobenthos. *Environ. Pollut.* 157:2751–58.

Ruus, A., et al. 2005. Experimental results on bioaccumulation of metals and organic contaminants from marine sediments. *Aquat. Toxicol.* 72:273–92.

Scaps, P., and O. Borot. 2000. Acetylcholinesterase activity of the polychaete *Nereis diversicolor*: Effects of temperature and salinity. *Comp. Biochem. Physiol.* 125C:377–83.

Semlitsch, R. D., C. M. Bridges, and A. M. Welch. 2000. Genetic variation and a fitness tradeoff in the tolerance of gray treefrog (*Hyla versicolor*) tadpoles to the insecticide carbaryl. *Oecologia* 125:179–85.

Simboura, N., and A. Zenetos. 2002. Benthic indicators to use in ecological quality classification of Mediterranean soft bottom marine ecosystems, including a new biotic index. *Mediterr. Mar. Sci.* 3:77–111.

Slobodkin, L. B., and H. L. Sanders. 1969. On the contribution of environmental predictability to species diversity. *Brookhaven Symp. Biol.* 22:82–95.

Smolders, R., L. Bervoets, and R. Blust. 2004. *In situ* and laboratory bioassays to evaluate the impact of effluent discharges on receiving aquatic ecosystems. *Environ. Pollut.* 132:231–43.

Sokołowski, A., et al. 2004. Abnormal features of *Macoma balthica* (Bivalvia) in the Baltic Sea: Alerting symptoms of environmental adversity? *Mar. Pollut. Bull.* 49:17–22.

Solé, M., J. Kopecka-Pilarczyk, and J. Blasco. 2009. Pollution biomarkers in two estuarine invertebrates, *Nereis diversicolor* and *Scrobicularia plana*, from a marsh ecosystem in SW Spain. *Environ. Int.* 35:523–31.

Spurgeon, D. J., et al. 2000. Relative sensitivity of life-cycle and biomarker responses in four earthworm species exposed to zinc. *Environ. Toxicol. Chem.* 19:1800–08.

Stohler, R. A., J. Curtis, and D. J. Minchella. 2004. A comparison of microsatellite polymorphism and heterozygosity among field and laboratory populations of *Schistosoma mansoni*. *Int. J. Parasitol.* 34:595–601.

Sun, F. H., and Q. X. Zhou. 2008. Oxidative stress biomarkers of the polychaete *Nereis diversicolor* exposed to cadmium and petroleum hydrocarbons. *Ecotoxicol. Environ. Saf.* 70:106–14.

Takamura, N., F. Kasai, and M. M. Watanabe. 1989. Effects of Cu, Cd and Zn on photosynthesis of freshwater benthic algae. *J. Appl. Phycol.* 1:39–52.

Tanguy, A., et al. 2002. Polymorphism of metallothionein genes in the Pacific oyster *Crassostrea gigas* as a biomarker of response to metal exposure. *Biomarkers* 7:439–50.

Tanguy, A., et al. 2003. Metallothionein genes in the European flat oyster *Ostrea edulis*: A potential ecological tool for environmental monitoring? *Mar. Ecol. Prog. Ser.* 257:87–97.

Teixeira, H., et al. 2009. Quality assessment of benthic macroinvertebrates under the scope of WFD using BAT, the benthic assessment tool. *Mar. Pollut. Bull.* 58:1477–86.

TGD. 2003. Technical Guidance Document on Risk Assessment in support of Commission Directive 93/67/EEC on Risk Assessment for new notified substances, Commission Regulation (EC) n° 1488/94 on Risk Assessment for existing substances and Directive 98/8/EC of the European Parliament and of the Council concerning the placing of biocidal products on the market. EUR 20418 EN/2. European Commission, Joint Research Centre.

Truchet, M., R. Martoja, and B. Berthet. 1990. Conséquences histologiques de la pollution métallique d'un estuaire sur deux mollusques, *Littorina littorea* L. et *Scrobicularia plana* Da Costa. *C. R. Acad. Sci. Paris* 311:261–68.

van der Oost, R., J. Beyer, and N. P. E. Vermeulen. 2003. Fish bioaccumulation and biomarkers in environmental risk assessment: a review. *Environ. Toxicol. Pharmacol.* 13:57–149.

van der Oost, R., C. Porte-Visa, and N. W. van den Brink. 2005. Biomarkers in environmental assessment. In *Ecotoxicological testing of marine and freshwater ecosystems: Emerging techniques, trends and strategies*, ed. P. J. den Besten and M. Munawar, 87–152. Boca Raton, FL: CRC Press.

Vermeire, T., et al. 1999. Assessment factors for human health risk assessment: A discussion paper. *Crit. Rev. Toxicol.* 29:439–90.

Whalon, M. E. 2008. Introduction. In *Global pesticide resistance in arthropods*, ed. M. E. Whalon, D. Mota-Sanchez, and R. M. Hollingworth, 1–4. Wallingford, UK: CABI.

Wirgin, I., and J. R. Waldman. 2004. Resistance to contaminants in North American fish populations. *Mutation Res.* 552:73–100.

4 Microbial Pollution-Induced Community Tolerance

Ahmed Tlili and Bernard Montuelle

CONTENTS

4.1 INTRODUCTION: FROM THE SIMPLE TO THE COMPLEX

The assessment of the ecological quality of natural ecosystems is partly based on chemical criteria, such as the chemical nature and concentrations of pollutants, and physical criteria, such as topography, geology, and hydrology (Carluer and De Marsily 2004; Lagacherie et al. 2006). Even if such a physicochemical approach is essential, it has many limitations (specific information, no predictive information on the ecotoxicological risks associated with contamination) and many constraints from an analytical point of view (large number of molecules and degradation products of these molecules to analyze, unknown molecules, detection and quantification limits, etc.). In addition, biological approaches measure the impact of exposure to toxic contaminants on biological communities in natural ecosystems. According to the scheme proposed by Lagadic and Caquet (1998), three strategic levels for the *in situ* evaluation of the biological quality of ecosystems should be considered. The first is the measurement of the concentrations of chemical substances in organisms after exposure to toxicants (absorption, bioaccumulation, etc.). The second level is a measurement of biomarkers (biochemical and physiological) such as enzyme activities (catalase, glutathione peroxidase, etc.). And finally, the third level is a measure of the effect of the toxic contaminant on populations via the use of bioindicators (species or species groups) such as the diatom biological index or the IPS (pollution sensitivity index) in aquatic ecosystems, for example. These indices are based primarily on species characteristics analysis (taxonomy, species abundance statistics, and identification of key species), but do not reflect functional changes (growth, metabolic activities) that can develop in biological communities in response to anthropogenic disturbances. In addition to obtaining ecological information from *in situ* organisms and communities, ecotoxicological methods and techniques have the potential to improve environmental assessment.

Mono-specific bioassays are often used to determine the degree of environmental contamination and the risk associated with toxic substances. These ecotoxicity tests are performed in the laboratory under controlled conditions on selected organisms based on different criteria (sensitivity to toxicants, ease of maintenance and *in vitro* culture, repeatability), and the impact is assessed by dose-response tests in relation to different parameters (survival, mortality, reproduction). A major shortcoming of these tests is their lack of environmental representativeness and the reductionist approach to effect issues. Indeed, ecotoxicity tests are generally applied to "strains" that are cultivated in artificial culture media in the laboratory, sometimes using genetically modified strains (e.g., several bacterial tests: Ames, Microtox). Moreover, this approach is performed with single species and therefore does not include interspecific interactions. To overcome this drawback, it is essential to develop ecotoxicological studies that take into account both the biodiversity and functioning of microbial communities. Biological and ecological specificities of microbial communities make them very interesting tools for environmental quality assessment and impact assessment whether at the *in situ* level or using a "cosms" techniques to develop causality studies. The twenty-year-old concept of pollution-induced community tolerance (PICT), first introduced by Blanck et al. (1988), is an ecotoxicological tool that provides a good alternative for environmental status characterization, useful for not only assessing an immediate impact but also taking into account the contamination history of the ecosystem at the community level. It is also a tool that allows a better understanding of the ecological drivers in adaptation processes. Finally, the PICT concept allows the introduction of "more ecology in ecotoxicology."

4.2 THE PICT CONCEPT IN BRIEF: PRINCIPLE, OBJECTIVES, AND APPLICATION

The large diversity of microbes in natural ecosystems (algae, bacteria, protozoa, fungi, viruses) and the associated multitude of biological and physicochemical processes induce high structural and functional complexity into microbial communities. This generates various sensitivities and a highly variable type of response to anthropogenic pressures. The presence of an increasing bioavailability of xenobiotics especially leads to such structural and functional disturbances of microbial communities. One result is adaptation and the selection of the most tolerant species or strains of microorganisms to these toxicants, resulting in an overall increase in tolerance toward the toxicants of the whole exposed communities. This process underlies the concept of pollution-induced community tolerance. Assessing the acquisition of tolerance is a method that tries to establish a cause-effect relationship between one pollutant and its impact on biological communities. Indeed, the PICT approach is primarily a measure of the sensitivity of biological communities to a toxicant, combined with taxonomic analysis (Figure 4.1). However, a lot of issues are still open: Is it possible to target the effect of a given compound in an environmental pollutant cocktail? What are the relevant endpoints for PICT studies? What is the threshold to conclude that a community is impacted?

Basically, the main goal of such a method is to highlight differences in tolerance toward a toxin of two comparable biological communities, with one of them being "naïve" with respect to this substance (the reference community). In principle, the PICT concept can be applied to all types of ecosystems, aquatic or terrestrial (Hjorth et al. 2006; Niklinska et al. 2006), and to all microbial communities such as bacteria (Boivin et al. 2005) or photosynthetic microorganisms (Dahl and Blanck 1996). Table 4.1 summarises a nonexhaustive list of microbial PICT studies (in aquatic ecosystems and different targeted organisms) (see Table 14.2 for more details on soil-bacterial PICT cases).

Besides impact assessment, and as a result of generation time and microbial dynamics specificity, the PICT approach could also be used for environmental recovery studies. However, this application is less developed than impact studies (Blanck and Dahl 1998; Larsen et al. 2003; Boivin et al. 2006; Demoling and Baath 2008a,b).

4.3 SELECTION PROCESS, DETECTION, AND VALIDATION OF THE PICT APPROACH

According to Molander and Blanck (1991), the PICT concept can be summarised in two phases: a selection phase and a detection phase based on the functional response of biological communities. The selection phase is based on how communities adapt to stress conditions. Several levels of biological responses could be involved, from strain-specific sensitivity, cellular detoxification capacity or contaminant storage, to selection of resistant strains and changes in diversity or in specific individual numbers. The detection phase concerns the expression of the acquired tolerance and the way to characterise it by short-term bioassays.

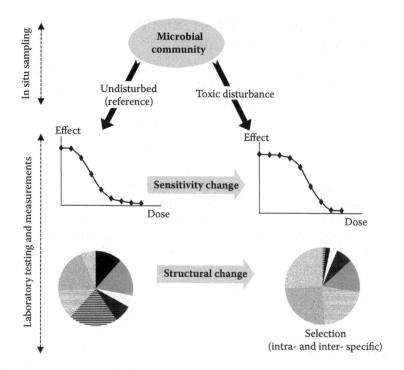

FIGURE 4.1   The PICT concept is based on the fact that exposure of a biological community to a toxicant will induce an inter- and intra-specific selection of the most tolerant organisms or the establishment of mechanisms for detoxification. Thus, the entire community may be restructured, present physiological alterations, and come finally to display an overall increase in tolerance to the toxicant compared to a reference community that has never been exposed. This tolerance difference, which is evaluated by *in vitro* short-term bioassays and with increasing concentrations of toxicant, may give an indication of past exposure (*in situ*) to toxicant of the different sampled communities.

4.3.1   Selection Phase or "Survival of the Fittest"

Basically, exposure of communities to toxicants leads to the selection of the most tolerant species. This tolerance could result from cellular metabolic processes that detoxify the contaminant. As examples, stress proteins are often involved in the tolerance to a metal, or changes may occur in the activity of an enzyme that modifies metal speciation (e.g., Cr^{6+} to the less toxic Cr^{3+}, resulting from reductase activity in *Methylococcus capsulatus* (Al Hasin et al. 2010)).

This section deals with other drivers of tolerance, such as selection processes and their role in PICT acquisition. Taken as a whole, tolerance acquisition depends on (1) inter- and intraspecific diversity and (2) the specific sensitivities of the strains or species that comprise the community.

4.3.1.1   Intra- and Interspecific Selection

Selection of a strain by a toxicant can lead to an increased tolerance in the species, population, and community concerned. There are many examples demonstrating

TABLE 4.1

Main Characteristics and Biological Targets of PICT Studies in Different Aquatic Ecosystems (river, lake, estuary, marine and coastal system)

Targeted Organisms	Selection (Structural Endpoint)	Detection (Functional Endpoint)	References
Photoautotrophic (algae and cyanobacteria) community in periphyton, phytoplankton, or sediment	Taxonomic analysis by microscopy; Molecular biology by DGGE, TTGE, or PFLA; High-performance liquid chromatography	Photosynthesis: ^{14}C incorporation and *in vivo*-induced fluorescence	Wangberg et al. 1991; Nyström et al. 2000; Paulsson et al. 2000; Soldo and Behra 2000; Bérard and Benninghoff 2001; Schmitt-Jansen and Altenburger 2005; Hjorth et al. 2006; Dorigo et al. 2007; Navarro et al. 2008; McClellan et al. 2008; Schmitt-Jansen and Altenburger 2008; Tlili et al. 2008; Eriksson et al. 2009; Blanck et al. 2009; Debenest et al. 2009; Pesce et al. 2010
Heterotrophic (bacteria and fungi) community in periphyton, phytoplankton, or sediment	Bacterial enumeration (DAPI-FISH, flow cytometry); Molecular biology (DGGE, TTGE, or ARISA); Community-level physiological profiling (CLPP)	Thymidine (Tdr) and leucine (Leu) incorporation techniques; Biolog ECO plates; Respiration (gas chromatography or MicroResp); Extracellular enzyme activity	Lehmann et al. 1999; Nyström et al. 2000; Barranguet et al. 2003; Blanck et al. 2003; Brandt et al. 2004; Massieux et al. 2004; Boivin et al. 2005; Boivin et al. 2006; Hjorth et al. 2006; Tlili et al. 2010

the phenomenon of intraspecific selection, especially for algal populations in aquatic ecosystems contaminated by metals or xenobiotics (Kasai et al. 1993; Bérard et al. 1998; Debenest et al. 2009). Hersh and Crumpton (1989) isolated and compared Chlorophyceae strains from different aquatic systems either highly contaminated by atrazine (agricultural areas) or uncontaminated. Chlorophyceae from the contaminated systems exhibited increased tolerance to atrazine compared to the Chlorophyceae isolated from the uncontaminated systems. Bérard et al. (1998) also tested the atrazine tolerance of two strains of *Chlorella vulgaris* (Chlorophyceae) isolated from two limnic systems: Lake Geneva (with low herbicide pollution) and Villaumur barrage (highly polluted by herbicides inhibiting Photosystem II). Results showed that the strain from the polluted site was more tolerant to atrazine than the strain from the unpolluted site, suggesting that there has probably been a genotypic selection by herbicides inhibiting photosynthesis, therefore confirming intraspecific selection.

The second selection level is interspecific, and most of the studies on PICT have shown that some species can survive at the expense of others, whether for bacteria (Dorigo et al. 2007; Demoling et al. 2009) or for algae (Schmitt-Janssen and Altenburger 2005; Dorigo et al. 2007; McClellan et al. 2008; Morin et al. 2010). A study conducted by Boivin et al. (2005) showed the selective effect of copper on bacterial communities in river biofilms, by using denaturing gradient gel electrophoresis (DGGE) analysis of the 16S rDNA gene. McClellan et al. (2008) studied interspecific algal community selection after exposure to diuron. They used *in vivo* algal-induced fluorescence (measured by PhytoPAM) and showed a 45% increase in the cyanobacterial proportion of the community, whereas the relative contribution of diatoms decreased to about 50% in the contaminated microcosms compared to the control microcosms. Serra (2009) also reported changes in the algal biofilm community exposed to copper, with the development of green algae and the decrease of diatoms and cyanobacteria, reflecting the copper tolerance of green algae.

Intra- and interspecific selection takes place simultaneously. Several contributing factors related to biological interactions and environmental factors (varying according to the studied sites and seasons) may affect the specific response of organisms to different selection pressures (DeLorenzo et al. 1999a,b; Espeland et al. 2001; Dorigo et al. 2004). Rimet et al. (1999) show that algal sensitivity increases with the complexity of the algal communities studied: the EC_{50} of nicosulfuron was higher for the algae from monocultures than for those studied in microcosms inoculated with phytoplankton collected from Lake Geneva. Hjorth et al. (2006), investigating the role of interaction complexity between marine phytoplankton and zooplankton under stress from the antifouling compound zinc pyrithione, provided evidence of effects on the function and structure of the marine plankton communities. Direct selective effects were observed immediately (such as reduced phytoplankton activity inducing a decrease in grazing rates of zooplankton), leading to cascading indirect effects throughout the community, such as a change in the algal community structure over time due to the decrease of the zooplankton predation pressure.

4.3.1.2　Ecosystem Characteristics and Their Role in the Selection Process

Many studies have shown the influence of environmental factors on the bioavailability of toxicants and therefore on the control of their toxicity (Lock and Janssen 2005; Brack et al. 2008). PICT reflects the accessibility to environmental chemicals by the local microorganisms and the bioavailability of these contaminants (metal speciation, adsorption of toxins onto organic matter, etc.). Measures of concentrations of *in situ* total contaminants are too widely inclusive, and can then appear poorly correlated with the PICT values of impacted communities.

For example, Almas et al. (2005) emphasised that localization in the soil matrix could influence organism exposure to trace metals, either by diffusion constraints or by metal sequestration onto soil colloids (providing a local environment with low chemical activity), or both. Neither of these mechanisms provides permanent protection if high metal concentrations occur in macropores for prolonged periods, but they would protect against transient peaks in macropore concentrations. To confirm this hypothesis, the authors used a set of Zn- and Cd-contaminated soils to explore the relationship between the binding strength of microorganisms to surfaces and their decreased PICT. They demonstrated that free and loosely attached cells had developed a strong PICT in response to eighty years of Zn and Cd pollution, whereas strongly attached cells were virtually unaffected. It appears that the position of strongly attached cells in biofilms and micropores of soil has effectively protected them against toxic metal selection. In an aquatic ecosystem study, Guasch et al. (2002) evaluated the effects of chronic copper exposure on natural periphyton in a nonpolluted calcareous river. In this study, copper exposure for sixteen days under variations in pH similar to those observed in the stream (i.e., diurnal changes from 8.2 to 8.6) affected the biomass and community structure of the stream algal community, by the replacement of metal-sensitive species with metal-tolerant ones. The authors concluded that the water alkalinity was important in affecting the bioavailability of the metals, and therefore their selection pressure on the biological community. On a broader study scale, Blanck et al. (2003) showed in a study on fifteen European rivers that Zn tolerance of the microbial community was driven by different factors (e.g., pH, calcium, or bicarbonate concentration) and that only regional models could describe this tolerance. Variability in Zn tolerances was high, reaching 1.5 to 2.5 orders of magnitude, ranging, for example, from 25 to 8,145 μM Zn for responses in photosynthesis.

Environmental factors (e.g., temperature, light, or nutrient availability), which interact with toxicants, may also influence microorganism sensitivity and thus the selection process within a community exposed to pollutants (Bérard and Benninghoff 2001; Guasch et al. 2003; Villeneuve et al. 2010). Effects of antibiotics on the PICT of the soil bacterial community could be modulated by the organic matter and nutrient content of soil. Schmitt et al. (2005) studied the microbial-induced tolerance in response to increasing sulfonamide sulfachloropyridazine concentrations in bulk soil and in nutrient-amended soil. They found that the antibiotic-induced tolerance (measured with the single carbon source microplate Biolog® technique) of the microbial community was increased if the soils were amended with nutrients. In the same way, Guasch et al. (2004) investigated the influence of phosphorus limitation on copper toxicity in periphytic communities, and they observed an enhancement of the phototrophic community tolerance to Cu in the P-enriched microcosm, coupled with a shift in the community structure between the exposed and the nonexposed periphyton communities.

More globally, Brack et al. (2008) considered that an effect-directed analysis (EDA) that combines chemical fractionation, chemical analysis, and biotesting could help to identify hazardous compounds in complex environmental mixtures, as this combined approach avoids artifacts in establishing relevant cause-effect relationships. The PICT approach is well adapted to such an EDA approach.

4.3.1.3 Exposure Time and Species Succession Duration

The type and the duration of exposure are obviously of major interest for the development of tolerance. PICT studies have been mainly conducted on small organisms with fast life cycles, mainly algae and bacteria, facilitating the analysis of short-term contamination effects (pulse contamination) (Tlili et al. 2008). The short generation time of bacteria and microalgae is a real advantage for comprehensive studies on the mechanism of tolerance acquisition at the community level. One of the most important parameters for the selection process in a PICT approach is the exposure time to the toxicant, since it must exceed the generation time of a nonexposed community. Moreover, the contact between toxicants and community must be established (Schmitt et al. 2005). The rate at which PICT develops has rarely been directly investigated. However, Ivorra et al. (1999) compared Zn and Cd tolerance in young biofilms and old biofilms (two and six weeks old, respectively) sampled in a Zn- and Cd-polluted river to the tolerance of naïve biofilms. These authors showed that two weeks' exposure to Zn and Cd was not sufficient to observe PICT development, unlike the six weeks' exposure. In a field study of irgarol tolerance in marine periphyton conducted over a decade, Blanck et al. (2009) showed that despite the fact that the irgarol concentrations had not changed in the polluted sites, PICT development was observed between the beginning and the end of the study, supporting the hypothesis that the PICT potential was low initially and that a persistent selection pressure was required to favor the development of irgarol-tolerant species. Such time-dependent acquisition of PICT is supported by the generation and growth time of algal or bacterial species, but also by the rate of immigration/emigration of cells in the environment, especially in biofilms.

The taxonomic analysis of biological communities highlights these structural differences and diversity evolution between impacted and nonimpacted communities, and subsequently supports the species selection hypothesis, for example, in the case of algae at a global level (Bérard and Benninghoff 2001; Guasch et al. 2004), or more specifically in the case of diatoms (Morin et al. 2008). Nontaxonomic methods also provide interesting evidence for microbial diversity changes. Fatty acid analysis (PLFA) was used especially to detect structural changes in microbial communities in contaminated soils (Baath et al. 1998; Demoling et al. 2009). Molecular biology tools are also used, such as polymerase chain reaction (PCR) on 16S rDNA (prokaryotes) and 18S (eukaryotes), followed by denaturing gradient gel electrophoresis (DGGE) or temperature gradient gel electrophoresis (TGGE) (Dorigo et al. 2002, 2007).

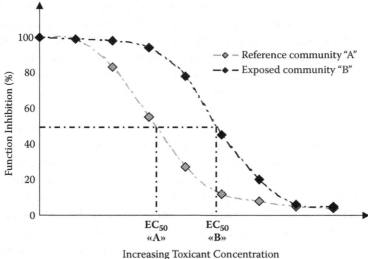

FIGURE 4.2 Dose-response curves for PICT characterisation. The tolerance level of biological community to a toxicant is measured by estimating the effective concentration of this toxic reducing x% of the selected functional descriptor intensity (EC_x). This estimation is based on the analysis of dose-response curves established by means of short-term bioassays. Comparison of EC_x obtained for different communities therefore allows evaluation of their level of tolerance for a toxicant: the highest EC_x value means a greater resistance to a pollutant. Thus, in the figure, community B is characterised by an EC_{50} higher than the EC_{50} of community A, which means that the exposed community is more tolerant than the reference community to the tested pollutant.

High-pressure liquid chromatography (HPLC), which allows the identification of photosynthetic pigments of major algal groups, allows investigation of the structural changes of photosynthetic communities (Dorigo et al. 2007).

4.3.2 MEASURING PICT (DETECTION PHASE)

To detect tolerance acquisition, short-term bioassays are applied, allowing the establishment of a dose-response curve for the toxicant-exposed and reference communities. Tolerance is subsequently expressed as an EC_x applied to the selected endpoint parameter. The difference between the EC_x obtained for the toxicant-exposed and reference communities allows the assessment of PICT (Figure 4.2).

The initial composition of a community that reflects environmental conditions before exposure to any stress (time reference) or under reference conditions (spatial reference) should be considered for PICT assessment. This was shown for phytoplankton, particularly in connection with seasonal events and algal succession (Bérard and Benninghoff 2001). Similarly, differences in periphyton microbial diversity between lotic and lentic areas in rivers (an effect of flow speed) resulted in differences in sensitivity and tolerance to a similar toxic exposure (Villeneuve 2010). The choice of the reference community is therefore an important step for the validation of a PICT approach. According to Blanck et al. (1988), a reference community is a community able to acquire significant tolerance after exposure to a toxicant: the tolerance acquisition depends not only on the community itself but also on its interactions with the toxicant. A highly diversified community with a large range of specific sensitivities toward a toxicant has an important selection potential, and thus an important potential for tolerance (so-called background tolerance) to toxicants.

The choice of a functional parameter in the induced tolerance assessment is also crucial, because its relevance as an ecotoxicological endpoint depends first on the mode of action of the toxicant (and its bioavailability) and second on the mechanisms involved in the community tolerance. In a recent study, Tlili et al. (2010) investigated the potential effect of a phosphorus gradient on biofilm tolerance to copper and diuron in indoor microcosms. To assess biofilm sensitivity to Cu and diuron, various physiological parameters were used in short-term inhibition tests, targeting the autotrophic and heterotrophic communities. It was concluded that induced tolerance to Cu or diuron varied according to the considered toxicant and the functional endpoint chosen, confirming the need to focus on functional diversity using complementary indicators for ecotoxicological investigations in river biofilms. Moreover, the combination of assessment tools for different activities (heterotrophic and autotrophic) can highlight possible interactions between different communities. Thus, Nyström (1997) showed a replacement of autotrophic communities by heterotrophic communities in atrazine-contaminated microcosms.

The most commonly used tests (Table 4.1) are the incorporation of labeled elements, such as carbon, as labeled ^{14}C, to measure primary production, considered as a relevant endpoint in the tolerance assessment in aquatic photosynthetic organisms in both the phytoplankton and the phytobenthos (Dahl and Blanck 1996; Bérard et al. 2003), and the incorporation of tritiated thymidine or leucine to measure bacterial activity, mainly in soil studies (Almas et al. 2005; Niklinska et al. 2006) and increasingly in aquatic environments (Pesce et al. 2006). Other tests, such as the induction of algal fluorescence by photosynthesis inhibitors or the use of Biolog plates and measures of respiration (concentration of CO_2 or O_2), are also applied (Witter et al. 2000; Niklinska et al. 2006; Tlili et al. 2010) (see also Section 14.3.2 for bacterial PICT endpoints).

4.3.3 GAPS IN PICT: NOBODY IS PERFECT

As explained above, the advantage of the PICT approach is, on the one hand, to give information about the history of exposure to a toxicant (not an occasional and brief contamination of the environment), focusing then on the link between selection pressure and impact, and, on the other hand, to generally provide a faster and complementary response than available from a taxonomic analysis of communities. However, to validate the PICT approach as a bioindicator tool of ecosystem pollution, many issues still need to be considered.

4.3.3.1 Short- or Long-Term Endpoints: Which Are the Most Relevant?

Most of the experimental studies on PICT use and compare the classical parameters of long-term stress (biomass, chlorophyll concentration, bacterial total density, etc.) and primary production (e.g., ^{14}C, bacterial leucine or thymidine incorporation, *in vivo* algal-induced fluorescence) used in short-term bioassays. The PICT approach can therefore be well validated, and tolerance results are in agreement with biomass or activity measurements realised during a study. However, in some cases, conventional endpoint measures of long-term stress appear to be more sensitive than those typically used in the PICT approach. Molander and Blanck (1992) showed, in a study on the pollution-induced tolerance in marine periphyton exposed to different concentrations of diuron, that some long-term parameters (increase of diatom species richness and increase in chlorophyll-*a*) were more sensitive than the lowest detection of PICT at 40 nM diuron, while PICT was more sensitive than other long-term parameters, such as a decrease of carbon incorporation rate. In a recent study, Tlili et al. (2008) highlighted the ecotoxicological potential impact of the herbicide diuron on biofilms during flooding events in a small river in the Beaujolais vineyard area (France). The authors investigated the responses of chronically contaminated biofilms exposed to short-term pulses of diuron. Biofilms were grown in two groups of indoor microcosms: one being noncontaminated and another contaminated by a chronic low-level of 1 µg.L^{-1} of diuron. Then these two groups of microcosms were splitted into lots with different experimental modalities: non exposed or exposed to pulse contamination (single pulse or double pulses, at two environmental concentration of diuron, 7 or 14 µg.L^{-1}). Exposure to pollution and its impact on biofilms were assessed by using various sets of long-term stress parameters, such as biomass parameters (chl *a*, ash-free dry weight (AFDW)) and community structure (using 18S and 16S rDNA gene analysis by DGGE, and HPLC pigment analysis to target eukaryotes, bacteria, and photoautotrophs, respectively), and also by performing photosynthesis inhibition bioassays (by ^{14}C assimilation). In this study, long-term stress parameters appeared to be more sensitive and discriminatory than PICT measurement. Whatever the exposure situation, no PICT development was observed, whereas biomass, fluorescence, and structural analysis showed a diuron pulse effect on biofilms.

Molander and Blanck (1992) proposed a conceptual three-stage model to explain the impact of diuron on marine periphyton as a function of water diuron concentrations. During the first stage, exposure to diuron has no long-term effect, but during the second stage there is a slight long-term effect on biomass and structure, and during the third and final stage, the diuron stress is so severe that sensitive species are eliminated, resulting in a decline in biomass diversity and chlorophyll-*a* in parallel with an increase in community tolerance, giving an easily detected PICT.

4.3.3.2 Environmental Factors and PICT

In the global context of assessing the impacts of complex pollution on microbial community diversity and function in natural ecosystems, one of the most critical points remains the distinction between the relative effects of selective pressures resulting from xenobiotic pollution and those resulting from environmental factors. As explained above, PICT is detected by comparing the tolerances of a selected community and an unselected naïve community, the latter having an initial composition reflecting environmental conditions before application of stress (time reference) or reference conditions (spatial reference) and an original sensitivity (baseline tolerance; Molander 1991). This baseline tolerance probably fluctuates with the environmental conditions (such as nutrients, temperature, and light) of the reference community, inducing limited variations in biomass, species composition, and phenotypic adaptations (Bérard and Benninghoff 2001; Dorigo et al. 2004; Guasch et al. 2003; Schmitt et al. 2005). When there are significant variations of the baseline tolerance, it will be difficult to prove an induced tolerance exclusively caused by the toxicant selection pressure. It is therefore necessary to have not only a reference site that is globally comparable to the study site, but also a sampling strategy (spatial and temporal) minimizing the interference of environmental factors in the detection of PICT. In a study of the effect of copper and temperature on aquatic bacterial communities, Boivin et al. (2005) showed that the copper tolerance at 10 and 14°C increased about three times, whereas copper tolerance at 20°C increased about six times. Temperature therefore had an effect on tolerance development, indicating that the effect of exposure to copper was enhanced at the higher temperature. In a soil study, Demoling and Baath (2008a) investigated the toxicity of different phenols on bacterial communities at different temperatures, by using leucine incorporation in bioassays as an endpoint, and they observed that the induced tolerance to phenols increased at lower temperature. Some studies have also focused on the effect of nutrients on tolerance acquisition (Ivorra et al. 2002; Brandt et al. 2009). Guasch et al. (2004) observed a clear influence of phosphorus limitation on copper toxicity. Biofilm communities that were previously fertilised for eighteen days were three times more tolerant than control communities, indicating that phosphorus limitation enhanced Cu toxicity, and tolerance induction was probably related to the higher phosphorus availability (negative correlation between Cu toxicity and algal biomass, which is stimulated by phosphorus supply). On the other hand, in a more recent study, these authors (Guasch et al. 2007) tested the influence of phosphorus on the tolerance of periphytic communities to the herbicide atrazine, and, conversely to copper, tolerance induction to atrazine did not require a phosphorus supply (because of the fast sorption kinetics and low rate of bioaccumulation of atrazine, which make its toxicity independent of the water chemistry, and thus of phosphorus enrichment). In contrast, light history seems to influence the sensitivity of periphytic algae to atrazine (Guasch and Sabater 1998).

Finally, considering the variability of terrestrial and aquatic environments, PICT assessment should take into account the mosaic of microhabitats to establish a reliable sampling strategy (Almas et al. 2005 in soil; Dorigo et al. 2009 in watercourses). These small-scale spatial, but also temporal, variabilities are in fact of major concern for population succession during PICT acquisition. They are sometimes considered stochastic events (Brandt et al. 2004).

4.3.3.3 The Phenomenon of Cotolerance

Cotolerance (Table 4.2) may occur when communities that have been exposed to one toxicant, but not to another, become tolerant to both toxicants. Occurrence of cotolerance depends on the means of acquiring tolerance and on the tolerance mechanisms involved (Blanck et al. 1988). This phenomenon is mainly caused by substances having

TABLE 4.2

Examples of Cotolerance between Toxicants as a Confounding Factor for PICT Assessment

Long-Term Exposure	Community	Endpoint: 1. Selection 2. Detection	Cotolerance with	References
Diuron	Phototrophic marine periphyton	1. Taxonomic analysis by microscopy 2. Photosynthesis (^{14}C incorporation)	Atrazine Bromacil Metribuzin	Molander al. 1991
Zinc	Soil bacteria	1. Plate count method (CFU) 2. Thymidine incorporation	Copper Cadmium Nickel Lead	Diaz-Ravina et al. 1994
Ultraviolet radiation	Phototrophic freshwater periphyton	1. Taxonomic analysis by microscopy 2. Photosynthesis (fluorimetry by MINI-PAM)	Cadmium	Navarro et al. 2008
Oxytetracycline	Soil bacteria	1. Plate count method (CFU) 2. Biolog ECO plates	Tetracycline Tylosin	Schmitt et al. 2005
Copper	Freshwater phytoplankton	1. Taxonomic analysis by microscopy 2. Photosynthesis (^{14}C incorporation)	Zinc	Gustavson and Wängberg 1995
Arsenate	Phototrophic marine periphyton	1. Taxonomic analysis by microscopy 2. Photosynthesis (^{14}C incorporation)	Thiophosphate Low cotolerance with arsenic and diuron	Blanck and Wängberg 1991
Phenol	Soil bacteria	1. Phospholipid fatty acid pattern 2. Leucine incorporation	2-Chlorophenol 2,4-Dichlorophenol 2,3,6-Trichlorophenol No cotolerance with copper and zinc	Demoling and Baath 2008c
Atrazine	Freshwater phytoplankton	1. Taxonomic analysis by microscopy 2. Photosynthesis (fluorimetry by Toxy PAM)	Isoproturon No cotolerance with diuron	Knauer et al. 2010
Diuron			Atrazine Isoproturon	
Copper	Phototrophic freshwater periphyton	1. Taxonomic analysis by microscopy 2. Photosynthesis (^{14}C incorporation)	Zinc Nickel Silver	Soldo and Behra, 2000

a similar mode of action (Molander and Blanck 1991; Soldo and Behra 2000) or inducing a similar detoxification mechanism (Gustavson and Wängberg 1995). According to
Molander (1991), two classes of mechanisms at the biochemical level could induce cotolerance. The first concerns tolerance mechanisms related to the uptake, translocation, or
metabolization/excretion of the toxicant, while the second is related to modifications of
the target site or of bypass reactions. Cotolerance is a phenomenon that should not be
confused with the multiple tolerance of a community subjected to exposure by several
toxicants simultaneously. Some studies have reported a cotolerance of biofilm communities between copper and some metals, such as Zn or Ni (Gustavson and Wängberg 1995;
Ivorra et al. 1999). Soldo and Behra (2000) have shown a strong cotolerance to silver in
autotrophic biofilm communities exposed to 5 µM copper. In contrast, Tlili (personal
communication) did not detect any cotolerance between copper and silver in a study
based on the use of the MicroResp™ method to assess PICT to metals for lotic biofilms.
These contradictory results may be explained by the fact that most of the studies that have
detected a cotolerance were based on photosynthesis as an endpoint to measure PICT
(targeting the photoautotroph community in the biofilm), whereas Tlili (personal communication) used the substrate-induced respiration endpoint, targeting the heterotrophic
community in the biofilm. Thus, depending on the kind of targeted activities and associated communities (phototrophic or heterotrophic), cotolerance assessment can be variable, and the combination of different endpoints may provide a better approach. Knauer
et al. (2010) examined the specificity of PICT by evaluating the cotolerance pattern for
three Photosystem II (PS II) inhibitor herbicides (atrazine, diuron, and isoproturon) to
the phytoplankton community, exposed separately over the long term to these three herbicides. These authors demonstrated that preexposure to diuron induced similar tolerance to all three herbicides. While it is difficult to distinguish the specific PICT effect
of two toxicants, the PICT method can separate molecule groups with common effects
that are associated with different pollution types (herbicides inhibitors of PS II, groups
of heavy metals, etc.). The PICT method, used with particular caution with respect to the
phenomenon of cotolerance, might, for example, be proposed as a preliminary approach
in the assessment of ecosystem pollution, choosing a model molecule for each different
group of pollutants.

4.4   THE PICT CONCEPT AND ITS POSITION
IN ECOSYSTEM FUNCTIONING

Even though the PICT concept has been proposed as a methodological approach
based on comparative community ecotoxicology, it has the potential to also be integrated into a broader vision of ecosystem functioning. Any investigation of the ecological basis of PICT needs to have a wide knowledge of the ecosystem itself.

4.4.1   PICT CONCEPT, REDUNDANCY, AND RESILIENCE:
WHAT IS THE CONNECTION?

One important and up-to-date question in microbial ecology concerns the nature
of the relationship that may exist between biodiversity and ecosystem functioning.

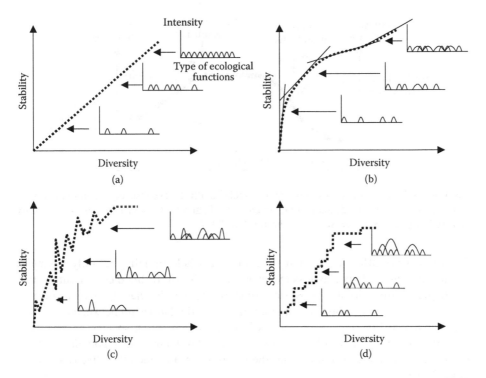

FIGURE 4.3 Four models showing the relationship between species richness and functional stability in communities. (a) Linear model: Functional stability decreases linearly with the reduction of species number. (b) Rivet model: A functional redundancy protects the system from loss of species until a threshold level. (c) Idiosyncratic model: Species disappearance effect is dependent on species interaction and hardly predictive. (d) Drivers and passengers model: Stability depends on the species ecological importance. Disappearance of driver species (keystone species) has more effect than disappearance of passenger species. (Modified from Figure 6.2 in Clements, W. H., and M. C. Newman, eds., *Community Ecotoxicology*, J. Wiley & Sons, Chichester, UK, 2002.)

Several theories attempt to explain the impact of diversity on ecosystem functions, such as matter and energy flows like photosynthesis, gene flow, or nutrient cycling. On the basis of assumptions about the functional role of diversity, there is the necessary assessment of the unequal ecological roles of various species in the ecosystem: the concepts of keystone species, guilds, or functional redundancy. It is generally admitted that large species richness allows a greater stability of the community, but the link between these two characteristics might vary, thereby leading to applicability of different models (Clements and Newman 2002; Figure 4.3).

One of the theories regarding the impact of species number on ecosystem function is based on the assumption of redundancy, which considers that the number of functions in an ecosystem increases with the number of species, but only to a certain threshold. Beyond this threshold, species that are added can be considered superfluous because they do not enrich ecosystem functions. According to this theory, the disappearance of the most sensitive species (as one of the processes underpinning

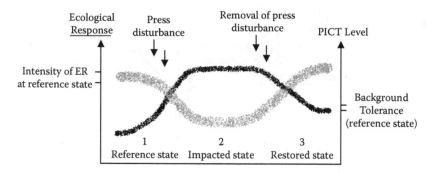

FIGURE 4.4 Ecological response and associated PICT response of a community to a stress or a recovery. (Modified from Clements, W. H., and M. C. Newman, eds., *Community Ecotoxicology*, J. Wiley & Sons, Chichester, UK, 2002.)

PICT) does not affect ecosystem function as a whole initially, but only beyond a certain threshold of species loss (Schwartz et al. 2000). If this theory assumes that the degree of diversity and ecosystem function are closely related, another one estimates that these two entities do not maintain a fixed relationship between them, and that ecosystem functions are a result of interactions between species. In this case, it is not only the number of species that counts, but their relative abundance, their organization in the community, and the environment in which they live (Ekschmitt et al. 2001).

Biological diversity appears therefore to have two major roles. The first is to ensure essential ecosystem function, and the second is to allow adaptation to changing environmental conditions (e.g., stress conditions). The concept of an "insurance policy," which stipulates that each species has advantages that enable it to deal successfully with certain environmental circumstances, explains why a more diversified system should be more stable. If a community contains a large number of redundant species, the probability that a functional role can still be exercised even after a strong disturbance is greater, through the survival of at least one of the more tolerant species (Mertz et al. 2007). In this case, PICT could correlate with a reduction of diversity or with changes in species evenness. This restructuring finally leads to changes or losses in ecological functions of the community, and so to a lowering of community stability (Figure 4.4).

The notion of ecosystem stability is also related to other concepts, which are a matter of discussion: resilience (return of ecosystems to a quasi-initial state following a disturbance), robustness (resistance to interference), and elasticity (ability to restore quickly the biomass to its initial value after disturbance). All these concepts are more or less involved in a comprehensive analysis of PICT. Barbault (1997) highlights "the probable importance of biodiversity to ecosystem resilience, which means not only their recovery ability after disturbance, but also their resistance to invasive species and therefore their long-term persistence" (p. 159). "Ecophysiological" demands of species are never identical, even when the species provide the same functions (identical ecological niches, redundant species, etc.). The numerical balance (abundance and richness) between these species may

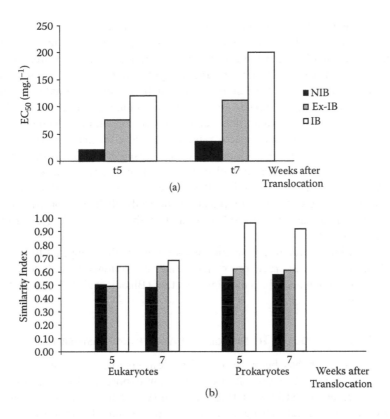

FIGURE 4.5 Periphytic community response after translocation for an impacted site (IB) to a nonimpacted site (NIB). The translocated biofilm (ex-IB) has a PICT intermediate between NIB tolerance (reference or background PICT) and the IB tolerance. Similarity Indices at 5 and 7 weeks after translocation of IB to the reference site indicate a low similarity with IB diversity for prokaryotes and an incomplete one for eukaryotes. (Modified from Dorigo, U., et al., *Aquat. Toxicol.*, 98:396–406, 2010.)

vary, depending on environmental changes, while the interactions between them remain unchanged.

These phenomena should also be considered for environmental recovery processes after a disturbance, and the PICT approach could also be used for studying the dynamics of environmental recovery. However, this application is less developed than impact studies (Blanck and Dahl 1998; Larsen et al. 2003; Boivin et al. 2006; Demoling and Baath 2008a). Kinetics of recovery at the community level after disturbances are variable, and physiological recovery from a chemical stress is generally more efficient at high microbial diversity and at high functional redundancy. However, recovering a function after a stress does not necessarily mean that a community returns to its original composition and tolerance (Figure 4.4). Dorigo et al. (2010) studied the structural and functional recovery of bacterial and eukaryote communities in biofilms naturally grown on stones for nine weeks in a pesticide-impacted site, after transferring them from the pesticide-polluted downstream site to a noncontaminated upstream site (Figure 4.5). Their results revealed that toxicant-induced changes between nonimpacted biofilm (NIB) and impacted biofilm (IB) communities remained present after translocation throughout the recovery period. The transfer of biofilm from impacted site to reference site did not break up structural differences between NIB and IB, and brought on only weak functional changes. Several causes could explain the maintained structural and functional differences. It is likely that persistent differences are preserved via long-lasting effects of the toxicant on the periphyton communities, particularly of copper, considering its adsorption onto biofilm and its incomplete release from biofilm matrices (Dorigo et al. 2010). This is in accordance with the results of Larsen et al. (2003) indicating that phytoplankton PICT was maintained over more than two weeks after the biocide Sea-Nine 211 had been degraded. It is also suggested that the limited recovery could also be due to limited new cell immigration and to the difficulty for microbial species to invade an already established community rather than invade non- or recently colonised substrata.

The use of biofilm recovery capacity assessed by the PICT appears to be an interesting bioindicator of the recovery of freshwater ecosystems; it is ecologically relevant and could potentially be a management tool (Blanck et al. 1988; Demoling and Baath 2008b).

4.4.2 COST OF INDUCED TOLERANCE AND ADDITIONAL DISTURBANCES

The PICT approach has been employed to demonstrate that community structural changes are the direct result of contaminant exposure. However, few studies have investigated the effects of this community restructuring on its sensitivity to other stressors. Acclimating or adapting to one set of environmental stressors may increase community sensitivity to novel stressors, suggesting a potential cost associated with tolerance acquisition (Wilson 1988; Clements and Rohr 2009). Mature stable ecosystems are characterised by a preponderance of organisms referred to as k-strategists, species that succeed by being well adapted to their environment. Earlier stages in a succession may have a greater proportion of r-strategists, organisms with broad environmental tolerances that do not survive so well in stable habitats where they cannot compete favorably against the more adapted k-species. By contrast, r-strategists have high reproductive rates, flood the environment with their cells, and are ready to colonise opportunistically any habitat space that may become available. In stressful environments, under either man-made stress or naturally harsh conditions, tolerance to abiotic factors becomes a greater determinant of community composition than biotic interactions and r-strategists predominate.

Kashian et al. (2007) showed that benthic communities subjected to long-term metal pollution were generally more tolerant to metals but more sensitive to UV-B radiation than communities from a reference site. The authors speculated that the increase of metal tolerance and the increase of sensitivity to UV-B were probably related to population-level responses (acclimation and adaptation) and shifts in community composition. The elimination of the most sensitive genotypes by pollution in a population, causing a loss of genetic variability, could reduce the ability of these populations to respond to future disturbances. Likewise, the same phenomenon could be possible with elimination of the most sensitive species from a community, inducing an alteration of the ecosystem's functions and augmentation of the sensitivity of the restructured community to additional perturbation (Paine et al. 1998). In 2004, Montuelle (personal communication) showed that the impact of 2,4-D on denitrification and respiration of a microbial soil community was more marked after thermal stress. This loss of functional tolerance due to an additive effect of both stresses was incidental to a change in diversity, as indicated by 16S rDNA fingerprints.

Most studies dealing with the cost of tolerance have shown that the ability of organisms to tolerate a disturbance has variable consequences if tolerant genotypes are favored and genetic diversity will be reduced (Wilson 1988; Courtney and Clements 2000; Zuellig et al. 2008). Moreover, apparent structural recovery, based on abundance or community composition, does not necessarily mean a functional recovery since similar communities with different histories of exposure can show divergent response trajectories after the removal of the stressors (Kashian et al. 2007). The cost of the induced tolerance may therefore explain the nonreturn to the initial state of a community after the removal of stress.

4.4.3 PICT AND MANAGEMENT IMPLICATIONS IN ECOLOGICAL RISK ASSESSMENT

Toxicants can disturb the sustainability of natural ecosystems by a variety of effects on species, populations, communities, and ecosystem processes. However, such systems have some capacity to absorb potentially toxic substances because of their dynamic stability. Toxicity testing has limitations in predicting ecological effects, and chemical measurement of environmental toxicant concentrations must be accompanied by ecological monitoring. A complementary approach is needed to distinguish between ecological effects resulting from the pollution and those due to naturally occurring environmental conditions.

Ecological risk assessment (ERA) aims to evaluate the likelihood that adverse ecological effects may occur, or are occurring, as a result of exposure to one or more stressors (USEPA 1992). The process is used to systematically evaluate and organise data, information, assumptions, and uncertainties in order to help understand and predict the relationships between stressors and ecological effects in a way that is useful for environmental decision making. Such assessment may involve chemical, physical, or biological stressors, and one stressor or many stressors may be considered.

The value of risk is obtained by comparing predicted environmental concentrations (PECs) of the toxic substance in the environment with a predicted no-effect concentration (PNEC) on exposed organisms. Whereas the PEC is determined from actual measurement of the toxic substance in the environment or estimated from environmental fate models, the PNEC is commonly derived from concentration-response curves obtained with single-species laboratory bioassays. But PNEC and PEC are not based on a homogenous data set, and this risk assessment ratio PEC/PNEC could then be questionable. The PICT approach can be easily integrated into an ERA process and applied as an analytical tool in the analysis phase. In this way, McClellan et al. (2008) showed, for example, that the chronic community-level effects of the herbicide diuron were not predictable from single-species tests. PICT methodology allows the minimizing of such uncertainty in an ERA, because it takes into account the species interactions by working at the community level. Furthermore, Efroymson and Suter (1999) emphasise that the use of small organisms is often omitted in ERA endpoints, and the PICT method is mainly based on small organisms. In their review of the significance of PICT, Boivin et al. (2002) proposed the connection of the PICT method to the conceptual model of the ecosystem (e.g., inclusion of higher trophic levels, community structure analysis, and food web modeling) in the planning stage of the ERA to overcome this problem. An active integration of PICT measurements as a complementary tool and as a microbial community endpoint in environmental assessment schemes would give more ecological relevance to the present battery of bioindicators.

REFERENCES

Al Hasin, A., et al. 2010. Remediation of chromium(VI) by a methane-oxidizing bacterium. *Environ. Sci. Technol.* 44:400–5.

Almas, A. R., J. Mulder, and L. R. Bakken. 2005. Trace metal exposure of soil bacteria depends on their position in the soil matrix. *Environ. Sci. Technol.* 39:5927–32.

Baath, E., et al. 1998. Microbial community-based measurements to estimate heavy metal effects in soil: The use of phospholipid fatty acid patterns and bacterial community tolerance. *Ambio* 27:58–61.

Barbault, R. 1997. *Biodiversité. Introduction à la biologie de la conservation.* Paris: Hachette.

Barranguet, C., et al. 2003. Copper-induced modifications of the trophic relations in riverine algal-bacterial biofilms. *Environ. Toxicol. Chem.* 22:1340–49.

Bérard, A., and C. Benninghoff. 2001. Pollution-induced community tolerance (PICT) and seasonal variations in the sensitivity of phytoplankton to atrazine in nanocosms. *Chemosphere* 45:427–37.

Bérard, A., et al. 1998. Caractérisation du phytoplankton de deux systèmes limniques vis-à-vis d'un herbicide inhibiteur de la photosynthèse. La méthode PICT (pollution-induced community tolerance): application et signification. *Ann. Limnol.* 34:269–82.

Bérard, A., et al. 2003. Comparison of the ecotoxicological impact of triazines irgarol 1051 and atrazine on microalgal cultures and natural microalgal communities in Lake Geneva. *Chemosphere* 53:935–44.

Blanck, H., and Dahl, B. 1998. Recovery of marine periphyton communities around a Swedish marina after the ban of TBT use in antifouling paint. *Mar. Pollut. Bull.* 36:437–42.

Blanck, H., S. A. Wängberg, and S. Molander. 1988. Pollution-induced community tolerance—A new ecotoxicological tool. In *Functional testing of aquatic biota for estimating hazards of chemicals,* ed. J. Cairns Jr. and J. R. Pratt, 219–30. ASTM STP 988. Philadelphia: ASTM.

Blanck, H., et al. 2003. Variability in zinc tolerance, measured as incorporation of radio-labeled carbon dioxide and thymidine, in periphyton communities sampled from 15 European river stretches. *Arch. Environ. Contam. Toxicol.* 44:17–29.

Blanck, H., et al. 2009. A retrospective analysis of contamination and periphyton PICT patterns for the antifoulant irgarol 1051, around a small marina on the Swedish west coast. *Mar. Pollut. Bull.* 58:230–37.

Boivin, M. E. Y., et al. 2002. Determination of field effects of contaminants—Significance of pollution-induced community tolerance. *Hum. Ecol. Risk Assess.* 8:1035–55.

Boivin, M. E. Y., et al. 2005. Effects of copper and temperature on aquatic bacterial communities. *Aquat. Toxicol.* 71:345–56.

Boivin, M. E. Y., et al. 2006. Functional recovery of biofilm bacterial communities after copper exposure. *Environ. Pollut.* 140:239–46.

Brack, W., et al. 2008. How to confirm identified toxicants in effect-directed analysis. *Anal. Bioanal. Chem.* 390:1959–73.

Brandt, K. K., et al. 2004. Microbial community-level toxicity testing of linear alkylbenzene sulfonates in aquatic microcosms. *FEMS Microbiol. Ecol.* 49:229–41.

Brandt, K. K., et al. 2009. Increased pollution-induced bacterial community tolerance to sulfadiazine in soil hotspots amended with artificial root exudates. *Environ. Sci. Technol.* 43:2963–68.

Carluer, N., and G. De Marsily. 2004. Assessment and modelling of the influence of man-made networks on the hydrology of a small watershed: Implications for fast flow components, water quality and landscape management. *J. Hydrol.* 285:76–95.

Clements, W. H., and M. C. Newman, eds. 2002. *Community ecotoxicology*. Chichester, UK: J. Wiley & Sons.

Clements, W. H., and J. R. Rohr. 2009. Community responses to contaminants: Using basic ecological principles to predict ecotoxicological effects. *Environ. Toxicol. Chem.* 28:1789–800.

Courtney, L. A., and W. II. Clements. 2000. Sensitivity to acidic pH in benthic invertebrate assemblages with different histories of exposure to metals. *J. N. Am. Benthol. Soc.* 19:112–27.

Dahl, B., and H. Blanck. 1996. Toxic effects of the antifouling agent irgarol 1051 on periphyton communities in coastal water microcosms. *Mar. Pollut. Bull.* 32:342–50.

Debenest, T., et al. 2009. Sensitivity of freshwater periphytic diatoms to agricultural herbicides. *Aquat. Toxicol.* 93:11–17.

DeLorenzo, M. E., G. I. Scott, and P. E. Ross. 1999b. Effects of the agricultural pesticides atrazine, deethylatrazine, endosulfan, and chlorpyrifos on an estuarine microbial food web. *Environ. Toxicol. Chem.* 18:2824 35.

DeLorenzo, M. E., et al. 1999a. Atrazine effects on the microbial food web in tidal creek mesocosms. *Aquat. Toxicol.* 46:241–51.

Demoling, L. A., and E. Baath. 2008a. The use of leucine incorporation to determine the toxicity of phenols to bacterial communities extracted from soil. *Appl. Soil Ecol.* 38:34–41.

Demoling, L. A., and E. Baath. 2008b. No long-term persistence of bacterial pollution-induced community tolerance in tylosin-polluted soil. *Environ. Sci. Technol.* 42:6917–21.

Demoling, L. A., and E. Baath. 2008c. Use of pollution-induced community tolerance of the bacterial community to detect phenol toxicity in soil. *Environ. Toxicol. Chem.* 27:334–40.

Demoling, L. A., et al. 2009. Effects of sulfamethoxazole on soil microbial communities after adding substrate. *Soil Biol. Biochem.* 41:840–48.

Diaz-Ravina, M., E. Baath, and A. Frostegard. 1994. Multiple heavy-metal tolerance of soil bacterial communities and its measurement by a thymidine incorporation technique. *Appl. Environ. Microbiol.* 60:2238–47.

Dorigo, U., A. Bérard, and J. F. Humbert. 2002. Comparison of eukaryotic phytobenthic community composition in a polluted river by partial 18S rRNA gene cloning and sequencing. *Microb. Ecol.* 4:372–80.

Dorigo, U., et al. 2004. Seasonal changes in the sensitivity of river microalgae to atrazine and isoproturon along a contamination gradient. *Sci. Total Environ.* 318:101–14.

Dorigo, U., et al. 2007. Lotic biofilm community structure and pesticide tolerance along a contamination gradient in a vineyard area. *Aquat. Microb. Ecol.* 50:91–102.

Dorigo, U., et al. 2009. Influence of sampling strategy on the assessment of the impact of pesticides on periphytic microbial communities in a small river. *FEMS Microbiol. Ecol.* 67:491–501.

Dorigo, U., et al. 2010. *In-situ* assessment of periphyton recovery in a river contaminated by pesticides. *Aquat. Toxicol.* 98:396–406.

Efroymson, R. A., and G. W. Suter. 1999. Finding a niche for soil microbial toxicity tests in ecological risk assessment. *Hum. Ecol. Risk Assess.* 5:715–27.

Ekschmitt, K., et al. 2001. Biodiversity and functioning of ecological communities—Why is diversity important in some cases and unimportant in others? *Z. Pflanz. Bodenkunde* 164:239–46.

Eriksson, K. M., et al. 2009. Community-level analysis of psbA gene sequences and irgarol tolerance in marine periphyton. *Appl. Environ. Microb.* 75:897–906.

Espeland, E. M., S. N. Francoeur, and R. G. Wetzel. 2001. Influence of algal photosynthesis on biofilm bacterial production and associated glucosidase and xylosidase activities. *Microb. Ecol.* 42:524–30.

Guasch, H., W. Admiraal, and S. Sabater. 2003. Contrasting effects of organic and inorganic toxicants on freshwater periphyton. *Aquat. Toxicol.* 64:165–75.

Guasch, H., M. Paulsson, and S. Sabater. 2002. Effect of copper on algal communities from oligotrophic calcareous streams. *J. Phycol.* 38:241–48.

Guasch, H., and S. Sabater. 1998. Light history influences the sensitivity to atrazine in periphytic algae. *J. Phycol.* 34: 233–41.

Guasch, H., et al. 2004. Phosphate limitation influences the sensitivity to copper in periphytic algae. *Freshwater Biol.* 49:463–73.

Guasch, H., et al. 2007. Influence of phosphate on the response of periphyton to atrazine exposure. *Arch. Environ. Contam. Toxicol.* 52:32–37.

Gustavson, K., and S. A. Wängberg. 1995. Tolerance induction and succession in microalgae communities exposed to copper and atrazine. *Aquat. Toxicol.* 32:283–302.

Hersh, C. M., and W. G. Crumpton. 1989. Atrazine tolerance of algae isolated from two agricultural streams. *Environ. Toxicol. Chem.* 8:327–32.

Hjorth, M., I. Dahllof, and V. E. Forbes. 2006. Effects on the function of three trophic levels in marine plankton communities under stress from the antifouling compound zinc pyrithione. *Aquat. Toxicol.* 77:105–15.

Ivorra, N., et al. 1999. Translocation of microbenthic algal assemblages used for *in-situ* analysis of metal pollution in rivers. *Arch. Environ. Contam. Toxicol.* 37:19–28.

Ivorra, N., et al. 2002. Responses of biofilms to combined nutrient and metal exposure. *Environ. Toxicol. Chem.* 21:626–32.

Kasai, F., N. Takamura, and S. Hatakeyama. 1993. Effects of simetryne on growth of various fresh-water algal taxa. *Environ. Pollut.* 79:77–83.

Kashian, D. R., et al. 2007. The cost of tolerance: Sensitivity of stream benthic communities to UV-B and metals. *Ecol. Appl.* 17:365–75.

Knauer, K., et al. 2010. Co-tolerance of phytoplankton communities to photosynthesis II inhibitors. *Aquat. Toxicol.* 96:256–63.

Lagacherie, P., et al. 2006. An indicator approach for describing the spatial variability of artificial stream networks with regard to herbicide pollution in cultivated watersheds. *Ecol. Indic.* 6:265–79.

Lagadic, L., and T. Caquet. 1998. Invertebrates in testing of environmental chemicals: Are they alternatives? *Environ. Health Persp.* 106:593–611.

Larsen, D. K., et al. 2003. Long-term effect of Sea-Nine on natural coastal phytoplankton communities assessed by pollution induced community tolerance. *Aquat. Toxicol.* 62:35–44.

Lehmann, V., G. M. J. Tubbing, and W. Admiraal. 1999. Induced metal tolerance in microbenthic communities from three lowland rivers with different metal loads. *Arch. Environ. Contam. Toxicol.* 36:384–91.

Lock, K., and C. R. Janssen. 2005. Influence of soil zinc concentrations on zinc sensitivity and functional diversity of microbial communities. *Environ. Pollut.* 136:275–81.

Massieux, B., et al. 2004. Analysis of structural and physiological profiles to assess the effects of Cu on biofilm microbial communities. *Appl. Environ. Microbiol.* 70:4512–21.

McClellan, K., R. Altenburger, and M. Schmitt-Jansen. 2008. Pollution-induced community tolerance as a measure of species interaction in toxicity assessment. *J. Appl. Ecol.* 45:1514–22.

Mertz, O., et al. 2007. Ecosystem services and biodiversity in developing countries. *Biodivers. Conserv.* 16:2729–37.

Molander, S. 1991. Detection, validity and specificity of pollution-induced community tolerance (PICT). PhD thesis, University of Göteborg, Sweden.

Molander, S., and H. Blanck. 1991. *Cotolerance pattern of marine periphyton communities restructured by the herbicide diuron.* Göteborg, Sweden: University of Göteborg, Department of Plant Physiology.

Molander, S., and H. Blanck. 1992. Detection of pollution-induced community tolerance (PICT) in marine periphyton communities established under diuron exposure. *Aquat. Toxicol.* 22:129–44.

Morin, S., et al. 2008. Long-term survey of heavy-metal pollution, biofilm contamination and diatom community structure in the Riou Mort watershed, South-West France. *Environ. Pollut.* 151:532–42.

Morin, S., et al. 2010. Recovery potential of periphytic communities in a river impacted by a vineyard watershed. *Ecol. Indic.* 10:419–26.

Navarro, E., C. T. Robinson, and R. Behra. 2008. Increased tolerance to ultraviolet radiation (UVR) and cotolerance to cadmium in UVR-acclimatized freshwater periphyton. *Limnol. Oceanogr.* 53:1149–58.

Niklinska, M., M. Chodak, and R. Laskowski. 2006. Pollution-induced community tolerance of microorganisms from forest soil organic layers polluted with Zn or Cu. *Appl. Soil Ecol.* 32:265–72.

Nyström, B. 1997. Metabolic indicators of ecotoxicological effects in freshwater periphyton communities. PhD thesis, University of Göteborg, Sweden.

Nyström, B., et al. 2000. Evaluation of the capacity for development of atrazine tolerance in periphyton from a Swedish freshwater site as determined by inhibition of photosynthesis and sulfolipid synthesis. *Environ. Toxicol. Chem.* 19:1324–31.

Paine, R. T., M. J. Tegner, and E. A. Johnson. 1998. Compounded perturbations yield ecological surprises. *Ecosystems* 1:535–45.

Paulsson, M., B. Nyström, and H. Blanck. 2000. Long-term toxicity of zinc to bacteria and algae in periphyton communities from the river Gota Alv, based on a microcosm study. *Aquat. Toxicol.* 47:243–57.

Pesce, S., C. Margoum, and B. Montuelle. 2010. *In-situ* relationships between spatio-temporal variations in diuron concentrations and phototrophic biofilm tolerance in a contaminated river. *Water Res.* 44:194–9.

Pesce, S., et al. 2006. Effects of phenylurea herbicide diuron on natural riverine microbial communities in an experimental study. *Aquat. Toxicol.* 78:303–14.

Rimet, F., et al. 1999. Atrazine and nicosulfuron effects on phytoplanktonic communities of Lake of Geneva. *Arch. Sci.* 52:111–22.

Schmitt, H., H. Haapakangas, and P. Van Beelen. 2005. Effects of antibiotics on soil microorganisms: Time and nutrients influence pollution-induced community tolerance. *Soil Biol. Biochem.* 37:1882–92.

Schmitt-Jansen, M., and R. Altenburger. 2005. Predicting and observing responses of algal communities to Photosystem II-herbicide exposure using pollution-induced community tolerance and species-sensitivity distributions. *Environ. Toxicol. Chem.* 24:304–12.

Schmitt-Jansen, M., and R. Altenburger. 2008. Community-level microalgal toxicity assessment by multiwavelength-excitation PAM fluorometry. *Aquat. Toxicol.* 86:49–58.

Schwartz, M. W., et al. 2000. Linking biodiversity to ecosystem function: Implications for conservation ecology. *Oecologia* 122:297–305.

Serra, A. 2009. *Fate and effects of copper in fluvial ecosystems: The role of periphyton.* PhD thesis, University of Girona, Spain, pp. 110–33.

Soldo, D., and R. Behra. 2000. Long-term effects of copper on the structure of freshwater periphyton communities and their tolerance to copper, zinc, nickel and silver. *Aquat. Toxicol.* 47:181–89.

Tlili, A., et al. 2008. Responses of chronically contaminated biofilms to short pulses of diuron—An experimental study simulating flooding events in a small river. *Aquat. Toxicol.* 87:252–63.

Tlili, A., et al. 2010. PO_4^{3-} dependence of biofilm autotrophic and heterotrophic communities tolerance to copper and diuron. *Aquat. Toxicol.* 98:165–177.

U.S. Environmental Protection Agency. 1992. Framework for ecological risk assessment. EPA/630/R-92/001. Washington, DC: Risk Assessment Forum, U.S. Environmental Protection Agency.

Villeneuve, A., A. Bouchez, and B. Montuelle. 2010. Influence of slight differences in environmental conditions (light, hydrodynamics) on the structure and function of the periphyton. *Aquat. Sci.* 72:33–44.

Wangberg, S. A., U. Heyman, and H. Blanck. 1991. Long-term and short-term arsenate toxicity to fresh water phytoplankton and periphyton in limnocorrals. *Can. J. Fish. Aquat. Sci.* 48:173–82.

Wilson, J. B. 1988. The cost of heavy-metal tolerance—An example. *Evolution* 42: 408–13.

Witter, E., et al. 2000. A study of the structure and metal tolerance of the soil microbial community six years after cessation of sewage sludge applications. *Environ. Toxicol. Chem.* 19:1983–91.

Zuellig, R. E., et al. 2008. The influence of metal exposure history and ultraviolet-B radiation on benthic communities in Colorado Rocky Mountain streams. *J. N. Am. Benthol. Soc.* 27:120–34.

5 Tolerance to Natural Environmental Change and the Effect of Added Chemical Stress

Herman Hummel, Adam Sokolowski, Christiaan Hummel, and Sander Wijnhoven

CONTENTS

5.1 INTRODUCTION

It has only been since the nineteenth century, after the major scientific expeditions of the *Beagle*, with Charles Darwin onboard from 1831 to 1836, and of the *Challenger*, from 1872 to 1876 under the supervision of Charles Thomson, that researchers started to study the variability of the marine environment and not use the sea only as a transport mechanism to explore other continents. These expeditions laid the foundation for the understanding that the sea was not a homogenous environment, and that the variation and changes in this environment underlie the huge variability and suite of adaptations found in nature.

However, it is also since that period that humanity started to influence its environment in an unprecedented way and increasingly exposed nature to strong unnatural changes. The impact of these anthropogenic changes nowadays leads to major changes in the functioning of ecosystems. In extreme cases, this means complete modification of the abiotic environment. Because of their sudden character, the changes are often much faster than genetic adaptations can keep pace with, and can lead to inhibition of biological processes as well as the extinction of endemic species.

Organisms living in naturally changing environments such as the brackish intertidal zone of estuaries, or those living under only moderately disturbed conditions, can survive due to a suite of adaptations and defence mechanisms to cope with the various continuously fluctuating factors, like salinity, water level, and temperature (Newell 1979). A range of adaptive capacities in response to moderate chemical disturbances whereby the organisms become more suited to the environment or better fitted to survive is described in the other chapters of this book.

In this chapter we will focus on the additional effect of chemical stresses on organisms that are already living under strong or fast changing natural conditions as occur in estuaries or at the limit of distribution of a species. For those organisms "familiar" to extreme environmental conditions and to multiple spatial and temporal changes in their environment, it might be expected that because of strongly developed (instantaneous) acclimation and (seasonal) acclimatisation capabilities, these species could better tolerate the additional stress of chemical disturbances. Below we will examine whether this assumption can be found to be true.

To assess the potential impact of chemical stressors on the (ecophysiological) performance of organisms, mostly laboratory experiments have been carried out. Additionally, indirect evidence has been collected by field observations, including long-term field studies on species composition and community changes in relation to environmental factors and pollutants.

Most of the laboratory experiments test the organism under the influence of one variable factor. Multifactorial experiments with multiple stress factors, especially in the case of a combination of natural factors and a chemical stressor, have been carried out rarely (see reviews of McLusky et al. 1986; Heugens et al. 2001; Heugens 2003 and references therein).

Moreover, in exposure experiments in the laboratory, chemical stressors are mostly used over a couple of days at concentrations far higher than those found in the field (Hummel and Patarnello 1994), in order to gain results from which tolerance levels can be abstracted through interpolation. It is thus strongly questionable whether these results are field relevant because of their short exposure periods and irrelevant high concentrations. However, longer-term tests with field-relevant concentrations of chemical stressors are again rare.

Another complicating factor is that it is not only the direct effect of a naturally changing environmental parameter together with an added chemical stress that can lead to multiple stress for organisms. Environmental factors can also exert an indirect effect, as they can have a large influence on the physicochemical characteristics, and therefore on the bioavailability, of chemicals in the environment (e.g., Chapman and Mann 1999; Allen 2002; Wijnhoven et al. 2006). Moreover, environmental factors influence the susceptibility of organisms, particularly soft-bodied animals and those living in direct contact with those chemicals (e.g., organisms in aquatic environments or living in direct contact with pore water in soils), by influencing the natural barrier of those animals (e.g., epidermal, epithelial, and cuticular structures or mucus) to chemical exposure and uptake (e.g., Bervoets et al. 1995). Furthermore, environmental parameters can influence the either passive or active uptake processes of animals by which contaminants can enter organisms instead of, or together with, nutrients (e.g., Bervoets et al. 1995; Roessink et al. 2008). And, even when no direct contact between organisms and pollutants occurs, environmental factors indirectly influence the organisms exposed to chemicals by food web transfer (e.g., Chapter 13), as uptake at lower trophic levels might be affected (e.g., Wijnhoven et al. 2006, 2007; Rozema et al. 2008).

All these flaws mentioned above together mean that it remains largely unknown whether the interactions between a multiple set of stressful natural factors and a chemical stressor on the performance and tolerance of an organism under field-relevant conditions are additive, synergistic, or as more likely to be expected, subtractive or antagonistic.

Moreover, Heugens et al. (2001) concluded that in ecotoxicological studies with multiple stressors, where the impact of chemical pollutants in combination with stress by natural factors was assessed, only in rare cases were the underlying ecophysiological causes studied. These rare studies have suggested that the underlying mechanism of a decreased tolerance might be an increased energy demand (see Chapter 12).

In studies on marine bivalves, Hummel (2006) has indeed found under multiple stresses an increased respiration rate together with a loss of weight and a decrease of the tolerance to the chemical disturbance. The mechanisms involved in tolerance to chemical disturbances thus do cost metabolic energy at the expense of other activities and processes, including reproduction and growth. This would indicate that a combination of natural and anthropogenic stressors would cost additional energy and, in very dynamic areas such as estuaries and at the distribution limit of species, would lead more quickly to a hampered performance of organisms, and in extreme cases would cause higher mortality and disappearance of species than in environmentally more stable areas.

Therefore, in the following sections we will summarise results on the effects of multiple stresses by a combination of natural and chemical stressors as found by Heugens et al. (2001) for mainly crustaceans and fish, and those by us and others on mainly marine bivalves.

To study the long-term effects of multiple stressors on bivalves, we have taken care, as advised by Hoffmann and Parsons (1991), along with classic exposure experiments in the laboratory, to use translocation experiments carried out over several months in the field.

5.2 TOLERANCE UNDER MULTIPLE STRESSES BY NATURAL FACTORS AND CHEMICAL STRESSORS

5.2.1 TEMPERATURE

In their review based on experiments mainly carried out with fish and crustaceans, Heugens et al. (2001) concluded that organisms at their thermal tolerance limits, when exposed to toxicants, experience stronger adverse thermal effects.

Heugens et al. (2001) suggested that the underlying mechanism of a decreased tolerance might be an increased energy demand, and we have indeed shown this to be true in mussels *Mytilus edulis* and clams *Macoma balthica* (Hummel 1999, 2006; Hummel et al. 2000a). The respiration rates of these marine bivalves increased more when they were exposed to polluted sediments and deviating temperatures (Figure 5.1). At the southern limit of its distribution, a species will suffer at elevated temperatures more from pollution, due to higher energy demands, than when living at more optimal conditions.

The stronger impact of, and thereby lower tolerance to, pollutants in mussels and clams at higher temperatures might also be related to the observation that ambient temperatures can directly affect metal uptake and accumulation in marine invertebrates. Wang et al. (2005) reported enhanced metal absorption with increasing temperature for the green mussel *Perna perna* in subtropical regions, as did

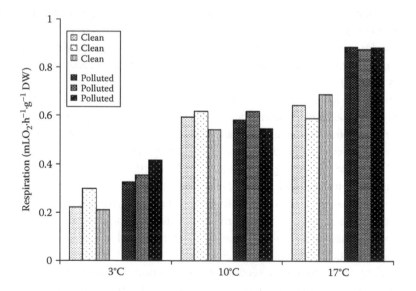

FIGURE 5.1 The respiration of clams *Macoma balthica* at different temperatures with and without polluted sediments. At optimum temperature, clams can tolerate polluted sediments, but at deviating temperatures the polluted sediments do cause additional higher respiration. (Exposure experiment after Hummel, H., et al., *J. Exp. Mar. Biol. Ecol.*, 251, 85–102, 2000a; Hummel, H., The Impact of Biodiversity Changes in Coastal Marine Benthic Ecosystems, BIOCOMBE EC contract EVK3-CT-2002-00072, final report, Netherlands Institute of Ecology, Yerseke, Netherlands, 2006.)

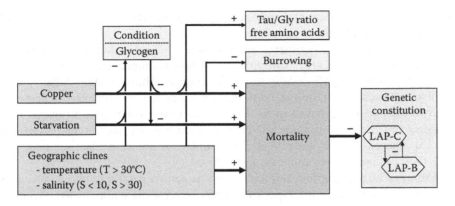

FIGURE 5.2   The interaction of natural and chemical stress factors on ecophysiological and genetic variables in the clam *Macoma balthica*. (After Hummel, H., *Océanis*, 25, 563–79, 1999; Hummel, H., et al., *J. Exp. Mar. Biol. Ecol.*, 191, 133–50, 1995.)

Wilkowska et al. (1994) for another mussel, *Mytilus trossulus*, in the brackish Baltic Sea. Similarly, a temperature-dependent absorption efficiency of dietary metals has been shown for mussels *Mytilus edulis* in the Arctic (Baines et al. 2005). A positive relationship between uptake of the nonessential metals Cd and Pb and temperature was shown experimentally also for *Mytilus edulis* from the Eastern Scheldt (the Netherlands). The enhanced metal accumulation in soft tissues of these bivalves was presumably related to the changes in solution chemistry and physicochemical kinetics of the elements, favouring higher absorption at higher temperature (Mubiana and Blust 2007).

5.2.2 NUTRITIONAL CONDITION

In previous studies, nutritional status in combination with exposure to chemicals has been found to have a strong effect in crustaceans, algae, and rotifers, with a decreased tolerance to chemicals at decreasing food or nutrient levels (see review by Heugens et al. 2001 and Chapter 12 in this book). The explanation was that at low food levels and a lower nutritional status there is a decreased ability to balance the energy costs for detoxification or repair mechanisms.

In marine clams *Macoma balthica* at the southern limit of their distribution, an extremely low condition index and a lower level of the energy reserve glycogen (1 to 2% of the body mass, whereas 5 to 6% is the normal lower limit) were also found to decrease the tolerance to copper (Hummel et al. 1996). In general, in this clam a strong decrease in its nutritional condition made the animals more sensitive to added stress (Hummel et al. 1995; see also Figure 5.2).

5.2.3 SALINITY

In previous studies on fish and crustaceans (Heugens et al. 2001) and on marine invertebrates in general (McLusky et al. 1986), the tolerance to most chemicals in general decreased as salinity declined and when the organisms were near their salinity limits. These responses were generally explained by increased physiological costs for osmoregulation.

From our exposure experiments with bivalves, it indeed became clear that higher copper concentrations in the environment may have an impact on the osmotic acclimation of mussels, *Mytilus edulis*, to salinity changes (Hummel 2006; Silber 2006). In response to changes in salinity (±10 PSU), mussels started to increase or decrease their free amino acid concentrations, which act as osmolytes in these bivalves. In the presence of Cu, the iso-osmotic volume regulation was impacted in the mussels. In the Cu-exposed mussels, the rate of change in free amino acid concentrations, and thus their osmoregulatory capacity, was reduced.

From this we may conclude that it is not only a decrease in salinity, but any deviation from the optimal salinity for a species that may increase the sensitivity to pollutants.

5.2.4 OXYGEN CONCENTRATION

The effect of environmental contaminants on marine organisms can also be markedly altered by dissolved oxygen concentration and redox potential, which affect directly the chemical form and activity (i.e., biological availability) of chemical stressors (Hughes 1981; Turetta et al. 2005). The problem of (lower) environmental oxygen levels and their potential impact on marine biota, especially benthic biota, is particularly persistent in inshore semienclosed water basins with limited water exchange or in the vicinity of nutrient and pollutant discharge points (Rees and Eleftheriou 1989). A number of studies clearly demonstrate the synergistic interaction of low oxygen saturation and some trace metals on benthic organisms in estuarine and coastal ecosystems.

Eriksson and Weeks (1994) showed in a series of laboratory experiments that hypoxia and increased copper concentration in the water resulted in elevated metal accumulation and lowered reproduction success rate (number of eggs produced) in the sediment-dwelling amphipod *Corophium volutator*. Under conditions of low oxygen saturation, the crustaceans constructed mud tubes protruding from the oxygen-deficient sediment surface, presumably reducing absorption of dissolved copper, which is less bioavailable in oxic water. Low oxygen concentrations increased the net accumulation of mercury in gill, kidney, liver, and muscle of (freshwater) carp *Cyprinus carpio* (Yediler and Jacobs 1995). Similarly, exposure to low oxygen conditions and dissolved copper also induced ecophysiological responses in the Baltic clam *Macoma balthica* (Neuhoff 1983). At low oxygen levels, copper concentrations in the range of 8 to 15 $\mu g\ dm^{-3}$ caused increased respiration rate and enhanced energy demand of the bivalve. In consequence, reserve materials with a fast metabolic turnover rate, like glycogen, reduced markedly, leading subsequently to the loss of soft tissue weight and a reduction of the adenylate energy charge (AEC) (Neuhoff 1983). In an extreme case, prolonged hypoxic and anoxic conditions combined with metal pollution in the ambient environment can bring about increased mortality of sedentary molluscs, as shown for mussels *Mytilus edulis* (Magni 1993).

The cumulative effect of oxygen depletion and copper pollution was also observed in field investigations on the dynamics of the process of colonisation of soft-bottom sediments in the inner Oslofjord, western Baltic Sea (Olsgard 1995; Trannum et al. 2004). Hypoxia affected adversely the establishment of recolonising fauna on a copper-polluted seafloor, and the benthic community composition was dominated by opportunistic and r-selected species (Trannum et al. 2004). In addition, Olsgard (1995) reported that variation in faunal structure and spatial distribution in the fjord was apparently controlled by a combination of oxygen content and the concentration of cadmium.

In the seriously eutrophicated Gulf of Gdańsk (southern Baltic Sea), oxygen deficiency in the overlying bottom water and surface sediments has been suggested to cause morphological deformations of the shell in *Macoma balthica* (Hummel et al. 2000b; Sokolowski et al. 2002, 2008). The prevalence of the deformed clams tended to increase with depth, reaching up to 65% of the total clam abundance in deep muddy and organic-rich sediments. Bivalves in deep-water regions are simultaneously exposed to the elevated metal concentrations both in dissolved form in the water column and bound to sediment particles, thus providing a good environment in which to test the tolerance of bivalves to concurrent environmental and chemical stress. The morphologically modified phenotypes exhibited higher tissue accumulated concentrations of some trace metals (As, Ag, Cd, Pb, Cu, and Zn) than bivalves with regular shell shape, indicative of a different ability of metal handling. It has been hypothesised that adverse hydrological conditions in the deeper waters of the Baltic (i.e., oxygen deficiency and raised concentrations of hydrogen sulphide—common phenomena in organically rich Baltic sediments), together with elevated bioavailabilities of metals (e.g., Cu, Pb, Zn; Sokolowski et al. 2007) and other potentially toxic contaminants (e.g., TBT), induce phenotypical changes (morphological irregularities) in part of the population and, in parallel, alterations of physiological mechanisms. As a result of the combined interactions of low oxygen concentration and the exposure of the bivalves to additional chemical stressors, deformed *Macoma balthica* absorb and accumulate metals to a higher extent than the animals not affected by shell deformation (Sokolowski et al. 2002).

The synergistic effect of oxygen concentration and trace metal contaminants on the tolerance of marine bivalves provides important clues for establishing monitoring programmes of environmental quality and for environmental impact assessment procedures. Sediment quality criteria based solely on concentrations of contaminants (e.g., trace metals, organic contaminants) either in a dissolved form in the water column or bound to surface sediments should be used with caution.

5.2.5 COMBINATION OF SEVERAL NATURAL STRESSORS

Few studies have investigated the influence of combinations of environmental factors on the toxicity of contaminants. Heugens et al. (2001) concluded that, for crustaceans, the combined action of temperature and salinity seemed to intensify their separate effects on contaminant toxicity.

Similar results were found for the Baltic clam *Macoma balthica*. Differences in the response of this marine bivalve could be related to the interaction of chemical pollutants present in the habitat, the initial condition (weight) of the bivalves prior to the experiment, as well as high temperatures and low food conditions, and salinities deviating from the optimum (Hummel et al. 1995). The combined effects of all (deviations of) natural environmental variables and pollution factors can be visualised in a model (Figure 5.2) showing their impact on the ecophysiology and survival of the animals, and ultimately on the genetic constitution of the population of the clams. The effect of pollutants was in these cases most clear at the lower pollutant concentrations. High pollutant concentrations were instantaneously lethal, and ecophysiological or genetic effects could not come to the foreground (Hummel et al. 1995). At the sublethal pollution concentrations, the essential tolerance mechanisms can be active.

It was concluded that the interaction of several natural factors and pollutants could account for the changes in, i.e., disappearance of, the southern distribution area of the clam *M. balthica* (Hummel 1999; Hummel et al. 2000a). In northern Spain and southwestern France, the primary factors for their disappearance (resulting in a northeastward shift in distribution) that could be indicated were a high level of pollution (particularly in Spanish estuaries) and higher temperatures (associated with global change). Similarly, the blue mussels *Mytilus edulis* and *M. galloprovincialis* presently show a strong northeastward shift in their distribution along the western European coast (Sanjuan et al. 1994, 1997; Hummel et al. 2001; Hummel 2006). For both bivalves in west and northern Spain and southwestern France the endemic ecotypes have already disappeared. The Mediterranean mussel (*M. galloprovincialis*) has replaced the Atlantic mussel (*M. edulis*), and the clam *M. balthica* has disappeared from these areas.

5.3 TOLERANCE IN COMMUNITIES UNDER EXTREME FIELD CONDITIONS OR SPECIES WITH SPECIFIC TRAITS

In any search for marine and brackish species with large environmental variable tolerance ranges, one might expect to find them among the species endemic to extreme environments, such as lagoons and estuaries, or among the species that can be found seemingly everywhere, cosmopolitan species. In this section we will discuss some such situations where species might be expected to possess a wide tolerance.

5.3.1 COMMUNITIES IN TRANSITIONAL WATERS

Searching for species of extreme environments, one comes to the group of species typical for transitional waters, such as lagoons, estuaries, and other paralic environments (waters at the landward side of the coast; Guelorget and Perthuisot 1989).

Such environments as lagoons are often characterised by a typical fauna of species independent of salinity level. Species like *Cerastoderma glaucum*, *Abra ovata*, *Nereis diversicolor*, *Gammarus locusta*, and *Sphaeroma hookeri* and several chironomid midge larvae can be present in either hypohaline or hyperhaline environments, as long as they are to some extent confined from the open sea. The salinity tolerance of these species is huge. Moreover, extreme high temperatures can occur in lagoons, and in these confined waters stratification is a common phenomenon, so low oxygen conditions may occur (Guelorget and Perthuisot 1989).

On the other hand, estuaries are characterised by their dynamics, with, in addition to fluctuating salinities, often strongly fluctuating currents, wave action, and intertidal regimes. Together with stress, estuaries also provide benefits for the residential fauna, like a large food supply, good dissolved oxygen conditions, and high water turnover rates removing faeces and dissolved pollutants. Are estuarine species better adapted to pollutants, or are "stressful" estuaries more susceptible to pollution? It is well known that estuaries are typified by much fewer species than either fully fresh or salt waters, where it is not the salinity in the brackish range causing this difference in species numbers, but the salinity fluctuations that can be considered stressful (Attrill 2002). Yet, whether these salinity-fluctuation-tolerant species are also more tolerant to pollutants is doubtful.

In lagoons, a clear bell-shaped total macrofauna biomass curve is observed across a scale representing confinement or degree of isolation from the adjacent seawater (Guelorget and Perthuisot 1989; Escaravage and Hummel 2010). Low biomass is found near the open sea, that is, at low nutrient levels and under highly exposed conditions (i.e., stronger currents, low water residence periods). A high biomass is found in slightly confined waters with high nutrient levels and less exposed conditions. Again, a lower biomass, but also low diversity, is found in the more confined waters with even higher nutrient conditions, yet low water quality. This suggests a trade-off, at which some species can survive in extreme environments with conditions where they have to cope with poor water quality conditions and associated stress, which is energy consuming, but where plenty of energy in the form of food is available and competition is low. Macrofaunal diversity in such environments is generally low (Pearson and Rosenberg 1978). The reason that these species are doing well in such environments would then be that there is little competition and plenty of food. In environments with more competition (where these species can also be found), their tolerance to environmental variability is probably reduced due to a (relative) lack of food and a higher energy demand (higher operational costs) due to competitive actions.

Another characteristic of confined transitional waters is their susceptibility to environmental enrichment and possible accumulation of pollutants as the water exchange of the system with more open systems is low. The question is whether these highly flexible species, especially in the more confined regions of transitional waters, are also very tolerant toward contaminants.

When the typical species for confined waters are weighted along a pollution gradient, via classification into ecological groups characteristic for a certain community health measure, as done in the AMBI classification (Borja et al. 2000), it can be observed that the species in confined regions of transitional waters are typically tolerant to excess organic matter enrichment, and that they may occur under normal conditions, but are promoted in slightly unbalanced situations. Most species in the confined regions are not the typical pollution-insensitive species indicative of heavily polluted aquatic systems (except for the chironomids, which are so-called second-order opportunistic species that are typical for slight to pronounced unbalanced situations (Borja et al. 2000)). In fact, the species typical of heavily polluted communities, the so-called first-order opportunistic species, are not common in confined regions. These are generally deposit feeders, which can proliferate in reduced sediments, like the polychaete worms *Capitella capitata* and *Malacoceros fuliginosus*, or oligochaetes, including *Tubificoides* spp. These species can cope with relatively heavy pollution, but exist in low dynamic, very stable environments, with very little other stress and competition. Typically the opportunistic species are the first to settle and maintain populations at recently polluted sites, but generally a pattern of succession of communities can be observed in time (Pearson and Rosenberg 1978). The opportunistic species to some extent help to detoxify the environment. Due to their action, pollutants are gradually bound by organic compounds, either in the environment itself or by passing through gut systems, or concentrations get lower by coverage with cleaner sediments or mixing and dilution processes by water currents or faunal activities (e.g., bioturbation). Thereby the opportunistic species make their environment gradually suitable for other species that are less pollutant tolerant but under natural/clean circumstances, more competitive.

It is thus not the most flexible species from confined regions that can cope better with pollutants, but those (opportunistic) species adapted to highly enriched but very stable conditions, for which there is no need to challenge other stresses. The typical species in confined regions of transitional waters are indeed relatively tolerant (to large amplitudes of natural factors) but cannot persist in (polluted) environments when there are too many competitors.

5.3.2 EXOTIC AND COSMOPOLITAN SPECIES

Organisms that might be expected to be tolerant toward pollutants are the successful exotic or invasive species and widespread cosmopolitan species.

The first group of species is found to be particularly successful in deteriorated environments, either enriched or impacted by pollutants. Invaders are successful as they get the chance to fill in empty niches in deteriorated systems. It seems, however, that it is not because of its invasive nature that a species is more tolerant toward pollution. It has, for instance, been found that some of the highly successful species in polluted environments do not really have relatively broad tolerance ranges, or that they are, in fact, less tolerant to changes in other environmental conditions (Rahel 2002; Wijnhoven et al. 2003). On the other hand, experiments with native and exotic gammarid amphipods seemed to indicate that the successful invader *Dikerogammarus villosus* is more resistant to the more polluted waters of the river Rhine in the Netherlands than the native species (Wijnhoven et al. 2003). Yet, the competitive advantage of *D. villosus* over the native gammarids seems to be particularly related to the elevated ionic contents of the water (the slightly brackish conditions), in combination with the fact that the invading species preys on the native gammarids. *Dikerogammarus villosus* appeared to be less tolerant toward, and less competitive under, ion-poor conditions, as occurring in the smaller, more natural streams entering the river. Another invading gammarid in the Rhine system, *Gammarus tigrinus*, appears to be better adapted to higher temperatures or lower oxygen contents. This, in combination with its more aggressive behaviour toward native species, can explain its success after introduction in the river Rhine during the 1980s (Van Riel et al. 2009), when hypoxic conditions were a common phenomenon. During the 1990s, when water quality gradually improved, other invaders, such as *D. villosus*, took over.

Furthermore, successful exotic species that become invasive often change their direct surroundings and make them suitable for their own kind, and less suitable for the original communities. Examples are the change of sediment structure by fixation of suspended matter (e.g., through faeces production), or the creation of hard substrate in soft sediment environments (for instance, the shells of the bivalve molluscs *Ensis directus* and *Crassostrea gigas* or the tubes of the amphipod crustacean *Chelicorophium curvispinum*; Kerckhof et al. 2007; Leuven et al. 2009; Troost 2009). Another way of becoming successful is by eating the competitors (or their larvae) or overgrowing them (for instance, *Styela clava*, *Dikerogammarus villosus*, *Crassostrea gigas*; Lützen 1999; Van Riel et al. 2006; Troost 2009).

In general, invaders are not successful because of a higher tolerance to pollutants than endemic species, yet (a combination with) other traits, like tolerance to brackish conditions, predatory behaviour, or transforming their surroundings, are crucial.

Cosmopolitan species are seemingly without a limit in their distribution, and this might imply that they possess a high tolerance to a wide range of environmental factors. Indeed, most cosmopolitan species (such as grasses, *Poa* spp., and mussels, *Mytilus* spp.) are plastic species, meaning that they are able to adapt themselves to a new environment (Hummel 2009). The plasticity of cosmopolitan species gives rise to the expectation that they might also be able to withstand chemical stress better, because they can adapt themselves, up to a certain level, to the presence of a chemical stressor. This in turn could mean that a cosmopolitan species might be more tolerant to contaminants than other species would be. Indeed, for ostracod crustaceans in a polluted Turkish lake, it has been shown that more cosmopolitan species occurred at polluted locations, indicating their higher tolerance (Külköylüoğlu et al. 2007). This would make cosmopolitan ostracod species like mussels useful as bioindicators of environmental quality status. However, in other studies in Turkish lakes, based on species occurrences and ecological preferences, only for ostracods was a trend observed, indicating that cosmopolitan species tended to have wider tolerance ranges than other species (Yılmaz and Külköylüoğlu 2006; Dugel et al. 2008). Similarly, among Australian chironomids there were only subtle differences among species according to biogeographic affinity, with tolerances of Gondwanan species being slightly narrower than those of cosmopolitan species (McKie et al. 2004).

Moreover, in tests on the sensitivity of Australian collembolans to phenol and heavy metals some cosmopolitan species were, in comparison to endemic species, more sensitive and others less sensitive (Greenslade and Vaughan 2003). Also, in germination tests with ten tropical dune plant species Martinez et al. (1992) could not demonstrate consistent differences in sensitivity to different natural stressors (salinity, temperature, nitrates) between species with distinct distribution ranges (endemics, pantropical, cosmopolites).

Therefore, at this moment we have to conclude, as for invaders, that there is, besides some circumstantial evidence of a slightly higher tolerance in cosmopolites, no clear proof for differences in tolerance to added chemical stressors between cosmopolitan and endemic species.

5.3.3 COMMONALITIES FOR SPECIES WITH SPECIFIC TRAITS

Whether they are lagoonal or estuarine species or invasive species, each of these species that seems to be highly environmentally flexible appears to exist or be successful at extreme conditions, as competition is low and food availability is high. There are several examples of species occurring in systems deteriorated by pollutants, but even there they can exist since other conditions yield a profit. These species often decrease or disappear when there is more competition or when conditions become more stable. In these cases initial benefits or advantages above other species are gradually disappearing and competitors that needed more time to establish are gradually entering the system.

However, it remains difficult to obtain strong evidence from field observations as multiple factors might interfere. Therefore, we need more proof from field-relevant experiments where species are tested under various conditions and in species combinations to define if certain species are more tolerant to stressful conditions in general.

5.4 CONCLUSIONS AND EPILOGUE

- For marine bivalves, aquatic crustaceans, and fish, it was concluded that toxicological stress interacts with environmental stress factors like salinity, temperature, or food condition, and produces significant effects through this interaction, even under conditions in which effects of the toxicant alone cannot be detected in the laboratory (Hummel et al. 1995, 2000a; Hummel 1999; Heugens et al. 2001; Heugens 2003; Hansen 2006). In general, tolerance to chemical stress decreased more at increasing temperature and decreasing food availability level, and at salinities deviating from the optimum.
- Organisms living under conditions close to their environmental tolerance limits appeared to be more vulnerable to additional chemical stress than their conspecifics living under more optimal conditions. Contaminant concentrations that are sublethal to organisms living at optimal environmental conditions may become lethal at suboptimal conditions.
- Results indicate that interactions between natural and chemical stressors have the ability to alter the ecological niche of a species and potentially the distribution of the species (Hummel 1999; Hummel et al. 2000a; Hansen 2006).
- The final conclusion has to be that the interaction between a combination of stressful natural factors and a chemical stressor on the performance and tolerance of an organism under field-relevant conditions is synergistic.
- In the context of major concern about the impact of global change on the environment, it is important to obtain a proper description of the synergistic effects of higher temperatures together with other changes in environmental factors, like salinity or oxygen, on the toxic effect of pollutants, since all changes together may yield an unequivocal acceleration of the disappearance of species.

- Especially in regions with several species at the edge of their distribution, as toward the Arctic, the synergistic action of natural and anthropogenic stressors, together with the lower tolerance of the organisms, may have a strong impact. The consequences of such changes for the functioning of marine ecosystems are not known. Further studies toward the synergistic effects of natural environmental changes and chemical stressors are thus urgently needed.

ACKNOWLEDGMENTS

This is publication 4742 of the Netherlands Institute of Ecology (NIOO-KNAW) and Monitor Taskforce Publication Series 2010-02.

REFERENCES

Allen, H. E. 2002. *Bioavailability of metals in terrestrial ecosystems: Importance of partitioning for bioavailability to invertebrates, microbes, and plants.* Pensacola, FL: Society of Environmental Toxicology and Chemistry (SETAC).

Attrill, M. J. 2002. A testable linear model for diversity trends in estuaries. *J. Anim. Ecol.* 71:262–69.

Baines, S. B., N. S. Fisher, and E. L. Kinney. 2005. Influence of temperature on dietary metal uptake in Arctic and temperate mussels. *Mar. Ecol. Progr. Ser.* 289:201–13.

Bervoets, L., R. Blust, and R. Verheyen. 1995. The uptake of cadmium by the midge larvae *Chironomus riparius* as a function of salinity. *Aquat. Toxicol.* 33:227–43.

Borja, A., J. Franco, and V. Pérez. 2000. A marine biotic index to establish the ecological quality of soft-bottom benthos within European estuarine and coastal environments. *Mar. Pollut. Bull.* 40:1100–14.

Chapman, P. M., and G. S. Mann. 1999. Sediment quality values (SQVs) and ecological risk assessment (ERA). *Mar. Pollut. Bull.* 38:339–44.

Dugel, M., O. Kulkoyluoglu, and M. Kilic. 2008. Species assemblages and habitat preferences of Ostracoda (Crustacea) in Lake Abant (Bolu, Turkey). *Belg. J. Zool.* 138:50–59.

Eriksson, S. P., and J. M. Weeks. 1994. Effects of copper and hypoxia on two populations of the benthic amphipod *Corophium volutator* (Pallas). *Aquat. Toxicol.* 29:73–81.

Greenslade, P., and G. T. Vaughan, 2003. A comparison of Collembola species for toxicity testing of Australian soils. *Pedobiologia* 47:171–79.

Guelorget, O., and J.-P. Perthuisot. 1989. *The paralic realm. Geological, biological and economic expressions of confinement.* Rome: Food and Agriculture Organization of the United Nations (FAO-UN).

Hansen, M. E. 2006. Risk assessment—The role of combined environmental and toxic stress. MSc thesis, Roskilde University.

Heugens, E. H. W. 2003. Predicting effects of multiple stressors on aquatic biota. PhD thesis, University of Amsterdam.

Heugens, E. H. W., et al. 2001. A review of the effects of multiple stressors on aquatic organisms and analysis of uncertainty factors for use in risk assessment. *Crit. Rev. Toxicol.* 31:247–84.

Hoffmann, A. A., and P. A. Parsons. 1991. *Evolutionary genetics and environmental stress.* New York: Oxford University Press.

Hughes, G. M. 1981. Effects of low oxygen and pollution on the respiratory system of fishes. In *Stress and fish*, ed. A. D. Pickering, 121–46. New York: Academic Press.

Hummel, C. A. 2009. Cosmopolite marine species. How do they do it? A literature review. MSc thesis, Free University, Amsterdam.

Hummel, H. 1999. Limits to adaptations in marine organisms: The model *Macoma balthica*. *Océanis* 25:563–79.

Hummel, H. 2006. The impact of biodiversity changes in coastal marine benthic ecosystems. BIOCOMBE EC contract EVK3-CT-2002-00072. Final report. Yerseke, Netherlands: Netherlands Institute of Ecology.

Hummel, H., and T. Patarnello. 1994. Genetic effects of pollutants on marine and estuarine invertebrates. In *Genetics and evolution of aquatic organisms,* ed. A. R. Beaumont, 425–34. London: Chapman & Hall.

Hummel, H., et al. 1995. Uniform variation in genetic traits of a marine bivalve related to starvation, pollution and geographic clines. *J. Exp. Mar. Biol. Ecol.* 191:133–50.

Hummel, H., et al. 1996. Sensitivity to stress of the estuarine bivalve *Macoma balthica* from areas between the Netherlands and its southern limits (Gironde). *J. Sea Res.* 35:315–21.

Hummel, H., et al. 2000a. The respiratory performance and survival of the bivalve *Macoma balthica* at the southern limit of its distribution area: A translocation experiment. *J. Exp. Mar. Biol. Ecol.* 251:85–102.

Hummel, H., et al. 2000b. Ecophysiological and genetic traits of the Baltic clam *Macoma balthica* in the Baltic: Differences between populations in the Gdansk Bay due to acclimatization or genetic adaptation? *Int. Rev. Hydrobiol.* 85:621–37.

Hummel, H., et al. 2001. Genetic traits in the bivalve *Mytilus* from Europe, with an emphsis on Arctic populations. *Polar Biol.* 24:44–52.

Kerckhof, F., J. Haelters, and S. Gollasch. 2007. Alien species in the marine and brackish ecosystem: The situation in Belgian waters. *Aquat. Invasions* 2:243–57.

Külköylüoğlu, O., D. Muzaffer, and K. Mustafa. 2007. Ecological requirements of Ostracoda (Crustacea) in a heavily polluted shallow lake, Lake Yeniçağa (Bolu, Turkey). *Dev. Hydrobiol.* 197:119–33.

Leuven, R. S. E. W., et al. 2009. The River Rhine: A global highway for dispersal of aquatic invasive species. *Biol. Invasions* 11:1989–2008.

Lützen, J. 1999. *Styela clava* Herdman (Urochordata, Ascidiacea), a successful immigrant to North West Europe: Ecology, propagation and chronology of spread. *Helgoländer Meeresun.* 52:383–91.

Magni, P. 1993. The effect of oxygen concentration on the bioavailability of copper for the bivalve *Macoma balthica*. Stage Report June 1992–February 1993 of Nederlands Instituut voor Ocecologisch Onderzoek.

Martinez, M. L., T. Valverde, and P. Moreno-Casasola. 1992. Germination response to temperature, salinity, light and depth of sowing of ten tropical dune species. *Oecologia* 92:343–53.

McKie, B. G., P. S. Cranston, and R. G. Pearson. 2004. Gondwanan mesotherms and cosmopolitan eurytherms: Effects of temperature on the development and survival of Australian Chironomidae (Diptera) from tropical and temperate populations. *Mar. Freshw. Res.* 55:759–68.

McLusky, D. S., V. Bryant, and R. Campbell. 1986. The effects of temperature and salinity on the toxicity of heavy metals to marine and estuarine invertebrates. *Oceanogr. Mar. Biol. Ann. Rev.* 24:481–520.

Mubiana, V. K., and R. Blust. 2007. Effects of temperature on scope for growth and accumulation of Cd, Co, Cu and Pb by the marine bivalve *Mytilus edulis*. *Mar. Environ. Res.* 63:219–35.

Neuhoff, H. G. 1983. Synergistic physiological effects of low copper and various oxygen concentrations on *Macoma balthica*. *Mar. Biol.* 77:39–48.

Newell, R. C. 1979. *Biology of intertidal animals*. Faversham, UK: Marine Ecological Surveys Ltd.

Olsgard, F. 1995. Monitoring the state of pollution of the inner Oslofjord. Studies of the soft-bottom fauna in 1993 [in Norwegian]. Report 622/95. Department of Biology, University of Oslo, Program for Pollution Monitoring, the Norwegian Pollution Control Authority.

Pearson, T. H., and R. Rosenberg. 1978. Macrobenthic succession in relation to organic enrichment and pollution of the marine environment. *Oceanogr. Mar. Biol. Annu. Rev.* 16:229–311.

Rahel, F. J. 2002. Homogenization of fresh water faunas. *Annu. Rev. Ecol. Syst.* 33:291–315.

Rees, H. L., and A. Eleftheriou. 1989. North Sea benthos: A review of field investigations into the biological effects of man's activities. *ICES J. Mar. Sci.* 45:284–305.

Roessink, I., A. A. Koelmans, and T. C. M. Brock. 2008. Interactions between nutrients and organic micro-pollutants in shallow freshwater model ecosystems. *Sci. Total. Environ.* 406:436–42.

Rozema, J., et al. 2008. Do high levels of diffuse and chronic metal pollution in sediments of Rhine and Meuse floodplains affect structure and functioning of terrestrial ecosystems? *Sci. Total Environ.* 406:443–48.

SanJuan, A., et al. 1994. *Mytilus galloprovincialis* and *M. edulis* on the coasts of the Iberian peninsula. *Mar. Ecol. Progr. Ser.* 113:131–46.

SanJuan, A., et al. 1997. Genetic differentiation in *Mytilus galloprovincialis* Lmk. throughout the world. *Ophelia* 47:13–31.

Silber, Y. 2006. Einfluss von Schwermetallen (Kupfer) auf die ökophysiologische Anpassungsfähigkeit von Miesmuscheln, *Mytilus* spp., verschiedener europäischer Küsten an kurzfristige Salzgehaltsänderungen. Diplom thesis, Rostock University.

Sokolowski, A., M. Wolowicz, and H. Hummel. 2007. Metal sources to the Baltic clam *Macoma balthica* (Mollusca: Bivalvia) in the southern Baltic Sea (the Gulf of Gdansk). *Mar. Environ. Res.* 63:236–56.

Sokolowski, A., et al. 2002. The relationship between metal concentrations and phenotypes in the Baltic clam *Macoma balthica* (L.) from the Gulf of Gdansk, southern Baltic. *Chemosphere* 47:475–84.

Sokolowski, A., et al. 2008. Shell deformations in the Baltic clam *Macoma balthica* from southern Baltic Sea (the Gulf of Gdansk); hypotheses on environmental effects. *AMBIO* 37:93–100.

Trannum, H. C., et al. 2004. Effects of copper, cadmium and contaminated harbour sediments on recolonisation of soft-bottom communities. *J. Exp. Mar. Biol. Ecol.* 310:87–114.

Troost, K. 2009. Pacific oysters in Dutch estuaries. Causes of success and consequences for native bivalves. PhD thesis, University of Groningen, the Netherlands.

Turetta, C., et al. 2005. Benthic fluxes of trace metals in the lagoon of Venice. *Microchem. J.* 79:149–58.

Van Riel, M. C., G. Van der Velde, and A. Bij de Vaate. 2009. Interference competition between alien invasive gammaridean species. *Biol. Invasions* 11:2119–32.

Van Riel, M. C., et al. 2006. Trophic relationships in the Rhine food web during invasion and after establishment of the Ponto-Caspian invader *Dikerogammarus villosus*. *Hydrobiologia* 565:39–58.

Wang, J. F., C. Y. Chuang, and W.-X. Wang. 2005. Metal and oxygen uptake in the green mussel *Perna viridis* under different metabolic conditions. *Environ. Toxicol. Chem.* 24:2657–64.

Wijnhoven, S., M. C. Van Riel, and G. Van der Velde. 2003. Exotic and indigenous freshwater gammarid species: Physiological tolerance to water temperature in relation to ionic content of the water. *Aquat. Ecol.* 37:151–58.

Wijnhoven, S., et al. 2006. Metal accumulation risks in regularly flooded and non-flooded parts of floodplains of the river Rhine: Extractability and exposure through the food chain. *Chem. Ecol.* 22:463–77.

Wijnhoven, S., et al. 2007. Heavy metal concentrations in small mammals from a diffusely polluted floodplain: Importance of species- and location-specific characteristics. *Arch. Environ. Contam. Toxicol.* 52:603–13.

Wilkowska, I., J. Kożuch, and J. Pempkowiak. 1994. The influence of selected abiotic factors (salinity, temperature) on the accumulation of cadmium from sea water by the blue mussel *Mytilus edulis. Biul. MIR* 1(131):61–65.

Yediler, A., and J. Jacobs. 1995. Synergistic effects of temperature; oxygen and water flow on the accumulation and tissue distribution of mercury in carp (*Cyprinus carpio* L.). *Chemosphere* 31:4437–53.

Yılmaz, F., and O. Külköylüoğlu. 2006. Tolerance, optimum ranges, and ecological requirements of freshwater Ostracoda (Crustacea) in Lake Aladag (Bolu, Turkey). *Ecol. Res.* 21:165–73.

Section

Mechanisms of Defence and the Acquisition of Tolerance to Chemical Stress

6 Biodynamic Parameters of the Accumulation of Toxic Metals, Detoxification, and the Acquisition of Metal Tolerance

Philip S. Rainbow and Samuel N. Luoma

CONTENTS

6.1 INTRODUCTION

Tolerance may be defined as the ability of organisms to cope with stress to which they are exposed in their environment (Amiard-Triquet et al. 2009). The stressor may be a natural factor, such as a large change in salinity or temperature, or anthropogenic disturbance, including the raised bioavailability of a toxic metal. There are two methods by which organisms can become resistant to a toxic contaminant (Klerks and Weis 1987; Amiard-Triquet et al. 2009): (1) Resistance may be gained by physiological acclimation during exposure of the individual organism to a sublethal bioavailability of the pollutant at some period; this resistance is not transferable to future generations. (2) Tolerance may also be acquired as a consequence of genetic adaptation in populations exposed over generations to the toxic contaminant, through the action of natural selection on genetically based individual variation in resistance; this tolerance is transferable to future generations, as usually confirmed by laboratory breeding of at least the F1 generation (e.g., Grant et al. 1989; Levinton et al. 2003). Klerks and Weis (1987) referred to the former as *acclimation* and the latter as *adaptation*. This adaptation may be lost in the absence of continuing exposure to the contaminant, again by natural selection, if, as appears to be usual, the genetically based tolerance has a metabolic cost bringing a selective disadvantage in the absence of a threshold bioavailability of the contaminant (e.g., Grant et al. 1989). The presence of such a tolerant population is direct evidence that the bioavailability of the toxic contaminant in the local environment is sufficient to be ecotoxicologically relevant (Luoma 1977).

In this review we have concentrated on examples of genetically based adaptation, highlighting when an example quoted is one of acclimation.

6.2 BIODYNAMIC MODELLING

The recent development of the biodynamic modelling of trace metal accumulation has advanced our understanding of metal ecophysiology and ecotoxicology, and allowed us to interpret the comparative physiological basis of the different metal concentrations accumulated by different aquatic organisms (Wang et al. 1996; Wang and Fisher 1999; Luoma and Rainbow 2005). Biodynamic modelling involves the quantification of the biodynamic parameters that contribute to comparative metal bioaccumulation, coupled with relevant geochemical measurements in the field. Biodynamic modelling can identify the critical parameters influencing the accumulated metal concentration in an animal, and can be used to separate the exposure pathways of metals to particular species (Luoma and Rainbow 2005).

In a biodynamic model (Wang et al. 1996; Wang and Fisher 1997), the steady-state accumulated metal concentration (C_{ss}, μg g^{-1}) in an aquatic animal is described by the equation

$$C_{SS} = (K_u \times C_W)/(K_{ew} + g) + (AE \times IR \times C_F)/(K_{ef} + g)$$

where K_u is the metal uptake rate constant from solution (μg g^{-1} d^{-1} per μg L^{-1} or L g^{-1} d^{-1}), C_W is the metal concentration in solution (μg L^{-1}), AE is the assimilation efficiency of the metal from ingested food with metal concentration C_F ($\mu g\ g^{-1}$), IR is the ingestion rate ($g\ g^{-1}\ d^{-1}$), and g is the growth rate constant (d^{-1}). K_e is the efflux rate constant (d^{-1}) of metal taken up, K_{ew} is for metal taken up from solution, and K_{ef} is for metal taken up from food. In many cases efflux rate constants after uptake from water and food are assumed (or have been shown) to be the same, expressed as K_e (Luoma and Rainbow 2005). The equation is essentially in two parts that are additive, describing first uptake of metal from solution and its subsequent efflux, and second uptake from ingested food and its subsequent efflux, allowing partitioning of the total accumulated concentration into metal derived from each route. Any effect of growth, as reflected in the growth rate constant, affects metal taken up from either source.

The comparative use of biodynamic modelling has the potential to identify differences between populations of the same species in the key bioaccumulation processes affecting metal uptake and subsequent accumulation, with consequences for comparative differences in susceptibility to metal toxicity (Wang and Rainbow 2006). Thus, biodynamic modelling can deconstruct rates of accumulation into the rates of metal uptake and efflux in different populations of a species, including, for example, a metal-tolerant population, any net metal accumulated necessarily being detoxified at the appropriate rate. If the difference between rates of uptake and efflux exceeds a maximum detoxification rate, then metal will accumulate in nondetoxified form and toxicity will ensue (Rainbow 2002, 2007). Biodynamic modelling can therefore identify which (if any) of the physiological parameters—uptake rate (from solution or diet), efflux rate (after uptake from solution or diet), or (by subtraction) detoxification rate (e.g., Croteau and Luoma 2009)—is significantly altered in that metal-tolerant population, thereby identifying the mechanistic basis of the selection for metal tolerance in a particular population.

As will be apparent in other chapters of this volume, there have been several excellent reviews of different aspects of the tolerance of invertebrates to trace metals (e.g., Klerks and Weis 1987; Posthuma and van Straalen 1993; Morgan et al. 2007). In this review we specifically assess whether the literature evidence available in (inevitably) selected case studies will allow us to make any generalisations as to which of the physiological processes of toxic metal bioaccumulation has been particularly susceptible to the selective processes driving the evolution of a metal-tolerant population in habitats in which the local trace metal bioavailability is sufficiently high to be of ecotoxicological significance.

6.3   CASE STUDIES

6.3.1   Restronguet Creek (Cornwall, UK): The Polychaete Worm *Nereis diversicolor*

Restronguet Creek in Cornwall, southwest England, receives discharge from the Carnon River, which drains a catchment with a long history of mining (Dines 1969), and correspondingly Restronguet Creek sediments contain extraordinarily high levels of As, Cu, Fe, Mn, and Zn (Bryan et al. 1980; Bryan and Gibbs 1983; Rainbow et al. 2009a,b). Restronguet Creek was the subject of intense investigation by Geoff Bryan and colleagues in the 1970s and 1980s, and they demonstrated

TABLE 6.1

Nereis diversicolor: Comparison of Ninety-Six-Hour LC$_{50}$
Dissolved Concentrations of Zn and Cu at 13°C between a
Population of Polychaete Worms from Restronguet Creek
and a Control Population from the Nearby Avon Estuary

| Metal | Salinity | LC$_{50}$ Concentration (µg L^{-1}) | |
		Restronguet Creek	Avon
Zn	17.5	94,000	55,000
	3.5	14,600	11,000
	0.35	2,300	1,500
Cu	17.5	2,300	540

Source: After Bryan, G. W., and P. E. Gibbs, *Occ. Publ. Mar. Biol. Assoc. UK*, 2, 1–112, 1983.

that Restronguet Creek housed populations of seaweeds and invertebrates tolerant to copper and zinc (Bryan and Hummerstone 1971, 1973; Bryan and Gibbs 1983). Metal-tolerant populations included three sediment-ingesting burrowing invertebrates, the polychaete *Nereis diversicolor* (Cu and Zn tolerant—Bryan and Hummerstone 1971, 1973; Bryan and Gibbs 1983—see Table 6.1), the bivalve *Scrobicularia plana* (Cu), and the amphipod crustacean *Corophium volutator* (Cu) (Bryan and Gibbs 1983). Other metal-tolerant invertebrate populations in Restronguet Creek were of the crab *Carcinus maenas* (Cu and Zn) and the polychaete *Nephtys hombergi* (Cu), and the seaweed *Fucus vesiculosus* also showed Cu tolerance (Bryan and Gibbs 1983). Grant et al. (1989) and Mouneyrac et al. (2003) confirmed the continuing presence at the top of Restronguet Creek of the copper- and zinc-tolerant population of *Nereis diversicolor* extending into the twenty-first century, substantiating the ongoing ecotoxicological significance of high local Cu and Zn bioavailabilities in the local sediments. Grant et al. (1989) and Hateley et al. (1989) confirmed that the copper and zinc tolerances of this upper Restronguet Creek population of *N. diversicolor* are genetically based, inheritable down generations, and bring a selective disadvantage beneath a threshold sediment concentration of each metal. The metabolic costs of the metal resistance in this population result in a scope for growth that is 46 to 62% less than that of nonresistant populations, equivalent to a mean energy cost of 1.31 mJ h^{-1} mg DW^{-1} (Pook et al. 2009).

Comparative studies of the bioaccumulation of metals by the metal-tolerant *Nereis diversicolor* population from Restronguet Creek and populations of the worm from other less metal-contaminated local estuaries have provided some elucidation of the physiological processes involved in the tolerance (Bryan and Hummerstone 1971, 1973; Bryan and Gibbs 1983).

Bryan and Hummerstone (1973) carried out pioneering laboratory experiments measuring the accumulation of radiolabelled zinc from solution by different populations of *N. diversicolor*, including the zinc-tolerant population from Restronguet Creek, and also compared body concentrations of zinc in different local populations of the worms. They concluded that the tolerance of the Restronguet Creek worms to zinc appeared to be based on a lower permeability of the worms to zinc, and perhaps greater excretion of the zinc taken up, enabling the worms to regulate the body Zn concentration fairly efficiently despite the high ambient levels of zinc (Bryan and Hummerstone 1973), a conclusion supported by Bryan and Gibbs (1983). Bryan and Hummerstone (1973) exposed worms in the laboratory to radiolabelled zinc over periods of seven to fifteen days, and based their conclusions on comparative permeabilities on measured net zinc accumulation rates, which they termed (net) absorption rates. Thus, the Zn-tolerant *N. diversicolor* from Restronguet Creek showed relatively low net accumulation rates of zinc taken up from solution in comparison with other local worm populations (Bryan and Hummerstone 1973). It is not, however, possible from their results to conclude with certainty whether this reduced net accumulation rate is attributable to reduced uptake (lower permeability) or increased excretion, or a combination of both.

More recently, Mouneyrac et al. (2003) confirmed that this Restronguet population of *N. diversicolor* remains Cu and Zn tolerant. Rainbow et al. (2009a) separately determined the rates of Zn uptake from solution, Zn uptake from food, and efflux of Zn taken up by either route in a comparison of this Restronguet Creek *N. diversicolor* population against other populations of *N. diversicolor* not tolerant to zinc. The comparative data are shown in Table 6.2. There was no evidence of decreased rate of uptake of Zn by the Restronguet Creek worms, either from solution (K_u) or from ingested sediment (AE); there was also no evidence for increased rates of efflux of zinc by these worms after uptake from either solution (K_{ew}) or ingested sediment (K_{ef}) (Table 6.2). *Nereis diversicolor* typically regulates the accumulated body concentration of Zn to an approximately constant level by matching the efflux rate to the total uptake rate (Bryan and Hummerstone 1973; Bryan et al. 1980; Amiard et al. 1987). At very high concentrations, regulation breaks down as uptake rate exceeds the efflux rate achievable and net accumulation ensues (Rainbow 2002). The accumulated Zn body concentration data of Bryan and Gibbs (1983) and Berthet et al. (2003) for *N. diversicolor* from Restronguet Creek indicate that the worms sampled were still just about able to regulate their Zn body concentrations to levels typical of this species in general (<400 µg Zn g⁻¹dry weight). However, in the case of the worms sampled from Restronguet Creek in 2007 by Rainbow et al. (2009a), body regulation of accumulated zinc had clearly broken down (Table 6.3). Thus, these Restronguet Creek Zn-tolerant worms survive the increased rates of Zn uptake locally (mostly via sediment ingestion; Rainbow et al. 2009b) by necessarily being able to match the rate of detoxification of excess incoming Zn to the rate of uptake from all sources (Rainbow et al. 2009a). The basis of Zn tolerance in the Restronguet Creek population of *N. diversicolor*, therefore, appears to be an inheritable increased rate of storage detoxification of accumulated zinc if and when Zn regulation has broken down. The major detoxified stores of Zn in this population are spherocrystals (Figure 6.1) in the gut wall (Mouneyrac et al. 2003; Rainbow et al. 2004).

The population of *Nereis diversicolor* at the top of the Restronguet Creek population is also tolerant to copper in comparison to local conspecific populations

TABLE 6.2
Nereis diversicolor: Biodynamic Accumulation Parameters for Zinc of Polychaete Worms from Restronguet Creek and Five Comparative Populations (means ± standard error SE)

	Parameter	SE	*n*	A posteriori ANOVA
Zn Uptake Rate Constant, K_u (L g^{-1} d^{-1})				
Blackwater	0.1021	0.0346	17	A
Gannel	0.0859	0.0297	14	A
Tavy	0.0752	0.0208	10	A
Restronguet Creek	*0.0660*	*0.0734*	*7*	*A*
West Looe	0.0336	0.0100	16	A
East Looe	0.0173	0.0366	7	A
Zn Efflux Rate Constant after Uptake from Solution, K_{ew} (d^{-1})				
Restronguet Creek	*0.0393*	*0.0034*	*5*	*A*
East Looe	0.0389	0.0043	7	A
Gannel	0.0386	0.0038	14	A
Blackwater	0.0359	0.0024	17	A
West Looe	0.0321	0.0018	17	A
Tavy	0.0235	0.0018	10	B
Zn Assimilation Efficiency from Ingested Sediment AE (%)				
Restronguet Creek	*65.3*	*4.4*	*4*	*A*
West Looe	58.8	9.0	3	A
East Looe	46.1	7.2	5	A
Blackwater	45.4	14.8	4	A
Tavy	38.6	6.5	6	A
Gannel	30.0	10.3	4	A
Zn Efflux Rate Constant after Uptake from Ingested Sediment, K_{ef} (d^{-1})				
Tavy	0.0712	0.0229	6	A
Restronguet Creek	*0.0441*	*0.0193*	*4*	*A,B*
West Looe	0.0292	0.0103	5	A,B
Gannel	0.0170	0.0147	4	A,B
Blackwater	0.0059	0.0098	4	A,B
East Looe	+0.0037	0.0104	5	B

Source: After Rainbow, P. S., B. D. Smith, and S. N. Luoma, *Mar. Ecol. Prog. Ser.*, 376, 173–84, 2009a.

Note: Parameters of worms from sites sharing a common letter for one metal in the *a posteriori* ANOVA column do not differ significantly ($p > 0.05$).

TABLE 6.3

Concentrations (mean ± 1 standard deviation, µg g⁻¹ dry weight) of Cu and Zn in Oxic Surface Sediment and *Nereis diversicolor* (after gut depuration) from Restronguet Creek and Five Comparative Estuaries

	Cu	Zn
	Sediment	
Blackwater	24.2 ± 3.5	93 ± 17
East Looe	16.1 ± 4.2	66 ± 15
Gannel	44.0 ± 2.8	251 ± 20
Restronguet Creek	3,390 ± 290	3,270 ± 260
Tavy	170 ± 16	211 ± 18
West Looe	48.9 ± 8.3	103 ± 4
	Nereis diversicolor	
Blackwater	24.5 ± 7.3	237 ± 37
East Looe	46.1 ± 24.6	294 ± 83
Gannel	48.8 ± 26.2	258 ± 60
Restronguet Creek	3,850 ± 1,770	1,925 ± 566
Tavy	214 ± 43	472 ± 90
West Looe	71.9 ± 28.8	366 ± 64

Source: 2007 data adapted from Rainbow, P. S., B. D. Smith, and S. N. Luoma, *Mar. Ecol. Prog. Ser.*, 376, 173–84, 2009a.

(Bryan and Hummerstone 1971; Bryan and Gibbs 1983; Mouneyrac et al. 2003; see Table 6.1). In contrast to zinc, accumulated copper is not regulated to an approximately constant body concentration by *N. diversicolor* (Bryan and Hummerstone 1971; Bryan et al. 1980; Bryan and Gibbs 1983; Amiard et al. 1987), with body copper concentrations rising in proportion to copper bioavailability, as modelled, for example, by local sediment copper concentrations (Bryan and Hummerstone 1971; Bryan et al. 1980). Bryan and Hummerstone (1971) exposed worms from Restronguet Creek and from the (control) Plym and Tamar estuaries to high dissolved copper concentrations in experiments lasting seventeen or thirty-seven days. In the absence of statistical comparisons of highly variable data, there was no evidence in these experiments that the Restronguet Creek worms had lower accumulation rates of copper than the other worms, but arguably higher copper accumulation rates (Bryan and Hummerstone 1971). Bryan and Gibbs (1983) used the rarely available short-lived ⁶⁴Cu radioisotope of copper in short-term experiments to show that Restronguet Creek *N. diversicolor* had significantly higher accumulation rates of copper from solution than worms from the (control) Avon estuary at 5 and 25 µg Cu L⁻¹ in 50% seawater, but lower rates of Cu accumulation from solution at the environmentally very high dissolved concentrations of 100 and 200 µg Cu L⁻¹ in 50% seawater.

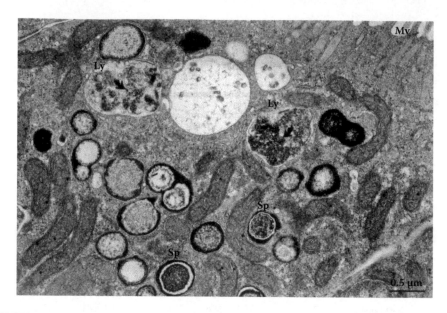

FIGURE 6.1 *Nereis diversicolor.* Ultrastructure of intestinal cells of Zn-tolerant worms from Restronguet Creek. Ly: lysosomes; M: mitochondrion; Mv: microvilli of intestinal cell; Sp: spherocrystals (containing Zn). (After Mouneyrac, C., et al., *Mar. Biol.*, 143, 731–44, 2003. With permission.)

The evidence is therefore ambiguous as to whether the Restronguet Creek population of *N. diversicolor* relies on reduced uptake from solution as the basis of its tolerance to copper. Rainbow et al. (2009a) were unable to include copper in their measurement of biodynamic accumulation parameters of the Cu-tolerant Restronguet Creek population of *N. diversicolor* because of the lack of an available suitable radio-isotope of Cu (in spite of the achievement of Bryan and Gibbs (1983) to obtain ^{64}Cu). The development of stable isotope techniques does, however, offer great potential for future investigation of this ecotoxicologically significant metal (Croteau and Luoma 2005). More than 99% of the Cd and 98% of the Zn accumulated by *N. diversicolor* is derived from sediment ingestion (Rainbow et al. 2009b). Given that copper shares similar chemical binding properties to these two metals (Luoma and Rainbow 2008), and that body concentrations of copper in the polychaete are proportional to sediment Cu concentrations (Bryan and Hummerstone 1971; Bryan et al. 1980), bioaccumulation parameters associated with the accumulation of copper from sediment may be worth examining for further understanding of the basis of the copper tolerance of the Restronguet Creek *N. diversicolor*.

Nevertheless, given the very high accumulated copper concentrations in the Restronguet Creek worms (Bryan and Hummerstone 1971; Bryan et al. 1980; Bryan and Gibbs 1983; Table 6.3), detoxification of accumulated copper must be occurring at an atypically high rate in these polychaetes. Since the copper tolerance of this Restronguet Creek population has a genetic basis (Grant et al. 1989; Hateley et al. 1989), it would appear that there has been an inheritable change in the rate of action of some part of the physiological detoxification mechanism employed for copper. Berthet et al. (2003)

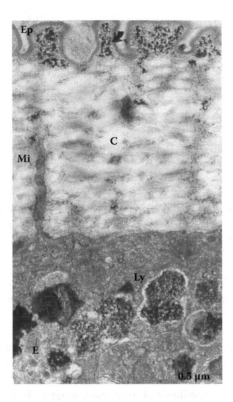

FIGURE 6.2 *Nereis diversicolor.* Ultrastructure of tegument of Cu-tolerant worms from Restronguet Creek. Ep: epicuticle; arrow: extracellular granules; C: collagen; E: epidermal cells; Ly: mineralised lysosomes; Mi: microvilli of epidermal cells. (After Rainbow, P. S., et al., *Mar. Ecol. Prog. Ser.*, 271, 183–91, 2004. With permission.)

found that 91% of accumulated copper was in insoluble form in these Restronguet Creek worms, of which 75% was present in metal-rich granules (68% of total body copper), with the remaining insoluble copper being bound to cellular debris. As described by Mouneyrac et al. (2003) and confirmed by Rainbow et al. (2004), extensive copper and sulphur-rich granules are present in two body locations of Restronguet Creek *N. diversicolor*—extracellular granules in the epicuticle of the body wall (apparently the major copper store) and granules of identical composition in lysosomes of the epidermal cells below the epicuticle of the body wall (Figures 6.2 and 6.3). One or both of these insoluble copper-rich deposits are probably derived from the lysosomal breakdown of copper-rich metallothionein (Luoma and Rainbow 2008). Metallothioneins are inducible cysteine-rich cytosolic proteins with a high binding affinity for particular trace metals, including cadmium, copper, mercury, silver, and zinc (Amiard et al. 2006).

6.3.2 RESTRONGUET CREEK (CORNWALL, UK): THE CRAB *CARCINUS MAENAS*

Bryan and Gibbs (1983) also showed that Restronguet Creek crabs (*Carcinus maenas*) had increased tolerance to dissolved zinc and copper in comparison with crabs from

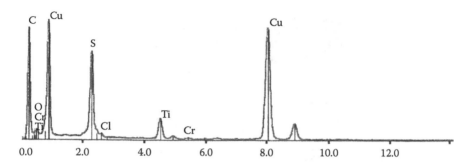

FIGURE 6.3 *Nereis diversicolor.* Energy-dispersive spectrophotometry microanalysis of epidermal Cu- and S-rich lysosomes in the tegument Cu-tolerant worms from Restronguet Creek. The peaks of C and O at the left of the spectrum originate from the organic matrix. Ti is emitted by the grid, Cl by the resin (epon). (After Rainbow, P. S., et al., *Mar. Ecol. Prog. Ser.*, 271, 183–91, 2004. With permission.)

other local estuaries. Because the crabs are mobile, it is almost certainly not valid to consider these Restronguet Creek crabs as one population sharing the same physiological adaptations to handle toxic metals. *C. maenas* zoea larvae settle out from the plankton after a pelagic phase of development, and it is quite likely that individual crabs entering Restronguet Creek (whether as larvae or as juveniles after settlement) are derived from parents from different parts of the Fal estuary system and beyond, with no history of exposure to raised toxic metal bioavailabilities. Thus, selection to survive the raised bioavailabilities of zinc and copper in Restronguet Creek would act on individual crabs as they live in or migrate to the Creek, and surviving crabs might be expected to have different mechanisms of metal tolerance based on different physiological bioaccumulation parameters with no common pattern across the crab population in Restronguet Creek (Rainbow et al. 1999). Moreover, any physiological basis of metal resistance in the crabs might well be down to physiological acclimation of individuals as opposed to the genetic adaptation of a population shown for both zinc and copper tolerance in Restronguet Creek *Nereis diversicolor.* This polychaete possesses a short-lived motile larval stage remaining close to the sediment and not straying far, rendering more likely any selective action on the physiological processes of an arguably more permanent local population.

Indeed, Bryan and Gibbs (1983) found that the resistance of Restronguet Creek *Carcinus maenas* to high dissolved zinc concentrations increased with the size of the crabs to support the above interpretation of individual selection during growth. The net accumulation rate of zinc by Restronguet Creek crabs was lower than that of nontolerant crabs from Falmouth and Plymouth, when exposed to the dissolved concentration of 10,000 µg Zn L^{-1} (Bryan and Gibbs 1983). Restronguet Creek crabs surviving thirty-eight days' exposure to this concentration had raised tissue concentrations of zinc in tissues compared to unexposed control Tamar crabs, especially in the gills and antennary gland, perhaps reflecting the role of these two tissues in zinc excretion (Bryan and Gibbs 1983). Bryan and Gibbs (1983) concluded that the Zn-tolerant crabs from Restronguet Creek may be less permeable to zinc or better able to excrete zinc than other crabs, but the net accumulation data presented do not allow discrimination between these two possibilities. An earlier study by Bryan (1966) had employed ^{65}Zn as a radiotracer in comparing uptake of dissolved zinc by Restronguet Creek and Tamar crabs over periods of twenty-four and thirty-three days. The Restronguet Creek crabs accumulated less radiolabelled zinc than the Tamar ones, but again, it is difficult to be sure whether this is a result of reduced zinc uptake (reduced permeability) or increased zinc excretion, or both.

Rainbow et al. (1999) measured the comparative uptake rates of zinc from solution by *C. maenas* from Restronguet Creek and other European estuaries of differential degrees of metal contamination. The dissolved uptake rate of zinc from 50 μg Zn L^{-1} by the Restronguet Creek crabs was not significantly different from that of crabs from a control site (Millport, Scotland), nor, in contrast to the results of Bryan (1966) and Bryan and Gibbs (1983), was the net accumulation rate of zinc significantly different between the crabs from the two sites (Rainbow et al. 1999).

Thus, it is not possible to draw concrete general conclusions about the dominant physiological mechanism underlying the zinc tolerance of Restronguet Creek crabs, perhaps expectedly given the probable individual distinctiveness of any zinc resistance between crabs.

Carcinus maenas from Restronguet Creek also showed tolerance to dissolved copper in comparison to crabs from a control estuary, the Tamar, with evidence for a reduced net copper accumulation rate in the Restronguet Creek crabs (Bryan and Gibbs 1983). Although Bryan and Gibbs (1983) suggested that the Creek crabs are therefore less permeable to copper, increased excretion cannot be ruled out. Evidence was also presented of decreased accumulation of Cu from ingested food by Restronguet Creek crabs in comparison to Tamar crabs (Bryan and Gibbs 1983).

For both zinc and copper, therefore, the physiological basis of any metal tolerance by Restronguet Creek crabs, whether it is acclimation or inheritable genetic adaptation, awaits further elucidation. Biodynamic modelling techniques have much to offer in this context, given the increasing availability of appropriate tracer techniques for copper (Croteau and Luoma 2005). Similarly, the physiological bases of the copper tolerances of Restronguet Creek populations of *Scrobicularia plana*, *Corophium volutator*, *Nephtys hombergi*, and *Fucus vesiculosus* await further such investigation.

6.3.3 HAYLE AND GANNEL RIVERS (CORNWALL, UK): THE ISOPOD CRUSTACEAN *ASELLUS MERIDIANUS*

The freshwater isopod crustacean *Asellus meridianus* is a common inhabitant of the streams and rivers of southwest England, and populations of *A. meridianus* that are tolerant to copper and lead have been identified in metal-rich streams of Cornwall, England (Brown 1977). One population in the river Hayle was found to be tolerant to both copper and lead, although the river Hayle contains only trace amounts of lead together with high concentrations of copper; the second population, tolerant only to lead, was found in the lead-rich river Gannel, which lacks high copper concentrations (Brown 1977). The implication here is that the mechanism for copper tolerance in the river Hayle population has simultaneously achieved lead tolerance (in the absence of raised lead bioavailability), but that the mechanism for lead tolerance in the river Gannel population is not associated with simultaneous copper tolerance in the absence of raised copper bioavailability.

Accumulation rates of copper from a high dissolved copper concentration (500 µg Cu L⁻¹) over eight days were similar in these two populations and in a third non-tolerant population of *A. meridianus* from elsewhere in the Hayle. The Cu-tolerant Hayle population reached higher final accumulated copper concentrations (2,200 µg Cu g⁻¹) over this exposure, simply because these isopods had higher initial concentrations (1,030 µg Cu g⁻¹). Most of the newly accumulated copper was stored in insoluble form in the hepatopancreatic caeca (Brown 1977). The copper-tolerant isopods also accumulated up to 6,800 µg Cu g⁻¹ after feeding on a copper-enriched diet for twelve days, whereas nontolerant animals showed no evidence of copper accumulation from food and no survivors remained after eight days (Brown 1977). Increased detoxification therefore appears to be the basis of copper tolerance in the copper-tolerant Hayle population of *A. meridianus*, with the copper and sulphur-rich intracellular deposits described by Brown (1977) probably being derived from the lysosomal breakdown of copper-rich metallothionein (Luoma and Rainbow 2008).

Accumulation of lead from eight days of exposure of the three populations to 500 µg Pb L⁻¹ was raised in the isopod population that was not tolerant to lead (no survival after nine days); a possible basis of lead tolerance is therefore reduced lead accumulation from solution in each of the two lead-tolerant populations (both survived fourteen days' exposure) (Brown 1977). Isopods from both lead-tolerant populations did, however, strongly accumulate lead during a fourteen-day exposure to a lead-enriched diet (up to 28,000 µg Pb g⁻¹), while animals not tolerant to lead showed no accumulation of lead from food, and no live animals remained after ten days (Brown 1977). Thus, when lead is delivered via the diet, both lead-tolerant populations can withstand strong accumulation of lead, necessarily associated with storage detoxification (Rainbow 2002; Luoma and Rainbow 2008). In contrast to the situation for copper accumulated in the copper-tolerant Hayle population (1,030 µg Cu g⁻¹ body concentration), the hepatopancreatic caeca did not appear to be important storage sites for lead in field-collected samples of the lead-tolerant Gannel population of *A. meridianus* (220 µg Pb g⁻¹ body concentration) (Brown 1977). The relatively low accumulated body concentrations of lead in these latter field specimens from the Gannel suggest a limited role for diet as an uptake route for lead in these isopods, and that their lead tolerance is based on reduced accumulation of lead from solution. Whether uptake from solution is reduced or excretion is enhanced awaits the application of biodynamic modelling, best achieved for lead by stable isotope labelling, as in the case of copper, given the lack of availability of suitable lead radiotracers.

6.3.4 Foundry Cove (New York): The Oligochaete Worm *Limnodrilus hoffmeisteri*

Foundry Cove is a tidal freshwater marsh on the Hudson River, New York, 87 km upriver from the mouth of the Hudson, which received waste from a nickel-cadmium battery factory from 1953 to 1971 (Klerks and Levinton 1991; Levinton et al. 2003). Dredging in 1972 and 1973 removed the most contaminated sediment, but cadmium concentrations ranging from 40,000 to 225,000 µg Cd g^{-1} dry sediment were still recorded as late as 1983, and the cadmium in Foundry Cove was bioavailable to plants, fish, and invertebrates (Klerks and Levinton 1991; Levinton et al. 2003). Klerks and Levinton (1989) showed that the Foundry Cove population of the aquatic deposit-feeding oligochaete *Limnodrilus hoffmeisteri* was tolerant to cadmium, and that this cadmium tolerance was genetically determined. The majority of the cadmium-contaminated sediments were removed in a major remediation effort in 1994–1995; this removed the selective force for cadmium tolerance in the local population of *L. hoffmeisteri*, for there then ensued a rapid loss of cadmium resistance in as few as nine to eighteen generations of the worm, indicating the strength of any trade-off between Cd tolerance and some life history trait (Levinton et al. 2003). The cadmium-tolerant oligochaetes, for example, had very slow somatic growth rates compared to nontolerant conspecifics (Levinton et al. 2003).

The physiological mechanism of the cadmium tolerance was not reduced cadmium accumulation from either sediment or solution, for in fact the tolerant worms had a higher cadmium accumulation rate from solution than nontolerant conspecific worms (Klerks and Bartholomew 1991). Cadmium-tolerant worms accumulated more cadmium in the cytosol of their tissues than the control worms; this cytosolic cadmium was associated with a metallothionein-like protein (MTLP) and is therefore considered detoxified (Klerks and Bartholomew 1991; Amiard et al. 2006). Elevated MTLP production was shown to be genetically determined and is the physiological method of cadmium tolerance proposed by Klerks and Bartholomew (1991). Highest concentrations of cadmium were found in chloragogenous tissue around the gut, where the greatest concentration of Cd-rich high-density granules (probably cadmium sulphide) and the unique presence of low-density cadmium-rich granules (apparently phosphate based) were also detected (Klerks and Bartholomew 1991). It can be concluded then that the basis of cadmium tolerance in this oligochaete population was an increased rate of storage detoxification of absorbed cadmium, involving a genetically based increased rate of MTLP synthesis, which probably in turn leads to incorporation of the cadmium into insoluble granules.

6.3.5 CUMBRIA (UK): THE EARTHWORMS *LUMBRICUS RUBELLUS* AND *DENDRODRILUS RUBIDUS*

Related oligochaete annelids, the earthworms, are of course dominant inhabitants of terrestrial soils, and inevitably earthworm populations have come into contact with soils rich in toxic metals, particularly as a result of mining contamination, with the potential for the associated development of metal tolerance. Earthworms have a freshwater ancestry (Barrington 1967), and therefore the potential to take up metals from the soil solution in a manner similar to that of their freshwater oligochaete relatives.

In fact, several studies designed to detect the presence of metal tolerance in field populations of earthworms from metal-contaminated sites have failed to find pronounced interpopulation differences in sensitivity for metals, including cadmium, copper, lead, and zinc (Bengtsson et al. 1992; Marino and Morgan 1999; Spurgeon and Hopkin 1999).

Nevertheless, Langdon et al. (2001) identified copper-tolerant populations of the earthworms *Lumbricus rubellus* and *Dendrodrilus rubidus* in metal-contaminated mine spoil soils at Carrock Fell, Cumbria, in northwest England, but the results presented on comparative bioaccumulation patterns between tolerant and nontolerant populations of earthworms were inconclusive. A copper-tolerant population of the earthworm *Dendrodrilus rubidus* has also been identified by Arnold et al. (2008a,b) from a former copper mining area, Coniston Mines, in Cumbria. Earthworms from the abandoned copper mine site tolerated significantly higher soil Cu concentrations (LC_{50} 437 mg Cu kg^{-1}, 95% CL 219–874) than a control population (LC_{50} 145 mg Cu kg^{-1}, 95% CL 45–462), and exhibited significantly less change in weight and a lower loss in condition during copper exposure (Arnold et al. 2008a). Laboratory attempts to establish any genetic component to the copper tolerance by comparing parent and F1 generation earthworms were unsuccessful, although the relatively small size of the mine site earthworms in comparison with control earthworms does suggest a metabolic cost of tolerance compatible with the tolerance being a genetic adaptation (Arnold et al. 2008a). The Cu mine earthworms accumulated more new copper than the control earthworms on exposure to Cu-enhanced soil for fourteen days (Arnold et al. 2008b), eliminating reduced accumulation as the mechanism of any copper tolerance. Comparative subcellular partitioning of accumulated copper in the two earthworm populations showed that the Cu mine worms accumulated more copper in existing storage detoxification reservoirs without change of proportions of accumulated copper in the different subcellular binding pools (Arnold et al. 2008b). Thus, the apparent copper tolerance of the Coniston Copper Mine population of *Dendrodrilus rubidus* is based on increased detoxification via existing copper detoxification mechanisms, associated with increased accumulation of copper compared to controls.

6.3.6 RIVER DOMMEL (BELGIUM): THE MIDGE LARVA *CHIRONOMUS RIPARIUS*

Larvae of the midge *Chironomus riparius* are common in rivers of western Europe, and Postma et al. (1995) identified cadmium-adapted populations of this midge larva from the river Dommel, which flows from Belgium to the Netherlands. The growth of the cadmium-adapted larvae was less sensitive to the toxic effects of cadmium than that of control larvae, and F1 generation larvae were also cadmium tolerant, indicating that the cadmium tolerance is by genetic adaptation (Postma et al. 1995). Comparative studies on cadmium bioaccumulation kinetics on first-generation laboratory-reared midge larvae from cadmium-adapted and nonadapted field populations showed that the Cd-tolerant larvae had decreased net accumulation rates when exposed to 19.8 μg Cd L^{-1}, associated with increased cadmium elimination rates from the gut (Postma et al. 1996).

6.3.7 METAL-TOLERANT CLADOCERAN CRUSTACEANS *DAPHNIA MAGNA*

The water flea *Daphnia magna*, a freshwater cladoceran crustacean that can reproduce asexually by parthenogenesis, has proved to be an ideal laboratory animal with which to investigate the development of metal tolerance over generations. Metal-tolerant field populations of this species have also been identified (Muyssen et al. 2002).

Two populations of *D. magna* from ponds on the site of a zinc smelter in Budel, the Netherlands, were shown to be tolerant to acute dissolved zinc toxicity, exhibiting a higher reproduction rate and carapace length than control water fleas (Muyssen et al. 2002). Clones of the zinc-tolerant populations gradually lost their zinc tolerance when cultured without added zinc, and Muyssen et al. (2002) concluded that the zinc tolerance was attributable to physiological acclimation without a genetic basis. There were no differences in the zinc concentrations accumulated upon zinc exposure by the different populations compared (Muyssen et al. 2002).

Physiological acclimation figures strongly in studies of the resistance of *Daphnia magna* to toxic metals. LeBlanc (1982) demonstrated that *D. magna* preexposed to copper, lead, or zinc for twenty hours became significantly resistant to the toxic effects of these metals. This resistance was determined to be the result of physiological acclimation and not genetic adaptation (LeBlanc 1982). Similarly, Bodar et al. (1990) showed that this water flea developed resistance to cadmium exposure during a single generation, again as a result of physiological acclimation. Net accumulation of cadmium from solution was raised in preexposed water fleas, without increased concentrations of the cadmium detoxificatory protein, metallothionein (MT) (Bodar et al. 1990). It remains possible that increased MT synthesis is associated with the increased cadmium accumulation, because increased lysosomal breakdown of MT may compensate for any increased MT synthesis with no change in standing concentration of MT (Ng et al. 2008).

Physiological acclimation was also the basis of the development of trace metal tolerance in two clones of *Daphnia magna* preexposed to dissolved Cd, Zn, and a Cd-Zn mixture by Stuhlbacher et al. (1992). Preexposure to Zn and Zn-Cd resulted in a significant increase in Cd resistance in both clones, while only one clone gained increased Cd resistance from Cd preexposure; preexposure to Zn and Cd-Zn, but not Cd, caused increased Zn resistance in both clones (Stuhlbacher et al. 1992). In contrast to the results of Bodar et al. (1990), the increased resistance to elevated cadmium was associated with an increase in metallothionein-like protein in the one clone investigated (Stuhlbacher et al. 1992). The two clones, however, varied in rates of cadmium accumulation from solution, with one, but not the other, showing increased cadmium accumulation (Stuhlbacher et al. 1992).

There are more recent studies of the kinetics of metal bioaccumulation by *Daphnia magna* exposed to raised bioavailabilities of toxic metals (Guan and Wang 2004, 2006, Tsui and Wang 2007). Guan and Wang (2006) examined the effects of exposing a sensitive and a tolerant clone of *D. magna* to Cd in three successive generations on subsequent toxicity and bioaccumulation kinetics. After exposure to 3 µg Cd L^{-1} for two consecutive generations, both clones improved their resistance to a lethal Cd challenge, but the high comparability of the two clones in individual fitness parameters and Cd biokinetics suggested that Cd resistance was due to physiological acclimation and not genetic adaptation (Guan and Wang 2006). Guan and Wang (2006) found that continuous Cd exposure increased the Cd uptake rate from solution and diet (assimilation efficiency) for each clone, and they concluded that it was the metallothionein concentration in the Cd-exposed daphnids that was the critical factor in handling lethal Cd stress.

Tsui and Wang (2007) have provided a detailed review of the kinetics of bioaccumulation of several toxic metals (Ag, Cd, Hg, MeHg, Se, Zn) and the development of metal tolerance by *Daphnia magna*. Generally this cladoceran consistently shows remarkable physiological acclimation to toxic metal exposure, relying heavily on the induction of metallothionein (probably increased MT turnover if not increased MT concentration) in the case of metals like cadmium that can induce MT synthesis (Amiard et al. 2006).

6.3.8 FINNISS RIVER (NORTHERN TERRITORY, AUSTRALIA): THE FISH *MELANOTAENIA NIGRANS*

Given the relative motility of fish in comparison with many of the invertebrate examples given above, fish populations are less likely to remain in position in a stream, river, or estuary of high trace metal bioavailability of ecotoxicological significance long enough to enable any natural selection for genetically adapted metal tolerance. An exceptional habitat might be an isolated lake receiving high metal input, be it from a feeder stream or by atmospheric deposition.

In fact, a copper-tolerant population of fish has been described from an acid-rich freshwater stream draining a uranium and copper mine (Gale et al. 2003). The Rum Jungle mine is located on the East Branch of the Finniss River in the Northern Territory, Australia, and dissolved copper concentrations as high as 1.36 mg Cu L^{-1} were measured 2.5 km downstream of the mine in 1993. This site was occupied by a population of the black-banded rainbow fish *Melanotaenia nigrans* that had potentially been exposed to elevated copper bioavailabilities for more than forty years (Gale et al. 2003). These fish were copper tolerant with a 96 h EC$_{50}$ more than eight times higher than that of a reference population from Coomalie Creek draining a separate catchment in the Northern Territory (Gale et al. 2003). Results of the exposure of these two populations to radiolabelled ($^{64/67}$Cu) dissolved copper showed reduced bioaccumulation of labelled copper from solution by the copper-tolerant population; this latter population had a modelled steady-state concentration after twenty-four hour exposure to 30 µg Cu L^{-1} of 167 ± 15 (SE) ng Cu g^{-1} wet weight, compared to 89 ± 12 ng Cu g^{-1} wet weight in the reference fish (Gale et al. 2003). It was concluded that the reduced accumulation had been brought about by reduced uptake of dissolved copper in the gills with no evidence for increased efflux (Gale et al. 2003). Allozyme electrophoresis on the two populations showed significantly different allozyme frequencies at aspartate amino transferase (AAT) and glucose-6-phosphate isomerase (GPI) loci, and reduced heterozygosity in the copper-tolerant population, suggesting that exposure to copper has resulted in some selection of copper-tolerant allozymes at some loci (Gale et al. 2003), i.e., that there is a genetic basis to the copper tolerance (genetic adaptation).

6.3.9 LABORATORY SELECTION: THE FISH *HETERANDRIA FORMOSA*

The least killifish *Heterandria formosa* is a freshwater fish widely distributed in the southeastern United States, and is well suited for laboratory study given its viviparity, small size, relatively short life span, and ease of maintenance (Xie and Klerks 2004b). Xie and Klerks (2003) conducted an artificial selection experiment for six generations of *H. formosa*, successfully selecting for cadmium resistance. There was a genetic basis to this cadmium tolerance that was accompanied by copper resistance, but had a metabolic cost in reduced heat tolerance, reduced brood size, longer time to first reproduction, and shorter female life expectancy than control populations (Xie and Klerks 2003, 2004b). Xie and Klerks (2004c) compared cadmium accumulation from solution in Cd-tolerant lines and control lines of *H. formosa*. The Cd-tolerant lines had reduced short-term (two-hour) uptake of dissolved Cd, and a lower rate of Cd accumulation from solution over fourteen hours' exposure to the very high dissolved concentration of 6 mg Cd l^{-1} (Xie and Klerks 2004c). Xie and Klerks (2004c) concluded that differences in Cd uptake and accumulation biodynamics were the major mechanism underlying Cd tolerance in the laboratory-selected Cd-tolerant line, with elevated levels of metallothionein-like protein playing a contributory but relatively minor role (Xie and Klerks 2004a).

6.3.10 PLOMBIÈRES (BELGIUM): THE SPRINGTAIL *ORCHESELLA CINCTA*

The case studies considered so far have concerned aquatic animals, mostly invertebrates, even in the case of earthworms, which can crudely be considered freshwater oligochaetes living in the water of terrestrial soils. In all the above cases, uptake of metals from solution can play a part in the delivery of ecotoxicologically significant metal loads to the animal. The final two case studies considered here concern fully terrestrial invertebrates, obtaining all metal from the diet—a springtail *Orchesella cincta* and a woodlouse *Porcellio scaber*, a close isopod crustacean relative of the freshwater *Asellus meridianus* highlighted above.

In the first terrestrial example, cadmium-tolerant populations of the collembolan *Orchesella cincta* have been described from litter samples collected from a mining area at Plombières in Belgium and near a lead smelter works at Stolberg, Germany, as recognised by resistance to cadmium-induced growth reduction on exposure to a cadmium-enriched diet (van Straalen et al. 1987; Posthuma 1990). The cadmium tolerance in each case persisted in first-generation laboratory-bred springtails, indicating the presence of a genetic component to the tolerance (Posthuma 1990). In *O. cincta*, excretion of accumulated metals takes place through exfoliation at every moult of the midgut epithelium, which stores accumulated trace metals in detoxified form (van Straalen et al. 1987). Springtails of both cadmium-tolerant populations showed significantly higher cadmium excretion efficiencies than reference populations, and the exfoliation of the midgut epithelium is considered an important component of the cadmium tolerance and adaptation (van Straalen et al. 1987; Posthuma et al. 1992). The basis of the cadmium tolerance can therefore be interpreted as increased storage detoxification in the midgut epithelium, a temporary store that is excreted at every moult, leading to enhanced excretion (van Straalen et al. 1987; Posthuma et al. 1992; Sterenborg and Roelofs 2003). Moreover, estimates of the genetic variation in the springtail populations indicate that genetically adapted tolerance may occur within a few (tens of) generations, and at least within the time span of operations like mines and smelters emitting metals (Posthuma and van Straalen 1993; Posthuma et al. 1993). More recent work has shown that the increased cadmium tolerance in the Plombières population can be related to changes in the gene regulation of metallothionein (Sterenborg and Roelofs 2003), with increased production of metallothionein being brought about by selection acting on allelic polymorphism in the promotor of the MT gene (van Straalen and Roelofs 2005), leading to overexpression of the single copy cadmium-inducible MT gene (Roelofs et al. 2006). Thus, cadmium assimilated into the midgut epithelium of *O. cincta* induces Cd-MT (at an enhanced rate in the Cd-tolerant population at Plombières); this MT and bound Cd is autolysed in midgut epithelium lysosomes, and Cd-rich lysosomal residual bodies are released back into the gut lumen and excreted at each moult as a new midgut epithelium is formed (van Straalen and Roelofs 2005). Some of the expelled cadmium may even be reabsorbed and reintroduced into this cycle, albeit with a net loss of accumulated cadmium via loss in the faeces (van Straalen and Roelofs 2005).

6.3.11 PLOMBIÈRES (BELGIUM): THE WOODLOUSE *PORCELLIO SCABER*

The second fully terrestrial example quoted here concerns the terrestrial isopod *Porcellio scaber*, going under a plethora of common names, including woodlouse, pillbug, and sowbug, from the same mining site at Plombières in Belgium. F1 generation offspring of the population of *P. scaber* from Plombières did not show reduced growth in response to a cadmium-enhanced diet in contrast to a reference population, indicating the presence of genetically adapted cadmium tolerance (Donker and Bogert 1991). Populations assimilated the same amount of cadmium, and it was concluded that the tolerance mechanism is based on differences in detoxification mechanisms (Donker and Bogert 1991).

6.4 CONCLUSIONS

We set out to attempt to identify which, if any, of the physiological processes of toxic metal bioaccumulation has been particularly susceptible to the selective processes driving the evolution of a metal-tolerant population of an animal in the field. Table 6.4 therefore brings together the available evidence from the case studies of genetically adapted metal tolerance quoted above, in an attempt to address this question. We accept that this choice of case studies cannot be comprehensive, but we have no reason to believe that it is not representative. Established cases of genetically based toxic metal tolerance in the field are not prolific, testament to the fact that organisms have a remarkable ability to cope with raised trace metal bioavailabilities without selection for enhanced physiological mechanisms for metal resistance (Rainbow et al. 1999). Furthermore, much relevant data on bioaccumulation parameters are simply not available, even for well-established metal-tolerant field populations. The application of biodynamic modelling has so much to offer in this context (Luoma and Rainbow 2005).

We go further in some cases in Table 6.4 by interpreting the available bioaccumulation data to propose the particular biodynamic parameter that might be of paramount importance in the field. We are well aware that, in the last decade, there has been increasing awareness of the relative importance of diet as a source of toxic metals to animals (Wang 2002; Luoma and Rainbow 2008), particularly in the case of sediment-ingesting invertebrates like *Nereis diversicolor* (Casado-Martinez et al. 2009; Rainbow et al. 2009a,b). This appreciation has affected our interpretation of the significance of the role of dissolved uptake in the ecotoxicology of metals in field-exposed populations, given that this uptake route can no longer be assumed by default to be the most ecotoxicologically significant.

If any one biodynamic parameter stands out from Table 6.4, it is an enhanced detoxification rate of the metal taken up at a high rate under conditions of increased bioavailability from solution or diet. Thus, we conclude that it is an enhanced rate of detoxification that is often the result of selection in the field to bring about metal tolerance in a particular animal population. In many cases an enhanced detoxification rate is attributable to an enhanced induction of metallothionein—often referred to as metallothionein-like protein if the final biochemical details of absolute MT identification have not been carried out (Amiard et al. 2006)—remembering that induction of MT may be reflected in increased synthesis and turnover of MT without necessarily increased MT concentration (Ng et al. 2008). It is relevant that metallothionein induction is also often the basis of the physiological acclimation of individuals to produce metal tolerance, well exemplified by the case studies on the cladoceran *Daphnia magna* described above. Increased MT induction can be achieved by a single gene duplication. Metallothionein gene duplications are by no means unknown and have been described in the literature, for example, in the case of natural populations of the fruit fly *Drosophila melanogaster* (Maroni et al. 1987). Similarly, MT promotor genes have been shown to be susceptible to selection for increased activity, as in the case of the collembolan *Orchesella cincta* (Roelofs et al. 2006). Increased tissue concentrations of MT can also be associated physiologically with increases in the assimilation efficiency of the metal that is bound by MT in aquatic invertebrates (Wang and Rainbow 2005), perhaps explaining the apparently anomalous but common observation that a metal-tolerant population may have a higher metal accumulation rate than a control population (Table 6.4). The implication here is that the added benefit gained by an enhanced detoxification rate more than offsets the toxic metal challenge imposed by any associated increased uptake rate of the metal from the diet.

In hindsight it can be argued that a changed detoxification rate based on single gene changes such as duplication may be more easily achieved than a reduction in metal uptake rate or an increase in metal efflux rate. Metal uptake, be it from solution across an epidermal epithelium in contact with the surrounding medium, or across an alimentary epithelium in contact with the contents of the gut lumen, involves multiple routes of varying significance physiologically and taxonomically (Luoma and Rainbow 2008). Any single gene change may simply not be significant enough to total metal uptake rates to be available for natural selection.

To conclude, we propose as a testable hypothesis that it is the rate of detoxification of incoming metals that is the parameter of bioaccumulation most likely to be

TABLE 6.4

Proposed Mechanisms of Genetically Adapted Metal Tolerance in Populations of Animals Identified to Be Metal Tolerant

		Field Concentration	Accumulation from Water				Accumulation from Diet			
			Accumulation Rate	Uptake Rate	Efflux Rate	Detoxification Rate	Accumulation Rate	Uptake Rate	Efflux Rate	Detoxification Rate
Annelids										
Polychaetes										
Nereis diversicolor										
Restronguet Creek	Zn	Regulated	Low	Low?	High?		?			
	Zn	High after regulation breakdown	Not changed	Not changed	Not changed	High	Not changed	Not changed	Not changed	**High**
Restronguet Creek	Cu	High	High?			High	High?			**High**
Oligochaetes										
Limnodrilus hoffmeisteri										
Foundry Creek	Cd	High	High			High	Not changed			**High**
Dendrodrilus rubidus										
Coniston Mines	Cu	High					High			**High**
Crustaceans										
Isopods										
Asellus meridianus										
River Hayle	Cu	High	Not changed			**High**	High			**High**
River Hayle	Pb	Low	Low				High			High
River Gannel	Pb	Not high	**Low**				High			High
Porcellio scaber										
Plombières	Cd						Not changed			**High**

	Metal						
Crabs							
Carcinus maenas							
Restronguet Creek (acclimation?)	Zn	Some low	Low?	High?			
Restronguet Creek (acclimation?)	Cu	**Low**	Low?	High?	Low?		
Insects							
Chironomus riparius							
Midge larvae							
River Dommel	Cd	Low		High			
Collembolans							
Orchesella cincta							
Plombières	Cd					**High**	**High**
Fish							
Teleosts							
Melanotaenia nigrans							
Finniss River	Cu	**Low**	Low	Not changed			

Note: Suggested dominant mechanisms in the field are presented in bold. The adjectives *high* and *low* are in comparison with nontolerant populations, as identified and discussed in the text, which provides the relevant references. The accumulation rate is the net difference between rates of uptake and efflux.

changed in the evolution of a metal-tolerant animal population. We also propose that the application of biodynamic modelling of metal bioaccumulation has much to offer to our understanding of the physiological basis of toxic metal tolerance. We hope, therefore, that this review will form a first step in future progress to this end.

REFERENCES

Amiard, J. C., et al. 1987. Comparative study of the patterns of bioaccumulation of essential (Cu, Zn) and non-essential (Cd, Pb) trace metals in various estuarine and coastal organisms. *J. Exp. Mar. Biol. Ecol.* 106:73–89.

Amiard, J. C., et al. 2006. Metallothioneins in aquatic invertebrates: Their role in metal detoxification and their use as biomarkers. *Aquat. Toxicol.* 76:160–202.

Amiard-Triquet, C., et al. 2009. Tolerance in organisms chronically exposed to estuarine pollution. In *Environmental assessment of estuarine ecosystems. A case study*, ed. C. Amiard-Triquet and P. S. Rainbow, 135–57. Boca Raton, FL: CRC Press, Taylor & Francis Group.

Arnold, R. E., M. E. Hodson, and C. J. Langdon. 2008a. A Cu tolerant population of the earthworm *Dendrodrilus rubidus* (Sanigny, 1862) at Coniston Copper Mines, Cumbria, UK. *Environ. Pollut.* 152:713–22.

Arnold, R. E., et al. 2008b. Comparison of subcellular partitioning, distribution, and internal speciation of Cu between Cu-tolerant and naive populations of *Dendrodrilus rubidus* Savigny. *Environ. Sci. Technol.* 42:3900–5.

Barrington, E. J. W. 1967. *Invertebrate structure and function*. London: Nelson.

Bengtsson, G., H. Ek, and S. Rundgren. 1992. Evolutionary response of earthworms to longterm metal exposure. *Oikos* 63:289–97.

Berthet, B., et al. 2003. Responses to metals of the polychaete annelid *Hediste diversicolor*, a key species in estuarine and coastal sediments. *Arch. Environ. Contam. Toxicol.* 45:468–78.

Bodar, C. W. M., et al. 1990. Cadmium resistance in *Daphnia magna*. *Aquat. Toxicol.* 16:33–40.

Brown, B. E. 1977. Uptake of copper and lead by a metal-tolerant isopod *Asellus meridianus* Rac. *Freshwater Biol.* 7:235–44.

Bryan, G. W. 1966. The metabolism of Zn and ^{65}Zn in crabs, lobsters and fresh-water crayfish. In *Radioecological concentration processes*, ed. B. Aberg and F. P. Hungate, 1005–16. Oxford: Pergamon Press.

Bryan, G. W., and P. E. Gibbs. 1983. Heavy metals in the Fal Estuary, Cornwall: A study of long-term contamination by mining waste and its effects on estuarine organisms. *Occ. Publ. Mar. Biol. Assoc. UK* 2:1–112.

Bryan, G. W., and L. G. Hummerstone. 1971. Adaptation of the polychaete *Nereis diversicolor* to estuarine sediments containing high concentrations of heavy metals. I. General observations and adaptation to copper. *J. Mar. Biol. Assoc. UK* 51:845–63.

Bryan, G. W., and L. G. Hummerstone. 1973. Adaptation of the polychaete *Nereis diversicolor* to estuarine sediments containing high concentrations of zinc and cadmium. *J. Mar. Biol. Assoc. UK* 53:839–57.

Bryan, G. W., W. J. Langston, and L. G. Hummerstone. 1980. The use of biological indicators of heavy metal contamination in estuaries. *Occ. Publ. Mar. Biol. Assoc. UK* 1:1–73.

Casado-Martinez, M. C., et al. 2009. Pathways of trace metal uptake in the lugworm *Arenicola marina*. *Aquat. Toxicol.* 92:9–17.

Croteau, M. N., and S. N. Luoma. 2005. Delineating copper accumulation pathways for the freshwater bivalve *Corbicula* using stable copper isotopes. *Environ. Toxicol. Chem.* 24:2871–78.

Croteau, M. N., and S. N. Luoma. 2009. Predicting dietborne metal toxicity from metal influxes. *Environ. Sci. Technol.* 43:4915–21.

Dines, H. G. 1969. *The metalliferous mining region of South-west England.* Vol. 1. London: Her Majesty's Stationery Office.

Donker, M. H., and C. G. Bogert. 1991. Adaptation to cadmium in three populations of the isopod *Porcellio scaber. Comp. Biochem. Physiol. C* 100:143–46.

Gale, S. A., et al. 2003. Insights into the mechanisms of copper tolerance of a population of black-banded rainbow fish (*Melanotaenia nigrans*) (Richardson) exposed to mine leachate, using $^{64/67}$Cu. *Aquat. Toxicol.* 62:135–53.

Grant, A., J. G. Hateley, and N. V. Jones. 1989. Mapping the ecological impact of heavy metals in the estuarine polychaete *Nereis diversicolor* using inherited metal tolerance. *Mar. Pollut. Bull.* 20:235–38.

Guan, R., and W. X. Wang. 2004. Cd and Zn uptake kinetics in *Daphnia magna* in relation to Cd exposure history. *Environ. Sci. Technol.* 38:6051–58.

Guan, R., and W. X. Wang. 2006. Comparison between two clones of *Daphnia magna*: Effects of multigenerational cadmium exposure on toxicity, individual fitness, and biokinetics. *Aquat. Toxicol.* 76:217–29.

Hateley, J. G., A. Grant, and N. V. Jones. 1989. Heavy metal tolerance in estuarine populations of *Nereis diversicolor*. In *Reproduction, genetics and distribution of marine organisms, Proceedings 23rd European Marine Biological Symposium*, ed. J. S. Ryland and P. A. Tyler, 379–85. Fredensborg, Denmark: Olsen and Olsen.

Klerks, P. L., and P. R. Bartholomew. 1991. Cadmium accumulation and detoxification in a Cd-resistant population of the oligochaete *Limnodrilus hoffmeisteri. Aquat. Toxicol.* 19:97–112.

Klerks, P. L., and J. S. Levinton. 1989. Rapid evolution of metal resistance in a benthic oligochaete inhabiting a metal-polluted site. *Biol. Bull.* 176:135–41.

Klerks, P. L., and J. S. Levinton. 1991. Evolution of resistance and changes in community composition in metal-polluted environments: A case study on Foundry Cove. In *Ecotoxicology of metals in invertebrates*, ed. R. Dallinger and P. S. Rainbow, 223–41. Boca Raton, FL: Lewis Publishers.

Klerks, P. L., and J. S. Weis. 1987. Genetic adaptation to heavy metals in aquatic organisms: A review. *Environ. Pollut.* 45:173–205.

Langdon, C. J., et al. 2001. Resistance to copper toxicity in populations of the earthworm *Lumbricus rubellus* and *Dendrodrilus rubidus* from contaminated mine wastes. *Environ. Toxicol. Chem.* 20:2336–41.

LeBlanc, G. A. 1982. Laboratory investigation into the development of resistance of *Daphnia magna* (Straus) to environmental pollutants. *Environ. Pollut. A* 27:309–22.

Levinton, J. S., et al. 2003. Rapid loss of genetically based resistance to metals after the cleanup of a Superfund site. *Proc. Natl. Acad. Sci.* 100:9889–91.

Luoma, S. N. 1977. Detection of trace contaminant effects in aquatic ecosystems. *J. Fish. Res. Board Canada* 34:436–39.

Luoma, S. N., and P. S. Rainbow. 2005. Why is metal bioaccumulation so variable? Biodynamics as a unifying concept. *Environ. Sci. Technol.* 39:1921–31.

Luoma, S. N., and P. S. Rainbow. 2008. *Metal contamination in aquatic environments. Science and lateral management.* Cambridge: Cambridge University Press.

Marino, F., and A. J. Morgan. 1999. The time-course of metal (Ca, Cd, Cu, Pb, Zn) accumulation from a contaminated soil by three populations of the earthworm, *Lumbricus rubellus. Appl. Soil Ecol.* 12:169–77.

Maroni, G., J. Wise, and E. Otto. 1987. Metallothionein gene duplications and metal tolerance in natural populations of *Drosophila melanogaster. Genetics* 117:739–44.

Morgan, A. J., P. Kille, and S. R. Stürzenbaum. 2007. Microevolution and ecotoxicology of metals in invertebrates. *Environ. Sci. Technol.* 41:1085–96.

Mouneyrac, C., et al. 2003. Trace-metal detoxification and tolerance of the estuarine worm *Hediste diversicolor* chronically exposed in their environment. *Mar. Biol.* 143: 731–44.

Muyssen, B. T. A., C. R. Janssen, and B. T. A. Bossuyt. 2002. Tolerance and acclimation to zinc of field-collected *Daphnia magna* populations. *Aquat. Toxicol.* 56:69–79.

Ng, T. Y.-T., et al. 2008. Decoupling of cadmium biokinetics and metallothionein turnover in a marine polychaete after metal exposure. *Aquat. Toxicol.* 89:47–54.

Pook, C., C. Lewis, and T. Galloway. 2009. The metabolic and fitness costs associated with metal resistance in *Nereis diversicolor*. *Mar. Pollut. Bull.* 58:1063–71.

Posthuma, L. 1990. Genetic differentiation between populations of *Orchesella cincta* (Collembola) from heavy metal contaminated sites. *J. Appl. Ecol.* 27:609–22.

Posthuma, L., R. F. Hogervost, and N. M van Straalen. 1992. Adaptation to soil pollution by cadmium excretion in natural populations of *Orchesella cincta* (L.) (Collembola). *Arch. Environ. Contam. Toxicol.* 22:145–56.

Posthuma, L., and N. M. van Straalen. 1993. Heavy-metal adaptation in terrestrial invertebrates: A review of occurrence, genetics, physiology and ecological consequences. *Comp. Biochem. Physiol. C* 106:11–38.

Posthuma, L., et al. 1993. Genetic variation and covariation for characteristics associated with cadmium tolerance in natural populations of the springtail *Orchesella cincta* (L.). *Evolution* 47:619–31.

Postma, J. F., M. Kyed, and W. Admiraal. 1995. Site specific differentiation in metal tolerance in the midge *Chironomus riparius* (Diptera: Chironomidae). *Hydrobiologia* 315:159–65.

Postma, J. F., P. van Nugteren, and M. B. Buckert-de Jong. 1996. Increased cadmium excretion in metal-adapted populations of the midge *Chironomus riparius* (Diptera). *Environ. Toxicol. Chem.* 15:332–39.

Rainbow, P. S. 2002. Trace metal concentrations in aquatic invertebrates: Why and so what? *Environ. Pollut.* 120:497–507.

Rainbow, P. S. 2007. Trace metal bioaccumulation: Models, metabolic availability and toxicity. *Environ. Int.* 33:576–82.

Rainbow, P. S., B. D. Smith, and S. N. Luoma. 2009a. Differences in the trace metal bioaccumulation kinetics among populations of the polychaete *Nereis diversicolor* from metal-contaminated estuaries. *Mar. Ecol. Prog. Ser.* 376:173–84.

Rainbow, P. S., B. D. Smith, and S. N. Luoma. 2009b. Biodynamic modelling and the prediction of Ag, Cd and Zn accumulation from solution and sediment by the polychaete *Nereis diversicolor*. *Mar. Ecol. Prog. Ser.* 390:145–55.

Rainbow, P. S., et al. 1999. Trace metal uptake rates in crustaceans (amphipods and crabs) from coastal sites in NW Europe differentially enriched with trace metals. *Mar. Ecol. Prog. Ser.* 183:189–203.

Rainbow, P. S., et al. 2004. Enhanced food chain transfer of copper from a diet of copper-tolerant estuarine worms. *Mar. Ecol. Prog. Ser.* 271:183–91.

Roelofs, D., et al. 2006. Additive genetic variation of transcriptional regulation: Metallothionein expression in the soil insect *Orchesella cincta*. *Heredity* 96:85–92.

Spurgeon, D. J., and S. P. Hopkin. 1999. Tolerance to zinc in populations of the earthworm *Lumbricus rubellus* from uncontaminated and metal-contaminated ecosystems. *Arch. Environ. Contam. Toxicol.* 37:332–37.

Sterenborg, I., and D. Roelofs. 2003. Field-selected cadmium tolerance in the springtail *Orchesella cincta* is correlated with increased metallothionein mRNA expression. *Insect Biochem. Mol. Biol.* 33:741–47.

Stuhlbacher, A., et al. 1992. Induction of cadmium tolerance in two clones of *Daphnia magna* Straus. *Comp. Biochem. Physiol. C* 101:571–77.

Tsui, M. T. K., and W. X. Wang. 2007. Biokinetics and tolerance development of toxic metals in *Daphnia magna*. *Environ. Toxicol. Chem.* 26:1023–32.

van Straalen, N. M., and D. Roelofs. 2005. Cadmium tolerance in a soil arthropod: A model of real-time microevolution. *Entomologische Berichten* 65:105–10.

van Straalen, N. M., et al. 1987. Efficiency of lead and cadmium excretion in populations of *Orchesella cincta* (Collembola) from various contaminated forest soils. *J. Appl. Ecol.* 24:953–68.

Wang, W. X. 2002. Interactions of trace metals and different marine food chains. *Mar. Ecol. Progr. Ser.* 243:295–309.

Wang, W. X., and N. S. Fisher. 1997. Modelling the influence of body size on trace element accumulation in the mussel *Mytilus edulis*. *Mar. Ecol. Prog. Ser.* 161:103–15.

Wang, W. X., and N. S. Fisher. 1999. Assimilation efficiencies of chemical contaminants in aquatic invertebrates: A synthesis. *Environ. Toxicol. Chem.* 18:2034–45.

Wang, W. X., N. S. Fisher, and S. N. Luoma. 1996. Kinetic determinations of trace element bioaccumulation in the mussel *Mytilus edulis*. *Mar. Ecol. Prog. Ser.* 140:91–113.

Wang, W. X., and P. S. Rainbow. 2005. Influence of pre-exposure on trace metal uptake in marine invertebrates. *Ecotoxicol. Environ. Safety* 61:145–59.

Wang, W. X., and P. S. Rainbow. 2006. Subcellular partitioning and the prediction of cadmium toxicity to aquatic organisms. *Environ. Chem.* 3:395–99.

Xie, L., and P. L. Klerks. 2003. Responses to selection for cadmium resistance in the least killifish *Heterandria formosa*. *Environ. Toxicol. Chem.* 22:313–20.

Xie, L., and P. L. Klerks. 2004a. Metallothionein-like protein in the least killifish *Heterandria formosa* and its role in cadmium resistance. *Environ. Toxicol. Chem.* 23:173–77.

Xie, L., and P. L. Klerks. 2004b. Fitness cost of resistance to cadmium in the least killifish (*Heterandria formosa*). *Environ. Toxicol. Chem.* 23:1499–503.

Xie, L., and P. L. Klerks. 2004c. Changes in cadmium accumulation as a mechanism for cadmium resistance in the least killifish *Heterandria formosa*. *Aquat. Toxicol.* 66:73–81.

7 Antioxidant Defenses and Acquisition of Tolerance to Chemical Stress

Francesco Regoli, Maura Benedetti, and Maria Elisa Giuliani

CONTENTS

7.1 ANTIOXIDANT DEFENSES

All aerobic cells produce reactive oxygen species (ROS) as an unavoidable consequence of their metabolism (Halliwell and Gutteridge 2007). Main sources include the four electron reduction of molecular oxygen to water coupled with oxidative phosphorylation, electron transport chains in mitochondria and microsomes, the activity of several enzymes that produce ROS as intermediates or final products, and immunological reactions such as active phagocytosis. The most commonly generated ROS are the singlet oxygen ($^1O_2$), superoxide anion (O_2^-), hydrogen peroxide (H_2O_2), and the hydroxyl radical ($\bullet OH$), which rapidly combine to form other molecules, like peroxynitrite (HOONO), hypochloric acid (HOCl), peroxyl radicals (ROO•), and alkoxyl radicals (RO•). Although all these molecules are generally termed as ROS, they greatly differ in terms of cellular reactivity and potential to cause toxic insults to lipids, proteins, and DNA (Regoli and Winston 1999).

To counteract the adverse effects of basal oxyradical formation and prevent ROS toxicity to cellular targets, aerobic cells have evolved a complex array of antioxidant defenses based on low molecular weight scavengers and antioxidant enzymes (Figure 7.1). The low molecular weight scavengers neutralize ROS by directly reacting with them, thus becoming oxidized or even "temporary prooxidant" radicals. After the reaction with oxyradicals, the oxidized scavengers are generally reconverted to their functionally active, reduced form by specific mechanisms. Scavengers comprise both water-soluble molecules like reduced glutathione, ascorbic acid,

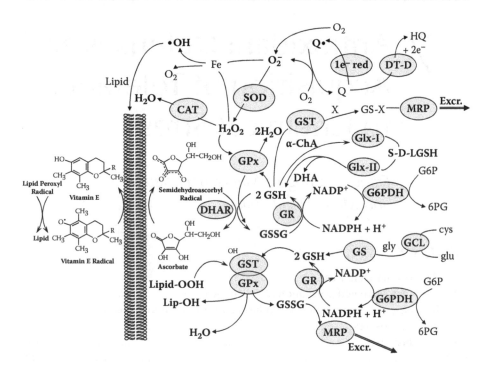

FIGURE 7.1 Main cellular antioxidant defenses and antioxidant pathways. CAT: catalase; SOD: superoxide dismutase; GPx: glutathione peroxidases; GR: glutathione reductase;
GSH: reduced glutathione; GSSG: oxidized glutathione; GST: glutathione S-transferases;
X: xenobiotic; Q: quinone; $1e^-$ red: one-electron reductases; DT-D: DT-diaphorase; MRP:
multidrug-resistance-related protein; DHAR: dehydroascorbate reductase; G6PDH: glucose
6-phosphate dehydrogenase; G6P: glucose 6-phosphate; 6PG: 6-phospogluconate; GCL:
γ-glutamylcysteine synthetase; GS: glutathione synthetase; Glx-I: glyoxalase I; Glx-II: glyoxalase II; α-ChA: α-keto aldehydes; S-D-LGSH: S-D-lactoylglutathione; DHA: d-hydroxy
acid.

and uric acid acting as antioxidants in the cytoplasm, and lipid-soluble defenses
(α-tocopherol, β-carotene, retinol) localized on the membranes to arrest the propagation of lipid peroxidation reactions.

The most abundant cytosolic scavenger is reduced glutathione (GSH), a tri-
peptide (γ-glutamyl-cysteinyl glycine) synthesized in a two-step reaction by
γ-glutamylcysteine synthetase, which binds glutamic acid and cysteine, and by
GSH synthetase, which adds glycine (Figure 7.1). Although predominantly cytosolic, mitochondrial and nuclear pools of glutathione are also present (Hayes and
McLellan 1999). GSH is a direct scavenger of ROS like singlet oxygen, hydroxyl
radical, and reactive nitrogen species, for which it is particularly effective in preventing formation of peroxynitrite. Furthermore, GSH has multiple protective roles
toward oxidative stress, acting also as a cofactor of several antioxidant enzymes: it
is the electron donor during the reduction of H_2O_2 and organic peroxides catalyzed
by glutathione peroxidases; it is conjugated to electrophilic compounds by glutathione S-transferases during xenobiotic detoxification reactions; and it is also used by

glyoxalase enzymes to neutralize highly toxic α-ketoaldehydes to d-hydroxy acids. Both nonenzymatic and enzymatic detoxification of ROS by GSH lead to oxidized glutathione, with two molecules forming a disulfide bond (GSSG). Oxidized GSSG is efficiently reconverted to GSH by glutathione reductase using NADPH as the reducing agent. The ratio GSH/GSSG can be a useful indicator of oxidative stress in cells, since GSH is normally present as 95 to 98% of total glutathione. The maintenance of the GSH/GSSG ratio, and therefore of the strong reducing environment of cells, is crucial in preventing oxidative damage (Meister 1989). When oxidative processes enhance formation of GSSG above the reducing capability of glutathione reductase, the ratio GSH/GSSG is temporarily altered before the excess of GSSG is excreted, mainly through multidrug-resistance-related protein 1 (MRP1) (Leier et al. 1996; Hayes and McLellan 1999) (see Chapter 10).

Ascorbic acid (vitamin C) is essential for several metabolic reactions, including noradrenalin biosynthesis, amino acid hydroxylation, and iron absorption (Hermes-Lima 2004). Animals, except for humans, can *de novo* synthesize ascorbate from glucose, or obtain it through the diet. Besides its metabolic functions, ascorbic acid has important antioxidant properties as a reducing agent for many ROS (H_2O_2, O_2^-, •OH, and lipid hydroperoxides); ascorbic acid is also essential in the recycling of oxidized α-tocopherol. During these reactions, ascorbic acid is converted to a radical intermediate, the semidehydroascorbyl radical, which is then recycled to ascorbic acid by dehydroascorbate reductase, an enzyme using GSH as cofactor (Figure 7.1).

Uric acid is formed during purine metabolism by oxidation of xanthine, a degradation product of both adenine and guanine; the reaction is catalyzed by xanthine oxidase or xanthine dehydrogenase. Upon oxidation, uric acid is converted to the urate radical, which can either be recycled to uric acid in a nonenzymatic reaction or converted to allantoin. Uric acid is found at high levels in some marine invertebrates like sea anemones (Regoli and Winston 1998), where it is a scavenger of activated oxygen species with an efficiency comparable to that of ascorbate (Muraoka and Miura 2003).

α-Tocopherols (vitamin E) are the major fat-soluble antioxidants localized within cell membranes. While plants can synthesize α-tocopherols, animals must take up these molecules through the diet (Hermes-Lima 2004). Antioxidant properties of vitamin E are conferred by the presence of a hydroxyl group that reacts with unpaired electrons, thus neutralizing ROS (i.e., peroxides), and protecting biological membranes from propagation of lipid peroxidation. The reaction of α-tocopherol with peroxyl radicals yields α-tocopherol radicals, which are reconverted to the functionally active form by other antioxidants, such as ascorbic acid, coenzyme Q, and carotenoids (Figure 7.1).

Carotenoids are lipid-soluble, membrane-associated antioxidants, produced by photoautotrophs as quenchers of ROS formed during photosynthetic processes. Animals do not synthesize but take up carotenoids, which represent important antioxidants against lipid peroxidation.

Compared to scavengers, which can interact with different ROS, enzymatic antioxidants catalyze specific reactions with specific substrates (Figure 7.1).

Superoxide dismutase (SOD) is a metal-dependent enzyme responsible for the removal of the superoxide anion ($2O_2^- \rightarrow O_2 + H_2O_2$). Despite being an antioxidant, SOD also represents a source of hydrogen peroxide, and it is thus necessary for the activity of this enzyme to be coordinated with that of H_2O_2 reducing enzymes. Various SOD isoforms differ for metals and subcellular localization; Mn-SOD is present in mitochondria, while Cu/Zn-SOD, mainly cytosolic, is also found in chloroplasts and peroxisomes (Nordberg and Arnér 2001).

Catalase (CAT) is a heme enzyme, mainly localized in peroxisomes, where it catalyzes the reduction of H_2O_2 to H_2O (Halliwell and Gutteridge 2007). At high H_2O_2 concentrations, the reaction involves two H_2O_2 molecules, which serve both as acceptor and as donor of hydrogen molecules, leading to the production of H_2O and O_2 ($2H_2O_2 \rightarrow 2H_2O + O_2$). This reaction proceeds at a very high rate, since catalase is an extremely active catalyst (Cadenas 1989). At low H_2O_2 concentrations, catalase also has a role in the detoxification of other substrates, as phenols and alcohols, through reactions coupled to H_2O_2 reduction. Since H_2O_2 is efficiently converted to hydroxyl radicals via the Fenton reaction, the antioxidant role of catalase is essential in lowering the formation of •OH radicals, which are poorly neutralized by other antioxidants. In this respect, catalase is considered to be one of the main defenses against the hydroxyl radical in several marine organisms, since it removes one of the main precursors. Due to its elevated conservation and wide distribution across evolutionary scale, catalase is considered to be a component of an essential system evolved to allow organisms to cope with aerobic environments (Halliwell and Gutteridge 2007).

Peroxidases are antioxidant enzymes involved in the removal of peroxides. Glutathione peroxidases (GPx) use GSH as an electron source to catalyze the reduction of H_2O_2 to H_2O and the reduction of lipid hydroperoxides to alcohols, with the concomitant oxidation of GSH to GSSG. GPx are present both in the cytosol and in the mitochondrial matrix, where catalase is absent, thus playing a critical role in removing H_2O_2 from mitochondria (Halliwell and Gutteridge 2007).

Glutathione S-transferases (GSTs) are phase II enzymes that catalyze the conjugation of GSH to various organic chemicals, both xenobiotics and endogenous aldehydic products of lipid peroxidation. Conjugated forms are more hydrophilic than their parent compounds, and thus more easily excreted through the bile or the urine. Some GST isoforms are highly reactive toward lipid peroxides, and modulated by prooxidant factors (Hayes and McLellan 1999).

Glutathione reductase (GR) is an NADPH-dependent flavoenzyme, which catalyzes the reduction of GSSG to GSH. Although GR is not a real antioxidant enzyme, nonetheless it is essential to maintain the correct GSH/GSSG ratio and the intracellular redox status in marine organisms. The reaction catalyzed by GR requires NADPH as an essential cofactor, which is mainly produced by glucose 6-phosphate dehydrogenase (G6PDH) during glucose metabolism via the pentose phosphate cycle. In this respect, the glycolytic enzyme G6PDH can also be considered to be involved in the wide system of antioxidant defenses (López-Barea et al. 1990).

DT-diaphorase, NAD(P)H: quinone oxidoreductase, is a two-electron transfer flavoprotein that catalyzes the conversion of quinones into hydroquinones, thus preventing redox cycling, and consequently quinone-dependent production of reactive oxygen species (Livingstone 2001).

Antioxidant responses have a great importance in organisms from polluted environments. In fact, the natural cellular balance between prooxidant factors and the efficiency of these defenses can be modulated or even overwhelmed by toxic chemicals, which, through different but interacting pathways, can both depress the antioxidant's capacity to remove oxyradicals and enhance the intracellular formation of ROS.

7.2 POLLUTANT-MEDIATED FORMATION OF OXYRADICALS

Among the several classes of pollutants released into the environment, trace metals can exert adverse effects through the generation of ROS as a result of their ability to lose electrons (Halliwell and Gutteridge 1984). Metals catalyze the Fenton reaction, in which a transition metal ion reacts with H_2O_2 producing $\bullet OH$ and the oxidized metal:

$$metal^{n+} + H_2O_2 \rightarrow metal^{n+1} + \bullet OH + OH^-$$

Several elements can catalyze this reaction with different efficiencies, depending on redox potential. Typical inducers are Fe(II), Cu(I), Cr(III), (IV), (V), (VI), and V(V), while in the presence of chelating agents, such as Gly-Gly-His and thiol-containing agents, Co(II), and Ni(II), also react with H_2O_2 and lipid peroxides to generate $\bullet OH$ and lipid radicals (Inoue and Kawanishi 1989; Shi et al. 1994).

In the Haber-Weiss reactions (Halliwell and Gutteridge 1984), an oxidized metal is reduced by O_2^- and then reacts with H_2O_2 to produce $\bullet OH$:

$$metal^{n+1} + O_2 \rightarrow metal^{n} + O_2$$

$$metal^{n} + H_2O_2 \rightarrow metal^{n+1} + \bullet OH + OH^-$$

Overall:

$$O_2^- + H_2O_2 \rightarrow + O_2 + \bullet OH + OH^-$$

with $(metal^{n+1}/metal^{n+})$

Haber-Weiss reactions are of particular importance during activated phagocytosis when a large amount of O_2^- is generated and a limited amount of metal is thus needed to act as catalyst (Freeman and Crap 1982).

Other mechanisms of trace-metal-mediated ROS production include the possibility that thiol radicals are generated by the reaction of some metals, e.g., Cr(VI), with cysteine. Such radicals may cause direct damage or react with other thiols to generate O_2^-, which in turn can lead to H_2O_2 (Shi and Dalal 1988).

Some metals (e.g., chromate and vanadate) are reduced by flavoenzymes, glutathione reductase, lipoil dehydrogenase, and ferroxin-NADP+ to generate active intermediates that react with O_2 to form O_2^-, while arsenite activates NADH oxidase and produces O_2^- (Halliwell and Gutteridge 1984; Livingstone 2001).

With an indirect action, metals like cadmium can represent a prooxidant stressor and increase oxyradical formation by interacting with –SH groups and depleting antioxidant defenses.

The induction of the cytochrome P450 system by aromatic compounds like polycyclic aromatic hydrocarbons (PAHs), polychlorinated biphenyls (PCBs), halogenated hydrocarbons, and dioxin and dioxin-like chemicals is another well-known source for intracellular ROS production (Stegeman and Lech 1991). Such substrates bind to P450, favoring the reduction of iron in the heme group by an electron transport chain ending with NADPH–cytochrome P450 reductase; the reaction with oxygen donates a second electron to the complex, forming a peroxide, while a substrate radical is formed, hydroxylated, and released (see Chapter 8). This process is generally very fast, but some chemicals, due to a slow oxidation, result in the uncoupling of electron transfer and oxygen reduction from substrate oxidation, thus greatly enhancing the release of ROS (Schlezinger et al. 2006). For some chemicals, reactive species are the final metabolites of the biotransformation process, as for benzo[a]pyrene, when 7,8-dihydrodiol-9,10-oxide is formed, which, adducting DNA, is responsible for the carcinogenicity of these compounds.

Several compounds, including PAH metabolites, quinones, nitroaromatics, nitroamines, and transition metal chelates, can increase the intracellular generation of ROS throughout the redox cycle (Livingstone 2001). In this cycle the compound is initially reduced to a radical intermediate with reducing equivalents typically provided by the activity of NAD(P)H reductases. The formed radical intermediates can give toxicity or, by reacting with oxygen, produce O_2^- regenerating the parent compound, which can go through another reaction cycle. The net products of these cyclic reactions are generation of O_2^- and consumption of NADPH.

Metabolism of quinones can include the activity of DT-diaphorase, which catalyzes a two-electron reduction of these compounds to hydroquinones (Burczynski and Penning 2000); however, quinones can also undergo univalent reductions that result in another pathway for ROS formation, like semiquinone radicals and the superoxide anion, particularly relevant in the metabolism of aromatic compounds in invertebrates.

Since organisms are exposed to complex mixtures of chemicals, it is highly important to figure out how various prooxidants can interact with each other and contribute to modulate the overall oxidative pressure (Figure 7.2). Some classes of polychlorinated biphenyls (PCBs) and pesticides elicit a response similar to that described for phenobarbitol (PB) and pregnenolone-16α-carbonitrile (PCN), in which two different receptors of the nuclear receptor family are involved: pregnane X receptor (PXR) and constitutively active receptor (CAR) (Honkakoski and Negishi 2000; Wei et al. 2000). Both PXR and CAR form a heterodimer with the retinoid X receptor (RXR) to activate transcription of genes of the cytochrome P450 system (CYP2B and CYP3A subfamilies), which enhance formation of ROS (Xie et al. 2000).

The majority of aromatic chemicals, e.g. dioxins (TCDD), dioxin-like compounds, polycyclic aromatic hydrocarbons (PAHs), and PCBs, interact with the cytosolic aryl

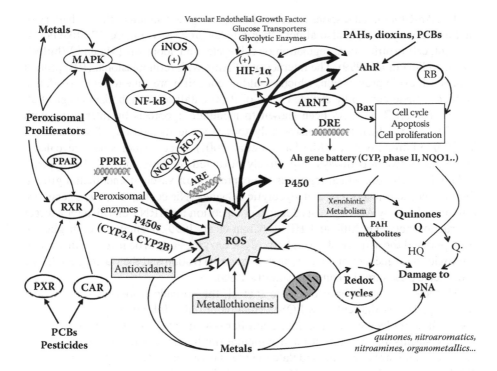

FIGURE 7.2 Main oxidative interactions between different classes of chemicals. ROS: reactive oxygen species; PAHs; polycyclic aromatic hydrocarbons; PCBs: polychlorinated biphenyls; AhR: aryl hydrocarbon receptor; ARNT: aryl hydrocarbon receptor nuclear translocator; DRE: dioxin-responsive element; NQO1: NADPH-quinone reductase 1; P450: cytochrome P450; RB: retinoblastoma protein; HIF-1α: hypoxia inducible factor 1α; ARE: antioxidant-responsive element; HO-1: heme oxygenase 1; PPAR: peroxisomal proliferator activated receptor; PPRE: peroxisomal proliferators-responsive element; RXR: retinoid X receptor; PXR: pregnane X receptor; CAR: constitutively active receptor; MAPK: mitogen-activated protein kinases; NF-kB: nuclear factor-kB; iNOS: inducible nitric oxide synthetase.

hydrocarbon (Ah) receptor, a ligand-activated transcription factor that, transported into the nucleus by the AhR nuclear translocator (ARNT), binds to specific regions of DNA (dioxin-responsive elements (DREs)) and initiates the transcription of the battery of Ah genes (Whitlock et al. 1996; Rowland and Gustafsson 1997). These genes include the P450s, CYP1A and CYP1B, UDP-glucoronosyl transferase 1AG, glutathione S-transferase Ya, class 3 aldehyde dehydrogenase, and NADPH-quinone reductase 1 (NQO1). Activation of CYP450 enzymes and of xenobiotics metabolism enzymes enhances production of ROS, semiquinone radicals, and PAH metabolites, which can form DNA adducts or generate ROS through the redox cycle, the same process also being activated by other chemical compounds (Wood et al. 1976; Di Giulio et al. 1995; Kim et al. 1998; Livingstone 2001); on the other hand, the induction of NQO1 limits the toxicity of PAH metabolites oxidized to carcinogenic quinones (Figure 7.2).

The AhR-mediated responses are not limited to the activation of the Ah battery of metabolic enzymes, but also affect other cellular processes (Figure 7.2), like development, cell proliferation, regulation of cell cycle, and apoptosis (Ma and Whitlock 1996; Kolluri et al. 1999). The interaction of AhR with retinoblastoma (RB) protein delays cell cycle progression, which would give a cell more time to repair oxidative adduct damage caused by elevated rates of P450-catalyzed metabolism (Puga et al. 1999). The AhR has also been shown to modulate apoptosis by interacting with the Bax gene, which is responsive only toward some ligands (i.e., PAHs and not TCDD) in human oocytes (Matikainen et al. 2001). Recent studies have revealed a cross-talk between pathways respectively mediated by AhR and HIF-1α (hypoxia inducible factor 1α), which regulates the expression of vascular endothelial growth factor, glucose transporters, and glycolytic enzymes (Nie et al. 2001); these genes are downregulated by AhR agonists due to a competition for ARNT and also for a direct interaction between AhR and HIF-1 (Chan et al. 1999). Deficits in both glucose transport and the activity of several glycolytic enzymes would be involved in some of the toxic consequences of dioxins, such as vascularization and the wasting syndrome after acute exposure (Liu and Matsumura 1995).

Moreover, we now have evidence that all the cascade responses of AhR and related pathways that enhance ROS formation are also inhibited by excessive levels of these molecules (Figure 7.2); a similar downregulation of cytochrome P450 by oxyradicals has important environmental implications, since other prooxidant chemicals and stressors can modulate the metabolism of AhR agonists (Barker et al. 1994; Morel and Barouki 1998). Metals, which frequently occur as co-contaminants with organic xenobiotics, can increase ROS formation through direct and indirect mechanisms (Figure 7.2), and can thus potentially reduce P450 biotransformation efficiency (Vakharia et al. 2001). Trace elements also induce expression of a battery of genes, characterized by the presence of antioxidant-responsive element (ARE), to limit oxidant challenge and oxidative damage (Dalton et al. 1999). Among these are NQO1 and heme oxygenase 1 (HO1), which transforms heme (prooxidant) in bilirubin (antioxidant), limiting the availability of heme groups for the CYP proteins (Carpenter et al. 2002); metals that induce HO1 can in this way affect the induction capability of CYP450 (Jacobs et al. 1999; Vakharia et al. 2001). A similar mechanism has been recently described in fish in which metals like cadmium and copper were shown to strongly depress the dioxin and PAH-mediated induction of cytochrome P450, mostly through oxidative effects at transcriptional, translational, and catalytic levels (Regoli et al. 2005b; Benedetti et al. 2007, 2009).

Metals modulate cellular redox status and P450 metabolism also through gene expression and activation of several signaling proteins (Figure 7.2), including MAP kinases, which induce HIF-1, the stress response transcription factor nuclear factor NF-kB, inducible nitric oxide synthase (i-NOS), HO1, and many other interacting pathways that enhance generation of ROS and negatively affect AhR-mediated gene transcription (Leonard et al. 2004).

Several organic xenobiotics like crude oil, lubricants, PAHs, PCBs, phthalate ester plasticizers, and alkylphenols have been shown to induce transcription of peroxisomal genes responsible for ROS formation (Cajaraville et al. 2003; Schrader and Fahimi 2006); these effects are mediated by peroxisomal proliferator activated receptor (PPARα), which belongs to the family of nuclear transcription factors (Figure 7.2). Furthermore, transition metals like iron and copper are abundant in peroxisomes and, released as an effect of xenobiotics, catalyze the generation of hydroxyl radicals through the Fenton reaction (Bacon and Britton 1989).

This elevated complexity of interactions between prooxidant pathways influences both the formation and the transformation rate of oxyradicals; thus, the efficiency of antioxidants is fundamental to prevent the onset of oxidative cellular damage.

7.3 RESPONSES OF ANTIOXIDANTS TO POLLUTION: SHORT-, MEDIUM-, AND LONG-TERM EXPOSURES

The variation of antioxidant concentrations can provide highly important information on the antioxidant status of an organism, and they have been often used as biomarkers of oxidative stress conditions. They can be considered early warning signals, very sensitive in revealing a prooxidant challenge, and useful in understanding mechanisms of action and relationship with prooxidant stressors. However, their variations are difficult to predict since antioxidants can be induced as a counteracting response toward an oxidative challenge, but they might also be inhibited, possibly anticipating a toxicity state, or be induced as a first phase of a response and then depleted at longer periods. All these possibilities can be expected and have been measured in different species or tissues, according to metabolic status, biological and environmental features, or the effects of other confounding factors.

Intensity and duration of exposure to environmental pollutants are key factors in determining different responses of antioxidants, with important implications when using similar biomarkers for assessing the evidence of biological disturbance in the short, medium, or long term.

Catalase, superoxide dismutase, glutathione S-transferases, and glutathione were shown to increase in various marine organisms during acute laboratory exposure to different organic and inorganic chemicals (Ahmad et al. 2004; Cheung et al. 2004; Damiens et al. 2004; Roméo et al. 2006; Ventura-Lima et al. 2007, 2009a,b; Sandrini et al. 2008; Faria et al. 2009; Silva et al. 2009; Monteiro et al. 2010), confirming the responsiveness of these parameters to cellular oxidative stress (Viarengo et al. 1989; Orbea et al. 2002, Regoli et al. 2002; Geret et al. 2003; Orbea and Cajaraville 2006). The short-term induction of catalase has been reported in early life stages of Antarctic silverfish *Pleuragramma antarcticum* exposed to 5 μg.L^{-1} benzo[a]pyrene for twenty-four hours (Regoli et al. 2005a), as well as in the Atlantic cod (*Gadus morhua*) and polar cod (*Boreogadus saida*) in response to PAHs (Sturve et al. 2006; Nahrgang et al. 2009).

On the other hand, the inhibition of the same antioxidant defenses has also been described in several laboratory exposures (Orbea et al. 2002; Vijayavel et al. 2004; Regoli et al. 2005b; Benedetti et al. 2007, 2009; Ahmad et al. 2009; Lushchak et al. 2009a,b). Vieira and coauthors (2008) indicated a different response of catalase, superoxide dismutase, and glutathione-dependent enzymes (GST, GR, GPx), when the common goby *Pamatoschistus microps* was exposed to benzo[a]pyrene, anthracene, or a complex mixture of fuel oil.

Other authors have reported contrasting results for different antioxidant parameters measured in mussels (*Mytilus galloprovincialis*) after very short-term exposures (few hours or few days) to several pollutants; while low molecular weight scavengers such as glutathione decreased in concentration, an increase of antioxidant enzyme activities was observed (Canesi 1999; Gravato et al. 2005).

Both inductions and decreases of antioxidant parameters were detected in organisms caged for relatively short periods (four to six weeks) at polluted sites, such as harbors, offshore platforms, or industrial areas (Cossu et al. 1997; Frenzilli et al. 2004; Regoli et al. 2004; Pampanin et al. 2005; Damiens et al. 2007; Bocchetti et al. 2008; Gorbi et al. 2008; Martín-Díaz et al. 2008; Pytharopoulou et al. 2008). In such conditions, organisms can react with a physiological adjustment of antioxidant levels to increased environmental oxidative pressure (Da Ros et al. 2000; Roméo et al. 2003a,b; Ahmad et al. 2006; Vlahogianni et al. 2007; Bocchetti et al. 2008; Maria et al. 2009a,b; Giarratano et al., 2010). Variations of antioxidant parameters in mussels, *M. galloprovincialis*, transplanted in a harbor area during dredging activities revealed various counteractive responses to prooxidant compounds in differently impacted areas or in different phases of such operations (Bocchetti et al. 2008).

The crab *Carcinus maenas* and the clam *Ruditapes philippinarum*, used as bioindicators to assess the effects of sediment contamination in four Spanish harbors, showed significant variations of oxidative parameters, with glutathione S-transferases and glutathione reductase discriminating different degrees of contamination (Martín-Díaz et al. 2008).

The sensitivity and activation sequence of various antioxidant defenses by pollutants were characterized in mussels, *M. galloprovincialis*, deployed for four weeks in a highly polluted harbor and collected on different days (Regoli et al. 2004). Results showed a biphasic trend for several antioxidants (catalase, glutathione S-transferases, glutathione peroxidases, glutathione reductase, total antioxidant capacity); no variations or increases of these defense biomarkers were observed during the first two weeks of exposure to the polluted site, followed by a progressive decrease up to severe depletion in the final part of the experiment, accompanied by the appearance of cell damage. These findings suggested an initial counteractive response of mussels to the enhanced prooxidant challenge, reflecting an adaptation or compensatory reaction toward ROS formation to overcome the stressful condition; however, antioxidants appeared overwhelmed over longer exposure periods (Regoli et al. 2004), making the organism more vulnerable to the toxicity of ROS and susceptible to oxidative stress (Regoli 2000).

Transient variations of antioxidants have also been observed in the gills of pearl oysters (*Pinctada fucata*) during acute exposure to copper (Jing et al. 2006); glutathione peroxidases and superoxide dismutase significantly decreased after twelve to twenty-four hours, while a recovery or even induction of these enzymes was observed after forty-eight to seventy-two hours. An antioxidant recovery was also reported in *M. galloprovincialis* caged in a metal-polluted area with a significant inhibition of antioxidant efficiency after four weeks, but a recovery after eight weeks (Regoli 2000).

All these results highlight that individual antioxidants do not vary in a synchronous way, and it is quite common that while some parameters may be induced, others do not vary or may even be depleted. Moreover, variations of antioxidants are rarely dose dependent, which is not surprising considering the high complexity of interactions with direct and indirect effects. It is impossible to identify a threshold of variation for these defenses above which an oxidative toxicity can be predicted. Alterations of individual antioxidants often do not correlate with the onset of cell damage, potentially suggesting an adaptation to contaminant exposure (Box et al. 2007; Lushchak et al. 2009b; Ventura–Lima et al. 2009b). In this respect, polychaetes exposed to 100 μg.L^{-1} of Cd revealed a significant enhancement of ROS formation and induction of antioxidants (SOD, GST), but no changes in lipid peroxidation (Sandrini et al. 2008). Analyses of antioxidants can be profitably integrated with the measurement of total oxyradical scavenging capacity (TOSC), which quantifies the overall cellular resistance toward different ROS like peroxyl radicals (ROO·), hydroxyl radicals (HO·), and peroxynitrite (HOONO) (Regoli and Winston 1999; Regoli 2000). Compared to individual defense biomarkers, TOSC is less sensitive, but has a greater prognostic value since an impaired capability to neutralize ROS has been associated with the onset of various forms of oxidative damage like lysosomal alterations and genotoxic damage (Regoli 2000; Frenzilli et al. 2001; Gorbi and Regoli 2003; Regoli et al. 2004). Similar relationships between multiple biomarkers allowed development of an integrated model to predict animal health condition by using lysosomal membrane stability as an index of cellular well-being (Moore et al., 2006).

Oxidative parameters respond also to temporary impacts in the medium term and can thus be useful for monitoring the progression or regression of a chemical impact. In the period 1997–2000, approximately 1,800,000 m^3 of materials dredged from the Port of Livorno, Italy, were discharged into the sea at a site 14 miles offshore. The red mullet (*Mullus barbatus*) was used as a bioindicator species for monitoring the onset of biological alterations caused by these discharges on a geographical and temporal scale (Regoli et al. 2002). Results obtained from several biomarkers indicated a biological impact in organisms sampled near the disposal site during the period of maximum intensity of dredging/disposal operations. These effects were mainly related to organic xenobiotics, as suggested by increased biotransformation activity, and caused significant variations in the levels and activities of several antioxidants. The analysis of total oxyradical scavenging capacity, however, revealed that the overall efficiency of tissues to absorb various oxidants was not seriously compromised when challenged by the increased prooxidant pressures. Variations of antioxidants were thus useful in revealing early warning and sensitive biological responses, while integration with TOSC analysis allowed evaluation of whether such changes could also determine a more integrated and functional biological effect with possible consequences at the organism level (Regoli et al. 2002).

Mussels (*Mytilus galloprovincialis*) sampled over eleven months, respectively before, during, and after harbor dredging operations, confirmed the possibility of biphasic responses of antioxidant defenses: after a significant induction of catalase and glutathione S-transferases in the first phase of operations, an inhibition of these antioxidant enzymes was observed during the second period of dredging, with a reduced total capability to neutralize ROS (Bocchetti et al. 2008). Similar results represent additional evidence that antioxidant responses are modulated by the duration and intensity of chemical disturbance (Stephensen et al. 2000; Regoli et al. 2002, 2004, 2005b).

Antioxidant parameters have been used as biomarkers even for monitoring the medium-term biological impact of oil spills. A follow-up study (Martínez-Gómez et al. 2009) was carried out on two demersal fish species (*Lepidorhmombus boscii* and *Callionymus lyra*) after the *Prestige* oil spill, which occurred in 2002 and affected the northern coast of Spain. Several biomarkers were monitored over three consecutive years (2003–2005) in both fish species, and various antioxidant enzymes showed significant variation in the period following the oil spill. In particular, catalase, glutathione reductase, and glutathione S-transferases in *L. boscii*, and glutathione reductase in *C. lyra*, were significantly decreased in the second part of the study, respectively two and three years after the accident, in most of the analyzed areas; such temporal variations of enzymatic activities were attributed to a decreasing oxidative challenge due to lower exposure to spilled PAHs, thus suggesting a recovery to baseline levels existing before the oil spill (Martínez-Gómez et al. 2009).

Responses of antioxidants after short-term or medium-term exposures (from days/weeks to a few years) may significantly differ from those obtained in field studies on organisms exposed to chronic pollution (several years to decades). Nonetheless, even in such populations, variable responses of these defenses can be expected. Levels of antioxidants constantly higher than in control organisms reflect the necessity to maintain greater protection, while recovery to comparable antioxidant efficiency between polluted and reference populations would indicate the occurrence of adaptive or counteractive mechanisms toward the environmental prooxidant challenge. Finally, when certain antioxidants remain depleted upon long-term exposures to prooxidant chemicals, two different scenarios can be hypothesized. In the first, the reduced antioxidants have only a limited effect on the overall antioxidant efficiency, and the organism's health condition will not be affected; second, if depressed defenses reflect an inability to counteract toxicity of ROS, then the occurrence of oxidative damage can be expected with potential consequences on organism or population performances.

The effects of chronically high levels of trace metals on antioxidant responses and the implications for the proper use of these biomarkers were investigated comparing two natural populations of mussels, *M. galloprovincialis*, from a polluted and a clean site, and mussels transplanted from the control to the polluted environment (Regoli and Principato 1995). Native mussels from this site were characterized by high tissue levels of As, Cu, Fe, Mn, Pb, and Se of industrial origin and, compared to control organisms, exhibited a lower content of glutathione and higher activities of glyoxalase enzymes: the same antioxidants showed similar results in caged mussels, suggesting the absence of time-dependent adaptations for chronic exposure (native polluted populations) compared to acute conditions (caged mussels). On the other hand, the activities of superoxide dismutase, catalase, glutathione peroxidases, and glutathione reductase were similar in chronically polluted and reference control organisms, a variation being observed during the acute exposure of translocated mussels (i.e., 75% inhibition of catalase after four weeks): these findings suggested the occurrence of adaptive mechanisms in chronically polluted environments where organisms may not give an expected biological response. Therefore, antioxidant defenses in chronically polluted organisms reveal both adapted and nonadapted responses (Regoli and Principato 1995).

The contemporary depletion and induction of different antioxidants in chronically polluted organisms were difficult to summarize in terms of health condition and susceptibility to oxidative stress. When the results on individual antioxidants were integrated with measurement of the total oxyradical scavenging capacity (TOSC), polluted organisms revealed a significantly lower capability to neutralize several reactive oxygen species, which correlated with higher levels of lipid peroxidation and lysosomal destabilization (Regoli 2000). The total antioxidant efficiency exhibited a sharp decline in control mussels transplanted for four weeks, while a certain recovery was observed after eight weeks, with comparable TOSC values in caged and natural polluted mussels. The greater initial reduction of TOSC values may reflect a high consumption of low molecular weight antioxidants during the first phase of exposure, confirming the temporary decrease observed for some antioxidant enzymes (including catalase) before compensatory mechanisms occur at longer (i.e., chronic) exposures (Regoli and Principato 1995). Overall, these data demonstrated that although some antioxidants possess a certain capability to recover after the acute phase, high levels of pollutants like trace metals may have a long-term adverse effect on exposed organisms that appear less protected from the toxicity of oxyradicals and are more susceptible to oxidative damage (Regoli and Principato 1995).

The previously described examples represent only a small percentage of the scientific evidence available on the crucial role of antioxidant defenses in determining resistance against pollution in organisms subjected to prolonged exposure; on the other hand, it is not easy to distinguish whether a similar resistance is acquired at physiological or genetic levels. If the increased tolerance is obtained via physiological mechanisms, the advantages obtained for the single individuals are not inheritable by the offspring, and must be acquired again through exposure. On the other hand, genetic adaptation occurs when the population has evolved resistance through the action of natural selection on genetically based individual variation (Klerks and Weis 1987): in these conditions, contaminants may act as selective agents, allowing the survival only of the most tolerant individuals and the formation of adapted populations.

To investigate physiologically and genetically acquired resistance, a field study has been carried out on Atlantic killifish (*Fundulus heteroclitus*) inhabiting a Superfund site on the Elizabeth River, Virginia (Meyer et al. 2003). The local population is tolerant of the acutely toxic sediments of the area, and such resistance appears based on both genetic adaptation, heritable for multiple generations, and physiological acclimation. Considering the elevated concentrations of prooxidant contaminants in the sediment, antioxidant defenses were analyzed in two generations of laboratory-raised descendants of killifish from Elizabeth River and from a reference site. In the resistant population, exposure to Elizabeth River sediments caused a significant increase of several antioxidant defenses (glutathione, glutathione reductase, glutathione peroxidases). These responses likely acted as protective agents against the acute toxicity of polluted sediments, thus representing part of the short-term process of physiological adaptation. However, other antioxidant parameters, particularly the total oxyradical scavenging capacity (TOSC), had constitutively higher levels in both the F1 and F2 generations of the Elizabeth River population than in the nonresistant population, thus identifying the tolerance as heritable. Such heritable (multigenerational) upregulation of the total antioxidant capacity, and the short-term induction of some antioxidant enzymes play an important role in the resistance to toxic effects of highly polluted sediments, contributing to survival of the killifish population in the Elizabeth River (Meyer et al. 2003) (see Chapter 8).

An additional example of antioxidant-mediated resistance to pollution has been reported in wild populations of brown trout (*Salmo trutta*), sampled from two rivers of the Roros area (Norway), chronically affected by metal contamination as a result of zinc and copper mining activities (Hansen et al. 2006). Activity and gene expression of various antioxidant enzymes were compared in these two populations and in a reference one. Superoxide dismutase and catalase exhibited tissue-dependent differences with generally higher enzymatic activities in the contaminated populations, potentially representing adaptive responses of brown trout tolerant to chronic metal pollution. Significant variations were measured also for gene expression of various antioxidant enzymes, although results on gene transcript levels were not consistent with those on enzymatic activities. Based on the absence of correlation between antioxidant variations at transcriptional and catalytic levels, the authors proposed as more probable a physiological mechanism for the oxidative tolerance to metals in these populations (Hansen et al. 2006).

An evident upregulation of antioxidant gene expression has also been detected in juvenile Coho salmon (*Oncorhynchys kisutch*) from Prince William Sound, Alaska (Roberts et al. 2006). This area has been affected by PAH contamination, resulting from the *Exxon Valdez* oil spill (which occurred fifteen years before the study) and other sources. The transcript levels of superoxide dismutase and glutathione S-transferases were significantly higher in gills of fish living in the contaminated sites than in control ones, providing further evidence of an inducible antioxidant-mediated adaptation mechanism (Roberts et al. 2006).

In some environments, organisms are naturally exposed to elevated and highly variable conditions of chemical-mediated oxidative challenge. Among these, the hydrothermal vents are characterized by extremely high temperature, low oxygen, and elevated concentrations of potentially harmful chemical species, mainly hydrogen sulfide, methane, and trace metals. The vent mussel, *Bathymodiolus azoricus*, is a typical inhabitant of the Mid-Atlantic Ridge hydrothermal vents, and antioxidant defenses were found to have an important role in adaptation to variable prooxidant pressure at various vent sites (Bebianno et al. 2005). Comparison of mussels from three different vent fields revealed the highest total antioxidant capacity and the lowest lipid peroxidation in organisms from Menez-Gwen field, exposed to relatively low environmental concentrations of metals. On the other hand, at Lucky Strike and especially Rainbow fields, particularly rich in Zn, Cd, Ag, Fe, and Mn, mussels exhibited reduced capability to neutralize reactive oxygen species, suggesting that organisms from these sites are more susceptible to oxidative stress (Bebianno et al. 2005). However, physiological adjustments of antioxidant levels may occur in *B. azoricus* upon variation of environmental prooxidant pressure, as demonstrated with transplantation experiments. Mussels from the more toxic Rainbow field, after two weeks' deployment at Lucky Strike, exhibited a significant induction of superoxide dismutase and glutathione peroxidases, and inhibition of catalase to values comparable to those of the resident population (Company et al. 2007). These results demonstrated the role of the environmental oxidative challenge in determining the physiological antioxidant characteristics of resident mussels, and also their capacity to adapt to external changes (Company et al. 2007).

Some species seem to be naturally more protected against oxidative damage as a result of distinct antioxidant features allowing more efficient adaptation to chronic pollution. Mussels (*Mytilus galloprovincialis*) and oysters (*Crassostrea angulata*) from the Spanish South Atlantic littoral are exposed to high levels of metals and have been shown to differ significantly in their response to such contamination (Funes et al. 2006). Relationships between metal levels, antioxidant defenses, and onset of oxidative damage revealed that the oysters are better adapted to copper and zinc pollution than the mussels. Despite the much higher accumulation of these elements, oysters are more efficiently protected from oxidative stress, showing increased antioxidant activities (superoxide dismutase, catalase, glutathione peroxidases, glutathione S-transferases) and less extensive oxidative damage (measured as malondialdehyde content). In contrast, the mussels accumulated much lower metal concentrations, but revealed an insufficient response of antioxidant defense mechanisms and a clear increase in oxidized biomolecules (malondialdehyde), indicating an impaired ability to scavenge ROS and protect cells from oxidative damage. These differences could explain the absence of the mussels from the most polluted sites, while the increase of antioxidant defenses in the oysters may confer an ecological advantage and explain the capability of these bivalves to accumulate metals at very high concentrations without apparent signs of oxidative damage (Funes et al. 2006).

7.4 CONCLUSIONS

Interactions between prooxidant factors and antioxidant defenses are very complex but fundamental in adaptation to stressful conditions and in modulating metabolism, biological effects, and toxicity of pollutants. Environmental chemicals enhance intracellular ROS formation, and exposures to complex mixtures cause reciprocal interactions between various oxidative pathways influencing both ROS generation and cascade oxidative responses.

Individual antioxidants are often sensitive in revealing a pollutant-mediated prooxidant challenge, and thus are useful as early warning biomarkers. However, their level variations are difficult to predict and are of questionable biological significance, especially when both induction and inhibition of different defenses occur simultaneously. Integration with analysis of total antioxidant capacity increases the ecological relevance of oxidative stress investigations since the reduced capability to neutralize ROS increases the susceptibility of an organism to oxidative stress and the appearance of adverse effects on an organism's health condition.

Intensity and duration of exposure to environmental pollutants (from days to decades) are key factors in adaptation to prooxidant conditions: the antioxidant status of exposed organisms can reveal various degrees of enhanced ROS toxicity or, conversely, reflect an acquired tolerance through physiological plasticity of antioxidant responses or a genetic modification of constitutive antioxidant efficiency.

REFERENCES

Ahmad, I., M. Pacheco, and M. A. Santos. 2004. Enzymatic and nonenzymatic antioxidants as an adaptation to phagocyte-induced damage in *Anguilla anguilla* L. following *in situ* harbour water exposure. *Ecotoxicol. Environ. Safe.* 57:290–302.

Ahmad, I., M. Pacheco, and M. A. Santos. 2006. *Anguilla anguilla* L. oxidative stress biomarkers: An *in situ* study of freshwater wetland ecosystem (Pateira de Fermentelos, Portugal). *Chemosphere* 65:952–62.

Ahmad, I., et al. 2009. Juvenile sea bass (*Dicentrarchus labrax* L.) enzymatic and nonenzymatic antioxidant responses following 17β-estradiol exposure. *Ecotoxicology* 18:974–82.

Bacon, B. R., and R. S. Britton 1989. Hepatic injury in chronic iron overload. Role of lipid peroxidation. *Chem. Biol. Interact.* 70:183–226.

Barker, C. W., J. B. Fagan, and D. S. Fasco. 1994. Down-regulation of P4501A1 and P450 1A2 in isolated hepatocytes by oxidative stress. *J. Biol. Chem.* 269:3985–90.

Bebianno, M. J., et al. 2005. Antioxidant systems and lipid peroxidation in *Bathymodiolus azoricus* from Mid-Atlantic Ridge hydrothermal vent fields. *Aquat. Toxicol.* 75:354–73.

Benedetti, M., et al. 2007. Oxidative and modulatory effects of trace metals on metabolism of polycyclic aromatic hydrocarbons in the Antarctic fish *Trematomus bernacchii*. *Aquat. Toxicol.* 85:167–75.

Benedetti, M., et al. 2009. Interactions between trace metals (Cu, Hg, Ni, Pb) and 2,3,7,8-tetrachlorodibenzo-*p*-dioxin in the Antarctic fish *Trematomus bernacchii*: Oxidative effects on biotransformation pathway. *Environ. Toxicol. Chem.* 28:818–25.

Bocchetti, R., et al. 2008. Contaminant accumulation and biomarker responses in caged mussels, *Mytilus galloprovincialis*, to evaluate bioavailability and toxicological effects of remobilized chemicals during dredging and disposal operations in harbour areas. *Aquat. Toxicol.* 89:257–66.

Box, A., et al. 2007. Assessment of environmental pollution at Balearic Islands applying oxidative stress biomarkers in the mussel *Mytilus galloprovincialis*. *Comp. Physiol. Biochem.* 146C:531–39.

Burczynski, M. E., and T. M. Penning. 2000. Genotoxic polycyclic aromatic hydrocarbon *ortho*-quinones generated by aldoketo reductases induce CYP1A1 via nuclear translocation of the aryl hydrocarbon receptor. *Cancer Res.* 60:908–15.

Cadenas, E. 1989. Biochemistry of oxygen toxicity. *Annu. Rev. Biochem.* 58:79–110.

Cajaraville, M. P., et al. 2003. Peroxisome proliferation as a biomarker in environmental pollution assessment. *Microsc. Res. Technol.* 61:191–202.

Canesi, L. 1999. Heavy metals and glutathione metabolism in mussel tissues. *Aquat. Toxicol.* 46:67–76.

Carpenter, D. O., K. Arcaro, and D. C. Spink. 2002. Understanding the human health effects of chemical mixtures. *Environ. Health Persp.* 110:25–42.

Chan, W. K., et al. 1999. Cross-talk between the aryl hydrocarbon receptor and hypoxia inducible factor signaling pathways—Demonstration of competition and compensation. *J. Biol. Chem.* 274:12115–23.

Cheung, C. C. C., et al. 2004. Antioxidant responses to benzo[*a*]pyrene and aroclor 1254 exposure in the green-lipped mussels, *Perna viridis*. *Environ. Pollut.* 128:393–403.

Company, R., et al. 2007. Adaptation of the antioxidant defence system in hydrothermal-vent mussels (*Bathymodiolus azoricus*) transplanted between two Mid-Atlantic Ridge sites. *Mar. Ecol.* 28:93–99.

Cossu, C., et al. 1997. Glutathione reductase, selenium-dependent glutathione peroxidase, glutathione levels, and lipid peroxidation in freshwater bivalves, *Unio tumidus*, as biomarkers of aquatic contamination in field studies. *Ecotoxicol. Environ. Safe.* 38:122–31.

Dalton, T. P., H. G. Shertzer, and A. Puga. 1999. Regulation of gene expression by reactive oxygen. *Annu. Rev. Pharmacol. Toxicol.* 39:67–101.

Damiens, G., et al. 2004. Evaluation of biomarkers in oyster larvae in natural and polluted conditions. *Comp. Biochem. Physiol.* 138C:121–28.

Damiens, G., et al. 2007. Integrated biomarker response index as a useful tool for environmental assessment evaluated using transplanted mussels. *Chemosphere* 66:574–83.

Da Ros, L., et al. 2000. Biomarkers and trace metals in the digestive gland of indigenous and transplanted mussels, *Mytilus galloprovincialis*, in Venice lagoon, Italy. *Mar. Environ. Res.* 50:417–23.

Di Giulio, R. T., et al. 1995. Biochemical mechanisms: Metabolism, adaptation, and toxicity. In *Fundamentals of aquatic toxicology. Effects, environmental fate and risk assessment*, ed. G. M. Rand, 523–61. Washington, DC: Taylor & Francis.

Faria, M., et al. 2009. Multi-biomarker responses in the freshwater mussel *Dreissena polymorpha* exposed to polychlorobiphenyls and metals. *Comp. Biochem. Physiol.* 149C:281–88.

Freeman, B. A., and J. D. Crap. 1982. Biology and disease: Free radicals and disease injury. *Lab. Invest.* 47:412–26.

Frenzilli, G., et al. 2001. DNA integrity and total oxyradical scavenging capacity in the Mediterranean mussel, *Mytilus galloprovincialis*: A field study in a highly eutrophicated coastal lagoon. *Aquat. Toxicol.* 53:19–32.

Frenzilli, G., et al. 2004. Time-course evaluation of ROS-mediated toxicity in mussels, *Mytilus galloprovincialis*, during a field translocation experiment. *Mar. Environ. Res.* 58:609–13.

Funes, V., et al. 2006. Ecotoxicological effects of metal pollution in two mollusc species from the Spanish South Atlantic littoral. *Environ. Pollut.* 139:214–23.

Geret, F., A. Serafim, and M. J. Bebianno. 2003. Antioxidant enzyme activities, metallothioneins and lipid peroxidation as biomarkers in *Ruditapes decussatus*. *Ecotoxicology* 12:417–26.

Giarratano, E., C. A. Duarte, and O. A. Amin. 2010. Biomarkers and heavy metal bioaccumulation in mussels transplanted to coastal waters of the Beagle Channel. *Ecotoxicol. Environ. Safe.* 73:270–79.

Gorbi, S., and F. Regoli. 2003. Total oxyradical scavenging capacity as an index of susceptibility to oxidative stress in marine organisms. *Comm. Toxicol.* 9:303–22.

Gorbi, S., et al. 2008. An ecotoxicological protocol with caged mussels, *Mytilus galloprovincialis*, for monitoring the impact of an offshore platform in the Adriatic Sea. *Mar. Environ. Res.* 157:3166–73.

Gravato, C., M. Oliveira, and M. A. Santos. 2005. Oxidative stress and genotoxic responses to resin acids in Mediterranean mussels. *Ecotoxicol. Environ. Safe.* 61:221–29.

Halliwell, B., and J. M. C. Gutteridge. 1984. Oxygen toxicity, oxygen radicals, transition metals, and disease. *J. Biochem.* 219:1–14.

Halliwell, B., and J. M. C. Gutteridge. 2007. *Free radicals in biology and medicine.* 4th ed. New York: Oxford University Press.

Hansen, B. H., et al. 2006. Antioxidative stress proteins and their gene expression in brown trout (*Salmo trutta*) from three rivers with different heavy metal levels. *Comp. Biochem. Physiol.* 143C:263–74.

Hayes, J. D., and L. I. McLellan. 1999. Glutathione and glutathione-dependent enzymes represent a co-ordinately regulated defence against oxidative stress. *Free Radical Res.* 31:273–300.

Hermes-Lima, M. 2004. Oxygen in biology and biochemistry: Role of free radicals. In *Functional metabolism: Regulation and adaptation*, ed. K. B. Storey, 319–68. Hoboken, NJ: John Wiley & Sons.

Honkakoski, P., and M. Negishi. 2000. Regulation of cytochrome P450 (CYP) genes by nuclear receptors. *Biochem. J.* 347:321–37.

Inoue, S., and S. Kawanishi. 1989. ESR evidence for superoxide, hydroxyl radicals and singlet oxygen produced from hydrogen peroxide and nickel(II) complex of glycylglycyl-L-histidine. *Biochem. Biophys. Res. Commun.* 159:445–51.

Jacobs, J. M., et al. 1999. Effect of arsenite on induction of CYP1A1, CYP2B, and CYP3A in primary cultures of rat hepatocytes. *Toxicol. Appl. Pharm.* 157:51–59.

Jing, G., et al. 2006. Metal accumulation and enzyme activities in gills and digestive gland of pearl oyster (*Pinctada fucata*) exposed to copper. *Comp. Biochem. Physiol.* 144C:184–90.

Kim, J. K., et al. 1998. Metabolism of benzo[a]pyrene and benzo[a]pyrene-7,8-diol by human cytochrome P450 1B1. *Carcinogenesis* 19:1847–53.

Klerks, P. L., and J. S. Weis. 1987. Genetic adaptation to heavy metals in aquatic organisms: A review. *Environ. Pollut.* 45:173–205.

Kolluri, S. K., et al. 1999. p27[Kip1] induction and inhibition of proliferation by intracellular Ah receptor in developing thymus and hepatoma cells. *Gene Dev.* 13:1742–53.

Leier, I., et al. 1996. ATP-dependent glutathione disulphide transport mediated by the MRP gene-encoded conjugated export pump. *Biochem. J.* 314:433–37.

Leonard, S. S., G. K. Harris, and X. Shi. 2004. Metal-induced oxidative stress and signal transduction. *Free Radical. Biol. Med.* 37:1921–42.

Liu, P. C. C., and F. Matsumura. 1995. Differential effects of 2,3,7,8-tetrachlorodibenzo-*p*-dioxin on the adipose-type and brain-type glucose transporters in mice. *Mol. Pharmacol.* 47:65–73.

Livingstone, D. R. 2001. Contaminant-stimulated reactive oxygen species production and oxidative damage in aquatic organisms. *Mar. Pollut. Bull.* 42:656–66.

López-Barea, J., et al. 1990. Structure, mechanism, functions and regulatory properties of glutathione reductase. In *Glutathione: Metabolism and physiological functions*, ed. J. Vina, 105–16. Boca Raton FL: CRC Press.

Lushchak, O. V., et al. 2009a. Chromium(III) induces oxidative stress in goldfish liver and kidney. *Aquat. Toxicol.* 93:45–52.

Lushchak, O. V., et al. 2009b. Low toxic herbicide Roundup induces mild oxidative stress in goldfish tissues. *Chemosphere* 76:932–37.

Ma, Q., and J. P. Whitlock Jr. 1996. The aromatic hydrocarbon receptor modulates the Hepa 1c1c7 cell cycle and differentiated state independently of dioxin. *Mol. Cell. Biol.* 16:2144–50.

Maria, V. L., M. A. Santos, and M. J. Bebianno. 2009b. Biomarkers of damage and protection in *Mytilus galloprovincialis* cross transplanted in Ria Formosa Lagoon (Portugal). *Ecotoxicology* 18:1018–28.

Maria, V. L., et al. 2009a. Wild juvenile *Dicentrarchus labrax* L. liver antioxidant and damage responses at Aveiro Lagoon, Portugal. *Ecotoxicol. Environ. Safe.* 72:1861–70.

Martín-Díaz, M. L., et al. 2008. Field validation of a battery of biomarkers to assess sediment quality in Spanish ports. *Environ. Pollut.* 151:631–40.

Martínez-Gómez, C., et al. 2009. Evaluation of three-year monitoring with biomarkers in fish following the *Prestige* oil spill (N Spain). *Chemosphere* 74:613–20.

Matikainen, T., et al. 2001. Aromatic hydrocarbon receptor-driven Bax gene expression is required for premature ovarian failure caused by biohazardous environmental chemicals. *Nat. Genet.* 28:355–60.

Meister, A. 1989. On the biochemistry of glutathione. In *Glutathione centennial. Molecular perspectives and clinical implications*, ed. N. Taniguchi et al., 3–21. San Diego: Academic Press.

Meyer, J. N., et al. 2003. Antioxidant defenses in killifish (*Fundulus heteroclitus*) exposed to contaminated sediments and model prooxidants: Short-term and heritable responses. *Aquat. Toxicol.* 65:377–95.

Monteiro, D. A., F. T. Rantin, and A. L. Kalinin. 2010. Inorganic mercury exposure: Toxicological effects, oxidative stress biomarkers and bioaccumulation in the tropical freshwater fish matrinxã, *Brycon amazonicus* (Spix and Agassiz, 1829). *Ecotoxicology* 19:105–23.

Moore, M. N., J. I. Allen, and A. McVeigh 2006. Environmental prognostics: An integrated model supporting lysosomal stress responses as predictive biomarkers of animal health status. *Mar. Environ. Res.* 61:278–304.

Morel, Y., and R. Barouki. 1998. Down-regulation of cytochrome P4501A1 gene promoter by oxidative stress-critical contribution of nuclear factor 1. *J. Biol. Chem.* 273:26969–76.

Muraoka, S., and T. Miura. 2003. Inhibition by uric acid of free radicals that damage biological molecules. *Pharmacol. Toxicol.* 93:284–89.

Nahrgang, J., et al. 2009. PAH biomarker responses in polar cod (*Boreogadus saida*) exposed to benzo[a]pyrene. *Aquat. Toxicol.* 94:309–19.

Nie, M., A. L. Blankenship, and J. P. Giesy. 2001. Interaction between aryl hydrocarbon receptor (AhR) and hypoxia signaling pathways. *Environ. Toxicol. Pharm.* 10:17–27.

Nordberg, J., and S. J. Arnér. 2001. Reactive oxygen species, antioxidants, and the mammalian thioredoxin system. *Free Radical Biol. Med.* 31:1287–312.

Orbea, A., and M. P. Cajaraville. 2006. Peroxisome proliferation and antioxidant enzymes in transplanted mussels of four Basque estuaries with different levels of polycyclic aromatic hydrocarbon and polychlorinated biphenyl pollution. *Environ. Toxicol. Chem.* 25:1616–26.

Orbea, A., M. O. Zarragoitia, and M. P. Cajaraville. 2002. Interactive effects of benzo[a]pyrene and cadmium and effects of di(2-ethylhexyl) phthalate on antioxidant and peroxisomal enzymes and peroxisomal volume density in the digestive gland of mussel *Mytilus galloprovincialis* Lmk. *Biomarkers* 1:33–48.

Pampanin, D. M., et al. 2005. Susceptibility to oxidative stress of mussels (*Mytilus galloprovincialis*) in the Venice lagoon (Italy). *Mar. Pollut. Bull.* 50:1548–57.

Puga, A., et al. 1999. Aromatic hydrocarbon receptor interaction with the retinoblastoma protein potentiates repression of E2F-dependent transcription and cell cycle arrest. *J. Biol. Chem.* 275:2943–50.

Pytharopoulou, S., et al. 2008. Translational responses of *Mytilus galloprovincialis* to environmental pollution: Integrating the responses to oxidative stress and other biomarker responses into a general stress index. *Aquat. Toxicol.* 89:18–27.

Regoli, F. 2000. Total oxyradical scavenging capacity (TOSC) in polluted and translocated mussels: A predictive biomarker of oxidative stress. *Aquat. Toxicol.* 50:351–61.

Regoli, F., and G. Principato. 1995. Glutathione, glutathione-dependent and antioxidant enzymes in mussels, *Mytilus galloprovincialis*, exposed to metals under field and laboratory conditions: Implications for the use of chemical biomarkers. *Aquat. Toxicol.* 31:143–64.

Regoli, F., and G. W. Winston. 1998. Application of a new method for measuring the total oxyradical scavenging capacity in marine invertebrates. *Mar. Environ. Res.* 46:439–42.

Regoli, F., and G. W. Winston. 1999. Quantification of total oxidant scavenging capacity of antioxidants for peroxynitrite, peroxyl radicals, and hydroxyl radicals. *Toxicol. Appl. Pharm.* 156:96–105.

Regoli, F., et al. 2002. Application of biomarkers for assessing the biological impact of dredged materials in the Mediterranean: The relationship between antioxidant responses and susceptibility to oxidative stress in the red mullet (*Mullus barbatus*). *Mar. Pollut. Bull.* 44:912–22.

Regoli, F., et al. 2004. Time-course variations of oxyradical metabolism, DNA integrity and lysosomal stability in mussels, *Mytilus galloprovincialis*, during a field translocation experiment. *Aquat. Toxicol.* 68:167–78.

Regoli, F., et al. 2005a. Antioxidant efficiency in early life stages of the Antarctic silverfish, *Pleuragramma antarcticum*: Responsiveness to pro-oxidant conditions of platelet ice and chemical exposure. *Aquat. Toxicol.* 75:43–52.

Regoli, F., et al. 2005b. Interactions between metabolism of trace metals and xenobiotic agonists of the aryl hydrocarbon receptor in the Antarctic fish *Trematomus bernacchii*: Environmental perspectives. *Environ. Toxicol. Chem.* 24:1475–82.

Roberts, A. P., J. T. Oris, and W. A. Stubblefield. 2006. Gene expression in caged juvenile Coho salmon (*Oncorhynchys kisutch*) exposed to the waters of Prince William Sound, Alaska. *Mar. Pollut. Bull.* 52:1527–32.

Roméo, M., et al. 2003a. Mussel transplantation and biomarkers as useful tools for assessing water quality in the NW Mediterranean. *Environ. Pollut.* 122:369–78.

Roméo, M., et al. 2003b. Multimarker approach in transplanted mussels for evaluating water quality in Charentes, France, coast areas exposed to different anthropogenic conditions. *Environ. Toxicol.* 18:295–305.

Roméo, M., et al. 2006. Responses of *Hexaplex* (*Murex*) *trunculus* to selected pollutants. *Sci. Total Environ.* 359:135–44.

Rowland, J. C., and J. A. Gustafsson. 1997. Aryl hydrocarbon receptor mediated signal transduction. *Crit. Rev. Toxicol.* 27:109–34.

Sandrini, J. Z., et al. 2008. Antioxidant responses in the Nereid *Laeonereis acuta* (*Annelida, Polychaeta*) after cadmium exposure. *Ecotoxicol. Environ. Safe.* 70:115–20.

Schlezinger, J. J., et al. 2006. Uncoupling of cytochrome P450 1A and stimulation of reactive oxygen species production by co-planar polychlorinated biphenyl congeners. *Aquat. Toxicol.* 77:422–32.

Schrader, M., and H. D. Fahimi. 2006. Peroxisomes and oxidative stress. *Biochim. Biophys. Acta* 1763:1755–66.

Shi, X. L., and N. S. Dalal. 1988. On the mechanism of the chromate reduction by glutathione: ESR evidence for the glutathionyl radical and an isolable Cr(V) intermediate. *Biochem. Biophys. Res. Commun.* 156:137–42.

Shi, X., et al. 1994. Reaction of Cr(VI) with ascorbate and hydrogen peroxide generates hydroxyl radicals and causes DNA damage: Role of a Cr(IV)-mediated Fenton-like reaction. *Carcinogenesis* 15:2475–78.

Silva, C. A., et al. 2009. Evaluation of waterborne exposure to oil spill 5 years after an accident in Southern Brazil. *Ecotoxicol. Environ. Safe.* 72:400–9.

Stegeman, J. J., and J. J. Lech. 1991. Cytochrome P-450 monooxygenase systems in aquatic species: Carcinogen metabolism and biomarkers for carcinogen and pollutant exposure. *Environ. Health Persp.* 90:101–9.

Stephensen, E., et al. 2000. Biochemical indicators of pollution exposure in shorthorn sculpin (*Myoxocephalus scorpius*), caught in four harbors in the southwest coast of Iceland. *Aquat. Toxicol.* 48:431–42.

Sturve, J., et al. 2006. Effects of North Sea oil and alkylphenols on biomarker responses in juvenile Atlantic cod (*Gadus morhua*). *Aquat. Toxicol.* 78(Suppl. 1):S73–78.

Vakharia, D. D., et al. 2001. Effect of metals on polycyclic aromatic hydrocarbon induction of CYP1A1 and CYP1A2 in human hepatocyte cultures. *Toxicol. Appl. Pharm.* 170:93–103.

Ventura-Lima, J., et al. 2007. Toxicological responses in *Laeonereis acuta* (*Annelida, Polychaeta*) after arsenic exposure. *Environ. Int.* 33:559–64.

Ventura-Lima, J., et al. 2009a. Effects of arsenic (As) exposure on the antioxidant status of gills of the zebrafish *Danio rerio* (Cyprinidae). *Comp. Biochem. Physiol.* 149C:538–43.

Ventura-Lima, J., et al. 2009b. Effects of different inorganic arsenic species in *Cyprinus carpio* (Cyprinidae) tissue after short-time exposure: Bioaccumulation, biotransformation and biological responses. *Environ. Pollut.* 157:3479–84.

Viarengo, A., et al. 1989. Lipid peroxidation and level of antioxidant compounds (GSH, vitamin E) in the digestive glands of mussels of three different age groups exposed to anaerobic and aerobic conditions. *Mar. Environ. Res.* 28:291–95.

Vieira, L. R., et al. 2008. Acute effects of benzo[a]pyrene, anthracene and a fuel oil on biomarkers of the common goby *Pomatoschistus microps* (*Teleostei, Gobiidae*). *Sci. Tot. Environ.* 395:87–100.

Vijayavel, K., ct al. 2004. Sublcthal cffcct of naphthalene on lipid peroxidation and antioxidant status in the edible marine crab *Scylla serrata*. *Mar. Pollut. Bull.* 48:429–33.

Vlahogianni, T., et al. 2007. Integrated use of biomarkers (superoxide dismutase, catalase and lipid peroxidation) in mussels *Mytilus galloprovincialis* for assessing heavy metals' pollution in coastal areas from the Saronikos Gulf of Greece. *Mar. Pollut. Bull.* 54:1361–71.

Wei, P., et al. 2000. The nuclear receptor CAR mediates specific xenobiotic induction of drug metabolism. *Nature* (London) 407:920–23.

Whitlock Jr., J. P., et al. 1996. Induction of cytochrome P4501A1: A model for analyzing mammalian gene transcription. *FASEB J.* 10:809–18.

Wood, A. W., et al. 1976. Metabolism of benzo[a]pyrene and benzo[a]pyrene derivatives to mutagenic products by highly purified hepatic microsomal enzymes. *J. Biol. Chem.* 251:4882–90.

Xie, W., et al. 2000. Reciprocal activation of xenobiotic response genes by nuclear receptors SXR/PXR and CAR. *Gene Dev.* 14:3014–23.

8 Biotransformation of Organic Contaminants and the Acquisition of Resistance

Michèle Roméo and Isaac Wirgin

CONTENTS

8.1   DAMAGE TO AQUATIC POPULATIONS FROM AROMATIC HYDROCARBON CONTAMINANTS

Contamination of the environment with organic compounds is a widespread phenomenon and occurs from both point and diffuse nonpoint sources. Aquatic ecosystems are often the final sink for contaminants from both sources. Acute toxicity of adult organisms from chemical exposures is rarely observed anymore in natural populations. However, more subtle and pernicious toxic outcomes may still be prevalent. Two toxic outcomes in fishes from exposure to these chemicals include neoplasia and early life stage toxicities. Fish populations in North American estuaries, European coastal waters, and tributaries of the Laurentian Great Lakes have suffered epizootics of preneoplastic and neoplastic hepatic and skin lesions presumably due to exposure to these contaminants. These examples include systems that are contaminated with polychlorinated biphenyls (PCBs), polychlorinated dibenzo-*p*-dioxins (PCDDs), polychlorinated dibenzofurans (PCDFs), and polyaromatic hydrocarbons (PAHs) (reviewed in Wirgin and Waldman 1998). Two species with documented elevated prevalences of hepatic tumors include Atlantic tomcod from the PCB-contaminated Hudson River, New York (Dey et al. 1993), and Atlantic killifish from the PAH-contaminated Elizabeth River, Virginia (Vogelbein et al. 1990). Studies with flatfishes in Puget Sound, Washington, have indicated that significant relationships exist between tumor levels and sediment concentrations of PAHs, and less strong associations with PCBs (Myers et al. 2003). Furthermore, increased disease prevalence may be exacerbated by other stressors, such as global warming, overharvest, and invasive species (Lafferty et al. 2004).

Because of the relationship between cancer and stressors, prevalences of preneoplastic and neoplastic hepatic and skin lesions in selected sentinel species have been used by managers to evaluate environmental quality and efficacy of remediation (Pinkney and Harshbarger 2006; Stentiford et al. 2009).

Additionally, evidence suggests that PAHs and dioxin-like chemicals have deleteriously affected early life stage viability of fishes at impacted localities and thereby decreased recruitment to adult populations. For example, direct *in situ* examination demonstrated that contamination with *Exxon Valdez* oil resulted in early life stage toxicities and genotoxicities to multiple generations of pink salmon *Oncorhynchus gorbuscha* (Bue et al. 1996) and Pacific herring *Clupea pallasi* in Prince William Sound, Alaska (Brown et al. 1996). Similarly, contamination of the Great Lakes with 2,3,7,8-tetrachlorodibenzo-*p*-dioxin (TCDD)-like chemicals is believed to have impaired reproduction of lake trout *Salvelinus namaycush* and led to their extirpation there (Tillitt et al. 2008). However, unlike the presence of tumors, the prevalence of toxicity to embryos and larvae is often difficult to observe, document, and quantify in natural populations. As a result, the sensitivity of target organisms to toxicants under controlled laboratory conditions often must be evaluated, thresholds determined that evoke toxicity, and body burdens of aromatic hydrocarbon (AH) contaminants in exposed populations compared to these thresholds. From these comparisons, conclusions can be reached on the likelihood of *in situ* toxicity under natural conditions.

Pollutants, which are present in the environment, interact with environmental and biological systems based upon their physicochemical properties and reactivity, yielding a characteristic pattern of environmental and internal exposure concentrations for each pollutant. Hydrophilic compounds can be excreted directly, whereas less hydrophilic and nonpolar compounds such as PAHs, halogenated aromatic hydrocarbons (HAHs; i.e., PCDDs and PCDFs), polyhalogenated biphenyls (PCBs and polybrominatedbiphenyls (PBBs)), and polyhalogenated diphenyl ethers (PCDEs and PBDEs) bioaccumulate and often biomagnify in the tissues of organisms. In fact, organic pollutants can be enzymatically biotransformed or metabolized at the cellular level before they can be excreted, thereby preventing cellular toxicity. However, at very high levels of chronic exposure, as sometimes experienced by populations at contaminated sites, these enzymes may become saturated, and therefore their defense activities are overwhelmed by the environmental chemicals. In that case, rather than protective, the activities of these enzymes on occasion lead to metabolic conversion of organics to forms that are damaging to both cellular proteins and DNA. The tolerance offered by the biotransformation process is therefore limited by saturation of the mechanism itself. Biotransformation is organized into two phases (I and II), which are followed by phase III (see Chapter 10 on multixenobiotic resistance) leading to excretion. The two aspects of biotransformation/resistance and the occurrence of cellular toxicity of organic chemicals will be discussed below.

8.2 RESISTANCE MECHANISMS: BIOTRANSFORMATION OF ORGANICS

8.2.1 PHASE I OF BIOTRANSFORMATION

8.2.1.1 Catalytic Activity of CYP1A

Phase I reactions consist of oxidation, reduction, and hydrolysis. A great number of enzymes, such as dehydrogenases, reductases, hydrolases, and monooxygenases, are involved in phase I. Functional groups are produced or exposed during phase I reactions, which render hydrophobic compounds more water soluble (Figure 8.1). Cytochrome P450 (CYP) is the main enzyme responsible for the metabolism of some endogenous compounds and the above-mentioned nonpolar xenobiotics, including the metabolic activation of many environmental toxic chemicals and carcinogens. CYPs are enzymes referred to as mixed-function oxidases (MFOs) (Di Giulio et al. 1995).

CYPs are hemoproteins, their molecular mass ranging between 43 and 60 kDa (Goksøyr and Förlin 1992). They are found associated with membranes in the

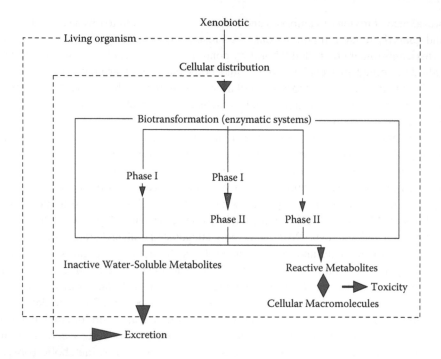

FIGURE 8.1   Phases I and II of xenobiotic biotransformation.

endoplasmic reticulum or mitochondria of different tissues: liver, lung, kidney, intestine, etc. (Stegeman and Hahn 1994). They catalyze the oxidation of a substrate RH (organic compound that becomes hydroxylated) by inserting one atom of molecular oxygen, of which the second atom is reduced to water following the equation

$$RH + O_2 + NADPH + H^+ \rightarrow ROH + NADP^+ + H_2O$$

The catalytic mechanism is cyclic and involves a series of reduction and oxidation reactions of cytochrome P450 (involving NADPH cytochrome P450 reductase) (Guengerich 1993). In the first step, cytochrome P450 binds RH; then cytochrome P450 reductase is reduced by NADPH. The first electron is transferred to cytochrome P450 by means of NADPH cytochrome P450 reductase, and a ferrous cytochrome P450–dioxygen complex is formed. A second electron is transferred by cytochrome P450 reductase or cytochrome b5, the oxygen-oxygen bond is cleaved, and a water molecule is formed.

8.2.1.2   Induction of CYP1A in Fishes

Much of our understanding of the characteristics and function of xenobiotic biotransformation enzymes in aquatic species has evolved from the vast mammalian literature. There are numerous similarities in the P450 enzymes between aquatic invertebrate and fish species and mammals. The *de novo* synthesis of P450 proteins by organisms exposed to xenobiotics, termed induction, leads to increased enzymatic activity. Induction has been well known for forty years in humans and other mammals, more recently in fishes and plants, and of late in invertebrates (Stegeman and Hahn 1994). Payne and Penrose (1975) were among the first to report elevated P450 activity in fish from petroleum-contaminated areas.

The induction of cytochrome P450 isoenzymes responds to exposure to xenobiotics by the way of a selective, receptor-mediated stimulation of the CYP1A gene transcription rate, resulting in increased levels of specific mRNA, new synthesis of cytochrome P450 isoenzymes, and an increase in their catalytic activities (for instance, ethoxyresorufin O-deethylase (EROD) for CYP1A). The receptor that mediates the regulation of the CYP1A gene expression is known as the aryl hydrocarbon (AH) receptor (AHR) (Poland and Glover 1975; Guengerich 1993). Studies have demonstrated that activation of the AHR pathway is necessary for benzo[a] pyrene (B[a]P)-induced hepatic carcinogenicity in mice (Shimizu et al. 2000) and TCDD- and PCB-induced early life stage toxicities in fishes (Antkiewicz et al. 2006). The functioning of the AHR pathway in fishes is almost identical to that in mammals except that fish have two or more forms of AHR (AHR1 and AHR2) due to genome duplication events (Hahn 2002a). After diffusing into the cell, the xenobiotic binds to a protein complex in the cytoplasm consisting of AHR, a dimer of heat shock protein 90 (Hsp90), p23, and ZAP2 (also known as ARA9 and AIP) (Figure 8.2). Upon ligand binding, ZAP2 is released, exposing the nuclear localization signal on AHR and leading to translocation of AHR from the cytoplasm to the nucleus. Within the

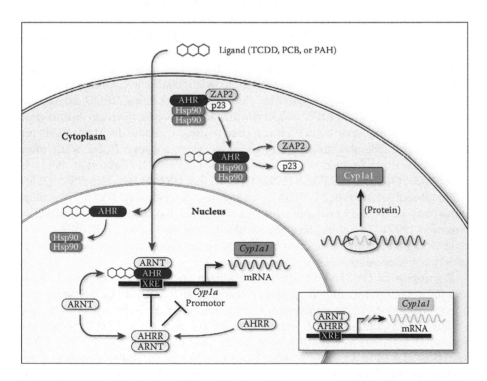

FIGURE 8.2 The functioning of the aryl hydrocarbon receptor (AHR) pathway in fishes.

nucleus, Hsp90s are released, and AHR heterodimerizes with another protein, the aryl receptor nuclear translocator (ARNT). The AHR-ARNT complex then binds to multiple enhancer elements in the promoter region of responsive genes in the AHR battery, such as CYP1A.

These enhancer elements are termed AHR-, dioxin-, or xenobiotic-responsive elements (AHREs, DREs, or XREs) and contain a core 5′-GCGTC-3′ DNA sequence that is recognized by the AHR-ARNT complex. As a result, coactivator complexes can access the promoter region of CYP1A (and other genes in the AHR battery), histone modifications occur, and transcription proceeds (reviewed in Denison et al. 2002). It should be noted that the activity of the AHR repressor (AHRR) serves as a negative feedback mechanism to downregulate activity of the AHR pathway in mammals and fishes (Hahn 2002a).

More than 8,100 distinct CYP gene sequences are already known (website of the P450 Nomenclature Committee: http://drnelson.utmem.edu/CytochromeP450.html, February 2010). The nomenclature used for cytochrome P450s is based on sequence homology (Nebert and Nelson 1991): two cytochrome P450s belong to the same family when their peptide sequence presents more than 45% amino acid homology, and to the same subfamily if homology is higher than 55%. The abbreviation CYP corresponds to a cytochrome P450 gene; it is completed with a number representing the family, then a letter indicating the subfamily (e.g., CYP4A) and a last number when there are several genes within the same subfamily (e.g., CYP4A1, CYP4A2, etc.). Conventionally, genes are written in italics, CYP1A1 (Goksøyr and Förlin 1992), whereas mRNA and proteins are in capitals. Nelson (1998) has developed a classification scheme where CYP families are classified into CLANS, i.e., clusters of higher-order groupings of P450 families.

CYP1As are induced by PAHs, coplanar PCBs, PCDDs, and PCDFs (Goksøyr and Förlin 1992), which are pollutants of the 3-methylcholanthrene (3-MC) type and are now considered AH receptor agonists. Three enzyme activities, EROD, ethoxycoumarin O-deethylase (ECOD), and arylhydrocarbon (benzo[a]pyrene) hydroxylase (AHH), are largely specific in their response to these compounds. Many PAHs are both inducers and substrates for CYP1A. In contrast, coplanar PCBs, while often good inducers, are frequently poor substrates for CYP1A (Di Giulio et al. 1995). In their review, Goksøyr and Förlin (1992) reported that CYP2B is induced by coplanar PCBs (phenobarbital type), CYP3A by endogenous steroids, and CYP4A by endogenous fatty acids and xenobiotics such as phthalates and peroxisome proliferators (Simpson 1997). Therefore, members of the cytochrome P450 family of monoxygenases can metabolize (and often bioactivate under toxic forms; see below) a wide variety of endogenous molecules and xenobiotics.

Regulation of CYP1A-like activities has been studied in the Atlantic killifish *Fundulus heteroclitus* for thirty years (Stegeman 1978). Nearly twenty years ago, *CYP1A* was cloned in the rainbow trout *Salmo gairdneri* (Heilmann et al. 1988), and a decade ago in the killifish *F. heteroclitus* (Morrison et al. 1998) and the Atlantic tomcod *Microgadus tomcod* (Roy et al. 1995). In addition to *F. heteroclitus*, CYPs have been cloned in a variety of other fishes (see Nelson's website). Recently, genes from a new vertebrate CYP1C family were found (Godard et al. 2005). Identification and cloning of the full-length sequences of three new *F. heteroclitus CYP1* genes,

CYP1B1, *CYP1C2*, and *CYP1D1*, were also recently reported by Zanette et al. (2009). At present, products of all steps of the CYP1A induction cascade, mRNA, protein, and catalytic activity, can be used to measure CYP1A expression in fishes with reverse transcriptase polymerase chain reaction (RT-PCR) or northern blotting (mRNA), western blotting using polyclonal and monoclonal antibodies raised against fish CYP1A, enzyme-linked immunosorbent assay (ELISA) and histochemistry for protein, and EROD, ECOD, and AHH enzyme activities. There is evidence that the three levels (mRNA, protein, enzymatic activity) are related to PAH presence in the animals and in the environment in a dose-dependent manner. Furthermore, in controlled laboratory experiments, levels of CYP1A expression are dose responsive to graded treatments of these chemicals. In fact, recent scans of the transcriptome of several fishes treated with TCDD or coplanar PCB mixes using microarray analysis have revealed that expression changes (upregulation) of CYP1A and CYP1B far exceed those of other differentially expressed genes (Handley-Goldstone et al. 2005; Carlson et al. 2009). This sensitivity and wide window of response provide a statistically robust system in which to evaluate the effects of these xenobiotics on impacted natural populations (see below).

8.2.1.3   Induction of CYP1A in Invertebrates

Measurement of CYP1A expression provides information relevant to exposure only in those organisms possessing the appropriate response mechanism (AHR activation of *CYP1A*). AH receptors exist in vertebrates, including fishes, birds, and mammals (Goksøyr et al. 1991; Nelson 1998). The presence of an AHR responding to PAH exposure has not been confirmed in aquatic invertebrates (Hahn 2002b), and the use of *CYP1A* induction as a biomarker of organic contaminant pollution has been problematic in these organisms. An AHR homolog with hallmark functional domains was identified in many tissues of soft-shell clam *Mya arenaria*, but it lacked TCDD and β-naphtoflavone (β-NF) binding abilities (Butler et al. 2001). Michel et al. (1994) found increased activity of NADPH cytochrome c reductase, and cytochrome P450 concentration in mussels (*Mytilus galloprovincialis*) exposed to BaP. Wootton et al. (1995) observed marked variation in levels of CYP1A-like mRNA in mussels (*Mytilus* spp.) collected from several sites in the Venice area (Italy), indicating modulation of gene expression with some correlation with the tissue levels of PAHs and PCBs. Snyder (1998) assigned six new cytochrome P450s from marine invertebrates belonging to the arthropod, mollusk, and echinoderm phyla to the CYP4 family. In northern blotting experiments with digestive gland tissues of the mussel *Mytilus galloprovincialis*, the expression of the CYP4Y1 gene was found to be inhibited by increasing concentrations of β-NF (a CYP1A-type inducer), and as a result, Snyder (1998) hypothesized that the reduction may be related to direct effects of a putative AHR on gene transcription. Chaty et al. (2004) identified CYP1A and CYP4 transcripts in the freshwater mussel *Unio tumidus* and compared their expression with those of the rainbow trout *Oncorhyncus mykiss* after treatment with the specific CYP1A1 and CYP4 vertebrate inducers, Aroclor 1254 and di(2-ethylhexyl)phthalate (DEHP). The authors could not amplify a CYP1A sequence from the cDNA of the digestive gland from either control or Aroclor-treated freshwater mussels *U. tumidus*. On the other hand, CYP4 transcripts were PCR amplified in digestive glands from control and DEHP-treated *U. timidus*, but no variation of CYP4 mRNA levels was found in mussels after phthalate treatment, as opposed to results in fishes. The regulation of *CYP4* gene expression appears to be different between vertebrates and invertebrates after an exposure to a vertebrate CYP4 inducer (phthalate). In 2005, mRNA sequences of the freshwater mussel *Dreissena polymorpha* AHR were published (EMBL: accession number DQ159188). New developments will advance the usefulness of these genes to understand the regulation and normal function of invertebrate cytochromes.

Polychaetes are important in terms of marine biomass and food webs. The presence of cytochrome P450 is well documented in different species of polychaetes, as well as the relationship between mixed-function oxygenase activity and cytochrome P450, and substrates metabolized by the annelid cytochrome P450 system (Lee 1998). Two partial CYP cDNA sequences, both of 453 bp, were characterized by Rewitz et al. (2004) from *Nereis virens*. Jørgensen et al. (2005) cloned two complete sequence genes in the same species, one named CYP342A1, the first member of a new family, and the other named CYP4BB1. Because both CYP enzymes catalyze the hydroxylation of pyrene to 1-hydroxypyrene, it is likely that they are involved in xenobiotic transformation.

8.2.1.4 Field Studies and Influence of Natural Factors on EROD Activity

Enhanced enzyme content or activity has been reported in a review dealing with field studies in fish collected in polluted sites from different areas (Bucheli and Fent 1995). EROD has been employed extensively as a biomarker of organic contaminants in the UK (Kirby et al. 1999, 2004), along the French Mediterranean coast (Garrigues et al. 1993), in other European countries (Beyer et al. 1996, Eggens et al. 1996), and throughout North America (Collier et al. 1995). In fact, there is a vast literature on the use of levels of CYP1A expression in fishes as a sensitive biomarker of exposure to environmentally relevant concentrations of contaminants such as PCDD/Fs, PCBs, and PAHs and as an early warning system of their biological effects (Oris and Roberts 2007). An overview of results of EROD activities in fish, monitored for four years in the Baltic Sea in the framework of the EU-funded BEEP project (2001–2004), was recently published (Lehtonen et al. 2006). In the analyzed fish species (pikeperch *Sander lucioperca* L., eelpout *Zoarces viviparus*, European flounder *Platichthys flesus*), EROD activity was highest in locations contaminated with PCDDs, PCBs, and PAHs. However, seasonal variation in EROD activity was commonly found due to the effects of temperature and sexual development (Lehtonen et al. 2006). In fact, Bucheli and Fent (1995) reported that temperature is probably the most dominant of the abiotic factors that influence cytochrome P450 expression. More recent literature points to the influence of other natural factors (seasonal variability in environmental hydrographic and climatic conditions) (Kopecka et al. 2006) and condition index, associated with "novel pollutants" such as flame retardants (Kuiper et al. 2007) on EROD activity in the flounder *P. flesus*. Lehtonen et al. (2006) recommended that if EROD activity is measured in fish collected from the sea, only one sex should be studied, or if both sexes are studied, they should be considered separately, and when using female fish, their gonad weight must be higher than 1% of the body weight. EROD activity is commonly analyzed in liver microsomes, but the gill, which is directly exposed to waterborne pollutants, can also be used (Abrahamson et al. 2007). The high sensitivity of the gill to enzyme induction compared with the liver and the kidney was confirmed in studies of fish caged in waters in urban and rural areas in Sweden.

A recent investigation demonstrated that estrogenic chemicals suppress PAH-mediated EROD induction in the flounder *P. flesus* (Kirby et al. 2007). The metabolism of estrogens involve some MFO activities (cytochrome aromatase, for instance). These researchers hypothesized that the extent of EROD induction in published monitoring data may be underestimated if it is assumed that estrogen-mediated MFO suppression is occurring in wild fish populations.

Greater research on system interactions and other factors such as the ones mentioned above is necessary to better understand the regulation of cytochrome P450s in organisms exposed to organic contaminants.

8.2.2 PHASE II OF BIOTRANSFORMATION

Phase II enzymes catalyze the conjugation between xenobiotic compounds containing electrophilic sites and endogenous molecules such as glutathione (glutathione S-transferases (GSTs)), sulfate (sulfo-transferases), glucuronic acid (UDP-glucuronosyltransferases (UDPGTs)), and epoxide (epoxide hydrolase) (Stegeman et al. 1992). Phase II enzymes serve to link metabolites from phase I with endogenous substrates, increasing their water solubility, expediting their excretion, and thereby reducing their toxicity (Figure 8.1). Phase II reactions take place in the liver of organisms in mammals and fishes (Förlin and Haux 1985; Wilce and Parker 1994; Bucheli and Fent 1995). Phase II enzymes can also directly metabolize molecules bearing –OH groups such as phenol, toluene, or trichlorethanol.

UDPGT enzyme activity seems to be moderately inducible (one- to threefold) in fishes by PAH-type inducers (Andersson et al. 1985). Nevertheless, Förlin et al. (1985) observed elevated activities in fish caught in the vicinity of a pulp-producing facility, and these UDPGT activities were associated with high CYP1A activities. UDPGT enzymes metabolize endogenous substrates such as steroid hormones, and increased activities could affect steroid levels (Förlin and Haux 1985). Neither epoxide hydrolase (EH) nor sulfotransferase (ST) expression seems to be affected by any of the PAH-type inducers (Foreman 1989). Few papers have concerned UDPGT, epoxide hydrolase, and sulfotransferase as phase II enzymes responding to organic pollution in the environment. On the contrary, the activities of GSTs have been widely studied since they are involved in the detoxification of many chemical compounds, including hydrocarbons, organochlorine insecticides, and PCBs (Clark et al. 1984).

GSTs are a multigenic family of enzymes named alpha, mu, pi, theta, sigma, zeta, and omega (Mannervick 1985; Sheehan et al. 2001). GST activities have been shown to increase in organisms such as the rainbow trout *Oncorhyncus mykiss* (Petrivalsky et al. 1997), the Mediterranean fish *Mullus barbatus* (Burgeot et al. 1996), mollusks (freshwater mussel *Sphaerium corneum*: Boryslawskyj et al. 1988), and in other invertebrates such as crabs (blue crab *Callinectes sapidus*: Lee 1988). Because phase I enzyme activities cannot be reliably used in invertebrates, GST activity has been used as a biomarker of organic contamination and biotransformation in mollusks collected from the field or caged in specific areas. However, GST activity in mollusks was shown to increase when salinity decreases (Bebianno et al. 2007). GST activity is generally measured with 1-chloro-2,4-dinitrobenzene (CDNB), a relatively non-specific GST reference substrate (Gallagher et al. 2000) following the reaction:

$$GSH + CDNB \xrightarrow{\text{GST}} GS - DNB + HCl$$

GST-CDNB activity reflects all GST isoenzyme activities. Activities determined with other substrates (i.e., ethacrynic acid) may evoke isoforms that are induced in response to different molecules (Fitzpatrick et al. 1995). In the clam *Ruditapes decussatus* exposed to pp′DDE and methoxychlor (Hoarau et al. 2004), several isoforms of GSTs belonging to the alpha and pi classes were identified using immunoblotting with rabbit anti-alpha and anti-pi class GST, followed by immunoelectrofocusing. On the other hand, a GST-pi gene was cloned in *Mytilus galloprovincialis* and the mRNA levels were evaluated by semiquantitative RT-PCR in mussels collected from the field, and were related to their pollutant burden (Hoarau et al. 2006).

8.3   RESISTANCE IN FISH POPULATIONS

Studies have demonstrated that organisms from polluted sites may exhibit an alternative strategy to cope with high levels of toxicants: the development of resistance. Resistance can occur by two different mechanisms: (1) physiological acclimation or (2) genetic adaptation, or in some cases by both. Physiological acclimation implies that tolerance develops when organisms are initially exposed to stressors and subsequently become desensitized to additional exposures. Reduced sensitivity or resistance is lost when exposure is terminated. For example, this may occur when contaminated sites are remediated. Physiological acclimation can be considered a process that occurs on the organismal level. Adaptation is a second mechanism by which organisms can become resistant to chronic exposure to xenobiotics. Unlike physiological acclimation, adaptation is heritable and may persist indefinitely even in the event of site remediation, unless the presence of the resistance phenotype results in significant evolutionary cost to the population. Unlike physiological acclimation, adaptation is a process that occurs at the population level and is likely to occur only if certain life history requirements are met. It is important to recognize that the genetic variant(s) conferring resistance is present in the population, albeit at low frequency, before xenobiotic exposure occurs. Natural selection, success in the face of toxicant exposure, increases the frequencies of beneficial variant alleles in the population so that they eventually predominate. Development of adaptation to xenobiotics requires that the impacted population is highly exposed, reproductively isolated from other nonimpacted proximal populations, and will most likely occur in larger rather than smaller populations.

Populations of two fish species in U.S. Atlantic Coast estuaries provide dramatic examples of resistance to AHs; however, manifestations of resistance differ among these populations (reviewed in Wirgin and Waldman 2004), as do the

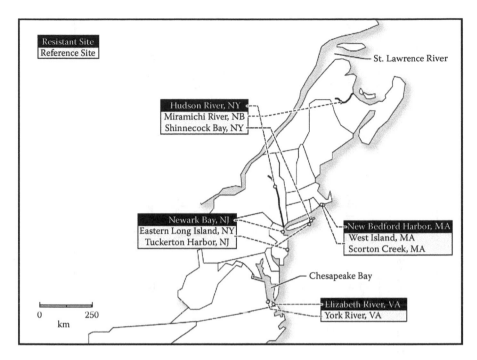

FIGURE 8.3 Location of populations (and reference) of two fish species (Atlantic killifish and Atlantic tomcod) in U.S. Atlantic coast estuaries.

mechanisms whereby resistance occurs. These include Atlantic killifish from PCB-contaminated New Bedford Harbor, Massachusetts (Nacci et al. 1999; Bello et al. 2001); PCB-contaminated Hudson River, New York (Elskus et al. 1999); TCDD-contaminated Newark Bay (western Hudson River Estuary, New Jersey) (Prince and Cooper 1995a); PAH-contaminated Elizabeth River, Virginia (Van Veld and Westbrook 1995; Owenby et al. 2002); and Atlantic tomcod from the Hudson River Estuary (Wirgin et al. 1992; reviewed in Wirgin and Chambers 2006) (Figure 8.3 and Table 8.1).

All of these populations share a common exposure to high levels of AHs, most certainly as mixtures among themselves and with a variety of other chemicals, including metals. Both species are abundant, bottom-dwelling fish, and their distributions are restricted to their natal estuaries. Killifish populations are modest in size, nonmobile, and localized in discontinuous demes. Tomcod are vagile within their natal estuaries, exhibiting seasonal within-estuary spawning migrations, but little migration among adjoining ecosystems. Both species exhibit moderate to high levels of genetic variation at allozyme loci and the mtDNA control region, and these genetic analyses support the reproductive isolation of populations. Empirical data suggest that these population characteristics in combination with chronic exposure allow for natural selection to operate rapidly to cause evolutionary change in highly impacted populations.

TABLE 8.1

Resistance Phenotype and Its Mechanistic Bases in North American Fish Populations

Population	Species	Major AH Contaminant	Resistance to	Evolutionary Cost	Reduced Genetic Diversity	Heritability	Mechanism—AHR Pathway Expression	Mechanism—AHR Polymorphism
New Bedford Harbor, Massachusetts	Atlantic killifish	PCBs	PCB126, 3-MC, TCDF	Unknown, not immunosuppression, growth, or condition	Not in mtDNA, AFLP, or allozymes	Yes, at least to F_2	No	Not AHR1
Newark Bay, New Jersey	Atlantic killifish	TCDD	TCDD, Aroclor 1248, Aroclor 1262, PCB77, B[a]P	Not tested	Not tested	Yes, at least to F_2	Not tested	Not tested
Elizabeth River, Virginia	Atlantic killifish	PAHs	3-MC, β-NF, PCB126	Immunocompromise, hypoxia	Not in mtDNA or allozymes	At least F_1 and maybe F_2	No	Not tested
Hudson River, New York	Atlantic tomcod	PCBs	Coplanar PCBs, TCDD; not B[a]P or β-NF	Not tested	Not in mtDNA	Yes, at least to F_2	No	Yes, AHR2

8.3.1 REDUCED CYP1A INDUCIBILITY AND SENSITIVITY TO EARLY LIFE STAGE TOXICITIES IN RESISTANT POPULATIONS OF ATLANTIC KILLIFISH

In all instances, environmentally collected adults and laboratory-reared offspring from resistant populations of killifish exhibited significantly decreased inducibility of hepatic and sometimes extra-hepatic CYP1A mRNA, CYP1A protein, and EROD enzyme activity compared to matched conspecifics from cleaner localities (Van Veld and Westbrook 1995; Prince and Cooper 1995b; Elskus et al. 1999; Nacci et al. 1999; Bello et al. 2001; Owenby et al. 2002). It is thought that modified inducibility of CYP1A in impacted populations may protect against the generation of bulky adducts and reactive oxygen species (Schlezinger et al. 2006), both of which may damage DNA. Interestingly, when killifish were collected from multiple sites along the Atlantic Coast of the United States with a gradient of PCB contamination, CYP1A inducibility by PCB126 in their progeny was negatively correlated with sediment concentrations of PCBs at sites where the collection of adults was made (Nacci et al. 2002a), strongly implicating PCB exposure as the cause of reduced expression. More importantly, the laboratory-bred embryonic and larval offspring of killifish collected from these resistant populations were significantly less sensitive to acute and chronic toxicities from coplanar PCBs and B[a]P than conspecifics from elsewhere (Nacci et al. 1999; Prince and Cooper 1995a; Owenby et al. 2002; Arzuaga and Elskus 2010). In addition to reduced survivorship, developmental toxicities observed in embryonic and larval offspring of sensitive populations included those typically observed in fishes treated under controlled laboratory conditions with TCDD or dioxin-like PCBs, including pericardial edema, craniofacial malformations, yolk sac edema, and abnormal spinal curvature—all of which probably resulted from altered vascular remodeling, decreased peripheral blood flow, reduced heart size, and compromised cardiac function (Antkiewicz et al. 2005), and all of which are incompatible with successful recruitment to adult populations. In an expanded study, Nacci et al. (2009) reported that the LC20 to PCB126 in F_1 killifish embryos from twenty-four estuarine sites in the United States from Maine to Virginia ranged over four orders of magnitude, and sensitivities of the response reflected sediment total PCB concentrations. Furthermore, one heretofore unknown resistant population was discovered in highly polluted Bridgeport Harbor, Connecticut (Nacci et al. 2009).

In these studies, sensitivities to TCDD or coplanar PCB induced teratogenicity cosegregated with inducibility of CYP1A, suggesting that they share a common mechanistic basis. Furthermore, studies in zebrafish embryos and larvae indicated that morpholino oligonucleotide antisense knockdown of AHR2 or ARNT1 expression (there are at least two forms of AHR and ARNT in fishes) resulted in reduced sensitivity to TCDD-induced CYP1A inducibility (Teraoka et al. 2003) and cardiac toxicities (reduced cardiac myocyte numbers, reduced heart size, and compromised function) (Antkiewicz et al. 2006), indicating that both share a common activating mechanistic basis, i.e., AHR pathway activation. Interestingly, morpholino knockdown of CYP1A expression by itself did not abrogate TCDD-induced cardiac toxicity (Antkiewicz et al. 2006).

8.3.2   Reduced CYP1A Expression and Sensitivity to Early Life Stage Toxicities in Atlantic Tomcod from Hudson River

As expected, environmentally exposed adult tomcod collected from the contaminated Hudson River and immediately sacrificed exhibited significantly higher levels of hepatic CYP1A mRNA expression, bile metabolites of PAHs, and bulky hepatic DNA adducts than adult tomcod collected from four cleaner North American estuaries (Kreamer et al. 1991; Wirgin et al. 1994). Interestingly, CYP1A mRNA expression in environmentally exposed tomcod from the Hudson River returned to basal levels within several days when fish were depurated in clean laboratory water, indicating that inducers were quickly depurated and suggesting that they were not persistent PCBs or PCDD/Fs, and more likely were PAHs or other rapidly metabolized environmental chemicals. However, environmentally exposed adult tomcod from the Hudson River that were returned to the laboratory depurated for 20 to 300 days, and treated with coplanar PCB 77, exhibited reduced CYP1A mRNA inducibility by two orders of magnitude compared to conspecifics from cleaner reference populations (Wirgin et al. 1992). Decreased gene inducibility was also observed in Hudson River tomcod treated with three different coplanar PCBs or TCDD, but not two PAHs, B[a]P, or β-NF (Courtenay et al. 1999; Yuan et al. 2006). Similarly, F_1 and F_2 laboratory-reared offspring (embryos and larvae) of tomcod of Hudson River descent exhibited significantly reduced inducibility of CYP1A mRNA by PCB 77 (but not B[a]P) compared to similarly treated tomcod offspring from cleaner reference populations (Wirgin and Chambers 2006) (Figure 8.4). Interestingly, maximal CYP1A mRNA expression in PCB- and TCDD-treated adult and younger tomcod from the Hudson River was similar to that achieved in age-matched tomcod from reference

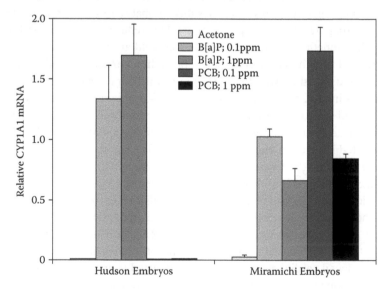

FIGURE 8.4   Heritability of CYP1A mRNA expression in F_2 Atlantic tomcod embryos treated with two doses of PCB77 or B[a]P or acetone vehicle control. Contaminated site: Hudson River; reference: Miramichi River.

populations, suggesting that the AHR pathway was functional, only less sensitive to activation in Hudson River tomcod. Finally, it was found that young tomcod collected over the species' entire range in the Hudson River (>140 km) were resistant to CYP1A induction by PCB77, but not B[a]P (Yuan et al. 2006). The investigators concluded that the Hudson River tomcod population was the most geographically extensive resistant vertebrate population ever reported.

Most importantly, young offspring of tomcod from the Hudson River were significantly less sensitive to TCDD and coplanar PCBs induced earlier life stage toxicities than tomcod from two different reference populations: the Miramichi River, New Brunswick, and Shinnecock Bay, New York (Wirgin and Chambers 2006). These toxicities included hatching success, a suite of morphometric alterations, and behavioral deficits. In summary, manifestation of resistance in tomcod differed from that in killifish in that tomcod from the Hudson River exhibited resistance to halogenated coplanar PCBs and TCDD, but not nonhalogenated PAHs. This suggests that the mechanistic basis of resistance in tomcod and killifish differs. Also, the fact that CYP1A expression was elevated in environmentally exposed tomcod from the Hudson River compared to those from elsewhere, but not inducible by controlled laboratory treatment with PCBs or TCDD, suggests that other AH compounds (likely PAHs) had induced CYP1A in environmentally exposed tomcod from the Hudson River.

8.3.3 Downstream Effects of the Refractory CYP1A Phenotype in Killifish

The downstream effects of reduced CYP1A inducibility were evaluated in killifish from resistant populations. Arzuaga and Elskus (2010) reported that levels of reactive oxygen species (ROS) (that are potentially DNA damaging) were less in PCB126- and 3-MC-treated killifish embryos from resistant populations in New Bedford Harbor and Newark Bay compared to embryo offspring of reference site descent. But, levels of oxidative DNA damage were not compared between PCB- or PAH-treated fish from the resistant and sensitive populations; thus, actual biological effects of elevated ROS are still not known. Nacci et al. (2002c) used the ^{32}P-postlabeling assay to evaluate the effects of the refractory CYP1A phenotype on the generation of bulky hepatic DNA adducts in resistant killifish adults from New Bedford Harbor and a reference site that were intraperitoneum (i.p.) injected with B[a]P. They found that levels of hepatic DNA adducts were three- to eightfold higher in treated killifish from the reference compared to the resistant population. Thus, in this case, the refractory CYP1A phenotype may have been beneficial to organism and population health. Further downstream, Jung et al. (2009) quantified and compared the extent of mitochondrial DNA (mtDNA) and nuclear DNA (nDNA) damage in adult killifish from the resistant Elizabeth River population and a reference site that were (1) environmentally exposed and depurated for four weeks in the laboratory; (2) environmentally exposed, depurated for four weeks in the laboratory, and treated with B[a]P; or (3) in their F_1 larval offspring, treated with B[a]P. DNA damage was quantified using an extra-long PCR approach that provides quantitative data on the number of DNA lesions per 10 kilobases of DNA. It is important to note that this assay is sensitive to the presence of bulky adducts and ROS-modified DNA bases. Jung et al. (2009) found that the environmentally exposed adults from the resistant Elizabeth River population exhibited higher levels of mtDNA and nDNA lesions in brain, liver, and muscle than those from the reference site. However, adults from the reference population treated with B[a]P exhibited significant increases in DNA damage in both genomes compared to basal levels, while those from the Elizabeth River did not. Furthermore, it was found that B[a]P treatment did not induce significant hepatic oxidative DNA damage in either population. This indicated that killifish from the resistant population were less vulnerable than those from the reference population to further damage from DNA adducts, perhaps due to reduced CYP1A inducibility.

PAHs are very rapidly metabolized in fish livers by CYP1A and other xenobiotic metabolizing enzymes. Wills et al. (2009) evaluated the effects of reduced CYP1A activities on cardiac deformities and metabolite patterns in B[a]P and fluoranthene-treated killifish embryos born to parents from the resistant Elizabeth River population and from a sensitive population. As expected, there was an absence of cardiac deformities in the resistant population, and its profile of B[a]P metabolism differed from that of the sensitive population. The half-life of the parent B[a]P compound was extended in Elizabeth River fish, and its metabolism was shifted toward BaP-9,10 dihydrodiol and away from the highly mutagenic BaP-7,8-dihydrodiol-9,10-epoxide (BPDE) product. Thus, in two killifish populations the refractory CYP1A phenotype co-occurred with reduced levels of DNA-damaging metabolites and the product of their interacting with genomic DNA, thereby affording some protection from the initiation of chemical carcinogenesis.

8.4 STUDIES OF MECHANISMS OF RESISTANCE IN FISH POPULATIONS

8.4.1 HERITABILITY OF RESISTANCE PHENOTYPES

The mechanistic basis of resistance to AHs has been explored in affected killifish and tomcod populations. The initial question posed was the extent of heritability of the resistance phenotype. It was important to evaluate heritability because it would provide initial insights into the mechanistic basis of resistance and predict its persistence in the event of site remediation. Heritability studies on the offspring of two populations of killifish—New Bedford Harbor (Nacci et al. 2002a) and Newark Bay (Nacci et al. 2010)—and tomcod from the Hudson River (Wirgin and Chambers 2006) revealed that resistance (in both CYP1A inducibility and teratogenicity) was transmitted to at least the F_2 generation (Figure 8.4). Similarly, Owenby et al. (2002) and Nacci et al. (2010) observed that resistance persisted through at least the F_2 generation in killifish from the Elizabeth River. But, Meyer and Di Giulio (2002) and Meyer et al. (2002) reported that resistance eroded in later developmental stages of the F_1 and subsequent generations of killifish from the Elizabeth River. It is possible that this was due to the maternal transfer of contaminants to F_1 embryos and their later dilution during growth, physiological acclimation, or perhaps to epigenetic silencing of CYP1A expression. The general trend of persistence of resistance beyond the F_1 generation in these populations indicated that its basis was primarily genetic, not physiological, because maternal transfer of contaminant burdens would be almost totally cleared by the F_2, but not the F_1 generation. The fact that resistance co-occurred for both modified CYP1A inducibility and teratogenicity suggests that they shared a common mechanistic basis.

8.4.2 OPEN-ENDED APPROACHES TO IDENTIFY GENES ASSOCIATED WITH RESISTANCE

Two different methods have been used to attempt to identify the genetic bases of resistance in these populations, one open-ended and the second a candidate gene(s) approach. Initially, Meyer et al. (2005) used differential display technology to identify genes in liver that were differentially expressed between environmentally exposed killifish from the Elizabeth River and a reference population that were immediately sacrificed after collection. Eleven genes were identified that were differentially expressed between the two populations and across both genders. However, it is unlikely that expression of any of these genes was the mechanistic basis of resistance in the Elizabeth River population, but more likely represented exposure to different environments at the two localities. Williams and Oleksiak (2008) used an amplified fragment length polymorphism (AFLP)-based population genetics approach to search for genetic polymorphisms at anonymous DNA loci that exhibited a signature of selection among killifish populations. They compared the frequencies of genetic variants at three hundred of these loci in three resistant (New Bedford Harbor, Newark Bay, and Elizabeth River) and six flanking sensitive populations. Cleverly, two neighboring sensitive populations (one to the north and one to the south) were paired with each resistant population. In comparisons between resistant and paired sensitive populations, they scanned for loci whose F_{ST} values (a measure of genetic differentiation between populations) were outliers from those expected of neutral evolution. In total, twenty-four loci (1 to 6% of loci in each of the three population comparisons) were identified as being under selection or linked to areas of the genome under selection. Four of these loci were shared between two resistant populations, suggesting that these populations may share common mechanisms of adaptation. Because the other twenty loci were not shared, this suggests that there also may be independent mechanisms of evolution that are not shared among these populations. These putative selected loci have yet to be identified and their significance functionally evaluated.

An alternative approach based on expression differences was used by the same investigators to identify genes that confer resistance (Fisher and Oleksiak 2007; Oleksiak 2008). The same nine populations (three population triads) and comparative strategy that were used in the population genetics study were analysed in a microarray study. A custom-designed microarray fabricated with cDNA sequences from 384 metabolic genes from killifish heart and liver libraries was spotted onto the arrays, of which 225 genes were successfully analysed in comparisons between the resistant and paired sensitive populations. Gene expression was analyzed in brain (Fisher and Oleksiak 2007) and liver (Oleksiak 2008) of environmentally exposed adult killifish that were collected from the nine populations, depurated, and maintained in clean laboratory water under common "garden conditions" for one year. Extensive depuration would ensure that expression pattern differences would be genetic and not reflect differential exposures. Many differentially expressed genes were identified in both tissues between killifish from the resistant and flanking sensitive populations. For brain, in total, seventy-six differentially expressed genes were identified whose expression differed between the three resistant and paired flanking sensitive populations. Of these, two genes in brain were shared among the three comparisons. More differentially expressed genes between resistant and paired sensitive populations were detected in liver. In total, 149 genes were differentially expressed in livers between the three resistant and paired sensitive populations. Interestingly, three genes were identified that were differentially expressed across the three comparisons between resistant and paired sensitive populations. However, it is not clear if these differentially expressed genes serve as the basis of the resistant phenotype or are downstream products of the genes that actually confer resistance. It is also curious that the RNAs analyzed were not from AH-treated fish. It will be interesting to see what functional evaluation of the three shared genes reveals. Similarly, Carlson et al. (2009) used a custom-designed tomcod cDNA microarray to explore for genes that were differentially expressed between coplanar PCB-treated tomcod embryos from the Hudson River and two reference populations. Although novel PCB-induced genes were identified, none was thought to serve as the basis of resistance. However, several transcripts representing biomarkers of cardiomyopathy in mammals (cardiac troponin T2, cathepsin L, and atrial natriuretic peptide) were differentially expressed by PCB treatment among tomcod from the three populations. The discovery of genes associated with cardiomyopathy is consistent with known sensitivity of fish heart to coplanar PCB teratogenicity. Not surprisingly, the strongest difference by far between resistant and sensitive populations of tomcod was for CYP1A expression.

8.4.3 AHR PATHWAY EXPRESSION AND RESISTANCE

A candidate gene approach was also used to identify the mechanistic resistance of these fishes to AHs. Several investigators hypothesized that alterations in expression or structure of AHR pathway genes was the singular basis of resistance. Although most fishes contain two AHR paralogues (and more in salmonids), it is believed that AHR2 is the more active of the two forms in mediating CYP1A expression (Jonsson et al. 2007) and early life stage toxicities (Prasch et al. 2003). Thus, these studies were driven by the hypothesis that decreased levels of AHR2 or ARNT, or increased levels of AHRR expression in resistant populations served to decrease their responsiveness to AH compounds. However, studies with resistant killifish from New Bedford Harbor and the Elizabeth River and tomcod from the Hudson River failed to support this hypothesis.

Levels of hepatic AHR2 and AHRR were inducible by β-NF in killifish from the King's Creek reference site, but not in those from the Elizabeth River. Neither hepatic AHR1 nor ARNT expression was inducible by β-NF in these two populations, and no differences in their expression were observed between populations. Most importantly, no differences in basal levels of AHR1, AHR2, or AHRR expression were detected between the resistant and sensitive populations (Meyer et al. 2003). Powell et al. (2000) reported similar results with killifish embryos from New Bedford Harbor and a sensitive population, Scorton Creek, Massachusetts. No discernible differences were observed in AHR1 or AHR2 expression between these two populations. These investigators did observe tissue differences in AHR1 expression between adult fish from these two populations, but these expression differences were not observed in the F_1 progeny from the two populations. Because previous studies had demonstrated that the resistant phenotype is transmitted intergenerationally in the New Bedford Harbor population, it is unlikely that this variation in AHR1 expression served as the mechanistic basis of resistance. Similarly, no differences in AHR2 or ARNT1 expression were observed in untreated, chemically treated, or environmentally exposed tomcod from the Hudson River and sensitive populations (Roy and Wirgin, unpublished data).

The AHR repressor (AHRR) is a recently discovered transcription factor for which orthologues have been detected in killifish, tomcod, and other fishes. AHRR expression is induced by AHs through activation of the AHR pathway. AHRR binds with ARNT and downregulates activity of the AHR pathway by a negative feedback loop (Figure 8.2), although the exact mechanism by which this occurs is largely unknown (Evans et al. 2008). It has been hypothesized that AHRR is overexpressed in fishes from resistant compared to sensitive populations and thereby downregulates AHR pathway activity. However, empirical studies have failed to substantiate this proposal. Karchner et al. (2002) found no difference in AHRR expression by treatment with TCDD or two Aroclors between killifish from New Bedford Harbor and a reference site across a variety of tissues. Meyer et al. (2003) reported that hepatic AHRR was down- rather than upregulated in wild-caught killifish from the resistant population in the Elizabeth River compared to the reference population. In fact, hepatic AHRR expression was inducible by β-NF in killifish from the King's Creek reference site, but not in those from the Elizabeth River. Roy et al. (2006) observed that hepatic AHRR expression was inducible by both B[a]P and PCB77 in juvenile tomcod from a reference population, but only by B[a]P and not PCB77 in the resistant Hudson River population. Interestingly, among environmentally exposed tomcod collected from natural populations, hepatic AHRR expression was highest in those from the Hudson River. In fact, expression of AHRR mRNA in chemically treated and environmentally exposed tomcod was positively correlated with CYP1A expression, a result consistent with the fact that expression of both genes is regulated by the AHR2 pathway. This finding provided more evidence that an impaired AHR2 pathway may underlie the resistance phenotype in tomcod. Results from these studies ruled out expression differences of AHR pathway components as the mechanistic basis of resistance in these fishes, but they did not address the importance of structural differences in AHR pathway molecules between resistant and sensitive populations.

8.4.4 AHR POLYMORPHISMS AND RESISTANCE

Studies with *in vivo* rodent (reviewed in Okey et al. 2005) and avian models (Head et al. 2008) have demonstrated that genetic polymorphisms (single nucleotide polymorphisms (SNPs), moderate-sized deletions, and splice variants) in AHR are the mechanistic basis of dramatic interspecific and intraspecific variation in sensitivity to TCDD- or B[a]P-induced toxicity. Thus, it was reasonable to propose that genetic polymorphisms that downregulate the functional activity of the AHR pathway may serve as the mechanistic basis of resistance in these fishes. These may include genetic polymorphisms in the structural genes themselves or *cis* and *trans* factors that regulate gene expression. In this regard, Hahn et al. (2004) compared the sequence of AHR1 between killifish from the resistant New Bedford Harbor and a nearby reference population. They found 25 SNPs in the AHR1 coding region, of which nine were nonsynonymous, and that frequencies of AHR1 haplotypes differed significantly between the resistant and sensitive populations. The functional significance of two of the AHR1 haplotypes whose frequencies differed between the two populations were tested in empirical assays with *in vitro* transcribed and translated variant killifish AHR1 proteins. However, the two AHR1 proteins did not differ in their binding capacities or affinities for TCDD or their abilities in transfection assays to activate TCDD induced expression of a plasmid construct containing the luciferase reporter gene under control of dioxin response elements. Therefore, it is unlikely that these AHR1 polymorphisms play any role in conferring resistance in this model. Because of the importance of AHR in normal development, it was surprising to find such a high level of diversity at this gene. Perhaps the focus in this study on AHR1 rather than AHR2, and the greater functional importance of AHR2 in fishes, is the reason for the abundance of genetic polymorphisms and the lack of their functional association with the resistant phenotype in this study. In fact, several studies subsequent to that of Hahn et al. (2004) using a morpholino knockdown approach have demonstrated that AHR2, rather than AHR1, activates PCB and PAH induced cardiac teratogenesis in several fishes including killifish from a sensitive population (Clark et al. 2010). This strongly suggests that associations in fishes between AHR polymorphisms and variability in sensitivity to AH agonists should focus on AHR2.

AHR1 has not been identified in Atlantic tomcod so fortuitously AHR2, rather than AHR1 sequence, was investigated for genetic polymorphisms associated with resistance in the Hudson River tomcod population. Using southern blot analysis, Yuan (2003) found moderate levels of genetic diversity at the 3′ end of AHR2 among tomcod from the Hudson River and three rivers in Canada. However, using this technique, it was impossible to determine the exact number or location of these polymorphic sites or whether they were in fact within AHR2 or downstream flanking sequences. In a followup study, Wirgin et al. (manuscript submitted) compared full-length AHR2 cDNA sequences between ten tomcod each from the resistant Hudson River and two sensitive populations, Shinnecock Bay and the Miramichi River. They found exceedingly low levels of variation at AHR2 within these populations (only three coding region polymorphisms detected) (Figure 8.5), but two AHR2 coding region polymorphisms exhibited fixed differences between the resistant and sensitive populations. This finding of extreme sequence conservation in AHR2 was consistent with expectations for a functionally important gene and contrasted with that found previously for AHR1 in killifish. Interestingly, the common AHR2 allele was only seen in heterozygotes in the Hudson River

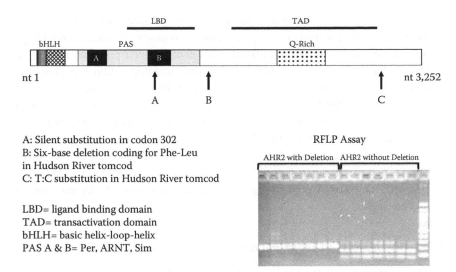

A: Silent substitution in codon 302
B: Six-base deletion coding for Phe-Leu
in Hudson River tomcod
C: T:C substitution in Hudson River tomcod

RFLP Assay

AHR2 with Deletion   AHR2 without Deletion

LBD= ligand binding domain
TAD= transactivation domain
bHLH= basic helix-loop-helix
PAS A & B= Per, ARNT, Sim

FIGURE 8.5 Functional domains and genetic polymorphisms in AHR2 among Atlantic tomcod populations.

population, and the variant AHR2 allele was only observed in heterozygotes from the Shinnecock Bay and Niantic River populations. The presence of the common AHR2 allele in low frequency in the Hudson River and the variant AHR2 allele in the two reference populations proximal to the Hudson River suggests that both alleles existed in all three populations prior to the relatively recent pollution of the Hudson River, but probably at a lower frequency in the Hudson River population. Restriction fragment length polymorphism (RFLP) assays were developed to rapidly screen for these variant AHR2 genotypes in many more tomcod specimens from these localities and additional reference populations. Given the dramatic difference in AHR2 allelic frequencies between the Hudson River and other less impacted tomcod populations, it will be informative to evaluate the functional significance of the variant AHR2 proteins in *in vitro* assays that compare the abilities of the two proteins to bind AHR agonists and to suport gene expression in chemically treated AHR deficient cells as described in Hahn et al. (2004).

8.4.4.1 Selection or Random Genetic Drift at AHR2 in Tomcod?

The hypothesis of strong directional selection at AHR2 in tomcod was empirically evaluated in further genetic studies of these populations (Wirgin et al. submitted). Sequence of the mitochondrial DNA (mtDNA) control region was analyzed because it is selectively neutral, and patterns of polymorphism among populations are believed due to stochastic factors (gene flow, population age, population size, and mutation rate) and therefore reflect the biogeographic histories of the populations rather than contemporary natural selection. Consistent with their life histories of year round residency in their natal estuaries, frequencies of mtDNA haplotypes differed significantly among U.S. estuaries. However, Canadian tomcod populations did not exhibit significant differences in mtDNA haplotype frequencies probably because of the effects of Pleistocene glaciation and the recent recolonization of Canadian rivers. As expected, genetic distance among populations was positively correlated with geographic distance, indicating that patterns of diversity among populations probably reflected stochastic factors. Although the Hudson River population was genetically distinct from its most proximal neighbors, the magnitude of its genetic differentiation was consistent with that among other U.S. tomcod populations. Interestingly, mtDNA haplotype diversity was higher in the Hudson River than at all other collection sites, indicating that the Hudson River population did not suffer a major genetic bottleneck as might be expected from such a dramatic shift in allelic frequencies as witnessed at AHR2. In the future, it would be informative to confirm the results observed with the mtDNA control region at selectively neutral nDNA markers such as microsatellites.

8.4.5   Do Changes in DNA Methylation Contribute to Resistance?

The role of epigenetic alterations in chemical carcinogenesis and developmental biology has gained increasing attention in recent years. Epigenetic changes are those that are transmitted among generations, but do not involve alterations in DNA sequence. Mechanisms of epigenetic change include alterations in levels of cytosine DNA methylation at CpG islands in the regulatory region of inducible genes or changes in the methylation, acetylation, phosphorylation, or ubiquitination levels of specific sites in histones. Epigenetic signatures at CpG islands and histones are known to regulate expression of downstream genes. It was hypothesized by two groups of investigators that epigenetic change, i.e., alterations of DNA methylation in the CYP1A promoter of killifish, was the mechanistic basis of its resistance. That is because hypermethylation of CpG islands in gene promoters is associated with downstream silencing of gene expression by cis-acting regulatory elements. However, that did not turn out to be the case.

Arzuaga et al. (2004) reported that treatment of resistant killifish from Newark Bay with the DNA demethylating agent, 5-azacytidine, did not restore CYP1A inducibility in PCB126-treated specimens. Timme-Laragy et al. (2005) used a more sensitive and direct approach to evaluate the extent of DNA methylation at CpG islands in the CYP1A promoter of killifish from the Elizabeth River. They hypothesized that methylation would be more extensive in killifish from the Elizabeth River than control populations. Using the bisulfite sequencing method, they found no evidence of methylation at 34 CpG sites (including three that are in the AHR binding DREs of CYP1A) in livers of wild-caught adult or F_1 generation embryo offspring of killifish from the Elizabeth River or a reference site. Roy and Wirgin (unpublished data), also using bisulfite sequencing, found evidence of cytosine methylation in the CYP1A promoter of environmentally exposed tomcod, but no difference in their levels between tomcod from the Hudson River and a reference site. Thus, cytosine methylation of the CYP1A promoter does not appear to be the mechanistic basis of refractory CYP1A inducibility in these resistant fish models.

8.4.6 Does Histone Modification Occur in Fish Populations and Contribute to Resistance?

It is possible that histone modifications may serve as an epigenetic mechanism to reduce CYP1A inducibility and vulnerability to downstream early life stage toxicities. Histone methylation is known to usually suppress gene transcription, while histone acetylation has the opposite effect. Studies in various taxa, including fishes, have shown that a variety of metals alter expression of CYP1A and other inducible genes at the transcriptional level. In adult tomcod, Sorrentino et al. (2005) observed that exposure to As, Cr, or Ni significantly reduced constitutive expression and inducibility of hepatic CYP1A mRNA by B[a]P, but the mechanism of this repression was unknown. Similar downregulation of constitutive and induced CYP1A mRNA expression was also observed in tomcod larve. Furthermore, Schnekenburger et al. (2007) reported that prior exposure to Cr significantly reduced CYP1A mRNA inducibility in mouse hepatoma cells by altering remodeling patterns of acetylation by B[a]P at the histone mark H3K14Ac. Roy and Wirgin (unpublished data), using western blot analysis and commercially available antibodies raised against modified mammalian histones, observed that exposure of tomcod larvae to Cr responsively reduced levels of global histone acetylation at H3K14Ac (Figure 8.6), and that prior exposure to Cr reduced levels of histone acetylation at this histone mark in B[a]P- or PCB77-exposed larvae. Reduction of global histone acetylation is consistent with downregulation of downstream genes such as CYP1A in resistant tomcod. It will be interesting in the future to examine changes in histone acetylation within the promoter of inducible genes such as CYP1A. We hypothesize that histone modification provides an additional mechanism to alter gene expression and toxic phenotypes of organisms that are exposed to complex contaminant mixtures in impacted natural environments. This hypothesis has yet to be evaluated in environmentally exposed fishes from natural populations.

8.5 EVOLUTIONARY COSTS ASSOCIATED WITH RESISTANCE

It is thought that genetic adaptation to chemical stressors is associated with evolutionary costs. Costs may include reduced fitness in population life history characteristics such as reproductive rate, growth rate, survivorship, and genetic variation. Alternatively, increased sensitivity to stressors (chemical or physical), other than that adapted to, may occur. For example, longer time to first reproduction, decreased fecundity, abbreviated female life expectancy (Xie and Klerks 2004), and reduced microsatellite DNA variation (Athrey et al. 2007) were observed in a laboratory-developed line of Cd-resistant least killifish *Heterandria formosa*. Investigations into possible fitness costs associated with resistance of killifish to PAHs and PCBs have yielded mixed results. Meyer and Di Giulio (2003) reported that F_1 and F_2 offspring of killifish from the Elizabeth River were less tolerant of low oxygen conditions than were F_1 offspring of the control population. Moreover, compared to the F_1 offspring of reference river parents, the F_1, but not F_2, offspring of Elizabeth River parents exhibited reduced growth and survivorship under clean conditions. This finding was consistent with the erosion of resistance in killifish between the

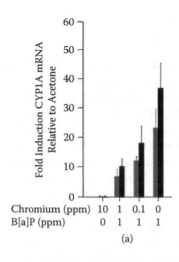
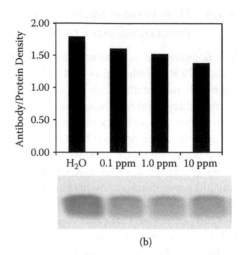

FIGURE 8.6 Effect of graded doses of Cr VI on (a) CYP1A mRNA expression, and (b) global histone acetylation at H4K16Ac in Atlantic tomcod larvae. Results in (a) are from two separate experiments in which tomcod larvae were coexposed to 1 ppm B[a]P or 10 ppm Cr VI alone or coexposed to two doses of Cr VI (1 and 0.1 ppm) and 1 ppm of B[a]P.

F_1 and F_2 generations reported by these investigators. This led these researchers to conclude that adaptation to PAH toxicity carried a cost of decreased fitness in other contexts. Furthermore, Frederick et al. (2007) compared the innate and acquired immune response of adult killifish collected from the Elizabeth River and a control site. They found that levels of circulating immunoglobulin M (IgM) and specific antibody responses against five ubiquitous marine bacteria were significantly lower in killifish from the Elizabeth River. Furthermore, killifish from the Elizabeth River failed to survive challenge with formalin-killed marine bacteria *Vibrio anguillarum*, while those from the control site thrived. However, because PAHs are known immunotoxicants in fishes, it is not clear if decreased capacity of immunosurveillance in Elizabeth River fish was due to within-generation exposure to contaminants or was a multigenerational evolutionary cost of resistance.

In similar studies of evolutionary costs, Nacci et al. (2009) hypothesized that resistance to PCBs in killifish from New Bedford Harbor was associated with increased susceptibility to challenge with a bacterial pathogen. However, they found that survival in F_2 laboratory-reared killifish or wild killifish from New Bedford Harbor was comparable to that in matched specimens from a control site that were i.p. injected with the marine bacteria *Vibrio harvyei*. In fact, greater survival was found in New Bedford Harbor than control site males. Furthermore, in earlier studies these investigators reported that New Bedford Harbor killifish exhibited normal reproductive output and high condition factors (Nacci et al. 2002b). They concluded that their results were inconsistent with the existence at these endpoints of evolutionary costs of resistance. It is interesting to note the differences between populations of killifish from the Elizabeth River and New Bedford Harbor in possible fitness costs associated with resistance. It is possible that they represent exposures to different toxicants with different mechanisms of toxicity and likely different mechanisms of adaptation.

8.5.1 Is Genetic Diversity Cropped in Resistant Populations?

Another cost of resistance is thought to be reduced genetic variation in resistant populations. Genetic diversity endows populations with increased fitness through the ability to adapt to changing environments, reduced sensitivity to disease, and is thought to covary with biodiversity. Genetic diversity in resistant populations may be cropped through intense directional selection for resistance alleles with a resulting reduction in effective population size. Several investigators have descriptively addressed this issue by comparing genetic diversity in resistant and matched control populations. Roark et al. (2005), who compared levels of allozyme diversity among eighteen populations of killifish, including that in New Bedford Harbor, failed to find an association between PCB exposure and levels of genetic diversity. Rather, patterns of genetic differentiation among populations were directly correlated with geographic distance, suggesting that stochastic rather than selective factors drove evolutionary change. Similarly, McMillan et al. (2006) failed to find significant differences in genetic diversity among six populations of killifish (including that from New Bedford Harbor) using a more sensitive DNA-based AFLP approach. Levels of allozyme (Mulvey et al. 2002) and mitochondrial DNA haplotype (Mulvey et al. 2003) diversity were also compared between the resistant-tolerant Elizabeth River and proximal control sites in the York River, Virginia. Both genetic analyses revealed that the Elizabeth River population was genetically distinct from other collections, and that the extent of genetic differentiation among sites was related to the extent of their sediment contamination with PAHs and not their geographic distance. However, no relationship was observed between levels of genetic diversity and contaminant exposure histories of these populations. This suggested that selection may have been operative to sculpt genetic relationships among populations, but not to reduce diversity in the resistant population. Wirgin et al. (manuscript submitted) also compared the mtDNA haplotype diversity in tomcod between the Hudson River and six reference populations that included two proximal and four distant from the Hudson. They also found that mtDNA diversity was not impoverished in the Hudson River population, and that spatial patterns of haplotype frequencies were directly correlated with geographic distance among populations. It is unlikely that any of the resistant populations have suffered from reduced genetic diversity as a result of contaminant exposure and resulting adaptation. It should be remembered that the focus of most of these studies was selectively neutral loci that would only exhibit decreased diversity if effective population sizes were severely reduced, a scenario that almost certainly did not occur.

8.6 CONCLUSIONS

Populations from polluted aquatic systems usually persist despite their bioaccumulation of toxic levels of PCBs, PCDDs, and PAHs singly, as mixtures, and in combination with other contaminants. Much of the ability of fishes to survive this challenge is through induction of phase I and phase II enzymes and their detoxifying activities. On occasion, the detoxifying capacities of these defenses are exceeded by high tissue burdens of these toxicants. In that case, the prevalence of environmental diseases, particularly hepatic and epithelial neoplasia and early life stage toxicities, may be elevated in impacted populations. It has also been demonstrated that populations from highly contaminated sites may persist through development of resistance through genetic adaptation or physiological acclimation. It is likely that both mechanisms ameliorate AH-induced toxicity through modulation of phase I and phase II enzyme activities. Studies in rodent knockout models have documented the essential role of AHR in the initiation and promotion of neoplasia in animals exposed to PAHs and PCBs. Presumably, the same applies to fishes, although the presence of duplicated AHR loci may complicate the story. Similarly, in fish knockdown models, the essentiality of AHR2 and downstream CYP1A activity has been confirmed in TCDD and some PAH-induced early life stage toxicities. Studies on resistant fish models have also demonstrated that downregulation of the expression and activities of these enzymes may prove beneficial to populations chronically exposed to high levels of these toxicants.

One remaining question is how extensive is resistance in populations from chemically impacted ecosystems? The discovery of resistant populations to date has been fortuitous. Do other species in these systems also exhibit a resistant phenotype? Is it the combination of high exposure levels, life history characteristics, and genetic profiles of these populations that renders them uniquely able to develop resistance? If not, how do other populations in these systems survive these harsh conditions?

The molecular mechanisms of resistance in these populations are largely unknown. One study has demonstrated that a single genetic polymorphism in AHR2 significantly diminishes CYP1A inducibility and presumably sensitivity to early life toxicities. Because these populations are largely reproductively isolated, there are probably multiple independent mechanisms through which resistance was developed and maintained, probably reflecting their exposure histories to different contaminants with differing modes of toxicity and their differing population demographics. It is likely that targets other than AHR pathway molecules and perhaps polygenic variation are the basis of heritable resistance in these populations. Based on the relationship between sediment PCB concentrations and LC20 values observed among many killifish populations, it is likely that physiological acclimation also occurs.

We predict that epigenetic modifications will prove an important factor in determining how populations persist in contaminated environments. To date, few studies have investigated epigenetic modifications of DNA or histones in natural populations, but their regular occurrence *in vivo* and *in vitro* in chemically treated mammalian models suggests their presence in natural populations. For example, the effects of several metals in altering global signatures have been reported at multiple histone marks in mammalian cells, and many of these probably result in modified downstream gene expression and phenotypes (Zhou et al. 2008; Sun et al. 2009). Once the presence of epigenetic alterations is confirmed globally in chemically treated organisms, technologies such as chromatin immunoprecipitation (ChIP) will allow for the localization of these altered histone patterns at regulatory elements of specific genes such as CYP1A and other toxicologically relevant genes in the AHR battery and elsewhere.

Paradoxically, high prevalences of hepatic neoplasms have been observed in two populations for which resistance to CYP1A induction and early life toxicities was also observed: killifish from the Elizabeth River and tomcod from the Hudson River (Wirgin and Waldman 1998). If these populations are resistant to PAHs and PCBs, respectively, why have epizootics of neoplasia developed in them? It is possible that their resistance developed recently and their prevalences of tumors today are significantly reduced compared to historic highs reported in the literature. It may be hypothesized that these populations have also developed resistance to chemically induced hepatic neoplasia. Given the demonstrated role of AHR2 in the development of tumors in B[a]P-treated rodents, it is possible that AHR pathway downregulation has reduced the vulnerability of tomcod to hepatic carcinogenesis. And indeed, while still elevated, the reported prevalence of hepatic neoplasms in tomcod is lower today than twenty-five years ago. It will be interesting in the future to empirically evaluate under controlled laboratory conditions the sensitivity of these resistant populations to PAH-initiated and PCB-promoted neoplasia. However, it is also possible that hepatic tumors may provide a selective advantage to populations challenged with high levels of chemical toxicants. Years ago, Farber and colleagues (Farber and Rubin 1991) proposed a resistant hepatocyte model in which rare hepatocytes become initiated by exposure to PAHs, express low levels of phase I bioactivating enzymes, high levels of phase II conjugation enzymes, and by this mechanism are provided a selective advantage allowing their development as hepatocyte nodules. Resistant hepatocytes are able to proliferate vigorously, resist cytotoxicity, and exhibit these unusual patterns of gene induction. On the organismal level, studies have demonstrated that rats with liver tumors initiated by this protocol develop a unique ability to express resistance to a variety of toxic agents (Yusuf et al. 1999). Because cancer is postreproductive in these fish models, environmental-induced hepatic neoplasia may provide another mechanism of resistance allowing populations to persist in the face of contaminant exposure. Hepatic neoplasia may provide a complementary mechanism for impacted populations to persist in some highly contaminated environments, albeit at the cost of an altered population age structure.

ACKNOWLEDGMENTS

I.W. acknowledges support of SBRP grant ES10344 and NIEHS Center Grant ES00260.

REFERENCES

Abrahamson, A., et al. 2007. Gill EROD in monitoring of CYP1A inducers in fish—A study in rainbow trout (*Oncorhynchus mykiss*) caged in Stockholm and Uppsala waters. *Aquat. Toxicol.* 85:1–8.

Andersson, T., M. Pesonen, and C. Johansson. 1985. Differential induction of cytochrome P-450-dependent monooxygenase, epoxide hydrolase, glutathione transferase and UDP glucuronosyl transferase activities in the liver of the rainbow trout by [beta]-naphthofla-vone or clophen A50. *Biochem. Pharmacol.* 34:3309–14.

Antkiewicz, D. S., et al. 2005. Heart malformation is an early response to TCDD in embryonic zebrafish. *Toxicol. Sci.* 84:368–77.

Antkiewicz, D. S., R. E. Peterson, and W. Heideman. 2006. Blocking expression of AHR2 and ARNT1 in zebra larvae protects against cardiac toxicity of 2,3,7,8-tetrachlorodibenzo-*p*-dioxin. *Toxicol. Sci.* 94:175–82.

Arzuaga, X., W. Calcano, and A. Elskus. 2004. The DNA de-methylating agent 5-azacytidine does not restore CYP1A induction in PCB resistant Newark Bay killifish (*Fundulus heteroclitus*). *Mar. Environ. Res.* 58:517–20.

Arzuaga, X., and A. Elskus. 2010. Polluted site killifish (*Fundulus heteroclitus*) embryos are resistant to organic pollutant mediated induction of CYP1A activity, reactive oxygen species, and heart deformities. *Environ. Toxicol. Chem.* 29:676–82.

Athrey, N. R. G., P. L. Leberg, and P. L. Klerks. 2007. Laboratory culturing and selection for increased resistance to cadmium reduce genetic variation in the least killifish *Heterandria formosa*. *Environ. Toxicol. Chem.* 26:1916–21.

Bebianno, M. J., et al. 2007. Glutathione S-tranferases and cytochrome P450 activities in *Mytilus galloprovincialis* from the south coast of Portugal: Effect of abiotic factors. *Environ. Int.* 33:550–58.

Bello, S. M., et al. 2001. Acquired resistance to Ah receptor agonists in a population of Atlantic killifish (*Fundulus heteroclitus*) inhabiting a marine Superfund site: *In vivo* and *in vitro* studies on the inducibility of xenobiotic metabolizing enzymes. *Toxicol. Sci.* 60:77–91.

Beyer, J., et al. 1996. Contaminant accumulation and biomarker responses in flounder (*Platichthys flesus* L.) and Atlantic cod (*Gadus morhua* L.) exposed by caging to pol-luted sediments in Sørfjorden, Norway. *Aquat. Toxicol.* 36:75–98.

Boryslawskyj, M., et al. 1988. Elevation of glutathione S-transferase activity as a stress response to organochlorine compounds, in the fresh water mussel, *Sphaerium corneum*. *Mar. Environ. Res.* 24:101–4.

Brown, E. D., et al. 1996. Injury to the early life stages of Pacific herring in Prince William Sound after the *Exxon Valdez* oil spill. In *Proceedings of the Exxon Valdez Oil Spill Symposium*, ed. S. D. Rice, R. B. Spies, D. A. Wolfe, and B. A. Wright, 448–62. Bethesda, MD: American Fisheries Society.

Bucheli, T., and K. Fent. 1995. Induction of cytochrome P450 as a biomarker for environmen-tal contamination in aquatic ecosystems. *Crit. Rev. Environ. Sci. Technol.* 25:201–68.

Bue, B. G., et al. 1996. Effects of the *Exxon Valdez* oil spill on pink salmon embryos and preemergent fry. In *Proceedings of the Exxon Valdez Oil Spill Symposium*, ed. S. D. Rice, R. B. Spies, D. A. Wolfe, and B. A. Wright, 619–27. Bethesda, MD: American Fisheries Society.

Burgeot, T., et al. 1996. Bioindicators of pollutant exposure in the northwestern Mediterranean Sea. *Mar. Ecol. Prog. Ser.* 131:125–41.

Butler, R. A., et al. 2001. An aryl hydrocarbon receptor (AHR) homologue from the soft-shell clam, *Mya arenaria*: Evidence that invertebrate AHR homologues lack 2,3,7,8-tetrachlorodibenzo-*p*-dioxin and β-napthoflavone binding. *Gene* 278:223–34.

Carlson, E. A., N. K. Roy, and I. I. Wirgin. 2009. Microarray analysis of polychlorinated biphenyl mixture-induced changes in gene expression among Atlantic tomcod popula-tions displaying differential sensitivity to halogenated aromatic hydrocarbons. *Environ. Toxicol. Chem.* 28:759–71.

Chaty, S., F. Rodius, and P. Vasseur. 2004. A comparative study of the expression of CYP1A and CYP4 genes in aquatic invertebrate (freshwater mussel, *Unio tumidus*) and vertebrate (rainbow trout, *Oncorhynchus mykiss*). *Aquat. Toxicol.* 69:81–94.

Clark, A. G., et al. 1984. Characterization of multiple glutathione transferases from the house fly, *Musca domestica* (L). *Pest. Biochem. Phys.* 22:51–59.

Clark, B. W., et al. 2010. AHR2 mediates cardiac teratogenesis of polycyclic aromatic hydrocarbons and PCB-126 in Atlantic killifish (*Fundulus heteroclitus*). *Aquat. Toxicol.* 99: 232–240.

Collier, T. K., et al. 1995. A field evaluation of cytochrome P4501A as a biomarker of contaminant exposure in three species of flatfish. *Environ. Toxicol. Chem.* 14:143–52.

Courtenay, S., et al. 1999. A comparison of the dose and time response of CYP1A1 mRNA induction in chemically treated Atlantic tomcod from two populations. *Aquat. Toxicol.* 47:43–69.

Denison, M. S., et al. 2002. Ligand binding and activation of the Ah receptor. *Chem.-Biol. Interact.* 141:3–24.

Dey, W. P., et al. 1993. Epizoology of hepatic neoplasia in Atlantic tomcod (*Microgadus tomcod*) from the Hudson River estuary. *Can. J. Fish. Aquat. Sci.* 50:1897–907.

Di Giulio, R., et al. 1995. Biochemical mechanisms of contaminant metabolism, adaptation, and toxicity. In *Fundamentals of aquatic toxicology*, ed. G. Rand, 523–61. 2nd ed. Bristol, PA: Taylor and Francis.

Eggens, M. L., A. Opperhuizen, and J. P. Boon. 1996. Temporal variation of CYP1A indices, PCB and 1-OH pyrene concentration in flounder, *Platichthys flesus*, from the Dutch Wadden Sea. *Chemosphere* 33:1579–96.

Elskus, A. A., et al. 1999. Altered CYP1A expression in *Fundulus heteroclitus* adults and larvae: A sign of pollution resistance? *Aquat. Toxicol.* 45:99–113.

Evans, B. R., et al. 2008. Repression of aryl hydrocarbon receptor (AHR) signaling by AHR repressor: Role of DNA binding and competition of AHR nuclear translocator. *Mol. Pharmacol.* 73:387–98.

Farber, E., and H. Rubin. 1991. Cellular adaptation in the origin and development of cancer. *Cancer Res.* 51:2751–61.

Fisher, M. A., and M. F. Oleksiak. 2007. Convergence and divergence in gene expression among natural populations exposed to pollution. *BMC Genomics* 8:108.

Fitzpatrick, P. J., D. Sheehan, and D. R. Livingstone. 1995. Studies on isoenzymes of glutathione S-transferase in the digestive gland of *Mytilus galloprovincialis* with exposure to pollution. *Mar. Environ. Res.* 39:241–44.

Förlin, L., et al. 1985. Effects of pulp bleach plant effluents on hepatic xenobiotic biotransformation enzymes in fish: Laboratory and field studies. *Mar. Environ. Res.* 17:109–12.

Förlin, L., and C. Haux. 1985. Increased excretion in the bile of 17[beta]-[3H]estradiol-derived radioactivity in rainbow trout treated with [beta]-naphthoflavone. *Aquat. Toxicol.* 6:197–208.

Foureman, G. 1989. Enzymes involved in metabolism of PAH by fishes and other aquatic animals: Hydrolysis and conjugation enzymes (for phase II enzymes). In *Metabolism of polycyclic aromatic hydrocarbons in the aquatic environment*, ed. U. Varanasi, 185–202. Boca Raton, FL: CRC Press.

Frederick, L. A., P. A. Van Veld, and C. D. Rice. 2007. Bioindicators of immune function in creosote-adapted estuarine killifish, *Fundulus heteroclitus*. *J. Toxicol. Environ. Health* 70A:1433–442.

Gallagher, E., et al. 2000. *In vitro* kinetics of hepatic glutathione S-transferase conjugation in largemouth bass and brown bullheads. *Environ. Toxicol. Chem.* 19:319–26.

Garrigues, P., et al. 1993. Banking of environmental samples for short-term biochemical and chemical monitoring of organic contamination in coastal marine environments: The GICBEM experience (1986–1990). *Sci. Total Environ.* 139–40:225–36.

Godard, C. A. J., et al. 2005. The new vertebrate CYP1C family: Cloning of new subfamily members and phylogenetic analysis. *Biochem. Biophys. Res. Commun.* 331:1016–24.

Goksøyr, A., et al. 1991. Immunochemical cross-reactivity of β-naphtoflavone-inducible cytochrome P450(P40IA) in liver microsomes from different fish species and rat. *Fish Physiol. Biochem.* 9:1–13.

Goksøyr, A., and L. Förlin. 1992. The cytochrome P-450 system in fish, aquatic toxicology and environmental monitoring. *Aquat. Toxicol.* 22:287–311.

Guengerich, F. P. 1993. Cytochrome P450 enzymes. *Am. Sci.* 81:440–47.

Hahn, M. E. 2002a. Aryl hydrocarbon receptors: Diversity and evolution. *Chem.-Biol. Interact.* 141:131–60.

Hahn, M. E. 2002b. Biomarkers and bioassays for detecting dioxin-like compounds in the marine environment. *Sci. Total Environ.* 289:49–69.

Hahn, M. E., et al. 2004. Aryl hydrocarbon receptor polymorphisms and dioxin resistance in Atlantic killifish (*Fundulus heteroclitus*). *Pharmacogenetics* 14:131–43.

Handley-Goldstone, H. M., M. W. Grow, and J. J. Stegeman. 2005. Cardiovascular gene expression profiles of dioxin exposure in zebrafish embryos. *Toxicol. Sci.* 85:683–93.

Head, J. A., M. E. Hahn, and S. W. Kennedy. 2008. Key amino acids in the aryl hydrocarbon receptor predict dioxin sensitivity in avian species. *Environ. Sci. Technol.* 42:7535–41.

Heilmann, L. J., et al. 1988. Trout P450IA1: cDNA and deduced protein sequence, expression in liver, and evolutionary significance. *DNA* 7:379–87.

Hoarau, P., et al. 2006. Cloning and expression of a GST-pi gene in *Mytilus galloprovincialis*. Attempt to use the GST-pi transcript as a biomarker of pollution. *Comp. Biochem. Physiol.* 143C:196–203.

Hoarau, P., et al. 2004. Effect of three xenobiotic compounds on glutathione S-transferase in the clam *Ruditapes decussatus*. *Aquat. Toxicol.* 68:87–94.

Jonsson, M. E., et al. 2007. Role of AHR2 in the expression of novel cytochrome P450 1 family genes, cell cycle genes, and morphological defects in developing zebra fish exposed to 3,3′,4,4′,5-pentachlorobiphenyl or 2,3,7,8-tetrachlorodibenzo-*p*-dioxin. *Toxicol. Sci.* 100:180–93.

Jørgensen, A., L. J. Rasmussen, and O. Andersen. 2005. Characterisation of two novel CYP4 genes from the marine polychaete *Nereis virens* and their involvement in pyrene hydroxylase activity. *Biochem. Biophys. Res. Commun.* 336:890–97.

Jung, D., et al. 2009. Effects of benzo[a]pyrene on mitochondrial and nuclear DNA damage in Atlantic killifish (*Fundulus heteroclitus*) from a creosote-contaminated and reference site. *Aquat. Toxicol.* 95:44–51.

Karchner, S. I., et al. 2002. Regulatory interactions among three members of the vertebrate aryl hydrocarbon receptor family: AHR repressor, AHR1 and AHR2. *J. Biol. Chem.* 77:6949–59.

Kirby, M. F., et al. 1999. Hepatic EROD activity in flounder (*Platichthys flesus*) as an indicator of contaminant exposure in English estuaries. *Mar. Pollut. Bull.* 38:676–86.

Kirby, M. F., et al. 2004. Hepatic ethoxyresorufin O-deethylase (EROD) activity in flounder (*Platichthys flesus*) from contaminant impacted estuaries of the United Kingdom: Continued monitoring 1999–2001. *Mar. Pollut. Bull.* 49:71–78.

Kirby, M. F., et al. 2007. Ethoxyresorufin O-deethylase (EROD) and vitellogenin (VTG) in flounder (*Platichthys flesus*): System interaction, crosstalk and implications for monitoring. *Aquat. Toxicol.* 81:233–44.

Kopecka, J., et al. 2006. Measurements of biomarker levels in flounder (*Platichthys flesus*) and blue mussel (*Mytilus trossulus*) from the Gulf of Gdansk (southern Baltic). *Mar. Pollut. Bull.* 53:406–21.

Kreamer, G.-L., et al. 1991. Cytochrome P450IA mRNA expression in feral Hudson River tomcod. *Environ. Res.* 55:64–78.

Kuiper, R. V., et al. 2007. Long-term exposure of European flounder (*Platichthys flesus*) to the flame-retardants tetrabromobisphenol A (TBBPA) and hexabromocyclododecane (HBCD). *Ecotoxicol. Environ. Saf.* 67:349–60.

Lafferty, K. D., J. W. Porter, and S. E. Ford. 2004. Are diseases increasing in the ocean? *Annu. Rev. Ecol. Evol. Syst.* 35:31–54.

Lee, R. 1988. Glutathione S-transferase in marine invertebrates from Langesundfjord. *Mar. Ecol. Prog. Ser.* 46:33–36.

Lee, R. F. 1998. Annelid cytochrome P-450. *Comp. Biochem. Physiol.* 121C:173–79.

Lehtonen, K. K., et al. 2006. The BEEP project in the Baltic Sea: Overview of results and outline for a regional biological effects monitoring strategy. *Mar. Pollut. Bull.* 53:523–37.

Mannervick, B. 1985. The isoenzymes of the glutathione S-transferase. *Adv. Enzymol.* 57:357–417.

McMillan, A. M., et al. 2006. Genetic diversity and structure of an estuarine fish (*Fundulus heteroclitus*) indigenous to sites associated with a highly contaminated urban harbor. *Ecotoxicology* 15:539–48.

Meyer, J. N., and R. T. Di Giulio. 2002. Patterns of heritability of decreased EROD activity and resistance to PCB 126-induced teratogenesis in laboratory-reared offspring of killifish (*Fundulus heteroclitus*) from a creosote-contaminated site in the Elizabeth River, VA, USA. *Mar. Environ. Res.* 54:621–26.

Meyer, J. N., and R. T. Di Giulio. 2003. Heritable adaptation and fitness costs in killifish (*Fundulus heteroclitus*) inhabiting a polluted estuary. *Ecol. Appl.* 13:490–503.

Meyer, J. N., D. E. Nacci, and R. T. Di Giulio. 2002. Cytochrome P4501A (CYP1A) in killifish (*Fundulus heteroclitus*): Heritability of altered expression and relationship to survival in contaminated sediments. *Toxicol. Sci.* 68:69–81.

Meyer, J. N., et al. 2005. Differential display of hepatic mRNA from killifish (*Fundulus heteroclitus*) inhabiting a Superfund estuary. *Aquat. Toxicol.* 73:327–41.

Meyer, J. N., et al. 2003. Expression and inducibility of aryl hydrocarbon receptor pathway genes in wild-caught killifish (*Fundulus heteroclitus*) with different contaminant histories. *Environ. Toxicol. Chem.* 22:2337–43.

Michel, X., et al. 1994. Benzo[a]pyrene hydroxylase activity in the marine mussel *Mytilus galloprovincialis*: A potential marker of contamination by polycyclic aromatic hydrocarbon-type compounds. *Mar. Environ. Res.* 38:257–73.

Morrison, H. G., et al. 1998. Molecular cloning of CYP1A from the estuarine fish *Fundulus heteroclitus* and phylogenetic analysis of CYP1 genes: Update with new sequences. *Comp. Biochem. Physiol.* 121C:231–40.

Mulvey, M., et al. 2003. Genetic structure and mtDNA diversity of *Fundulus heteroclitus* populations from polycyclic aromatic hydrocarbon-contaminated sites. *Environ. Toxicol. Chem.* 22:671–77.

Mulvey, M., et al. 2002. Genetic structure of *Fundulus heteroclitus* from PAH-contaminated and neighboring sites in the Elizabeth and York Rivers. *Aquat. Toxicol.* 61:195–209.

Myers, M. S., L. L. Johnson, and T. K. Collier. 2003. Establishing the causal relationship between polycyclic aromatic hydrocarbon (PAH) exposure and hepatic neoplasms and neoplasia-related liver lesions in English sole (*Pleuronectes vetulus*). *Hum. Ecol. Risk Assess.* 9:67–94.

Nacci, D. E., D. Champlin, and S. Jayaraman. 2010. Adaptation of the estuarine fish *Fundulus heteroclitus* (Atlantic killifish) to toxic pollutants. *Estuaries Coasts* 33:853–64.

Nacci, D., et al. 1999. Adaptations of wild populations of the estuarine fish *Fundulus heteroclitus* to persistent environmental contaminants. *Mar. Biol.* 134:9–17.

Nacci, D., et al. 2002a. Predicting the occurrence of genetic adaptation to dioxin-like compounds in populations of the estuarine fish *Fundulus heteroclitus. Environ. Toxicol. Chem.* 7:1525–32.

Nacci, D., et al. 2002b. Effects of environmental stressors on wildlife populations. In *Coastal and estuarine risk assessment: Risk on the edge*, ed. M. C. Newman, 247–72. Washington, DC: CRC Press/Lewis Publishers.

Nacci, D. E., et al. 2002c. Effects of benzo[a]pyrene exposure on a fish population resistant to the toxic effects of dioxin-like compounds. *Aquat. Toxicol.* 57:203–15.

Nacci, D., et al. 2009. Evolution of tolerance to PCBs and susceptibility to a bacterial pathogen (*Vibrio harveyi*) in Atlantic killifish (*Fundulus heteroclitus*) from New Bedford Harbor (MA, USA) harbor. *Environ. Pollut.* 157:857–64.

Nebert, D. W., and D. R. Nelson. 1991. P-450 gene nomenclature based on evolution. *Meth. Enzymol.* 206:3–11.

Nelson, D. R. 1998. Metazoan cytochrome P450 evolution. *Comp. Biochem. Physiol.* 121C:15–22.

Okey, A. B., et al. 2005. Toxicological implications of polymorphisms in receptors for xenobiotic chemicals: The case of the aryl hydrocarbon receptor. *Toxicol. Appl. Pharm.* 207(Suppl. 1):43–51.

Oleksiak, M. F. 2008. Changes in gene expression due to chronic exposure to environmental pollutants. *Aquat. Toxicol.* 90:161–71.

Oris, J., and A. Roberts. 2007. Statistical analysis of CYP1A biomarker measurements in fish. *Environ. Toxicol. Chem.* 26:1742–50.

Owenby, D. C., et al. 2002. Fish (*Fundulus heteroclitus*) populations with different exposure histories differ in tolerance of creosote-contaminated sediments. *Environ. Toxicol. Chem.* 21:1897–902.

Payne, J., and W. R. Penrose. 1975. Induction of aryl hydrocarbon (benzo[a]pyrene) hydroxylase in fish by petroleum. *Bull. Environ. Contam. Toxicol.* 14:112–16.

Petrivalsky, M., et al. 1997. Glutathione-dependent detoxifying enzymes in rainbow trout liver: Search for specific biochemical markers of chemical stress. *Environ. Toxicol. Chem.* 16:1417–21.

Pinkney, A. E., and J. C. Harshbarger. 2006. Tumor prevalence in mummichogs from the Delaware River watershed. *J. Aquat. Animal Health* 18:244–51.

Poland, A., and E. Glover. 1975. Genetic expression of aryl hydrocarbon hydroxylase by 2,3,7,8-tetrachloro dibenzo-p-dioxin: Evidence of a receptor mutation in genetically non-responsive mice. *Mol. Pharmacol.* 11:389–98.

Powell, W. H., et al. 2000. Developmental and tissue specific expression of AHR1, AHR2, and ARNT2 in dioxin-sensitive and -resistant populations of the marine fish *Fundulus heteroclitus*. *Toxicol. Sci.* 57:229–39.

Prasch, A. L., et al. 2003. Aryl hydrocarbon receptor 2 mediates 2,3,7,8-tetrachlorodibenzo-*p*-dioxin developmental toxicity in zebrafish. *Toxicol. Sci.* 76:138–50.

Prince, R., and K. R. Cooper. 1995a. Comparisons of the effects of 2,3,7,8-tetrachlordibenzo-*p*-dioxin on chemically impacted and nonimpacted subpopulations of *Fundulus heteroclitus*. I. TCDD toxicity. *Environ. Toxicol. Chem.* 14:579–588.

Prince, R., and K. R. Cooper. 1995b. Comparisons of the effects of 2,3,7,8-tetrachlorodibenzo-*p*-dioxin on chemically impacted and nonimpacted subpopulations of *Fundulus heteroclitus*. II. Metabolic considerations. *Environ. Toxicol. Chem.* 14:589–596.

Rewitz, K. F., et al. 2004. Identification of two *Nereis virens* (Annelida: Polychaeta) cytochromes P450 and induction by xenobiotics. *Comp. Biochem. Physiol.* 138C:89–96.

Roark, S. A., et al. 2005. Population genetic structure of a non-migratory estuarine fish (*Fundulus heteroclitus*) across a strong gradient of polychlorinated biphenyl contamination. *Environ. Toxicol. Chem.* 24:717–725.

Roy, N. K., et al. 2006. Characterization of the aryl hydrocarbon receptor repressor and a comparison of its expression in Atlantic tomcod from resistant and sensitive populations. *Environ. Toxicol. Chem.* 25:560–71.

Roy, N. K., et al. 1995. Characterization and prevalence of a polymorphism in the 3′ untranslated region of cytochrome P4501A1 in cancer-prone Atlantic tomcod. *Arch. Biochem. Biophys.* 322:204–13.

Schlezinger, J. J., et al. 2006. Uncoupling of cytochrome P450 1A and stimulation of reactive oxygen species production by co-planar polychlorinated biphenyl congeners. *Aquat. Toxicol.* 77:422–32.

Schnekenburger, M., G. Talaska, and A. Puga. 2007. Chromium cross-links histone deacetylase 1-DNA methyltransferase I complexes to chromatin, inhibiting histone-remodeling marks critical for transcriptional activation. *Mol. Cell. Biol.* 27:7089–101.

Sheehan, D., et al. 2001. Structure, function and evolution of glutathione transferases: Implications for classification of non-mammalian members of an ancient enzyme superfamily. *Biochem. J.* 360:1–16.

Shimizu, Y., et al. 2000. Benzo[a]pyrene carcinogenicity is lost in mice lacking the aryl hydrocarbon receptor. *Proc. Natl. Acad. Sci. USA* 97:779–82.

Simpson, A. E. C. M. 1997. The cytochrome P450 4 (CYP4) family. *Gen. Pharmacol.- Vasc.* S28:351–59.

Snyder, M. J. 1998. Cytochrome P450 enzymes belonging to the CYP4 family from marine invertebrates. *Biochem. Biophys. Res. Commun.* 249:187–90.

Sorrentino, C., et al. 2005. Co-exposure to metals modulates CYP1A1 mRNA inducibility in Atlantic tomcod *Microgadus tomcod* from two populations. *Aquat. Toxicol.* 75:238–52.

Stegeman, J. J. 1978. Influence of environmental contamination on cytochrome P-450 mixed function oxygenases in fish: Implications for recovery in the Wild Harbor Marsh. *J. Fish. Res. Board Can.* 35:668–74.

Stegeman, J. J., et al. 1992. Molecular responses to environmental contamination: Enzyme and protein systems as indicators of chemical exposure and effect. In *Biomarkers. Biochemical, physiological, and histological markers of anthropogenic stress*, ed. R. J. Huggett et al., 235–335. Boca Raton, FL: Lewis Publishers.

Stegeman, J. J., and M. Hahn. 1994. Biochemistry and molecular biology of monoxygenases: Current perspectives on forms, functions, and regulation of cytochrome P450 in aquatic species. In *Aquatic toxicology: Molecular, biochemical and cellular perspectives*, ed. G. Ostrander and D. Malins, 87–206. Boca Raton, FL: Lewis Publishers.

Stentiford, G. D., et al. 2009. Site-specific disease profiles in fish and their use in environmental monitoring. *Mar. Ecol. Prog. Ser.* 381:1–15.

Sun, H., et al. 2009. Modulation of histone modification and MLH1 gene silencing by hexavalent chromium. *Toxicol. Appl. Pharm.* 237:258–66.

Teraoka, H., et al. 2003. Induction of cytochrome P450 1A is required for circulation failure and edema by 2,3,7,8-tetrachlorodibenzo-*p*-dioxin in zebrafish. *Biochem. Biophys. Res. Commun.* 304:223–28.

Tillitt, D. E., et al. 2008. Reproductive impairment of Great Lakes lake trout by dioxin-like chemicals. In *The toxicology of fishes*, ed. R. T. Di Giulio and D. E. Hinton, 819–75. Boca Raton, FL: CRC Press.

Timme-Laragy, A. R., et al. 2005. Analysis of CpG methylation in the killifish CYP1A promoter. *Comp. Biochem. Physiol.* 141C:406–11.

Van Veld, P., and D. Westbrook. 1995. Evidence for depression of cytochrome P4501A in a population of chemically resistant mummichog (*Fundulus heteroclitus*). *Environ. Sci.* 3–4:221–34.

Vogelbein, W. G., et al. 1990. Hepatic neoplasms in the mummichog *Fundulus heteroclitus* from a creosote contaminated site. *Cancer Res.* 50:5978–86.

Wilce, M. C. J., and M. W. Parker. 1994. Structure and function of glutathione S-transferases. *BBA Protein Struct. Mol.* 1205:1–18.

Williams, L. M., and M. F. Oleksiak. 2008. Signatures of selection in natural populations adapted to chronic pollution. *BMC Evol. Biol.* 8:282.

Wills, L. P., et al. 2009. Effects of CYP1A inhibition on the biotransformation of benzo[a] pyrene in two populations of *Fundulus heteroclitus* with different exposure histories. *Aquat. Toxicol.* 92:195–201.

Wirgin, I. I., and R. C. Chambers. 2006. Atlantic tomcod (*Microgadus tomcod*): A model species for the response of Hudson River fish to toxicants. In *Hudson River fishes and their environment*, ed. J. R. Waldman, K. E. Limburg, and D. Strayer, 331–64. Bethesda, MD: American Fisheries Society.

Wirgin, I. I., et al. 1994. A biomarker approach in assessing xenobiotic exposure in cancer-prone Atlantic tomcod from the North American Atlantic coast. *Environ. Health Persp.* 102:764–70.

Wirgin, I. I., et al. 1992. Effects of prior exposure history on cytochrome P450IA mRNA induction by PCB congener 77 in Atlantic tomcod. *Mar. Environ. Res.* 34:103–8.

Wirgin, I. I., and J. R. Waldman. 1998. Altered gene expression and genetic damage in North American fish populations. *Mutation Res.* 399:193–219.

Wirgin, I., and J. R. Waldman. 2004. Resistance to contaminants in North America fish populations. *Mutation Res.* 552:73–100.

Wootton, A. N., et al. 1995. Evidence for the existence of cytochrome P450 gene families (CYP1A, 3A, 4A, 11A) and modulation of gene expression (CYP1A) in the mussel *Mytilus* spp. *Mar. Environ. Res.* 39:21–26.

Xie, L., and P. L. Klerks. 2004. Fitness costs of resistance to cadmium in the least killifish (*Heterandria formosa*). *Environ. Toxicol. Chem.* 23:1499–503.

Yuan, Z. 2003. Resistance of CYP1A mRNA induction to halogenated aromatic hydrocarbons in Atlantic tomcod (*Microgadus tomcod*) from the Hudson River—Its prevalence and mechanistic basis. PhD thesis, New York University.

Yuan, Z., et al. 2006. Evidence of spatially extensive resistance to PCBs in an anadromous fish of the Hudson River. *Environ. Health Persp.* 114:77–83.

Yusuf, A., et al. 1999. Development of resistance during the early stages of experimental liver carcinogenesis. *Carcinogenesis* 20:1641–44.

Zanette, J., et al. 2009. New cytochrome P450 1B1, 1C2 and 1D1 genes in the killifish *Fundulus heteroclitus*: Basal expression and response of five killifish CYP1s to the AHR agonist PCB126. *Aquat. Toxicol.* 93:234–43.

Zhou, X., et al. 2008. Arsenite alters global histone H3 modification. *Carcinogenesis* 29:1831–36.

9 Stress Proteins and the Acquisition of Tolerance

Catherine Mouneyrac and Michèle Roméo

CONTENTS

9.1 INTRODUCTION

Organisms in their ambient environment are exposed to various stressor agents, such as temperature, oxygen, pH levels, salinity, UV radiation, and chemical contaminants. Each of these abiotic factors may occur alone or in combination with others, including biotic factors (e.g., growth, reproduction). Abiotic stress usually brings about protein dysfunction. In order to survive under stress, it is very important for the cell to have mechanisms to maintain proteins in their functional conformations and to prevent nonnative proteins from aggregation. Now, it is well known that heat shock proteins play an important role in protein folding, assembly, translocation, and degradation in many normal cellular processes, stabilise proteins and membrane, and can assist in protein refolding under stress conditions (Wang et al. 2004). Heat shock proteins were given that name as their synthesis is induced when cultured cells or the whole organisms are exposed to elevated temperature. This response is observed in every organism in which it has been sought. However, there has been increasingly more data to show that heat shock proteins are not only induced by heat shock, but also by a wide variety of other stresses, and seem to have very general protective functions (Lindquist 1986; Almoguera et al. 1993; Alamillo et al. 1995; Sabehat et al. 1998; Harndahl et al. 1999; Hamilton and Heckathorn 2001; Iba 2002).

Heat shock proteins have been studied in many different organisms, from eubacteria to archaebacteria, from mice to soybeans. Most of the examples given in this chapter will deal with aquatic organisms since they can live in quickly changing environments. For example, temperature and salinity stresses are often combined and interact in nature, e.g., in intertidal regions, and the salinity tolerance of organisms is directly influenced by temperature (Kinne 1971). These factors have an impact directly or indirectly on physiological performance, and therefore may limit the distribution of a species. In response to cellular stress, so far the only known universal system is the production of stress proteins (Gross 2004). Different stress-protective molecular pathways, more or less specific for the different stressors and particularly heat shock proteins (HSPs), have evolved to cope with the molecular consequences of cell stress (Feder and Hofmann 1999). HSPs make up approximately 5 to 10% of the total protein content of cells under conditions of normal, healthy growth (Pockley 2003).

The final goal of a species' stress response might translate into an ecophysiological advantage, and the expression pattern of HSPs is directly related to thermal history (Tomanek and Somero 1999; Dong and Dong 2008). For example, Nakano and Iwama (2002) showed that two congeneric species of tidepool fishes of the genus *Oligocottus* that inhabit different thermal niches in the intertidal zone differ in their heat shock responses. The authors concluded that, compared with the thermally sensitive *O. snyderi* found only in lower tidepools or the subtidal zone with small changes of temperature, the less sensitive one, the tidepool sculpin *O. maculosus* distributed in upper tidepools with temperature extremes, may enhance its thermal tolerance by having a large pool of heat shock proteins (HSP70) that may have a close relationship with the thermal tolerance of these fish species. These sculpins are intertidal fishes that may be good models to study the effect of short-term changes in thermal history on HSP response since they are exposed to fluctuations in water environmental variables such as temperature.

Many papers deal with the induction of HSPs in both laboratory and field studies. Among HSPs, the HSP70 family members are the most extensively studied for their characterisation and induction in response to environmental stressors in a range of species. For instance, the production of HSP70 was investigated twenty years ago in a natural population of amphipods (*Gammarus* sp.) and HSP70s varied with ambient temperature (Bradley and Ward 1989); induction was assessed using immunoblotting techniques. In two species of limpets of the genus *Colisella*, *C. scabra* living in the upper part of the intertidal zone, and therefore subject to adverse temperatures, showed greater thermal tolerance as well as a larger response to acute thermal shock than did *C. pelta* living in the mid-intertidal zone (Sanders et al. 1991).

9.2   HEAT SHOCK PROTEIN FAMILY

The cellular response to stress consists of the induction of a family of ubiquitous proteins highly conserved in evolution, in particular heat shock proteins (now referred to as high stress proteins). The induction of the HSP family has been widely documented in different taxonomic groups (Pyza et al. 1997; Clayton et al. 2000; Wang et al. 2004; de Miguel et al. 2009; Slotsbo et al. 2009; Woolfson and Heikkila 2009).

The regulation of HSP70 gene expression occurs mainly at the transcriptional level (review in Basu et al. 2002). Analysis of heat shock protein genes and a comparison of their promoter sequences from a variety of organisms led to the identification of a palindromic heat shock element (HSE) CNNGAANNTTCNNG (Bienz and Pelham 1987). Heat shock protein induction results from the binding of an activated heat shock transcription factor (HSF) to HSEs upstream of *hsp* genes (Morimoto et al. 1992). Since most of the inducible *hsp* genes do not contain introns, the mRNA is rapidly translated into nascent protein within minutes of exposure to a stressor (Iwama et al. 1998). The involvement of HSPs in the cell response to thermal stress was reported for the first time by Ritossa (1962), who observed their increase as a function of temperature. The modulation in the expression of HSP mRNAs has developed through several isoforms showing a great versatility of the phenotype responding to thermal fluctuations (Hofmann and Somero 1995; Tomanek 2002, 2005; Tomanek and Somero 2002). A wide range of variations in the induction of HSP isoforms has been reported as a function of different stressors and different biological species (Yamashita et al. 2004; Ojima et al. 2005). HSPs are overexpressed in the case of chemical and physical stresses such as anoxia (Spector et al. 1986), salinity stress (Ramagopal 1987; Gonzalez and Bradley 1995), exposure to metals and xenobiotics (Sanders 1990), and more generally, oxidative stress (Freeman et al. 1999; Kalmar and Greensmith 2009). Recent evidence suggests that exposure of cells to microbial pathogens can also induce HSPs, which then modulate both innate and adaptive immune responses (Van Eden et al. 2005).

HSPs are classified as a function of their molecular weight (90, 70, 60 kDa, etc.) and there are three major stress protein families: HSP90, HSP70, HSP60. Proteins of low molecular weight (16 to 43 kDa) also belong to an HSP family (Feige et al. 1996; Sonna et al. 2002). The HSP with the lowest molecular weight (8 kDa) is called ubiquitin. Even within a specific heat shock protein family (i.e., HSP90), isoforms exist that are differentially regulated (Basu et al. 2002).

The stress proteins have a capacity to repair proteins harmed by stress or eliminating them when they cannot be repaired any further. They work as molecular "chaperones," accompanying, monitoring, and protecting other proteins (Frydman 2001; Hartl and Hayer-Hartl 2002). In response to cellular stresses like heat shock or oxidative damage, ubiquitin and other HSPs identify misfolded or unfolded proteins and target them for proteasomal degradation. More precisely, HSP70s have been recently shown to chaperone misfolded proteins to degradation via the ubiquitin-proteasome pathway (Park et al. 2007). They are also involved in the transfer of proteins to mitochondria and in the induction and regulation of apoptosis (Craig et al. 1994; Creagh et al. 2000; Frydman 2001). Some HSPs are constitutively expressed, since they are involved in protein homeostasis in normal cellular conditions and protein housekeeping (Sørensen et al. 2005). These are called heat shock cognates (HSCs).

The most studied of HSPs are the 70 kDa HSPs (Morimoto et al. 1992), which comprise two main forms: constitutive, which are present all the time (heat shock cognates, e.g., HSC70), and inducible, which are expressed in response to external stimuli (heat shock proteins, e.g., HSP70), working as molecular chaperones (Clark and Peck 2009).

HSP90 plays an important role in the aryl hydrocarbon receptor (AHR) pathway, which allows the biotransformation of organic xenobiotics (see Chapter 8) via the cytochrome P450 system (Hahn 2002). HSP90 is active in supporting various components of the cytoskeleton and steroid hormone receptors (Young et al. 2001).

Although plants synthesise a similar set of high molecular weight HSPs, most of the translation capacity is devoted to the synthesis of small molecular weight (12 to 40 kDa) HSPs (Vierling 1991). These small HSPs (sHSPs) have no known constitutive function and are only induced during stress (Ciocca et al. 1993). sHSPs (12 to 43 kDa) can be induced in plants; they are triggered by extreme temperatures (Lopez-Matas et al. 2004; Sanmiya et al. 2004; Charng et al. 2006), and the wide diversification and abundance of sHSPs in plants may reflect adaptation to temperature stress (Waters et al. 1996). Moreover, for instance, the heat-stress-associated 32 kDa protein (Hsa32), which is highly conserved in land plants, is absent in most other organisms (Charng et al. 2006). Calreticulins (CRTs) are calcium-binding proteins functioning as chaperones of Ca^{2+} regulation levels (Luana et al. 2007); an increase of their levels confers stress tolerance to the organism (Hofmann and Somero 1996; Chapple et al. 1998).

Another group of proteins, the glucose-regulated proteins (GRPs), have many similarities with HSPs (Welch 1991; Hightower 1993). They are also ubiquitously expressed and stimulated by glucose deprivation and other cellular stresses.

9.3 ACQUISITION OF TOLERANCE

The acquisition of tolerance by HSP induction is possible only up to a certain level and is specific to a species, and the levels of HSPs cannot increase unceasingly because the cost of HSP expression exceeds its benefits (Eckwert et al. 1997; Pyza et al. 1997), a phenomenon that is not specific to HSPs but common to any kind of protein. Induction of HSPs by different organisms indicates an adaptation to different stress conditions. This tolerance is often associated with damage, which can be at a low or high level, leading to necrosis (a passive and degenerative process) and apoptosis (an active process) in some cases. Since HSPs play a protective role against various stresses, is their induction linked to the development of a tolerance toward a subsequent stress? The most ancient example, which was demonstrated both *in vivo* and *in vitro*, concerns thermal tolerance, defined as the capacity of a cell or an organism to resist thermal shock after exposure to a sublethal temperature stress. The HSP induction threshold was clearly correlated with the level of thermal stress at which a species is naturally exposed in its environment. This could reflect the importance of thermal history of a species in evolution and suggests a possible role for HSPs in species differences in thermal tolerance (Fangue et al. 2006). The persistence of heat shock proteins in organisms long after the thermal shock is removed indicates that HSPs play a role in the long-term adaptation of animals to their environment (Parsell and Lindquist 1993; Morimoto and Santoro 1998). In *Arabidopsis thaliana*, Charng et al. (2006) suggest that heat-stress-associated 32 (Hsa32) is required not for induction, but rather for maintenance of acquired thermotolerance, a feature that could be important in plants.

Some authors have reported that threshold temperatures for heat shock protein induction and thermal tolerance may be strictly genetically controlled, while others have concluded that the threshold for HSP induction can be modified by acclimation (Basu et al. 2002). Dietz and Somero (1993) demonstrated that summer-acclimatised gobies (genus *Gillichthys*) had significantly higher levels of brain HSP90 than winter-acclimatised fish, and that the threshold induction temperature for HSP90 was 4°C higher in the summer-acclimatised group. Generally, the data suggest that HSP expression is subject to acclimatisation and that HSPs may fluctuate in response to normal variations in seasonal temperature.

Interestingly, thermal acclimation confers resistance not only against heat stress but also against other stresses, such as chemical contaminants (metals, organic compounds), but the contrary also occurs and will be discussed below. For instance, in the intertidal copepod *Tigriopus japonicus*, HSP70 has a conserved role of thermotolerance, and its expression in response to xenobiotics (metals, endocrine disruptors) exposure appears to be protective (Rhee et al. 2009).

The term *cross-tolerance* will therefore be used below when thermal and chemical stress act together. In plants, the overexpression of HSP70 genes associated with the acquisition of thermal tolerance (Lee and Schöffl 1996) leads to an enhanced tolerance to salt water (Leborgne-Castel et al. 1999). There have been many reports of an increase in HSP expression in plants in response to metals (see review in Hall 2002). However, no or very low amounts of HSPs were found in plants growing on metalliferous soils, suggesting that HSPs are not responsible for the heritable metal tolerance of the terrestrial plant *Silene vulgaris* (in Hall 2002). Small plant HSPs are also induced by salinity (Hamilton and Heckathorn 2001), drought (Sato and Yokoya 2008), osmotic pressure (Sun et al. 2001), and oxidative stress (Lee et al. 2000; Neta-Sharir et al. 2005; Volkov et al. 2006). The cellular response to hypoxia represents a more specific stress response mechanism than the general purpose HSP pathway (see Katschinski and Glueck 2003).

9.3.1 HSP INDUCTION BY ENVIRONMENTAL FACTORS AND CROSS-TOLERANCE WITH METALS

Examples of cross-tolerance acquired following thermal shock have been reported in relation to various other stress challenges. In cell cultures of the wild tomato *Lycopersicon peruvianum*, a short heat stress given prior to metal (Cd) stress induces a tolerance effect by preventing membrane damage, as judged by the localisation of HSP70 in the nucleus, cytoplasm, and at the plasma membrane (Neumann et al. 1994). In blue mussels (*Mytilus edulis*) from the Baltic Sea, exposure to elevated temperature led to increased tolerance to Cd. The increasing levels of HSP70 correlated with a maintained level of fitness, in terms of scope for growth, although the mussels showed increasing Cd body burdens (Tedengren et al. 2000). The effect of Cd on the accumulation of the small HSP (HSP30) in amphibian *Xenopus laevis* A6 kidney epithelial cells revealed that the combination of mild heat shock temperatures and cadmium chloride concentrations resulted in a synergistic accumulation of HSP30 protein and HSP30 mRNA. In contrast to heat shock, prior exposure of *Xenopus* A6 cells to cadmium chloride treatment, sufficient to induce the accumulation of HSPs, did not protect the cells against a subsequent thermal challenge (Woolfson and Heikkila 2009). It has been hypothesised that Cd and heat shock evoke distinctive patterns of damage compromising the health of the cell (Dokladny et al. 2008).

In contrast to metallothionein (MT) concentrations, increased HSP levels in the crab *Carcinus maenas* did not correlate with environmental metal concentrations along a Cu and Zn exposure gradient (Pedersen et al. 1997). Duffy et al. (1999) determined HSP60 and HSP70 levels in gills of Alaskan fish, i.e., living in a cold temperature environment: northern pike (*Esox lucius*), burbot (*Lota lota*), whitefish (*Coregonus nelsoni*), grayling (*Thymallus arcticus*), and sheefish (*Stenodus lencichthys*). They found that the major contaminant in the muscle tissue of the Alaskan fish was methyl mercury. No correlation was observed between HSP60 or HSP70 expression in gill tissue and mercury concentrations in muscle tissue.

The induction of HSP70 proteins in response to zinc exposure in a unicellular algae (*Raphidocelis subcapita*) grown under different environmental conditions (pH, temperature, humic acids, nitrates, phosphates) showed that only temperature and pH were able to induce acquired tolerance to subsequent zinc exposure (Bierkens et al. 1998). Adjustment of the pH and temperature in two physicochemically different natural surface waters was demonstrated to be sufficient to obtain similar induction patterns of HSP70 in the algae upon exposure of *R. subcapita* to zinc.

In contrast, mussels (*Mytilus edulis*) had to adapt to low salinity levels in the Baltic Sea, corresponding to the limit of their geographic range; they had levels of HSP70s lower than those of mussels from the North Sea and were more susceptible to exposure to cadmium (Brown et al. 1995). Similarly, in another study with the same species (*M. edulis*), Tedengren et al. (1999) showed that mussels from the Baltic Sea were more sensitive in terms of physiological responses and survival when exposed to contaminants, compared with populations from the North Sea. Could these differences between the two populations be explained by environmental factors or genetic differences in their ability to synthesise HSP70s? Immature specimens from the Baltic Sea were transplanted into the North Sea for a month and then exposed to copper in the laboratory. The results showed that physiological differences in performance between the two populations were largely explained by environmental factors, although a lower rate of induction of HSPs in mussels from the Baltic in comparison with those of the North Sea was observed.

Pyza et al. (1997) compared the levels of HSP70 in the centipede *Lithobius mutabilis* from woodlands differing in their degree of lead and zinc contamination and a reference site. No differences in HSP levels were observed between organisms originating from contaminated and reference sites, even among those differing in their degree of contamination. These authors concluded that the acquisition of tolerance by induction of HSPs is possible only to a certain level and is specific to one species. In a laboratory study, exposure to warm temperature (33°C) of prehatched embryos of a fish (*Danio rerio*) may have allowed them to generate more stress proteins, and this may have protected them initially from both Cd and heat stress (Hallare et al. 2005). In contrast, after hatching, the larvae showed increased sensitivity to Cd, probably due by the fact that the production of HSPs could have reached its maximum just before the time of hatching, and thus the toxic action of these stressors would have surpassed the ability of the cells to generate more HSPs (Hallare et al.

2005). This could be extrapolated to contaminated environments where survival of the fish could become problematic.

Cadmium tolerance was observed in seven clones of the cladoceran crustacean *Daphnia magna* originating from various geographical locations (Haap and Köhler 2009). The highest EC50 values, i.e., the lowest toxicity levels, were observed for the clones presenting the lowest HSP expression. These variations could not be related to differential accumulation of Cd. The association of stress insensitivity with low HSP70 induction, which has been reported, for example, for populations of different invertebrates under strong selection pressure, was suggested. This is particularly interesting since *Daphnia magna* is a largely parthenogenetic species.

In populations of terrestrial arthropods (the woodlice *Oniscus asellus* and *Porcellio scaber*, and the millipede *Julus scandinavius*) from sites impacted by metals, low levels of HSPs have been observed (Köhler et al. 1999). In contrast, *O. asellus* from a reference site were unable to maintain low levels in HSP70s when exposed in the laboratory to metals. These observations can be explained by the selection of phenotypes insensitive to metal-contaminated soils, better able to select less contaminated food, thus reducing the absorption of metals (Köhler et al. 1999). Natural populations of the fruit fly *Drosophila* exposed to stressful environments, such as climatic extremes for many generations, showed low HSP levels (Bettencourt et al. 1999; Lansing et al. 2000; Sørensen et al. 2001; Zatsepina et al. 2001).

In oysters *Crassostrea virginica* from populations adapted to different thermal regimes (Washington, North Carolina, and Texas), Ivanina et al. (2009) found that overall HSP and metallothionein induction patterns were similar in oysters from the three studied geographically distant populations. Levels of HSP69, and in some cases HSC72–77 and HSP90, were lower in Cd-exposed oysters than in their control counterparts during heat stress, indicating that simultaneous exposure to these two stressors may have partially suppressed the cytoprotective upregulation of molecular chaperones. These limitations of the stress protein response may contribute to the reduced thermal tolerance of oysters from metal-polluted environments. The mechanisms responsible for this impairment of stress protein response are unknown according to Ivanina et al. (2009); one of the possible explanations is the negative effect of Cd on the cell's biosynthetic capacity, such as DNA transcription or mRNA translation processes (Pytharopoulou et al. 2006). Similarly, Slotsbo et al. (2009) exposed the springtail *Folsomio candida* to a range of aqueous concentrations of $HgCl_2$ for twenty-four hours, and the same individuals were subsequently exposed to a range of high temperatures from 20 to 35.5°C. A highly significant synergistic interaction between the effects of mercury and heat was found, with a reduced tolerance to heat after exposure to sublethal concentrations of mercury. This reduced tolerance to heat in metal-exposed organisms could also be due to reduced aerobic scope and the resultant shift of thermal tolerance limits to lower values (see review by Sokolova and Lannig 2008).

The following studies often concern HSP responses to both metal and organic contamination. Both HSP70s and HSP90s were sensitive in *Daphnia magna* to concentrations of AgCl and sodium dodecyl sulphate (SDS) (Bradley 1993). Rhee et al. (2009) studied the expression of ten HSP transcripts in response to heat treatment in an intertidal marine copepod *Tigriopus japonicus* (TJ-HSP); the bacterial expression of TJ-HSP70 and its expression in response to metal and endocrine-disrupting chemical (EDC) exposures were also studied. Metals (copper, silver, and zinc) caused a concentration-dependent increase in the expression of HSP70 transcripts. The effect of EDCs on HSP70 expression was differential. While 4-nonylphenol (NP) and 4-*t*-octylphenol (OP) caused downregulation, bisphenol A (BPA) caused upregulation. The promoter region of the genomic HSP70 sequence contained putative xenobiotic response elements (XREs), indicating that TJ-HSP70 regulation was not only performed by temperature, but also by xenobiotics. These findings suggest that, in *T. japonicus*, HSP70 has a conserved role of thermal tolerance, and its expression in response to xenobiotic exposure appears to be a protective response.

9.3.2   HSP Induction by Environmental Factors and Cross-Tolerance with Organics

Several organic xenobiotics can increase HSPs in vertebrates and invertebrates. The nature of the dose-response curves depends on the form of stress protein investigated (HSP60, HSP70, or HSP90) and on the contaminant. Expression of HSP60 and HSP90 was examined in females and males of a fish, the medaka *Oryzias latipes*, exposed via the diet to a pyrethroid insecticide, esfenvalerate (Werner et al. 2002). HSP60 was significantly elevated in females in the middle- and high-exposure groups (21 and 148 mg.kg^{-1}), whereas HSP90 was significantly elevated in males of the highest exposed groups and showed an increasing trend in females of both groups. The authors assumed that dietary exposure to esfenvalerate may cause deleterious reproductive effects in the percentage of nonviable and dead larvae with increasing insecticide concentrations. Levels of members of the stress protein families HSP60 and HSP70 (Werner et al. 1998) in benthic estuarine amphipods (*Ampelisca abdita*) exposed to sediments from different sampling sites in northern San Francisco Bay were measured; HSP64 levels were positively correlated with concentrations of total polycyclic aromatic hydrocarbons (PAHs). Clams *Macoma nasuta* from the same area (Werner et al. 2004) accumulated metals, PAHs, and organochlorine pesticides (aldrin and p,p′-DDT and its metabolites p,p′-DDD and p,p′-DDE). Bioaccumulated PAH concentrations were negatively correlated with HSP70 expression in gill (dominantly 2,3-ring PAHs) and mantle (dominantly 4-ring PAHs), while tissue concentrations of other PAHs were positively linked to inflammation and germ cell necrosis in digestive gland and gonad. Although PAHs have been shown to inhibit transcription of specific genes in sea urchins (Pillai et al. 2003), mechanisms leading to a possible inhibition of HSP70 expression are presently unknown. Mortality in clams *Macoma nasuta* was not significantly correlated with any endpoints, but a weak association was seen with lysosomal membrane damage, which in turn was positively correlated with total lesion score, germ cell necrosis, and macrophage aggregates. This link merits further investigation according to Werner et al. (2004).

Daphnia magna is tolerant to exposure to normally lethal doses of the pesticide malathion following pretreatment to heat (Bond and Bradley 1997). Concentrations of both HSP60s and HSP70s were increased to approximately 2.5 to 3 times control levels at all tested doses in organotin (TBT)-exposed zebra mussels *Dreissena poly-morpha* (Clayton et al. 2000), whereas when mussels are exposed to copper, their HSP60 and HSP70 expression patterns differ, although both contaminants appear to share some similar cellular and subcellular effects. These last results are confirmed by the work of Solé et al. (2000), who showed HSP60 induction in the clam *Ruditapes decussata* exposed to various tributyltin (TBT) concentrations. The authors high-lighted the pronounced individual variability and the existence of a plateau in the stress response, limiting the applicability of HSPs as biomarkers in field studies.

Exposure to PAHs adsorbed to clay particles and to suspended field-contaminated sediments induced an HSP70 response in the oyster *Crassostrea virginica* (Cruz-Rodríguez and Chu 2002). The HSP70 expression for each treatment showed fluctua-tions at various time intervals, and no mortalities were recorded during the exposure experiments, indicating a protective role of HSP70. Following exposure of the clam *Ruditapes decussatus* to p,p'-DDE (dichlorodiphenyl dichloroethylene), Dowling et al. (2006) did not find any difference between control and treated samples except the appearance of HSP90 in the mantle of DDE-treated clams. Immunoblotting experiments with a dinitrophenol-specific antibody revealed extensive differences in both the extent and number of carbonylated proteins in mantle and digestive gland in response to p,p'-DDE, while gill was unaffected. These results demonstrate that DDE causes tissue-specific formation of reactive oxygen species in clams. Jonsson et al. (2006) evaluated expression of HSP70 in mussels (*Mytilus* sp.) exposed to a brominated flame retardant (BDE-47) and to crude oil, alone and spiked with alky-lphenols and PAHs. HSP70 was strongly induced in all exposure groups, which also included bisphenol A and diallylphthalate.

Developmental stage, which is influenced by temperature and salinity in natural environments, may have an influence upon HSP expression. The juvenile channel catfish *Ictalurus punctatus* was acutely exposed (intraperitoneal injection) to the aryl hydrocarbon receptor (AhR) agonists [beta]-naphthoflavone (BNF) or dimethyl benz[a]anthracene (DMBA) (Weber and Janz 2001). Ninety-six hours postinjection, BNF and DMBA significantly increased apoptosis and decreased HSP70 expression in juvenile catfish ovaries, whereas the liver did not exhibit increased apoptosis and instead showed increased hepatic HSP70 expression. However, DMBA had no effect on apoptosis or HSP70 levels in more reproductively mature juvenile fish that were housed at a lower water temperature. This may be due to a developmental or temper-ature-dependent component to these responses. In early life stages of the brown trout (*Salmo trutta* f. *fario*) exposed to a mixture of PAHs and pentachlorophenol (PCP), an induction of HSP70 was observed that is stage specific with increased response levels at advanced developmental stages (Luckenbach et al. 2003), but weight was reduced in embryos treated with the PCP-PAH mixture. Werner (2004) measured the heat shock protein response in the euryhaline clam, *Potamocorbula amurensis*, after exposure to a range of salinities reflecting normal and extreme environmental conditions. The ability to raise cellular HSP70 levels in response to heat shock was significantly impaired in this clam collected from a low (0.5 ppt)-salinity field site, and after fourteen days of exposure to low salinity in the laboratory.

9.3.3   HSP Induction Observed in Genomics and Proteomics Studies

The recent cloning of multiple heat shock transcription factor (HSF) genes in higher eukaryotes has revealed several novel features of their transcriptional response. Profiles of differentially expressed genes can be obtained via transcriptomics studies. Gornati et al. (2004) reported the coding sequences of HSP70 and HSP90 and a partial sequence of HSC70 in the fish *Dicentrarchus labrax*. According to Geist et al. (2007), exposure of the striped bass (*Morone saxatilis*) to esfenvalerate had tissue-specific effects on the transcription of HSP70, HSP90, and CYP1A1. The authors concluded that the stress response at the transcriptome level is a more sensitive indicator for esfenvalerate exposure at low concentrations than swimming behavior, growth, or mortality.

Heat shock (one hour at 35°C) caused rapid induction of HSC70 and HSP70 transcript levels in mussels *Mytilus galloprovincialis* (Franzellitti and Fabbri 2005). After exposure to Hg^{2+} or chromium Cr^{6+}, the authors reported that both the inducible and constitutive HSP70 members can be simultaneously and differentially modulated by these stressors.

In the Pacific oyster *Crassostrea gigas*, HSC70 and HSP70 genes have been sequenced and characterised (Boutet et al. 2003). HSP70 synthesis was inhibited by high metal concentration. Tanguy et al. (2008) reported an important number of new stress sequences in *Mytilus edulis*, among them two new HSP70 and HSP40 genes. The transcription of genes possibly involved in defence mechanisms (HSP70, HSP90, and ubiquitin) demonstrated only small and insignificant changes in the copepod *Calanus finmarchicus* exposed to low concentrations of naphthalene (Hansen et al. 2008), indicating a small degree of protein damage. Messenger RNA transcripts were measured by reverse transcriptase polymerase chain reaction (RT-PCR) in tissues of the deep-water mussel *Modiolus modiolus* collected from a reference site and a site near a preliminary municipal wastewater treatment outfall (Veldhoen et al. 2009). Significant elevation of HSP70 transcript levels was detected in adductor muscle and hepatopancreas, indicating some level of stress, whereas transcript levels for MDR, CYP4, rpS4, rpS9, and Ca^{2+}-ATPase were not different between sites in the tissues examined. Lyons et al. (2003) measured HSP mRNA levels in the gills of *Mytilus edulis* and found that the relative levels of individual HSPs followed the order HSP40 > HSP60 > HSP70 > HSP90. Lower HSP40 and 70 transcript levels were found in mussel samples from the Baltic Sea (low salinity reference site) than in those from the North Sea (high salinity and presence of pollutants). Salinity may slightly decrease HSP40 and 70, whereas HSP70 is upregulated approximately tenfold at the polluted site compared to the reference site; this was attributed to exposure to increased pollution levels.

A suppression subtractive hybridisation method was employed to characterise up- and downregulated genes in the oyster *Crassostrea gigas* during hydrocarbon exposure (Boutet et al. 2004). Genes differentially expressed (258) were involved in hydrocarbon detoxification (cytochrome P4501A1-like protein, cytochrome b5, flavin-containing monooxygenase 2, and glutathione S-transferase omega class), protection against oxidative stress (copper/zinc superoxide dismutase), and cell protection (heat shock protein 70 family). Moreover, an increase in the mRNA level of all genes studied was observed.

The increased availability and use of DNA microarrays (in model organisms, including fish and more recently bivalves) have allowed the characterisation of gene expression patterns associated with exposure to different toxicants. The effects of carbon tetrachloride (CCl^4) and pyrene on the transcriptome of juveniles have been considered in the kidney and liver of rainbow trout *Oncorhynchus mykiss*; expression of 1,273 genes was measured using a cDNA microarray (Krasnov et al. 2005). Metallothionein, HSP90, and mitochondrial proteins of oxidative phosphorylation were induced in both trout tissues. Stimulation of heat shock proteins was greater in the liver than in the kidney. Overall, pyrene suppressed a range of protective or acclimative reactions (defence and humoral immune response, protein binding, receptors, and signal transducers), many of which were stimulated with CCl_4. Using a commercially available Atlantic salmon and trout array, Hook et al. (2006) determined alteration in the gene expression profile in rainbow trout *O. mykiss* exposed to sublethal levels of a series of model toxicants. Benzo[a]pyrene (B[a]P) and Diquat, both of which exert toxicity via oxidative stress, upregulated twenty-eight of the same genes, of over one hundred genes altered by either treatment; for instance, HSP90 was upregulated in response to B[a]P.

Dondero et al. (2006) utilised a low-density oligonucleotide microarray, employing twenty-four different genes involved in both cellular homeostasis and stress-related responses in the mussel *Mytilus edulis* collected along a copper gradient. Heat shock genes were modulated along the copper pollution gradient, with HSP27 upregulated in mussels from the two most copper-polluted sites and HSP70 down regulated in individuals from a moderately contaminated site. As said before, the heat shock 70 gene family in mussels encompasses both constitutive and inducible isotypes; they can exhibit different (even opposite) expression patterns after exposure to various stressors (e.g., heat and metals). Overexpression of the HSP27 class of molecular chaperones has been demonstrated to provide resistance to both heat shock and oxidative stress in mammal cells, and this mechanism appears linked to the modulation of actin microfilament dynamics through phosphorylation of the HSP27 peptide (Lavoie et al. 1993).

Alteration of HSPs has been often observed in proteomics studies, presumably because of high degrees of sequence conservation, which allow for species-independent identification according to Monsinjon and Knigge (2007), who reported major protein classes identified in fifteen environmental proteomics studies. Petrak et al. (2008) pointed out that the same proteins (and first of all, HSP27s) seem to predominate in the "hit parade of repeatedly identified differentially expressed proteins." Abundance of HSPs can be explained by the fact that a significant portion of the experimental studies in proteomics concern nonphysiological conditions (Petrak et al. 2008). Using proteomics, Chora (2009) reported a downregulation of calreticulin 55 in the clam *Ruditapes decussatus* exposed to nonylphenol. This downregulation of calreticulin 55 was also observed by David et al. (2005) in gill and mantle of the oyster *Crassostrea gigas* submitted to hypoxia.

In mussels *Mytilus edulis* exposed to Aroclor or to fractions of sediments from PCB-impacted sites, Olsson et al. (2004), using one-dimensional electrophoresis gels, observed seven proteins, presumed to be from stress protein families (HSP60, HSP70, and HSP90); levels of all proteins except for three (62, 73, and 90 kDa) were significantly higher ($p < 0.05$) in treated than in control organisms. Of the physiological responses measured (changes in respiration, excretion, clearance rates, and scope for growth), only the last two differed significantly from control organisms. Olsson et al. (2004) also performed two-dimensional electrophoresis (2-DE) gels to determine protein expression changes by comparing the presence or absence of protein spots on gels from treated mussels with gels from control mussels. Between twenty-three and seventy-six protein spots were present and fifteen to twenty-three absent compared to controls. The authors concluded that the extensive reaction in terms of absent and present protein spots compared to controls affects energy turnover negatively and may reduce animal fitness.

De Wit et al. (2008) exposed two populations of adult zebrafish (*Danio rerio*) for fourteen days to 0.75 and 1.5 µM of the flame-retardant tetrabromobisphenol-A (TBBPA) with endocrine disruption properties. Differential expression of genes from the liver of *D. rerio* was observed, using oligonucleotide arrays, and the most obvious signal of stress response was an upregulation of the *hsp70* gene. The proteome of the liver of *D. rerio* was also analysed by means of differential in-gel electrophoresis (DiGE) and showed a stimulation of heat shock 70 kDa protein 5 in the liver of male fish exposed to the highest concentration; it was associated with an upregulation of the chaperone glucose-protein GP96, an endoplasmic-reticulum-resident stress protein belonging to the HSP90 family. TBBPA also elicited interferences with thyroid hormone and vitamin A homeostasis, which can lead to alteration of development, growth immune function, reproductive success, and metabolism that are not directly linked to HSP upregulation. These authors demonstrated the potential of a combined genome and proteome approach to generate detailed mechanistic ecotoxicological information.

9.4 CONCLUSIONS

HSP proteins, included in a test battery of complementary bioassays, may be very valuable as tier I biomarkers, i.e., broad response biomarkers that are used for preliminary screening of the environment (Bierkens 2000). In a recent study, Micovic et al. (2009) showed in mussels that stress proteins could be powerful biomarkers of marine pollution. Bradley (1993) highlighted that some HSPs have been associated with classes of stressors, others with specific diseases. The cellular stress response can vary, for example, according to tissue or organ, species, HSP family, and type of stressor. Because of the adaptation of organisms to naturally occurring fluctuations in the environment, HSP levels will not be suitable as unequivocal signatures for the presence of pollutants (De Pomerai 1996). The induction of heat shock proteins (HSPs) is considered an important protective, ecophysiologically adaptive, and genetically conserved response to environmental (natural and chemical) stress in all organisms. There is little direct evidence of correlation between HSP levels and toxic effects due to their role as chaperones. For instance, Glover and Lindquist (1998) demonstrated that HSP104, HSP70, and HSP40 constitute a chaperone system that rescues previously aggregated proteins. However, association of stress insensitivity with low HSP induction has been reported for numerous populations of invertebrates under strong selection pressure (Haap and Köhler 2009). An important question remains whether pollutant-induced changes, i.e., in *hsp* gene or HSP protein expression in organisms, are sufficiently diverse to allow for identification of specific modes of action or specific contaminants. De Wit et al. (2008) observed responses at the protein level of zebrafish exposed to tetrabromobisphenol-A that were confirmative of the oligonucleotide microarray results. As for gene expression analysis, proteome analysis revealed a stimulation of HSP70 in the liver of male fish. These findings and others reported by De Wit et al. (2008) imply that responses measured at the genome level can give indications about effect on the proteome. Proteins represent the biologically active molecules in an organism, and therefore carry an essential amount of information in their responses. The identification of some proteins specifically regulated by exposure to pollutants, in particular HSPs, allows us to move to a global view of the action of pollutants and might lead to the identification of the molecular mechanisms that are involved in the responses of animals. These molecular mechanisms may have further influence at the population level.

REFERENCES

Alamillo, J., et al. 1995. Constitutive expression of small heat shock proteins in vegetative tissue of the resurrection plant *Craterostigma plantagineum. Plant Mol. Biol.* 29:1093–99.

Almoguera, C., M. A. Coca, and J. Jordano. 1993. Tissue-specific expression of sunflower heat shock proteins in response to water stress. *Plant J.* 4:947–58.

Basu, N., et al. 2002. Heat shock protein genes and their functional significance in fish. *Gene* 295:173–83.

Bettencourt, B. R., M. E. Feder, and S. Cavicchi. 1999. Experimental evolution of HSP70 expression and thermotolerance in *Drosophila melanogaster. Evolution* 53:484–92.

Bienz, M., and H. R. Pelham. 1987. Mechanisms of heat shock gene activation in higher eukaryotes. *Adv. Gene.* 24:31–72.

Bierkens, J., W. Van de Perre, and J. Maes. 1998. Effect of different environmental variables on the synthesis of HSP70 in *Raphidocelis subcapitata. Comp. Biochem. Physiol.* 120A:29–34.

Bierkens, J. G. E. A. 2000. Applications and pitfalls of stress-proteins in biomonitoring. *Toxicology* 153:61–72.

Bond, J., and B. P. Bradley. 1997. Resistance to malathion in heat-shock *Daphnia magna. Environ. Toxicol. Chem.* 16:705–71.

Boutet, I., A. Tanguy, and D. Moraga. 2004. Response of the Pacific oyster *Crassostrea gigas* to hydrocarbon contamination under experimental conditions. *Gene* 329:147–57.

Boutet, I., et al. 2003. Molecular identification and expression of heat shock cognate 70 (hsc70) and heat shock protein 70 (HSP70) genes in the Pacific oyster *Crassostrea gigas. Cell Stress Chaperon.* 8:76–85.

Bradley, B. P. 1993. Are the stress proteins indicators of exposure or effect? *Mar. Environ. Res.* 35:85–88.

Bradley, B. P., and J. B. Ward. 1989. Detection of a major stress protein using a peptide antibody. *Mar. Environ. Res.* 28:471–75.

Brown, D. C., B. P. Bradley, and M. Tedengren. 1995. Genetic and environmental regulation of HSP70 expression. *Mar. Environ. Res.* 39:181–84.

Chapple, J. P., et al. 1998. Seasonal changes in stress-70 protein levels reflect thermal tolerance in the marine bivalve *Mytilus edulis* L. *J. Exp. Mar. Biol. Ecol.* 229:53–68.

Charng, Y.-Y., et al. 2006. *Arabidopsis* Hsa32, a novel heat shock potein, is essential for acquired thermotolerance during long recovery after acclimation. *Plant Physiol.* 140:1297–305.

Chora, S. C. 2009. Study of contaminant effects in a sentinel species, the clam *Ruditapes decussatus*, through protein expression analysis: Monitoring in marine environment. PhD thesis, University of Algarve, Faro, Portugal.

Ciocca, D. R., et al. 1993. Biological and clinical implications of heat shock protein 27,000 (HSP27): A review. *J. Natl. Cancer Inst.* 85:1558–70.

Clark, M. S., and L. S. Peck. 2009. HSP70 heat shock proteins and environmental stress in Antarctic marine organisms: A minireview. *Mar. Genomics* 2:11–18.

Clayton, M. E., R. Steinmann, and K. Fent. 2000. Different expression patterns of heat shock proteins HSP 60 and HSP 70 in zebra mussels (*Dreissena polymorpha*) exposed to copper and tributyltin. *Aquat. Toxicol.* 47:213–26.

Craig, E. A., J. S. Weissman, and A. L. Horwich. 1994. Heat shock proteins and molecular chaperones: Mediators of protein conformation and turnover in the cell. *Cell* 78:365–72.

Creagh, E. M., R. J. Carmody, and T. G. Cotter. 2000. Heat shock protein 70 inhibits caspase-dependent and -independent apoptosis in Jurkat T cells. *Exp. Cell Res.* 257:58–66.

Cruz-Rodríguez, L. A., and F.-L.E Chu. 2002. Heat-shock protein (HSP70) response in the eastern oyster, *Crassostrea virginica*, exposed to PAHs sorbed to suspended artificial clay particles and to suspended field contaminated sediments. *Aquat. Toxicol.* 60:157–68.

David, E., et al. 2005. Response of the Pacific oyster *Crassostrea gigas* to hypoxia exposure under experimental conditions. *FEBS J.* 272:5635–52.

de Miguel, N., et al. 2009. Structural and functional diversity in the family of small heat shock proteins from the parasite *Toxoplasma gondii*. *Biochim. Biophys. Acta* 1793:1738–48.

De Pomerai, D. I. 1996. Heat-shock proteins as biomarkers of pollution. *Hum. Exp. Toxicol.* 15:279–85.

De Wit, M., et al. 2008. Molecular targets of TBBPA in zebrafish analysed through integration of genomic and proteomic approaches. *Chemosphere* 74:96–105.

Dietz, T. J., and G. N. Somero. 1993. Species- and tissue-specific synthesis patterns for heat shock proteins HSP70 and HSP90 in several marine teleost fishes. *Physiol. Zool.* 66:863–80.

Dokladny, K., et al. 2008. Lack of cross-tolerance following heat and cadmium exposure in functional MDCK monolayers. *J. Appl. Toxicol.* 88:885–94.

Dondero, F., et al. 2006. Assessing the occurrence of a stress syndrome in mussels (*Mytilus edulis*) using a combined biomarker/gene expression approach. *Aquat. Toxicol.* 78(Suppl. 1):S13–24.

Dong, Y. W., and S. L. Dong. 2008. Induced thermotolerance and the expression of heat shock protein 70 in sea cucumber *Apostichopus japonicus* Selenka. *Fish. Sci.* 74:573–78.

Dowling, V., et al. 2006. Protein carbonylation and heat shock response in *Ruditapes decussatus* following p,p′-dichlorodiphenyldichloroethylene (DDE) exposure: A proteomic approach reveals that DDE causes oxidative stress. *Aquat. Toxicol.* 77:11–18.

Duffy, L. K., et al. 1999. Comparative baseline levels of mercury, HSP 70 and HSP 60 in subsistence fish from the Yukon-Kuskokwim delta region of Alaska. *Comp. Biochem. Physiol.* 124C:181–86.

Eckwert, H., G. Alberti, and H. R. Köhler. 1997. The induction of stress proteins (HSP) in *Oniscus asellus* (Isopoda) as a molecular marker of multiple heavy metal exposure. I. Principles and toxicological assessment. *Ecotoxicology* 6:249–62.

Fangue, N., M. Hofmeister, and P. Schulte. 2006. Intraspecific variation in thermal tolerance and heat shock protein gene expression in common killifish *Fundulus heteroclitus*. *J. Exp. Biol.* 209:2859–72.

Feder, M. E., and G. E. Hofmann. 1999. Heat-shock proteins, molecular chaperones, and the stress response: Evolutionary and ecological physiology. *Annu. Rev. Physiol.* 61:243–82.

Feige, U., et al. 1996. *Stress-inducible cellular responses*. Basel: Birkhauser Verlag.

Franzellitti, S., and E. Fabbri. 2005. Differential HSP70 gene expression in the Mediterranean mussel exposed to various stressors. *Biochem. Biophys. Res. Commun.* 336:1157–63.

Freeman, M. L., et al. 1999. On the path to the heat shock response: Destabilization and formation of partially folded protein intermediates, a consequence of protein thiol modification. *Free Radical Bio. Med.* 26:737–45.

Frydman, J. 2001. Folding of newly translated proteins *in vivo*: The role of molecular chaperones. *Annu. Rev. Biochem.* 70:603–47.

Geist, J., et al. 2007. Comparisons of tissue-specific transcription of stress response genes with whole animal endpoints of adverse effect in striped bass (*Morone saxatilis*) following treatment with copper and esfenvalerate. *Aquat. Toxicol.* 5:28–39.

Glover, J. R., and S. Lindquist. 1998. Hsp104, Hsp70 and Hsp40: A novel chaperone system that rescues previously aggregated proteins. *Cell* 94:73–82.

Gonzalez, C. R. M., and B. P. Bradley. 1995. Are there salinity stress proteins? *Mar. Environ. Res.* 39:205–8.

Gornati, R., et al. 2004. Rearing density influences the expression of stress-related genes in sea bass (*Dicentrarchus labrax*, L.). *Gene* 341:111–8.

Gross, M. 2004. Emergency services: A bird's eye perspective on the many different functions of stress proteins. *Curr. Protein Pept. Sci.* 5:213–23.

Haap, T., and H. R. Köhler. 2009. Cadmium tolerance in seven *Daphnia magna* clones is associated with reduced HSP70 baseline levels and induction. *Aquat. Toxicol.* 94:131–37.

Hahn, M. E. 2002. Aryl hydrocarbon receptors: Diversity and evolution. *Chem.-Biol. Interact.* 141:131–60.

Hall, J. L. 2002. Cellular mechanisms for heavy metal detoxification and tolerance. *J. Exp. Bot.* 53:1–11.

Hallare, A. V., et al. 2005. Combined effects of temperature and cadmium on developmental parameters and biomarker responses in zebrafish (*Danio rerio*) embryos. *J. Therm. Biol.* 30:7–17.

Hamilton, E. W., and S. A. Heckathorn. 2001. Mitochondrial adaptations to NaCl. Complex I is protected by anti-oxidants and small heat shock proteins, whereas complex II is protected by proline and betaine. *Plant Physiol.* 126:1266–74.

Hansen, B. H., et al. 2008. Effects of naphthalene on gene transcription in *Calanus finmarchicus* (Crustacea: Copepoda). *Aquat. Toxicol.* 86:157–65.

Harndahl, U., et al. 1999. The chloroplast heat shock protein undergoes oxidation-dependent conformational changes and may protect plants from oxidative stress. *Cell Stress Chaperon.* 4:129–38.

Hartl, F. U., and M. Hayer-Hartl. 2002. Molecular chaperones in the cytosol: From nascent chain to folded protein. *Science* 295:1852–58.

Hightower, L. E. 1993. A brief perspective on the heat-shock response and stress proteins. *Mar. Environ. Res.* 35:79–83.

Hofmann, G., and G. Somero. 1995. Evidence for protein damage at environmental temperatures: Seasonal changes in levels of ubiquitin conjugates and HSP 70 in the intertidal mussel *Mytilus trossolus*. *J. Exp. Biol.* 198:1509–18.

Hofmann, G., and G. Somero. 1996. Interspecific variation in thermal denaturation of proteins in the congeneric mussels *Mytilus trossolus* and *M. galloprovincialis*: Evidence from the heat-shock response and protein ubiquitination. *Mar. Biol.* 126:65–75.

Hook, S. E., et al. 2006. Gene expression patterns in rainbow trout, *Oncorhynchus mykiss*, exposed to a suite of model toxicants. *Aquat. Toxicol.* 77:372–85.

Iba, K. 2002. Acclimative response to temperature stress in higher plants: Approaches of gene engineering for temperature tolerance. *Annu. Rev. Plant Biol.* 53:225–45.

Ivanina, A. V., C. Taylor, and I. M. Sokolova. 2009. Effects of elevated temperature and cadmium exposure on stress protein response in eastern oysters *Crassostrea virginica* (Gmelin). *Aquat. Toxicol.* 91:245–54.

Iwama, G., et al. 1998. Heat shock protein expression in fish. *Rev. Fish Biol. Fish.* 8:35–56.

Jonsson, H., et al. 2006. Expression of cytoskeletal proteins, cross-reacting with anti-CYP1A, in *Mytilus* sp. exposed to organic contaminants. *Aquat. Toxicol.* 78(Suppl. 1):S42–48.

Kalmar, B., and L. Greensmith. 2009. Induction of heat shock proteins for protection against oxidative stress. *Ad. Drug. Deliv. Rev.* 61:310–18.

Katschinski, D. M., and S. B. Glueck. 2003. Hot worms can handle heavy metal. Focus on "HIF-1 is required for heat acclimation in the nematode *Caenorhabditis elegans.*" *Physiol. Genomics* 14:1–2.

Kinne, O. 1971. Salinity: Invertebrates. In *Marine ecology*, ed. O. Kinne, 821–996. London: Wiley Interscience.

Köhler, H.-R., et al. 1999. Interaction between tolerance and 70 kDa stress protein (HSP70) induction in collembolan populations exposed to long-term metal pollution. *Appl. Soil Ecol.* 11:43–52.

Krasnov, A., et al. 2005. Transcriptome responses to carbon tetrachloride and pyrene in the kidney and liver of juvenile rainbow trout (*Oncorhynchus mykiss*). *Aquat. Toxicol.* 74:70–81.

Lansing, E., J. Justesen, and V. Loeschcke, 2000. Variation in the expression of HSP70, the major heat-shock protein, and thermotolerance in larval and adult selection lines of *Drosophila melanogaster. J. Therm. Biol.* 25:443–50.

Lavoie, J., et al. 1993. Modulation of actin microfilament dynamics and fluid phase pinocytosis by phosphorylation of heat shock protein. *J. Biol. Chem.* 268:24210–14.

Leborgne-Castel, N., et al. 1999. Overexpression of BiP in tobacco alleviates endoplasmic reticulum stress. *Plant Cell* 11:459–70.

Lee, B. H., et al. 2000. Expression of the chloroplast-localized small heat shock protein by oxidative stress in rice. *Gene* 245:283–90.

Lee, J. H., and F. Schöffl. 1996. An HSP70 antisense gene affects the expression of HSP70/HSC70, the regulation of HSF and the acquisition of thermotolerance in transgenic *Arabidopsis thaliana. Mol. Gen. Genet.* 252:11–19.

Lindquist, S. 1986. The heat-shock response. *Annu. Rev. Biochem.* 55:1151–91.

Lopez-Matas, M. A., et al. 2004. Protein cryoprotective activity of a cytosolic small heat shock protein that accumulates constitutively in chestnut stems and is up-regulated by low and high temperatures. *Plant Physiol.* 134:1708–17.

Luana, W., et al. 2007. Molecular characteristics and expression analysis of calreticulin in Chinese shrimp *Fenneropenaeus chinensis. Comp. Biochem. Physiol.* 147B:482–91.

Luckenbach, T., et al. 2003. Developmental and subcellular effects of chronic exposure to sublethal concentrations of ammonia, PAH and PCP mixtures in brown trout (*Salmo trutta* f. *fario* L.) early life stages. *Aquat. Toxicol.* 65:39–54.

Lyons, C., et al. 2003. Variability of heat shock proteins and glutathione S-transferase in gill and digestive gland of blue mussel, *Mytilus edulis. Mar. Environ. Res.* 56:585–97.

Micovic, V., et al. 2009. Metallothioneins and heat shock proteins 70 in marine mussels as sensors of environmental pollution in Northern Adriatic Sea. *Environ. Toxicol. Pharmacol.* 28:439–47.

Monsinjon, T., and T. Knigge. 2007. Proteomic applications in ecotoxicology. *Proteomics* 7:2997–3009.

Morimoto, R. I., and M. G. Santoro. 1998. Stress-inducible responses and heat shock proteins: New pharmacological targets for cytoprotection. *Nat. Biotechnol.* 16:833–38.

Morimoto, R. I., K. D. Sarge, and K. Abravaya. 1992. Transcriptional regulation of heat shock genes. *J. Biol. Chem.* 267:21987–90.

Nakano, K., and G. K. Iwama. 2002. The 70-kDa heat shock protein response in two intertidal sculpins, *Oligocottus maculosus* and *O. snyderi*: Relationship of HSP70 and thermal tolerance. *Comp. Biochem. Physiol.* 133A:79–94.

Neta-Sharir, I., et al. 2005. Dual role for tomato heat shock protein 21: Protecting Photosystem II from oxidative stress and promoting color changes during fruit maturation. *Plant Cell* 17:1829–38.

Neumann, D., et al. 1994. Heat-shock proteins induce heavy-metal tolerance in higher plants. *Planta* 194:360–67.

Ojima, N., M. Yamashita, and S. Watabe. 2005. Quantitative mRNA expression profiling of heat-shock protein families in rainbow trout cells. *Biochem. Biophys. Res. Commun.* 329:51–57.

Olsson, B., et al. 2004. Physiological and proteomic responses in *Mytilus edulis* exposed to PCBs and PAHs extracted from Baltic Sea sediments. *Hydrobiologia* 514:15–27.

Park, S.-H., et al. 2007. The cytoplasmic HSP70 chaperone machinery subjects misfolded and endoplasmic reticulum import-incompetent proteins to degradation via the ubiquitin-proteasome system. *Mol. Biol. Cell.* 18:153–65.

Parsell, D. A., and S. Lindquist. 1993. The function of heat-shock proteins in stress tolerance: Degradation and reactivation of damaged proteins. *Annu. Rev. Genet.* 27:437–96.

Pedersen, S. N., A. K. Lundebye, and M. H. Depledge. 1997. Field application of metallo-thionein and stress protein biomarkers in the shore crab (*Carcinus maenas*) exposed to trace metals. *Aquat. Toxicol.* 37:183–200.

Petrak, J., et al. 2008. "Déjà vu" in proteomics. A hit parade of repeatedly identified differentially expressed proteins. *Proteomics* 8:1744–49.

Pillai, M. C., et al. 2003. Polycyclic aromatic hydrocarbons disrupt axial development in sea urchin embryos through a [beta]-catenin dependent pathway. *Toxicology* 186:93–108.

Pockley, A. G. 2003. Heat shock proteins as regulators of the immune response. *Lancet* 362:469–76.

Pytharopoulou, S., et al. 2006. Evaluation of the global protein synthesis in *Mytilus galloprovincialis* in marine pollution monitoring: Seasonal variability and correlations with other biomarkers. *Aquat. Toxicol.* 80:33–41.

Pyza, E., et al. 1997. Heat shock proteins (HSP70) as biomarkers in ecotoxicological studies. *Ecotox. Environ. Saf.* 38:244–51.

Ramagopal, S. 1987. Salinity stress induced tissue specific proteins in barley seedlings. *Plant Physiol.* 84:324–31.

Rhee, J.-S., et al. 2009. Heat shock protein (HSP) gene responses of the intertidal copepod *Tigriopus japonicus* to environmental toxicants. *Comp. Biochem. Physiol.* 149C:104–12.

Ritossa, F. 1962. A new puffing pattern induced by heat shock and DNP in *Drosophila*. *Experientia* 18:571–73.

Sabehat, A., S. Lurie, and D. Weiss. 1998. Expression of small heat-shock proteins at low temperatures. *Plant Physiol.* 117:651–58.

Sanders, B. 1990. Stress proteins: Potential as multitiered biomarkers. In *Environmental biomarkers*, ed. L. Shugart and J. McCarthy, 165–91. Chelsea, MI: Lewis Publishers.

Sanders, B. M., et al. 1991. Relationships between accumulation of a 60 kDa stress protein and scope-for-growth in *Mytilus edulis* exposed to a range of copper concentrations. *Mar. Environ. Res.* 31:81–97.

Sanmiya, K., et al. 2004. Mitochondrial small heat-shock protein enhances thermotolerance in tobacco plants. *FEBS Lett.* 557:265–68.

Sato, Y., and S. Yokoya. 2008. Enhanced tolerance to drought stress in transgenic rice plants overexpressing a small heat-shock protein, sHSP17.7. *Plant Cell Rep.* 27:329–34.

Slotsbo, S., et al. 2009. Exposure to mercury reduces heat tolerance and heat hardening ability of the springtail *Folsomia candida*. *Comp. Biochem. Physiol.* 150C:118–23.

Sokolova, I. M., and G. Lannig. 2008. Interactive effects of metal pollution and temperature on metabolism in aquatic ectotherms: Implications of global climate change. *Climate Res.* 37:181–201.

Solé, M., Y. Morcillo, and C. Porte. 2000. Stress-protein response in tributyltin-exposed clams. *Bull. Environ. Contam. Toxicol.* 64:852–58.

Sonna, L. A., et al. 2002. Molecular biology of thermoregulation: Invited review: Effects of heat and cold stress on mammalian gene expression. *J. Appl. Physiol.* 92:1725–42.

Sørensen, J. G., J. Dahlgaard, and V. Loeschcke. 2001. Genetic variation in thermal tolerance among natural populations of *Drosophila buzzatii*: Down regulation of HSP70 expression and variation in heat stress resistance traits. *Funct. Ecol.* 15:289–96.

Sørensen, J. G., et al. 2005. Altitudinal variation for stress resistant traits and thermal adaptation in adult *Drosophila buzzatii* from the new world. *J. Evol. Biol.* 18:829–32.

Spector, M., et al. 1986. Global control in *Salmonella typhimurium*: Two dimensional electrophoretic analysis of starvation, anaerobiosis, and heat shock-inducible proteins. *J. Bacteriol.* 168:420–24.

Sun, W., et al. 2001. At-HSP17.6A, encoding a small heat-shock protein in *Arabidopsis*, can enhance osmotolerance upon overexpression. *Plant J.* 27:407–15.

Tanguy, A., et al. 2008. Increasing genomic information in bivalves through new EST collections in four species: Development of new genetic markers for environmental studies and genome evolution. *Gene* 408:27–36.

Tedengren, M., et al. 1999. Heavy metal uptake, physiological response and survival of the blue mussel (*Mytilus edulis*) from marine and brackish waters in relation to the induction of heat-shock protein 70. *Hydrobiologia* 393:261–69.

Tedengren, M., et al. 2000. Heat pretreatment increases cadmium resistance and HSP 70 levels in Baltic Sea mussels. *Aquat. Toxicol.* 48:1–12.

Tomanek, L. 2002. The heat-shock response: Its variation, regulation and ecological importance in intertidal gastropods (genus *Tegula*). *Integr. Comp. Biol.* 42:797.

Tomanek, L. 2005. Two-dimensional gel analysis of the heat-shock response in marine snails (genus *Tegula*): Interspecific variation in protein expression and acclimation ability. *J. Exp. Biol.* 208:3133–43.

Tomanek, L., and G. Somero. 1999. Evolutionary and acclimatation-induced variation in the heat-shock responses of congeneric marine snails (genus *Tegula*) from different thermal habitats: Implications for limits of thermotolerance and biogeography. *J. Exp. Biol.* 202:2925–36.

Tomanek, L., and G. Somero. 2002. Interspecific and acclimation-induced variation in levels of heat-shock proteins 70 (HSP70) and 90 (HSP90) and heat-shock transcription factor-1 (HSF1) in congeneric marine snails (genus *Tegula*): Implications for regulation of HSP gene expression. *J. Exp. Biol.* 205:677–85.

Van Eden, W., R. Van der Zee, and B. Prakken. 2005. Heat-shock proteins induce T-cell regulation of chronic inflammation. *Nat. Rev. Immunol.* 5:318–30.

Veldhoen, N., et al. 2009. Gene expression profiling in the deep water horse mussel *Modiolus modiolus* (L.) located near a marine municipal wastewater outfall. *Aquat. Toxicol.* 93:116–24.

Vierling, E. 1991. The roles of heat shock proteins in plants. *Annu. Rev. Plant Phys.* 42:579–620.

Volkov, R. A., et al. 2006. Heat stress-induced H2O2 is required for effective expression of heat shock genes in *Arabidopsis*. *Plant Mol. Biol.* 61:733–46.

Wang, W., B. Vinocur, O. Shoeyov, and A. Altman. 2004. Role of plant heat-shock proteins and molecular chaperones in the abiotic stress response. *Trends Plant Sci.* 9:244–52.

Waters, E. R., G. J. Lee, and E. Vierling. 1996. Evolution structure and function of the small heat shock proteins in plants. *J. Exp. Bot.* 47:325–38.

Weber, L. P., and D. M. Janz. 2001. Effect of [beta]-naphthoflavone and dimethylbenz[a] anthracene on apoptosis and HSP70 expression in juvenile channel catfish (*Ictalurus punctatus*) ovary. *Aquat. Toxicol.* 54:39–50.

Welch, W. J. 1991. The role of heat-shock proteins as molecular chaperones. *Curr. Opin. Cell Biol.* 3(6):1033–38.

Werner, I. 2004. The influence of salinity on the heat-shock protein response of *Potamocorbula amurensis* (Bivalvia). *Mar. Environ. Res.* 58:803–7.

Werner, I., Kline, K. F., and J. T. Hollibaugh. 1998. Stress protein expression in *Ampelisca abdita* (Amphipoda) exposed to sediments from San Francisco Bay. *Mar. Environ. Res.* 45:417–30.

Werner, I., et al. 2002. Effects of dietary exposure to the pyrethroid pesticide esfenvalerate on medaka (*Oryzias latipes*). *Mar. Environ. Res.* 54:609–14.

Werner, I., et al. 2004. Biomarker responses in *Macoma nasuta* (Bivalvia) exposed to sediments from northern San Francisco Bay. *Mar. Environ. Res.* 58:299–304.

Woolfson, J. P., and J. J. Heikkila. 2009. Examination of cadmium-induced expression of the small heat shock protein gene, HSP30, in *Xenopus laevis* A6 kidney epithelial cells. *Comp. Biochem. Physiol.* 152A:91–99.

Yamashita, M., K. Hirayoshi, and K. Nagata. 2004. Characterization of multiple members of the HSP70 family in platyfish culture cells: Molecular evolution of stress protein HSP70 in vertebrates. *Gene* 336:207–18.

Young, J. C., I. Moarefi, and F. U. Hartl. 2001. HSP90: A specialized but essential protein-folding tool. *J. Cell Biol.* 154:267–73.

Zatsepina, O. G., et al. 2001. A *Drosophila melanogaster* strain from sub-equatorial Africa has exceptional thermotolerance but decreased HSP70 expression. *J. Exp. Biol.* 204:1869–81.

10 The Multixenobiotic Transport System

A System Governing Intracellular Contaminant Bioavailability

Gautier Damiens and Christophe Minier

CONTENTS

10.1 IMPORTANCE OF TRANSPORT SYSTEMS IN XENOBIOTIC BIOAVAILABILITY

Any adverse effect resulting from a chemical toxicant is primarily due to its interaction with a particular target within the organism or the cell. The availability and duration of the interaction of a xenobiotic with a given cellular target are dependent upon its toxicokinetic properties, i.e., its absorption, distribution, metabolism, and excretion (ADME). Any environmental contaminant that is not absorbed will not represent a particular toxicological risk for the organism (with the exception of contact toxicity). Distribution may lead to accumulation in nontarget organs, thus preventing its toxicological activity, or into organs specialised in metabolism and excretion. Metabolism can inactivate and eliminate the toxic compound and may lead to its excretion even if intermediate metabolites may have deleterious effects (see Chapter 8).

These kinetic processes may also be described at the cellular level. Uptake, sequestration, biotransformation, and efflux of xenobiotics will largely modulate the putative toxicological effects of a given xenobiotic. Although other mechanisms, such as alteration of targets or of signalling pathways, may also contribute to xenobiotic resistance, processes that alter xenobiotic bioavailability are of the outmost importance.

In the past, classical toxicology focussed on metabolism, dividing it into two phases (see also Chapter 8). Phase I represents the functionalisation of foreign organic compounds. Oxidation, hydrolysis, and sometimes reduction lead to transformation of the parent compounds, which then generally harbour a polar radical. Phase II involves the conjugation of partially detoxified metabolites to make them readily water soluble and excretable, and thus unable to reenter the cell by passive diffusion. Surprisingly, after phase II, it was generally considered that the xenobiotics were "detoxified" and no longer considered. However, accumulation of the metabolites may result in cell injury, and their excretion is of particular importance. By now it has been realised that the transport systems are just as important as the previously known processes (Cascorbi 2006; Lesli et al. 2005). Phase 0 and phase III, involved in the modulation of the cellular entry and exit, respectively, are key processes that result in the modulation of toxicological effects (Figure 10.1).

Since all tissue barrier functions, including absorption in the gut, export in the liver and kidney (or hepatopancreas), and transport through the blood-brain, blood-testis, or other specialised interfaces, involve specific and mostly polarised cell layers, phases 0 and III are directly relevant to the toxicological features of xenobiotics.

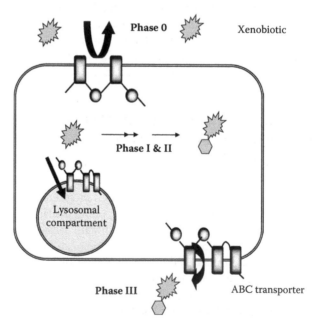

FIGURE 10.1 Schematic representation of the involvement of ABC proteins in xenobiotics transport. ABC transporters can pump untransformed xenobiotics out of the cell (phase 0), export conjugated metabolites (phase III), or may sequester xenobiotics in the lysosomal compartment.

Among transporters, ABC (ATP-binding cassette) proteins have been recognised as essential contributors to phases 0 and III of xenobiotic disposition (Leslie et al. 2005). Accordingly, this review will focus on transporters of the ABC family that are mainly involved in xenobiotic transport and, as such, are key players in the general defence system of organisms.

10.2   STRUCTURAL ORGANISATION OF ABC PROTEINS

In recent years, characterisation of prokaryotic and eukaryotic genome sequences has allowed the complete identification of ABC genes in genomes representative of all major phyla. (Decottignies and Goffeau 1997; van Veen and Konings 1998; Dean et al. 2001; Sanchez-Fernandez et al. 2001; Anjard et al. 2002; Roth et al. 2003; Garcia et al. 2004; Sheps et al. 2004; Sauvage et al. 2006). ABC proteins are encoded by the largest gene superfamily in all sequenced genomes to date and include a range of mostly membrane-bound proteins mainly involved in the transport of a wide variety of molecules. The importance of this class of proteins is illustrated by the fact that almost 5% of the *Escherichia coli* genome is occupied by genes encoding seventy-nine distinct ABC proteins (Blattner et al. 1997; Linton and Higgins 1998).

ABC transporters are characterised by the sequence and organisation of the ATP-binding domain. This domain, also known as the nucleotide-binding fold (NBF), contains two conserved motifs (the Walker A and Walker B motifs), separated by 90 to 120 amino acids. While Walker motifs are also found in all ATPases and ATP-binding proteins other than the ABC transporters, a characteristic fifteen-amino acid ABC signature sequence (also known as the C motif) in the nucleotide-binding domain (NBD) between the Walker motifs is used for identification and homology cloning of members of the ABC superfamily. This ATP-binding site is cytoplasmic and generally bound to a transmembrane domain (TMD) that contains several (six to eleven) α-helices. A full transporter such as the ABCB1 or ABCC1 proteins typically comprises two NBFs and two TDMs organised with the topology TMD-NBF-TMD-NBF, whereas half transporters such as the ABCG2 protein are restricted to the association of a TMD with a NBF (Figure 10.1). Nevertheless, it must be noted that half transporters may function as dimers rather than monomers (Hrycyna and Gottesman 1998).

The ABC family is divided into subgroups based on phylogenetic relationships between their nucleotide-binding domains, and members of each subfamily generally exhibit the same domain structure (Dean et al. 2001). Accordingly, eukaryotic ABC proteins can be divided into eight families, from ABCA to ABCH, seven of which (A to G) are found in mammalian genomes.

10.3   GENERAL CHARACTERISTICS OF THE
ABC TRANSPORT SYSTEM

Research on ABC transporters and the associated concept of multiple resistances to xenobiotics is largely due to investigations conducted on cancer resistance to chemotherapeutic agents. Clinical oncologists were the first to observe that some cancers treated with multiple different anticancer drugs tended to develop cross-resistance to many other cytotoxic agents to which they had never been exposed, effectively leading to failure of the chemotherapy. *In vitro* studies using cell lines have helped characterise this phenomenon. When exposed to low concentrations of a unique cytotoxic drug such as colchicine, vinca alkaloids, or anthacycline, cultured cells spontaneously developed, at low frequency, simultaneous resistance to a large group of structurally and functionally unrelated chemicals (Biedler and Riehm 1970). Drug transport experiments in multi-drug-resistant cell lines indicated that the emergence of multidrug resistance was linked to a decreased intracellular drug accumulation and to an increased drug efflux, both of which are strictly adenosine triphosphate (ATP) dependent (Danö 1973). The most common form of this multidrug resistance arises from the selection of cells with high amounts of a 170 KDa transport protein referred to as P-glycoprotein (P-gp), which later was identified as the ABCB1 protein. This is a membrane-spanning, ATP-binding protein that transports chemotherapeutic drugs out of the cells (Juliano and Ling 1976). An absence of definitive substrate specificity is the most characteristic and also the most enigmatic aspect of this transporter. Evidence that this multidrug resistance phenotype arises from overexpression of the ABCB1 protein was provided by transfection experiments with the *ABCB1* gene, which showed that overexpression of P-gp alone can confer multidrug resistance on an otherwise drug-sensitive cell line (Gros et al. 1986; Ueda et al. 1987). It is assumed that this protein acts to reduce accumulation of its substrates in the cell, and the mechanism of reduced accumulation involves the ATP-dependent transport of substrates out of the cell (Figure 10.2).

Although increased *ABCB1* gene expression was usually associated with poor response to chemotherapeutic agents, it later became evident that multidrug resistance could also occur in the absence of detectable *ABCB1* gene expression. Further investigations led to the discovery of two new families of ABC proteins (and also some non-ABC transporters). First, the involvement of a 190 kDa protein in multidrug resistance was identified (Cole et al. 1992) and was named the multidrug-resistance-associated protein (MRP). This protein proved to be distantly related to the ABCB1 protein (and is now known as the ABCC1 protein) and facilitated the discovery of eight more genes within the same MRP subfamily, of which at least four are potentially involved in mediating drug resistance (ABCC2, ABCC3, ABCC4, and ABCC5; reviewed in Leslie et al. 2005). A third drug transporter, encoded by the *ABCG2* gene, also distantly related to P-glycoprotein and the MRPs, was originally isolated from a multi-drug-resistant breast cancer cell in an effort to elucidate non-P-glycoprotein mechanisms of drug resistance (Doyle et al. 1998; Bates et al. 2001).

10.4 PHYSIOLOGICAL FUNCTIONS AND RELEVANCE OF ABC PROTEINS

In addition to their role in drug resistance, ABCB1, ABCC1, ABCC2, and ABCG2 are expressed in nonmalignant tissues and are involved in protecting tissues from xenobiotic accumulation and resulting toxicity. In mammals, ABCB1 is expressed on the mucosal surface of the gastrointestinal tract, which suggests a function to

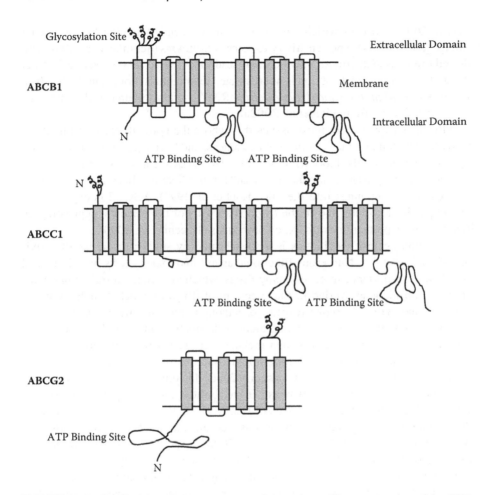

FIGURE 10.2 Predicted secondary structures of three drug efflux transporters of the ATP-binding cassette (ABC) family. ABCB1 protein consists of two transmembrane domains, each containing six transmembrane segments and two ATP-binding sites. It is *N*-glycosylated at the first extracellular loop. ABCC1 protein has an additional aminoterminal extension containing five transmembrane segments and is *N*-glycosylated near the N-terminus and at the sixth extracellular loop. ABCG2 protein is a half transporter consisting of one ATP-binding site and six transmembrane segments, and is generally *N*-glycosylated at the sixth extracellular loop. In contrast to the other transporters, the ATP-binding site of ABCG2 protein is at the aminoterminal end of the polypeptide. N denotes aminoterminal ends of the proteins.

prevent uptake of xenobiotics and perhaps to facilitate excretion (Thiebaut et al. 1987). ABCB1 is also present in the brush border of the proximal tubule of the kidneys and on the biliary face of the hepatocytes. These localisations are consistent with a role for ABCB1 in excretion of xenobiotics and endogenous metabolites into the urine and bile, respectively. The protein is also localised on the luminal surface of capillary endothelial cells in the brain and the testes and within the placenta, thus indicating a role in protection of these particular organs (Thiebaut et al. 1989;

Fromm 2004). Mice in which *Abcb1* genes have been inactivated by insertional mutagenesis show extreme sensitivity to some neurotoxins (Schinkel et al. 1994) and altered drug tissue distribution, which may lead to increased fetal exposure (Smit et al. 1999). This tissue distribution is not restricted to mammals. For example, in a fish, the turbot *Scophthalmus maximus Abcb1* mRNA is mainly found in the intestine, brain, and kidney (Tutundjian et al. 2002a).

ABCC1 is expressed in most tissues throughout the body. Its localisation at the basolateral cellular surface of lung, testis, kidneys, and liver suggests a role in excreting compounds into the blood (Flens et al. 1996). ABCC2 has a more limited distribution, but its presence in canalicular cells in the liver has been associated with the export of compounds into the bile (Kool et al. 1997). ABCG2 is expressed in hematopoetic stem cells and in the trophoblast cells of the placenta, suggesting an involvement in protection from toxic compounds (Rocchi et al. 2000).

Protection against xenobiotics is also indicated by studies on transport activities of ABC proteins. Numerous functional assays and toxicity tests have indicated that ABC transporters can efflux many drugs, which are often natural compounds. There is considerable overlap in the spectrum of drugs to which ABCB, ABCC1, ABCC2, and ABCG2 confer resistance, although some substrate specificity can be noted. ABCB1 transports moderately hydrophobic, amphiphilic, neutral, or positively charged substances, which include (but are not restricted to) anticancer drugs, analgesics, protease inhibitors, immunosuppressive agents, corticoids, pesticides, and antibiotics (reviewed in Cascorbi 2006). In addition to nonbiotransformed compounds, ABCC1 can export phase II metabolites, including glucuronides, sulphate, and glutathione conjugates (Jedlitschky et al. 1996). ABCC1 and ABCC2 also require reduced glutathione (GSH) as a cosubstrate to efflux compounds such as vincristine (Loe et al. 1996; Evers et al. 2000). ABCC2 is primarily an organic anion transporter and is thus involved in the cotransport of weakly basic compounds along with GSH. Notably, it can export GSH conjugates of inorganic arsenic (Kala et al. 2000). ABCG2 confers resistance to a narrower range of anticancer agents than ABCB1, ABCC1, and ABCC2. Like ABCB1, ABCG2 is not dependent on GSH for transport of its substrates.

In accordance with the wide substrate specificity of the ABC transporters, an increasing number of compounds have been identified as substrates. These include numerous compounds of environmental relevance, such as pesticides (reviewed in Buss and Callaghan 2008), polychlorobiphenyls (Galgani et al. 1996), polycyclic aromatic hydrocarbons (Yeh et al. 1992; Britvic and Kurelec 1999), alkylphenols polyethoxylates (Sturm et al. 2001), antibiotics (Eberl et al. 2007), dialyl phthalate (Minier et al. 2008), musks (Luckenbach et al. 2004), statins and fibrates (Caminada et al. 2008), detergents (Li-Blatter et al. 2009), and trace metals such as cadmium and zinc (Achard et al. 2004) or sodium arsenite (Eufemia and Epel 1998).

Apart from their role in xenobiotic excretion, there may be several other roles for ABC transporters that have not yet been fully explored. High ABCB1 protein expression is found in steroid-secreting glands, and inactivation of the mouse *Abcb1* gene results in reduced steroid secretion (Ueda et al. 1992; Altuvia et al. 1993). ABCB1 has been shown to play a role in the migration of dentritic cells (Randolph et al.

1998). There is also some evidence that ABCC1, ACC2, and ABCG2 are associated with cellular folic acid homeostasis (Hooijberg et al. 2003; Ifergan et al. 2004).

10.5 ABC PROTEINS AND MULTIXENOBIOTIC RESISTANCE

In contrast to medical or agricultural research, which aims to discover how circumventing or inhibiting the ABC protein activities might lead to better treatment of cancers or pests, environmental research aims to understand how these transporters keep xenobiotics out of the cells, and to what extent they ensure protection from environmental contaminants. Research on aquatic ecosystems has been pioneering in this respect, and the first insights of the existence of an ABC defence system have been reported in aquatic invertebrates. Membrane vesicles isolated from several organs of the freshwater mussel *Anodonta cygnea* and the marine mussel *Mytilus galloprovincialis* showed an ability to bind xenobiotics in a verapamil (an ABC transporter inhibitor)-sensitive manner (Kurelec and Pivčević 1989, 1991). Immuno-related proteins were first identified in the blue mussel *Mytilus edulis*, the oyster *Crassostrea gigas* (Minier et al. 1993), and the marine echiuran worm *Urechis caupo* (Toomey and Epel 1993). Transport activities have been demonstrated using model compounds in *U. caupo* (Toomey and Epel 1993) and *M. galloprovincialis* (Kurelec and Pivčević 1991). These first steps led Kurelec (1992) to propose that the ABC transport system is designated as the multixenobiotic resistance (MXR) mechanism, serving as a defence system against environmental xenobiotics. In accordance with this hypothesis, some environmental xenobiotics, including moderately hydrophobic pesticides and, to a lesser extent, highly hydrophobic xenobiotics such as PCBs, were shown to interact with the mussel ABC system (Cornwall et al. 1995; Galgani et al. 1996). Activity tests using the ability of compounds to inhibit or to compete for the transporters demonstrated that the herbicide simazine enhanced dye accumulation in mussel blood cells by 25% (Minier and Moore 1998). An increasing number of ABC genes are now described in many aquatic organisms (Table 10.1).

Demonstrating that the ABC system exists as a direct response to an exposure to xenobiotics was an important step. The relationship between the expression of the MXR phenotype and exposure to toxic compounds was first tested with *in vitro* experiments where mussels were exposed to increasing concentrations of the *Vinca* alkaloid vincristine (Minier and Moore 1996). Treatment with 5 µg.ml^{-1} vincristine enabled the blood cells to reduce rhodamine B accumulation by 35%, and those treated with 10 µg.ml^{-1} by 48%, compared to control samples. This indicates that ABC-like efflux activity is related to the concentration of the toxic compound. Nevertheless, when analysing all concentrations tested (at different time points), a direct linear relationship was not shown. Although the activity clearly increased with the dose, an effect of the repeated injections was noted; i.e., chronic exposure to the toxic compound contributed to the increased efflux activity. This result suggests that resistant cells are selected under toxic pressure. Accordingly, a decrease in blood cell concentration was noted at the end of the experiment (Minier and Moore 1996).

Furthermore, when comparing the dye exclusion activity of the cells exposed to vincristine for fifteen days to those that had been left untreated for fourteen days

TABLE 10.1
Partial or Complete Cloned Sequences of ABCB1, ABCC1, and ABCG2
Genes in Aquatic Organisms from the Swiss-Prot Database

Species	Access Number	Authors
	ABCB1	
Amphibian		
Xenopus laevis	Q91586	Castillo G., Shen H.J., Horwitz S.B.
Echinoderm		
Strongylocentrotus purpuratus	Q3T7C8	Hamdoun A., Cherr G., Roepke T., Chou S., Epel D.
Molluscs		
Venerupis philippinarum	C0KJQ5	Liu N., Pan L.Q., Miao J.J.
Aequipecten opercularis	C1L366	Lozano V., Mauriz O., Martinez-Escauriaza R., Sanchez J.L., Pazos A.J.
Brachidontes pharaonis	Q4VYA4	Feldstein T., Nelson N., Mokady O.
Crassostrea gigas	Q8WQ93	Minier C., Lelong C., Djemel N., Rodet F., Tutudjian R., Favrel P., Mathieu M., Leboulenger F.
Crassostrea virginica	Q7YZC2	Christl T.J., James E.R., Karnaky K.J.Jr., Scott G.I.
Dreissena polymorpha	Q8MPN1	Tutundjian R., Leboulenger F., Minier C.
Mimachlamys varia	C1L368	Lozano V., Mauriz O., Martinez-Escauriaza R., Sanchez J.L., Pazos A.J.
Mytilus californianus	B2WTH9	Luckenbach T., Epel D.
Mytilus edulis	Q9Y0C3	Alpermann T.J., Luedeking A., Mengedoht D.M., Koehler A.
Mytilus galloprovincialis	B7S5C3	Damiens G., Romeo M.
Pecten maximus	C1L367	Lozano V., Mauriz O., Martinez-Escauriaza R., Sanchez J.L., Pazos A.J.
Unio pictorum	Q5I3N1	Sauerborn Klobucar R., Zaja R., Smital T.
Fishes		
Barbus barbus	C3VPX3	Klobucar R.S., Zaja R., Smital T.
	Q4PS58	Sauerborn Klobucar R., Krca S., Zaja R., Smital T.
Carassius gibelio	Q4PS55	Sauerborn Klobucar R., Krca S., Zaja R., Smital T.
Chondrostoma nasus	Q58HM1	Sauerborn Klobucar R., Krca S., Zaja R., Smital T.
Cyprinus carpio	Q4VC52	Sauerborn Klobucar R., Krca S., Zaja R., Smital T.
Fundulus heteroclitus	Q9W6L9	Cooper P.S., Van Veld P.A., Reece K.S.
Leuciscus cephalus	Q4VC50	Sauerborn Klobucar R., Krca S., Zaja R., Smital T.
Mullus barbatus	Q5I6A4	Sauerborn Klobucar R., Zaja R., Smital T.
Oncorhynchus mykiss	Q5I1Z1	Sauerborn Klobucar R., Zaja R., Smital T.
Platichthys flesus	Q98TN5	Williams T.D., Gensberg K., Minchin S.D., Chipman J.K.
Poeciliopsis lucida	Q4PJT7	Sauerborn Klobucar R., Zaja R., Smital T.
Pseudopleuronectes americanus	Q08134	Chan K.M., Davies P.L., Childs S., Veinot L., Ling V.
Trematomus bernacchii	D0PRB1	Zucchi S., Corsi I., Bard S., Richards R., Focardi S.

(continued)

TABLE 10.1
Partial or Complete Cloned Sequences of ABCB1, ABCC1, and ABCG2
Genes in Aquatic Organisms from the Swiss-Prot Database (Continued)

Species	Access Number	Authors
	ABCC1	
Amphibian		
Xenopus tropicalis	B1H398	
Molluscs		
Mytilus californianus	B2WTH9	Luckenbach T., Epel D.
Mytilus edulis	Q962X4	Luedeking A., Alpermann T., Koehler A.
Mytilus galloprovincialis	A0SXG3	Franzellitti S., Fabbri E.
Fishes		
Barbus barbus	Q4PS57	Sauerborn Klobucar R., Krca S., Zaja R., Smital T.
Carassius gibelio	Q4PS56	Sauerborn Klobucar R., Krca S., Zaja R., Smital T.
Chondrostoma nasus	Q58HM2	Sauerborn Klobucar R., Krca S., Zaja R., Smital T.
Cyprinus carpio	Q6B520	Sauerborn Klobucar R., Stupin Polancec D., Brozovic A., Zaja R., Smital T.
Leuciscus cephalus	Q4VC51	Sauerborn Klobucar R., Krca S., Zaja R., Smital T.
Mullus barbatus	C3VPX3	Klobucar R.S., Zaja R., Smital T.
Platichthys flesus	Q711I6	Cutler C.P., Brown A., Cramb G.
Poeciliopsis lucida	Q4FAA8	Zaja R., Sauerborn Klobucar R., Smital T.
	ABCG2	
Amphibians		
Xenopus laevis	A1L2M4	
Xenopus tropicalis	Q28BS4	Amaya E., et al.
Fishes		
Oncorhynchus mykiss	A8IJF9	Zaja R., Munic V., Smital T.
Salmo salar	C0PUC7	Leong J., von Schalburg K., Cooper G., Moore R., Holt R., Davidson W.S., Koop B.F.
	C0HA23	Leong J., von Schalburg K., Cooper G., Moore R., Holt R., Davidson W.S., Koop B.F.

after these first injections, no significant difference was observed. Hence, no reversion of the enhanced MXR phenotype has occurred in this experiment. Similarly, a lasting effect of xenobiotic exposure was noted in the mussel *M. californianus*, where efflux activity was still partially inhibited forty-eight hours after the removal of the ABC substrate (Luckenbach and Epel 2005).

Western blot analysis using C219 monoclonal antibody was carried out with the same mussel samples. This antibody recognises a highly conserved sequence common to almost all known ABCB1 proteins whose sequence is known (Endicott and

Ling 1989). Concentrations of C219-immunodetected proteins were 1.5 to 2.5 times higher in all treated samples than in control samples. This suggests that the vincristine treatments have induced enhancement in the MXR protein expression, and that this protein may be responsible (at least partly) for the reduced rhodamine accumulation in cells (Minier and Moore 1996).

Thus, these *in vitro* experiments added consistency to the previously described MXR phenotype, notably the ability of cells to reduce toxic compound accumulation in a verapamil-sensitive manner and the expression of the C219 antibody-recognised protein. Following this study, other works demonstrated that the MXR phenotype was inducible by polycyclic aromatic hydrocarbons (Smital and Kurelec 1998), pentachlorophenol, chlorthal, sodium arsenite (Eufemia and Epel 1998), metals (Achard et al. 2004), or dialyl phthalate (Minier et al., 2008). As MXR activity is an inducible mechanism, it could be suitable as a biomarker of exposure to environmental contaminants (Minier et al. 1999).

Both quantification of ABC proteins or assessment of transport activity can be used for environmental studies and monitoring. The most extensive survey was performed along the French coast (Minier et al. 2006). Mussels, *Mytilus edulis* and *M. galloprovincialis*, and the oyster *Crassostrea gigas* were sampled from a total of forty-three sites along the French coast and analysed for their C219-recognised protein levels. High expression levels were found in animals from the major French estuaries (Seine, Loire, and Gironde), at a few sites in Brittany, and in nearly all sites from the Mediterranean mainland coasts. Multivariate analysis of the data indicated that expression of MXR protein was strongly associated with contaminant concentrations in mussels, and that polycyclic aromatic hydrocarbons (PAHs) and polychlorinated biphenyls (PCBs) were directly correlated with MXR protein concentration. Variation between sites showed that levels could be more than ten times higher at contaminated sites (e.g., estuaries) than at reference sites. Although the results do not infer a causal linkage between mussel MXR protein, PAHs, and PCBs, since many other chemical contaminants are also present in the environment, they do show clearly that ABC protein expression is related to xenobiotic exposure. Interestingly, correlations with some of the xenobiotics assessed in mollusc tissues indicated that C219-recognised proteins were rapidly reaching a threshold value, and that a further increase in the tissue burden was not leading to an increase of immunodetected protein levels (Figure 10.3). This could indicate that the molluscs were then relying on an increased transport activity or on another defence mechanism. Alternatively, the organisms' health might have already been compromised, so that they were unable to further increase their MXR defence mechanism.

10.6 ENVIRONMENTAL RELEVANCE OF ABC PROTEINS

Because the MXR system is a nonspecific mechanism, a number of environmental toxicants can interact with the ABC transporters and alter their activities by blocking their active sites or saturating them. The resultant competition or inhibition, termed chemosensitisation, can decrease transporter activities such that xenobiotics

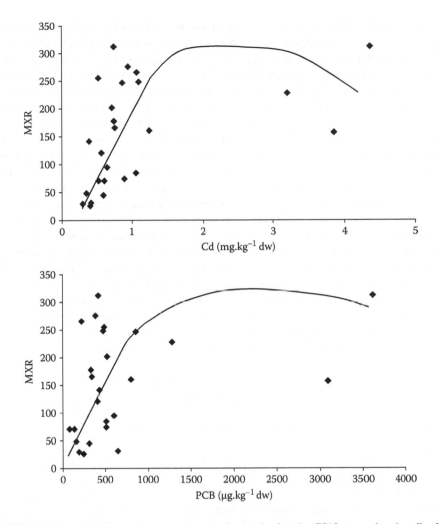

FIGURE 10.3 ABCB1 protein expression (as detected using the C219 monoclonal antibody) in mussels from the French coasts in relation with body burden of cadmium or polychlorobiphenyls. A significant linear relationship exists up to a concentration limit of circa 1.2 mg.kg^{-1} dw cadmium and 1 mg.kg^{-1} dw PCBs. (From Minier, C., personal data.)

normally excluded from the cell may then enter, accumulate, and exert toxicological effects (Higgins 2007).

Chemosensitisation may have important implications, especially for developing embryos. In early development stages, biotransformation activities (phases I and II) are generally not observed. Indeed, expression of enzyme systems such as cytochrome P450 generally does not appear until the differentiation of tissues. Embryos may then rely on other defence mechanisms, such as the ABC system, which appears as a first line of defence. Accordingly, upregulation of efflux transporter activity begins twenty-five minutes postfertilisation in sea urchin eggs, and at least three ABC mRNA (from *Abcb1*, *Abcc1*, and *Abcc2* genes) are present (Hamdoun et al. 2004).

Embryos of the echiuroid worm *Urechis caupo* have levels of ABCB1-related efflux activity even higher than those seen in drug-resistant cancer cells which overexpress ABC transporters (Toomey and Epel 1993), and high levels of ABC gene expression have also been reported in oyster embryos (Minier et al. 2002). Nevertheless, inhibition of MXR activity may have dramatic consequences. Presence of the competitive inhibitor verapamil during the first divisions of the *Urechis* embryo increases the toxicity of cytotoxic drugs, such as emetine and vinblastine, leading to a delay in cell division and production of abnormal cells upon division (Toomey and Epel 1993). Similarly, the median effective concentration of vinblastine drops by more than an order of magnitude in the presence of the inhibitor MK571 when compared to the control for sea urchin, *Strongylocentrotus purpuratus*, embryos (Hamdoun et al. 2004). Inhibition of the transporter also retards progression through the cell cycle, with major effects on anaphase. This effect on mitosis, and the timing of upregulation of activity during embryogenesis, points to a role for efflux transport in promoting entry or progression through the first cell division. In the amoeba *Dyctyostelium discoideum*, ABC transporters regulate cell differentiation (Good and Kuspa 2000), and in the nematode *Caenorhabditis elegans*, an ABC transporter is involved in larval development (Yabe et al. 2005). ABC proteins may be implicated in the transport of signalling molecules, and inhibition of the ABC transport system may have important consequences on development. Accordingly, developmental abnormalities have been reported in *Mytilus edulis* larvae exposed to ABC transporter inhibitors (McFadzen et al. 2000).

Protecting the brain from toxicants is crucial for the organisms and may even be ecologically relevant, as it may have profound consequences on behaviour. Activity of the ABC system has been studied in the killifish (*Fundulus heteroclitus*), and results showed that it contributes to active xenobiotic excretion at the blood-brain barrier (Miller et al. 2002; Bard and Stegeman 2002). However, ABC protein function and thus blood-brain barrier integrity can be compromised by chemosensitisers that inhibit transport activity. Fish treated with ivermectin (a pesticide used in veterinary medication) and the efflux inhibitor cyclosporine A are significantly more sensitive to the pesticide compared to fish treated with ivermectin only (Bard and Gadbois 2007). Inhibition of the ABC transporters is associated with significantly earlier onset and increased mortality in ivermectin-exposed fish. Furthermore, at lower doses, neurobehavioural effects can be monitored. These effects include lethargy, slowing of pectoral fin movement, altered posture, and decreased escape response (Bard and Gadbois 2007).

The ABC transport system can transport xenobiotics out of the cell even if this compound is present at very low concentration. Significant transport activity of rhodamine was measured using *Urechis* embryos incubated with concentrations as low as 10^{-8} M of the dye (Toomey and Epel 1993). Similar results were obtained with mussel gills (Minier, unpublished data). Sea urchin embryos can keep calcein-acetoxymethyl ester (calcein-AM) out of the cells at levels between 10^{-9} and 10^{-5} M, a 10^4 range of concentration (Hamdoun et al. 2004). These low effective concentrations may appear surprising considering the lack of specificity of the ABC transporters. Nevertheless, it must be noted that a number of ABC substrates accumulate within the cell membrane, from which they can be extruded by the ABC

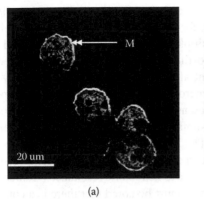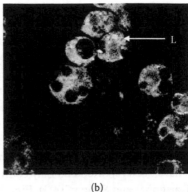

(a)   (b)

FIGURE 10.4 Confocal laser microscopy observations of turbot hepatocytes incubated with 1 mM rhodamine B for one hour without (a) or with (b) 20 mM verapamil. M means membrane and L lysosome. Incubation with verapamil led to a 26% increase of the cell fluorescence ($p < 0.001$, $n = 20$). Note the rhodamine B fluorescence within the cell membrane in the absence of the inhibitor (a). (From Tutundjian, R., et al., *Mar. Environ. Res.*, 54, 443–47, 2002b. With permission.)

transporters. Thus, concentrations of xenobiotics in direct proximity of the transporters may be considerably higher than the average medium concentration. This can be visualised using fluorescent substrates (Figure 10.4). Although the efficiency of ABC transporters may depend on the structure of the substrate (Hamdoun et al. 2004), their activity may be highly relevant in environmental conditions considering the concentration of xenobiotics in, for example, aquatic environments and their presence as a complex mixture of compounds. Complex interactions might occur between various chemicals and ABC transporters. The additive effects of several substrates may lead to chemosensitisation even if these particular compounds are not themselves in high individual amounts. Chemosentisation may then appear as a special risk, leading even nontoxic compounds to have detrimental effects (Luckenbach and Epel 2005).

10.7 CONCLUSIONS

It is becoming increasingly evident that the ABC transporters play a significant role in xenobiotic bioavailability. This role is revealed by the tissue distribution of ABC transporters, which are found highly expressed at important toxicological barriers (e.g., the blood-brain barrier), epithelia regulating absorption and excretory sites. As the activity of a xenobiotic ultimately depends on the ability of the compound to reach its target, ABC transporters constitute an effective toxicological barrier that could confer resistance to a wide range of compounds.

Genome sequencing programs have revealed that ABC transporters constitute a superfamily of numerous genes that are highly conserved throughout the evolution. The number of ABC transporters represents a significant proportion of the genes in the species for which the complete genome has been sequenced, and an increasing number of ABC members are being discovered in numerous species. This indicates the important role of these transporters for living organisms. As far as defence against xenobiotics is concerned, the multiple transporters that are expressed may form a functional network capable of exporting a very wide range of chemical compounds. Each member may have different although overlapping specificity, which may help the cells to cope with the numerous xenobiotics that they are exposed to. Whereas ABCB1 transports unmodified neutral or positively charged hydrophobic compounds, substrates of the ABCC subfamily members and ABCG2 extend to organic anions and phase II metabolites. This network may be spatially organised to include sequestration in compartments such as the lysosomes. Indeed, ABC transporters have been reported and may contribute to xenobiotic resistance (Köhler et al. 1998; Rajagopal and Simon 2003). Finally, it must be noted that there is a considerable overlap in the substrate specificity and regulation of the phase I and phase II enzymes, such as cytochrome P450 and glutathione transferase and the ABC transporters, thus indicating that they play complementary roles in xenobiotic resistance (Paetzold et al. 2009).

Because of its obvious *in vitro* effects, chemosentisation appears as a special risk for the organisms. Blocking the ABC system results in increased toxicity, and even a nontoxic compound may lead to detrimental effects because it could contribute to an enhanced accumulation of other toxic compounds. Given the wide specificity of the different ABC transporters and the complex mixture of compounds present in the environment, xenobiotic interaction in the membrane should be significant. This may result in unexpected toxicity, as is often reported for drugs in humans (Greiner et al. 1999; Imai et al. 2004; Zhang et al. 2006; Igel et al. 2007).

REFERENCES

Achard, M., et al. 2004. Induction of a multixenobiotic resistance protein (MXR) in the Asiatic clam *Corbicula fluminea* after heavy metals exposure. *Aquat. Toxicol.* 67:347–57.

Altuvia, S., et al. 1993. Targeted disruption of the mouse *mdr1b* gene reveals that steroid hormones enhance mdr gene expression. *J. Biol. Chem.* 268:27127–32.

Anjard, C., et al. 2002. Evolutionary analyses of ABC transporters of *Dictyostelium discoideum*. *Eukaryot. Cell* 1:643–52.

Bard, S. M., and S. Gadbois. 2007. Assessing neuroprotective P-glycoprotein activity at the blood-brain barrier in killifish (*Fundulus heteroclitus*) using behavioural profiles. *Mar. Environ. Res.* 64:679–82.

Bard, S. M., and J. J. Stegeman. 2002. Inhibition of rhodamine B transport in liver, brain and ovary by cyclosporine A: Development of an *in vitro* model for analysis of P-glycoprotein function. *Mar. Environ. Res.* 54:461–63.

Bates, S. E., et al. 2001. The role of half transporters in multidrug resistance. *J. Bioenerg. Biomembr.* 33:503–11.

Biedler, J. M., and H. Riehm. 1970. Cellular resistance to actinomycin D in Chinese hamster cells *in vitro*: Cross-resistance, radioautographic and cytogenetic studies. *Cancer Res.* 30:1174–84.

Blattner, F. R., et al. 1997. The complete genome sequence of *Escherichia coli* K-12. *Science* 277:1453–74.

Britvic, S., and B. Kurelec. 1999. The effect of inhibitors of multixenobiotic resistance mechanism on the production of mutagens by *Dreissena polymorpha* in waters spiked with premutagens. *Aquat. Toxicol.* 47:107–16.

Buss, D. S., and A. Callaghan. 2008. Interaction of pesticides with p-glycoprotein and other ABC proteins: A survey of the possible importance to insecticide, herbicide and fungicide resistance. *Pestic. Biochem. Physiol.* 90:141–53.

Caminada, D., et al. 2008. Human pharmaceuticals modulate P-gp1 (ABCB1) transport activity in the fish cell line PLHC-1. *Aquat. Toxicol.* 90:214–22.

Cascorbi, I. 2006. Role of pharmacogenetics of ATP-binding cassette transporters in the pharmacokinetics of drugs. *Pharmacol. Ther.* 112:457–73.

Cole, S. P. C., et al. 1992. Overexpression of a transporter gene in a multidrug-resistant human lung cancer cell line. *Science* 258:1650–54.

Cornwall, R., et al. 1995. Characterization of multixenobiotic/multidrug transport in the gills of the mussel *Mytilus californianus* and identification of environmental substrates. *Aquat. Toxicol.* 31:277–96.

Danö, K. 1973. Active outward transport of daunomycin in resistant Ehrlich ascites tumor cells. *Biochim. Biophys. Acta* 323:466–83.

Dean, M., A. Rzhetsky, and R. Allikmets. 2001. The human ATP-binding cassette (ABC) transporter superfamily. *Genome Res.* 11:1156–66.

Decottignies, A., and A. Goffeau. 1997. Complete inventory of the yeast ABC proteins. *Nat. Genet.* 15:137–45.

Doyle, L. A., et al. 1998. A multidrug resistance transporter from human MCF-7 breast cancer cells. *Proc. Natl. Acad. Sci. USA* 95:15665–70.

Eberl, S., et al. 2007. Role of p-glycoprotein inhibition for drug interactions: Evidence from *in vitro* and pharmacoepidemiological studies. *Clin. Pharmacokinet.* 46:1039–49.

Endicott, J. A., and V. Ling. 1989. The biochemistry of P-glycoprotein-mediated multidrug resistance. *Annu. Rev. Biochem.* 58:137–71.

Eufemia, N. A., and D. Epel. 1998. The multixenobiotic defence mechanism in mussels is induced by substrates: Implications for a general stress response. *Mar. Environ. Res.* 46:401–5.

Evers, R., et al. 2000. Vinblastine and sulfinpyrazone export by the multidrug resistance protein MRP2 is associated with glutathione export. *Br. J. Cancer* 83:375–83.

Flens, M. J., et al. 1996. Tissue distribution of the multidrug resistance protein. *Am. J. Pathol.* 148:1237–47.

Fromm, M. F. 2004. Importance of P-glycoprotein at blood–tissue barriers. *Trends Pharmacol. Sci.* 25:423–29.

Galgani, F., et al. 1996. Interaction of environmental xenobiotics with a multixenobiotic defense mechanism in the bay mussel *Mytilus galloprovincialis* from the coast of California. *Environ. Toxicol. Chem.* 15:325–31.

Garcia, O., et al. 2004. Inventory and comparative analysis of rice and *Arabidopsis* ATP-binding cassette (ABC) systems. *J. Mol. Biol.* 343:249–65.

Good, J. R., and A. Kuspa. 2000. Evidence that a cell-type-specific efflux pump regulates cell differentiation in *Dictyostelium. Dev. Biol.* 220:53–61.

Greiner, B., et al. 1999. The role of intestinal P-glycoprotein in the interaction of digoxin and rifampin. *J. Clin. Invest.* 104:147–53.

Gros, P., et al. 1986. Isolation and expression of a cDNA (*MDR*) that confers multidrug resistance. *Nature* (London) 323:728–31.

Hamdoun, A. M., et al. 2004. Activation of multidrug efflux transporter activity at fertilization in sea urchin embryos (*Strongylocentrotus purpuratus*). *Dev. Biol.* 276:452–62.

Higgins, C. F. 2007. Multiple molecular mechanisms for multidrug resistance transporters. *Nature* (London) 446:749–57.

Hooijberg, J. H., et al. 2003. The role of multidrug resistance proteins MRP1, MRP2 and MRP3 in cellular folate homeostasis. *Biochem. Pharmacol.* 65:765–71.

Hrycyna, C. A., and M. M. Gottesman. 1998. Multidrug ABC transporters from bacteria to man—An emerging hypothesis for the universality of molecular mechanism and function. *Drug Resist. Update* 1:81–83.

Ifergan, I., et al. 2004. Folate deprivation results in the loss of breast cancer resistance protein (BCRP/ABCG2) expression: A role for BCRP in cellular folate homeostasis. *J. Biol. Chem.* 279:25527–34.

Igel, S., et al. 2007. Increased absorption from digoxin the human jejunum due to inhibition of intestinal transporter-mediated efflux. *Clin. Pharmacokinet.* 46:777–85.

Imai, Y., et al. 2004. Phytoestrogen/flavonoids reverse breast cancer resistance protein/ ABCG2-mediated multidrug resistance. *Cancer Res.* 64:4346–52.

Jedlitschky, G., et al. 1996. Transport of glutathione, glucuronate and sulfate conjugates by the MRP gene-encoded conjugate export pump. *Cancer Res.* 56:988–94.

Juliano, R. L., and V. Ling. 1976. A surface glycoprotein modulating drug permeability in Chinese hamster ovary cell mutants. *Biochem. Biophys. Acta* 455:152–62.

Kala, S. V., et al. 2000. The MRP2/cMOAT transporter and arsenic-glutathione complex formation are required for biliary excretion of arsenic. *J. Biol. Chem.* 275:33404–8.

Köhler, A., et al. 1998. Detection of P-glycoprotein mediated MDR/MXR in *Carcinus maenas* hepatopancreas by immuno-gold-silver labeling. *Mar. Environ. Res.* 46:411–14.

Kool, M., et al. 1997. Analysis of expression of cMOAT (MRP2), MRP3, MRP4 and MRP5, homologues of the multidrug resistance-associated protein gene (*MRP1*), in human cancer lines. *Cancer Res.* 57:3537–47.

Kurelec, B. 1992. The multixenobiotic resistance mechanism in aquatic organisms. *Crit. Rev. Toxicol.* 19:291–302.

Kurelec, B., and B. Pivčević. 1989. Distinct glutathione-dependent enzyme activities and a verapamil-sensitive binding of xenobiotics in a fresh-water mussel. *Biochem. Biophys. Res. Commun.* 164:934–40.

Kurelec, B., and B. Pivčević. 1991. Evidence for a multixenobiotic resistance mechanism in the mussel *Mytilus galloprovincialis. Aquat. Toxicol.* 19:291–302.

Leslie, E. M., R. G. Deeley, and S. P. Cole. 2005. Multidrug resistance proteins: Role of P-glycoprotein, MRP1, MRP2, and BCRP (ABCG2) in tissue defense. *Toxicol. Appl. Pharmacol.* 204:216–37.

Li-Blatter, X., P. Nervi, and A. Seelig. 2009. Detergents as intrinsic P-glycoprotein substrates and inhibitors. *Biochim. Biophys. Acta* 1788:2335–44.

Linton, K. J., and C. F. Higgins. 1998. The *Escherichia coli* ATPbinding cassette (ABC) proteins. *Mol. Microbiol.* 28:5–13.

Loe, D. W., et al. 1996. Multidrug resistance protein (MRP)-mediated transport of leukotriene C4 and chemotherapeutic agents in membrane vesicles. Demonstration of glutathione-dependent vincristine transport. *J. Biol. Chem.* 271:9675–82.

Luckenbach, T., I. Corsi, and D. Epel. 2004. Fatal attraction: Synthetic musk fragrances compromise multixenobiotic defense systems in mussels. *Mar. Environ. Res.* 58:215–19.

Luckenbach, T., and D. Epel. 2005. Nitromusk and polycyclic musk compounds as long-term inhibitors of cellular xenobiotic defense systems mediated by multidrug transporters. *Environ. Health Persp.* 113:17–24.

McFadzen, I., et al. 2000. Multidrug resistance in the embryos and larvae of the mussel *Mytilus edulis. Mar. Environ. Res.* 50:319–23.

Miller, D. S., et al. 2002. Xenobiotic efflux pumps in isolated fish brain capillaries *Am. J. Physiol. Reg. I* 282:191–98.

Minier, C., F. Akcha, and F. Galgani. 1993. P-glycoprotein expression in *Crassostrea gigas* and *Mytilus edulis* in polluted seawater. *Comp. Biochem. Physiol.* 106B:1029–36.

Minier, C., N. Eufemia, and D. Epel. 1999. The multixenobiotic resistance phenotype as a tool to biomonitor the environment. *Biomarkers* 4:442–54.

Minier, C., and M. N. Moore. 1996. Rhodamine B accumulation and MXR protein expression in mussel blood cells: Effects of exposure to vincristine. *Mar. Ecol. Prog. Ser.* 142:165–73.

Minier, C., and M. N. Moore. 1998. Calcein accumulation in mussel blood cells. *Mar. Environ. Res.* 46:425–28.

Minier, C., et al. 2002. Expression and activity of a multixenobiotic resistance system in the Pacific oyster *Crassostrea gigas*. *Mar. Environ. Res.* 54:455–59.

Minier, C., et al. 2006. Multixenobiotic resistance protein expression in *Mytilus edulis*, *Mytilus galloprovinciallis* and *Crassostrea gigas* from the French coasts. *Mar. Ecol. Prog. Ser.* 322:155–68.

Minier, C., et al. 2008. Multixenobiotic resistance, acetyl-choline esterase activity and total oxyradical scavenging capacity of the Arctic spider crab, *Hyas araneus*, following exposure to bisphenol A, tetra bromo diphenyl ether and diallyl phthalate. *Mar. Pollut. Bull.* 56:1410–15.

Paetzold, C., et al. 2009. Up-regulation of hepatic ABCC2, ABCG2, CYP1A1 and GST in multixenobiotic-resistant killifish (*Fundulus heteroclitus*) from the Sydney Tar Ponds, Nova Scotia, Canada. *Mar. Environ. Res.* 68:37–47.

Rajagopal, A., and S. M. Simon. 2003. Subcellular localization and activity of multidrug resistance proteins. *Mol. Biol.* 14:3389–99.

Randolph, G. J., et al. 1998. A physiologic function for P-glycoprotein (MDR-1) during the migration of dentritic cells from skin via afferent lymphatic vessels. *Proc. Natl. Acad. Sci. USA* 95:6924–29.

Rocchi, E., et al. 2000. The product of the ABC half-transporter gene ABCG2 (BCRP/MXR/ ABCP) is expressed in the plasma membrane. *Biochem. Biophys. Res. Commun.* 271:42–46.

Roth, C. W., et al. 2003. Identification of the *Anopheles gambiae* ATP-binding cassette transporter superfamily genes. *Mol. Cell.* 15:150–58.

Sanchez-Fernandez, R., et al. 2001. The *Arabidopsis thaliana* ABC protein superfamily, a complete inventory. *J. Biol. Chem.* 276:30231–44.

Sauvage, V., et al. 2006. Identification and expression analysis of ABC protein-encoding genes in *Toxoplasma gondii*: *Toxoplasma gondii* ATP-binding cassette superfamily. *Mol. Biochem. Parasitol.* 47:177–92.

Schinkel, A. H., et al. 1994. Disruption of the mouse *mdr1a* P-glycoprotein gene leads to a deficiency in the blood-brain barrier and to increased sensitivity to drugs. *Cell* 77:491–502.

Sheps, J. A., et al. 2004. The ABC transporter gene family of *Caenorhabditis elegans* has implications for the evolutionary dynamics of multidrug resistance in eukaryotes. *Genome Biol.* 5:R15.

Smit, J. W., et al. 1999. Absence or pharmacological blocking of placental P-glycoprotein profoundly increases fetal drug exposure. *J. Clin. Invest.* 104:1441–47.

Smital, T., and B. Kurelec. 1998. The activity of multixebiotic resistance mechanism determined by rhodamine B-efflux method as a biomarker of exposure. *Mar. Environ. Res.* 46:443–47.

Sturm, A., J. P. Cravedi, and H. Segner. 2001. Prochloraz and nonylphenol diethoxylate inhibit an mdr1-like activity *in vitro*, but do not alter hepatic levels of P-glycoprotein in trout exposed *in vivo*. *Aquat. Toxicol.* 53:215–28.

Thiebaut, F., et al. 1987. Cellular localization of the multidrug-resistance gene product P-glycoprotein in normal human tissues. *Proc. Natl. Acad. Sci. USA* 84:7735–38.

Thiebaut, F., et al. 1989. Immunohistochemical localization in normal tissues of different epitopes in the multidrug transport protein P170: Evidence for localization in brain capillaries and crossreactity of one antibody with a muscle protein. *J. Histochem. Cytochem.* 37:159–64.

Toomey, B. H., and D. Epel. 1993. Multidrug resistance in *Urechis caupo* embryos: Protection from environmental toxins. *Biol. Bull.* 185:355–64.

Tutundjian, R., et al. 2002a. Genetic and immunological characterisation of a multixenobiotic resistance system in the turbot (*Scophthalmus maximus*). *Comp. Biochem. Physiol.* 132B:463–71.

Tutundjian, R., et al. 2002b. Rhodamine exclusion activity in primary cultured turbot (*Scophthalmus maximus*) hepatocytes. *Mar. Environ. Res.* 54:443–47.

Ueda, K., et al. 1987. Expression of a full-length cDNA for the human *MDR*1 gene confers resistance to colchicine, doxorubicin and vinblastine. *Proc. Natl. Acad. Sci. USA* 84:3004–8.

Ueda, K., et al. 1992. Human P-glycoprotein transports cortisol, aldosterone and dexamethasone but not progesterone. *J. Biol. Chem.* 267:24248–52.

van Veen, H. W., and W. N. Konings. 1998. The ABC family of multidrug transporters in microorganisms. *Biochim. Biophys. Acta* 1365:31–36.

Yabe, T., et al. 2005. Multidrug resistance-associated protein MRP-1 regulates dauer diapause by its export activity in *Caenorhabditis elegans*. *Development* 132:3197–207.

Yeh, G. C., et al. 1992. A new functional role for P-glycoprotein: Efflux pump for benzo(alpha) pyrene in human breast cancer MCF-7 cells. *Cancer Res.* 52:6692–95.

Zhang, L., et al. 2006. Scientific perspectives on drug transporters and their role in drug interactions. *Mol. Pharm.* 3:62–69.

Section

Ecological and Ecophysiological Aspects of Tolerance

11 Tolerance and Biodiversity

Judith S. Weis

CONTENTS

11.1 INTRODUCTION

The tolerance discussed in most chapters in this volume is a result of populations in contaminated environments being subjected to selection pressure and then becoming more tolerant to those contaminants. This evolutionary response may provide a measure of contaminant effects and can enable species to persist and perhaps even thrive, and thus contribute to the diversity of species. Discussed in Chapter 4 is pollution-induced community tolerance (PICT), wherein the community in the polluted environment becomes dominated by (1) species that have evolved greater tolerance and (2) species that are more tolerant to begin with. More tolerant species may be so because of reduced uptake of the toxicants, altered inducibility of cytochrome P-450 systems, differential binding to detoxifying molecules such as metallothioneins (see Chapter 6), or enhanced ability to store contaminants in intracellular structures that prevent damage to the cell. The inherent or induced tolerance of these species enables them to persist, and therefore contributes to increasing the overall diversity of the communities in which they live. However, tolerance comes with costs, and it has frequently been noted that populations adapted to one set of environmental stressors have increased susceptibility to other stressors.

The word *biodiversity* is a contracted version of *biological diversity*. The Convention on Biological Diversity (an international treaty signed by 150 government leaders at the 1992 Rio Earth Summit) (http://www.cbd.int/) defines *biodiversity* as "the variability among living organisms from all sources including, *inter*

249

alia, terrestrial, marine and other aquatic ecosystems and the ecological complexes of which they are a part; this includes diversity within species, between species, and of ecosystems" (UN Convention on Biodiversity 1992). Biodiversity therefore encompasses many levels of biological organization. It includes genetic diversity within a species, the diversity of species in an area, and the diversity of habitat types within a landscape. This chapter will discuss the first aspect, genetic diversity within a species, and will then focus on the second component, species diversity within communities.

11.2   GENETIC DIVERSITY AND POLLUTION TOLERANCE

Pollutants can act as powerful selective forces by altering genetic variability. If populations in uncontaminated areas have a wide variation in tolerance, and those inhabiting polluted areas are selected for tolerance, it would be expected that the genetic diversity in the polluted sites would be reduced. Reduced genetic diversity in contaminated sites can also be a result of low population density and population bottlenecks. On the other hand, if contaminants are mutagenic, one might predict that polluted populations would have greater genetic diversity. In studies on various taxa subjected to different kinds of pollutants, both kinds of responses have been seen. However, few of the studies that examined genetic diversity also examined tolerance. There is an underlying but often untested assumption in most of these studies that when reduction in genetic diversity occurs in polluted populations, the prevailing genotypes are associated with increased tolerance.

11.2.1   MARINE INVERTEBRATES AND METALS

Ross et al. (2002) studied marine prawns (*Leander intermedius*) and isopod crustaceans (*Platynympha longicaudata)* from the vicinity of a metal smelter and three reference sites. The genetic diversity of the prawns from the smelter discharge site (Port Pirie, South Australia) was lower than that in one reference population, but not significantly different from the other two. However, genetic diversity of the Port Pirie isopods was significantly lower than all reference populations. In the laboratory, preexposed and reference prawns showed no differences in response to metals. However, the preexposed isopods were more tolerant to metals than the reference population. Thus, the isopods, but not the prawns, exhibited both decreased genetic variation and increased tolerance to the metal stressors.

Xiao et al. (2000) studied metal-impacted mussels, *Mytilus galloprovincialis,* and barnacles, *Balanus glandula,* and found that the populations of both species from impacted sites had lower genetic diversity than those from clean sites. Individuals from impacted sites were more likely to share the same haplotypes than were those from clean sites, suggesting that pollution reduces genetic diversity. Tolerance was not examined, however.

Kim et al. (2003) studied periwinkle snails (*Littorina brevicula*) from metal-polluted and unpolluted environments along the southeast coast of Korea. No genetic differences within cyt-b mtDNA were detected between environments, but differences in haplotype diversity and structuring were found within ND6 mtDNA between polluted and unpolluted snails; genetic diversity was significantly lower in polluted environments.

In a very different approach to the issue of tolerance and genetic diversity, Nevo et al. (1986) examined the tolerance of populations that were initially more or less genetically diverse. They tested three pairs of species belonging to different genera of marine gastropods, *Monodonta turbinata*, *M. turbiformis* (Trochidae), *Littorina punctata*, *L. neritoides* (Littorinidae), *Cerithium scabridum*, and *C. rupestre* (Cerithiidae), for resistance to various inorganic (metals) and organic (detergents and oil) pollutants. Each pair consisted of one narrow-niche species with low genetic diversity and one broad-niche species with higher genetic diversity. In all three cases the species with a higher level of genetic diversity was more resistant to all pollutants than its counterpart.

11.2.2 Terrestrial Plants

Many studies of genetic diversity have been performed on terrestrial plants, especially in mining areas. Deng et al. (2007) examined genetic diversity in mining and reference populations of a metal hyperaccumulator plant, *Sedum alfredii*. Four of the sampled sites were heavily contaminated with metals (Zn, Cd, Pb), and extremely high concentrations of these metals were seen in tissues of the plants. Reduced genetic diversity was found in the mining area populations, which could be due to selection, a bottleneck effect, or the prevalence of vegetative reproduction in the mining populations.

Keane et al. (2005) studied six dandelion (*Taraxacum officinale*) populations (three urban and three rural), which showed patterns that appeared to be influenced by contaminants. Genetic similarity among individuals within a population was correlated with increasing levels of airborne particulate matter ($\le 10 \, \mu m$, PM_{10}) and soil metal concentrations. The mean genetic similarity was always significantly higher (diversity lower) at urban than at rural sites. There was a significant negative correlation between the number of genotypes at a site and the amounts of PM_{10} and soil metals. They concluded that the data were consistent with the hypothesis that pollution-induced selection has contributed to the lower genetic diversity at the urban sites, but no tests of tolerance were actually done.

Tolerance *per se* was considered in a study by Prus-Głowacki et al. (2006), who found significant differences in genetic composition between two groups of European Scots pine (*Pinus sylvestris*) trees that were either tolerant or sensitive to metals. Total and mean numbers of alleles and genotypes per locus were higher in the pollution-sensitive group, but heterozygosity was lower. Pollution-tolerant trees were in Hardy-Weinberg equilibrium, while the sensitive group had an excess of homozygosity. There were many chromosomal aberrations in meristematic tissue of seedlings grown from seeds collected from trees in the polluted area, suggesting mutagenic effects. Both allozymes and cytogenetic analyses therefore showed a significant influence of metals on genetic structure.

Fluoride pollution effects were studied by Kozyrenko et al. (2007) on larch (*Larix gmelinii*) (Pinaceae). Genetic variability in two populations from a region free of fluoride pollution was considerably higher than in populations near a fluorite quarry that were growing on soils with a high fluoride content. The genetic diversity of the fluoride-tolerant population was the lowest. It was presumed that the reduction of genetic diversity was a response to the high concentration of fluorides in the soil.

In contrast, higher genetic diversity was seen by Prus-Glowacki et al. (1999) in *Pinus sylvestris* populations near a copper smelter than at a control area. Based on data from ten enzymatic loci, they characterized the mean number of alleles and genotypes per locus, heterozygosity and genotype polymorphism index, genetic diversity among sites, and genetic similarity. Mean heterozygosity and genotype polymorphism indices were lower at the control site than at the polluted one; populations from the polluted site had higher genetic diversity than the controls. The authors felt that their results provided evidence of an adaptation strategy to pollution connected with increasing genetic variation, but did not discuss mutagenicity.

The effects of traffic were studied in annual bluegrass (*Poa annua*) by Li et al. (2004), who found that traffic pollution can dramatically change genotypic frequencies at some loci and produce deviations from Hardy-Weinberg equilibrium due to an excess of heterozygotes. The effective number of alleles per locus and the observed and expected heterozygosity were higher in polluted sites than at the control site, but the increase was not correlated with the extent of pollution. The same species was studied for genetic response along a transect from an organic reagents factory (Chen et al. 2003). Pollution dramatically changed genotypic frequencies at some loci. At polluted sites, significant deviations from Hardy-Weinberg equilibrium were observed at some loci due to an excess of heterozygotes, especially in the two sites nearest the pollution source where all the individuals were heterozygous at some loci. The authors suggest that heterozygotes were more tolerant to organic pollution than homozygotes. They also found that the effective number of alleles per locus and the observed and expected heterozygosity were much higher at the polluted sites than the control site. Thus, this species appeared to respond in a similar manner to both traffic and organic pollutants by increasing heterozygosity and genetic diversity.

Keane et al. (1999) studied populations of the cattail (bullrush) (*Typha latifolia*) and found significant genetic differentiation among five study sites that had different levels of contamination. Significantly higher genetic diversity was detected at sites that were the most severely impacted by pollutants. Although this correlation does not establish cause and effect, the results indicate that the analysis of genetic diversity may be a useful tool for monitoring anthropogenic-induced changes in the genetic structure of natural populations of plants.

11.2.3 OTHER TAXA

Parthenogenetic species have also been found to have differential tolerance in different populations. Lopes et al. (2004) investigated copper tolerance in clones of the freshwater cladoceran crustacean *Daphnia longispina* from mine drainage and reference sites and found that the stressed population, which was more tolerant, did not include the most sensitive lineages, although the most tolerant ones were also present in the reference population. This confirmed the genetic erosion hypothesis. In a follow-up study, populations (clones) of *Daphnia longispina*, differing in their sensitivity to a metal-rich mine drainage effluent, were subjected to different levels of metals as well as to culling (nonselective removal). Culling did not affect population density, but resulted in clonal diversity higher than the control. Populations stressed by mine drainage recovered to their initial densities, but the most sensitive genotypes disappeared under both low and high metal levels, indicating reduced genetic diversity (Lopes et al. 2009). However, the most resistant clones that survived the high stress were found to be the most sensitive to other chemical stressors. Because of these differences in sensitivity to different chemicals, the authors suggest that successive inputs of partially lethal concentrations of different chemicals could lead to the disappearance of the population. In addition, populations exposed to low levels of culling plus metals showed the same clonal diversity as controls, demonstrating the importance of low-level effects of nonselective stressors in the maintenance of clonal diversity. Haap and Kohler (2009) investigated the sensitivity to Cd of seven clones of *Daphnia magna* from different geographical regions. The clones showed major differences in their susceptibility to Cd, with the highest EC_{50} values obtained for the clone with the lowest hsp70 expression (see Chapter 9).

In a study that examined both genetic diversity and tolerance, Coors et al. (2009) explored the relationship between land use around ten ponds, population genetic diversity of *Daphnia magna* as measured by allozymes, and the tolerance of the various populations to the pesticide carbaryl and potassium dichromate. Genetic diversity was reduced at sites with greater agricultural land use, indicating that genetic bottlenecks may have occurred from increased exposures. Differences were found in susceptibility to both toxicants among the populations, carbaryl tolerance increasing with increasing agricultural land use. Because the experimental design excluded the possibility of physiological adaptation to the toxicants, the authors concluded that the differences in susceptibility were genetic and that selection pressure imposed by pollution reduced the genetic diversity.

Tapio et al. (2006) investigated whether genetic diversity in two small passerine birds, the great tit (*Parus major*) and the pied flycatcher (*Ficedula hypoleuca*), was changed near point sources of metals (copper smelters). They measured metal concentrations and nucleotide diversity in mitochondrial DNA in feather samples from nestlings in polluted and control sites. The *P. major* population near a smelter showed significantly higher nucleotide diversity than a control population, suggesting increased mutation rates in the polluted environment. On the other hand, *F. hypoleuca* showed reduced nucleotide diversity at smelter sites. The authors suggested that the different responses in the two bird species may be due to differential ability to handle toxic compounds.

A similar study was done with wood mouse (*Apodemus sylvaticus*) populations near smelters, with very different results. Genetic analyses revealed high levels of genetic variation in all populations, but those from the most polluted sites did not differ from those of less polluted sites in terms of heterozygosity or mean allelic richness. No correlation was found between measures of genetic diversity and the degree of metal pollution (Berckmoes et al. 2005).

An experimental laboratory approach was used by Nowak et al. (2009) in studying tributyltin (TBT) effects on genetic structure of chironomid midge populations. TBT-exposed groups showed increased larval mortality and delayed development, as well as reduced genetic variation (losing almost half of their initial heterozygosity after twelve generations). This study indicates that organotin pollution can lead to a rapid decrease in genetic diversity but did not investigate tolerance in the polluted population.

11.3   COMMUNITY DIVERSITY AND CONTAMINANTS

The connection between evolutionary/genetic responses and effects at higher levels of biological organization (individual organisms, populations, and communities) is not always clear. A number of studies have considered potential linkages and relationships among different levels of biological organization (Weis and Weis 1989; Clements 2000; Theodorakis et al. 2000; Medina et al. 2007). Biochemical effects may occur rapidly, but their ecological relevance is often unclear. When populations become resistant, they can persist in a contaminated environment, but there may be costs to overall fitness. These costs may involve a reduced ability to respond to additional disturbances. Since organisms, populations, and communities are exposed to a variety of stressors and disturbances, becoming more tolerant to one may reduce the ability to respond to others.

At the community level, community structure (including diversity) is a frequently used indicator of responses to contaminants or other anthropogenic stressors (Resh and Jackson 1993). Taxa that are more sensitive to particular contaminants will become reduced in numbers and perhaps disappear altogether, causing the community to become less diverse, and progressively more dominated by tolerant species. The more tolerant species may be so because of reduced contaminant uptake, differential binding to detoxifying molecules such as metal-binding proteins (see Chapter 6), or enhanced ability to store contaminants in intracellular structures that reduce damage to the cell. It has also been observed that metal-tolerant communities in polluted environments may be more susceptible to new stressors such as acidification, UV-B radiation, and climate change (Clements 1999; Clements and Rohr 2009).

11.3.1   FRESHWATER AND MARINE ECOSYSTEMS

Rodriguez et al. (2007) evaluated a molecular approach using polymerase chain reaction (PCR)–temperature gradient gel electrophoresis (TGGE) to measure changes in cyanobacterial diversity along a pollution gradient in a river and compared it with microscopic observations of field-fixed and cultured samples. The different 16S rDNA genes present in the cyanobacterial community of each sampling point were separated, giving a characteristic pattern of bands for each site. This pattern represents a fingerprint of the community, allowing direct comparisons of the different samples. The TGGE results revealed that the structure of the cyanobacterial community differed along the pollution gradient. Microscopic and molecular approaches both showed that cyanobacterial diversity decreased in a downstream (i.e., decreasing contamination) direction. Similar results were obtained by the two methods, as indicated by the high correlation between them.

Many studies have focused on diversity of benthic communities in streams. Benthic macroinvertebrates are commonly studied, as they are in contact with contaminated sediments and are much less mobile than planktonic or nektonic communities. The community diversity in sediments is frequently used as one measure of sediment toxicity, along with chemical and toxicological data, the "sediment triad" approach (Chapman 1996). Measures of community diversity include species richness and diversity indices, such as Simpson's or Shannon-Wiener indices, that combine species richness and evenness into a single calculated numerical index value. The presence or absence of certain indicator species can also contribute to assessing the health of communities (La Point and Fairchild 1992).

The macroinvertebrate community in a Montana stream downstream from a mine had elevated metal levels in their tissues and reduced species richness, due to the loss of metal-sensitive species (Beltman et al. 1999). Similarly, in New Zealand, a high correlation was found between toxicity tests and measures of benthic community structure, with reduced richness and disappearance of certain taxa in streams that were the most affected by metal pollution (Hickey and Clements 1998). Focusing only on chironomid larvae in subarctic lakes, Mousavi et al. (2002) found that species richness was lowest in the most contaminated lake and that Simpson's diversity index was highest in the least polluted site. Their data indicated that species that were rare or absent in the heavily polluted sites were sensitive to metals, and species that persisted in heavily polluted sites were more tolerant. Insight into why certain species are more sensitive than others was obtained by Cain et al. (2004), who studied subcellular partitioning of metals in different insect species. Mayflies, which were sensitive and were absent or rare in the polluted sites, tended to concentrate metals in subcellular components that the authors defined as metal sensitive (Cain et al. 2004). In contrast, tolerant taxa (caddisflies—*Hydropsyche* spp.) had less accumulation and put a higher proportion of accumulated metals into metal-binding proteins; they were dominant in the polluted sites. Another tolerant insect (mayfly—*Baetis* sp.) accumulated much less copper than the sensitive taxa and was also dominant in the polluted areas. Thus, differences in bioaccumulation rates and the subcellular partitioning of accumulated metals can explain the differential sensitivity of the various taxa.

In a study examining community-level responses to two different types of stressors, metals and UV-B, Kashian et al. (2007) found that communities from a metal-polluted site were more tolerant to metals (as expected) but less tolerant to UV-B than reference communities. UV-B radiation significantly reduced many taxa in the metal-polluted site, but had no effects on the benthic community from the reference site. These findings demonstrate the potential costs associated with tolerance.

In any study of pollution tolerance, having a good reference site is vital. Since reference sites may differ, Bailey et al. (1998) stressed the importance of having a set of reference sites with which to compare the test site. They encouraged the use of multiple descriptors, including species richness, Simpson's diversity index, Simpson's equitability, family biotic index, and Bray-Curtis distance to the median reference community.

Wetland plant communities associated with mine sites were studied (Batty 2005). Certain plant species are very tolerant to metals and may be referred to as metallophytes. These species can be used in treatment wetlands. While the diversity of wetland plants in contaminated mine sites in the UK is quite low (only ten species were identified), the author argues that these sites may support some rare and threatened species.

In urban estuaries, sediments are contaminated with toxic chemicals and also subject to hypoxia. The combination of these stressors can have major effects on benthic communities. A general pattern of community change is that larger-bodied, K-selected species with a long life span decrease, while opportunistic, small, R-selected species tend to become dominant in communities that become less diverse (Gaston and Young 1992; McGee et al. 1995). Marine meiofaunal diversity was examined in the Gulf of Taranto (Italy) in relation to organic pollution in the sediment (Sandulli et al. 2004). Abundance and diversity were significantly lower in the area most heavily contaminated by organic pollution, which was impacted by sewage and industrial discharge and anoxic. In a study of a contaminated brackish intertidal marsh community, Weis et al. (2004) found that while species richness was most closely associated with the height of the location relative to the tide, rather than with sediment metal concentrations, the Shannon-Wiener and Simpson's indices of diversity were significantly associated with the metal levels.

Tolerance can affect the success of invasive species in communities. Piola and Johnston (2008) sampled hard-substrate invertebrate taxa in coastal sites in Australia and found that increasing pollution decreased the diversity of native species, while there was no change in the diversity of nonindigenous species (see Chapter 2). At three of the contaminated sites, assemblages that had previously been dominated by natives became more dominated by nonindigenous species. In laboratory studies, Piola and Johnston (2009) found that nonindigenous bryozoan species were much more tolerant to copper than native species. In field-based larval assays, nonindigenous species showed strong recruitment and growth in high copper areas relative to native species. Thus, community diversity alone was inadequate to describe community response to contaminants when invasive species may be more tolerant than natives to the stressors.

11.3.2 TERRESTRIAL PLANT COMMUNITIES

Terrestrial plants can be stressed by both contaminants in the soil and gaseous pollutants in the air. There is much evidence for development of pollution tolerance in plant populations, with early work focusing on plants in contaminated mine spoils (Bradshaw and McNeilly 1981). Gallagher (2008) studied the vegetative assemblage that developed on a degraded brownfield area—a former industrial site in Jersey City, New Jersey. The plant community that developed was comprised primarily of species tolerant to elevated soil metals, but the diversity (Shannon-Wiener index) decreased as total soil metal loads increased. Air quality was not included in this study.

Plants can also become resistant to the air pollutants sulfur dioxide and ozone (Barnes et al. 1999). The selective force of air pollution can bring about alterations in the composition of plant communities through the elimination of sensitive individuals or species, and changes can occur quite rapidly. Communities in polluted environments generally have reduced diversity, and air pollution plays a role in changing the distribution of many plant species. However, different types of pollutants have different effects, and some environments are more susceptible, such as those with a low buffering capacity. A meta-analysis of data from eighty-seven studies on changes in plant species richness and air pollution was performed by Toivonen and Kozlov (2008), who found that while species richness generally decreased with pollution, effects were not uniform. Soil acidification appeared to have more severe effects on diversity than other types of pollution. Effects were greater at low latitudes than high; this was attributed to the initially greater diversity of the original community and mean summer temperatures. The latter suggests that climate change will cause existing pollutant loads to become more damaging.

11.3.3 PARASITE COMMUNITIES

Effects of pollution on parasite communities can be positive or negative. On one hand, hosts in contaminated areas might be stressed and weakened so that the amount of parasitism could rise. This might involve an increase in the types of parasites (more diversity) or increasing the prevalence of a certain parasite (less diversity). On the other hand, parasites themselves can be negatively affected by the contaminants, so that hosts would benefit by having fewer parasites. Some parasites accumulate more contaminants than their host (MacKenzie et al. 1995; Sures et al. 1997). Since some parasites have complicated life cycles with different hosts during different stages of their life cycle, pollution effects on one host species could affect the parasite communities in other hosts. Eutrophication generally increases parasitism (Lafferty 1997), while metal exposure generally decreases it. Parasites of fishes are generally negatively associated with toxic pollutants (Lafferty 1997). Yellow eels (*Anguilla rostrata*) had fewer parasite species (reduced diversity) in more acidified sites (Marcogliese and Cone 1996). Certain types of parasites are particularly sensitive, namely, digenean trematodes, which require mollusks as intermediate hosts, and these do not survive well in acidified conditions. Digeneans, normally infesting the liver of the mummichog or killifish (*Fundulus heteroclitus*), were missing from a contaminated mummichog population in a site with no snails (Schmalz et al. 2002). Other parasites, like ciliates and acanthocephalans, can become more abundant in acidic (Halmetoja et al. 2000) or metal-contaminated environments (Lafferty 1997). The increase in ciliates seems to be a response to increased susceptibility of the host. Blanar et al. (2009) performed a meta-analysis of the literature that revealed consistent, strong negative effects of metal pollution on digenean and monogenean trematode populations and community richness. Ectoparasites were more susceptible to a wide range of pollutants than internal parasites. So, as with free-living communities, it is not just a change in overall diversity, but a change in species distribution that occurs, in which the more tolerant (parasite) species become more dominant.

Lafferty (1993) found that populations of the snail *Cerithidea californica* in an area receiving runoff were uninfected by larval digenean trematodes, while in an adjacent marsh they were infected; so the absence of parasites was an indicator of a disturbed environment. Similarly, Santiago Bass and Weis (2008) found that restored marshes in the Hackensack Meadowlands of New Jersey had significantly more snails (*Littoridinops tenuipes*), and the mummichog (*F. heteroclitus*) had more digenean trematodes than in nearby degraded marshes.

11.3.4 MULTIMETRIC INDICES

Since species richness and the various diversity indices by themselves did not appear to be adequate to evaluate the ecological health of an area, there has been a need to develop more holistic measures. *Ecological integrity* has been defined as "the ability to support and maintain a balanced, integrated, adaptive community of organisms having a species composition, diversity, and functional organization comparable to that of natural habitats of the region" (Karr 1991 p. 69). The approach of such an index is to combine various metrics at different levels of biological organization into a single index that characterizes the biotic integrity in a way similar to that of a battery of medical tests to evaluate individual health. Karr (1991) developed the index of biological integrity (IBI) for fish in streams, to evaluate anthropogenic effects on individuals, populations, communities, and the ecosystem. The metrics that go into an IBI include six measures of species richness and composition, three of which are taxon specific for midwestern United States. The three general metrics are total number of native fish species, number and identity of intolerant species, and percent of individuals of tolerant species. The total species richness must be compared with reference sites in a region and must consider the size of the stream involved. IBI scores can be used to evaluate current conditions, determine trends over time, and compare sites. There is a possibility of it helping to identify the source of the degradation. The IBI has been useful in detecting biological changes associated with pollution in a variety of locations. Karr (1991) emphasized that IBIs could be developed for other geographic regions and for other taxonomic groups, such as invertebrates.

Since then, IBIs have been developed for a wide variety of taxa and geographic areas, including the Pacific Northwest rivers (Mebane et al. 2003), palustrine wetlands (Simon and Stewart 1998), fish in rivers (in India, Ganasan and Hughes 1998; in the Seine, France, Oberdorff and Hughes 1992; and in Australia, Harris and Silveira 1999), macroinvertebrates in mid-Atlantic Highlands streams (Klemm et al. 2003), plants in Pennsylvania wetlands (Miller et al. 2006), estuarine fishes (Harrison and Whitfield 2004), and estuarine benthos (van Dolah et al. 1999; Paul et al. 2001; Alden et al. 2002; Llanso et al. 2002). The benthic index developed for the Gulf of Mexico (Engle et al. 1994) has three major components: the Shannon-Wiener diversity index (H′), the total abundance of tubificid oligochaetes (indicators of organic pollution), and the proportion of bivalve mollusks (indicative of undegraded conditions). The index portrays an unimpacted site as diverse, having a large proportion of upper trophic level organisms and a small proportion of opportunistic species. However, a more complex taxon-based approach rates the sensitivity to pollution of each individual taxon and its relative abundance, with particular emphasis on amphipod crustacean and shrimp taxa richness, but also includes total number of taxa (richness) (Eaton 2001). While the vast majority of indices have been developed for aquatic environments, indices have also been developed for bird communities (O'Connell et al. 1998; Canterbury et al. 2000).

There is disagreement over whether the multimetric approach is always preferable to focusing either on diversity alone or on certain taxa that are particularly sensitive or tolerant. In some cases (Carlisle and Clements 1999; Alden et al. 2002), diversity or species richness alone was as good as, if not better than, the complicated multimetric approaches in classifying the environmental quality of the sites.

Other approaches are taxon based and examine abundance of particular groups that have differential sensitivity/tolerance to environmental stresses. For example, soft-bottom marine communities can be comprised of (1) species very sensitive to organic enrichment, including specialist carnivores and some deposit-feeding polychaetes; (2) species indifferent to enrichment, including less selective deposit feeders, suspension feeders, and scavengers; (3) species tolerant to excess organic matter whose populations are stimulated by enrichment; (4) second-order opportunistic species, mostly small polychaetes; and (5) first-order opportunistic species, deposit feeders that proliferate in reduced sediments (Borja et al. 2000). A marine biotic index has been constructed based on the proportions of abundance of members of these groups that does not consider diversity at all and has been used in many areas of Europe (Borja et al. 2003).

11.4 CONCLUSIONS

No unifying scheme can be drawn from this review of the literature (Figure 11.1). The responses to stress associated with tolerance are highly variable, depending on the species and, within the same species, the type of stressor. The role of tolerance in the conservation of communities is limited to a relatively narrow gradient of contamination since increasing levels of stress result in contrasting responses of susceptible versus resistant species/populations.

In some cases, tolerance may not be beneficial for environmental conservation since some invasive species are better able to cope with stress than native ones.

Consequent changes in communities thus need to be carefully evaluated. Since there has been a bewildering array of indices developed, there is a need to focus on those that are (1) most widely successful, (2) establish criteria for validation to demonstrate accuracy and reliability, (3) compare and intercalibrate methods to achieve uniform assessment scales across geographic areas and habitats, and (4) integrate indices across ecosystem elements (fish and plankton in addition to benthos) (Borja et al. 2009). Validation is particularly important for the middle range of disturbance, as most indices work well for the extremes of highly polluted and reference conditions. Ranasinghe et al. (2009) compared five indicators used in the same California estuaries and found that none of the individual indices performed as well as the average professional judgment of nine benthic experts, but several combinations of indices outperformed the average expert. Reaching scientific consensus on simplifying approaches is a challenge for the future.

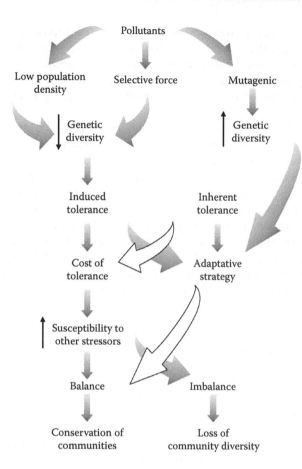

FIGURE 11.1 Exploration of the links between chemical contamination, genetic variability, tolerance, and consequences at the community level.

REFERENCES

Alden, R. W., et al. 2002. Statistical verification of the Chesapeake Bay benthic index of biological integrity. *Environmetrics* 13:473–98.

Bailey, R. C., et al. 1998. Biological assessment of freshwater ecosystems using a reference condition approach: Comparing predicted and actual benthic invertebrate communities in Yukon streams. *Freshwater Biol.* 39:765–74.

Barnes, J., et al. 1999. Natural and man-made selection for air pollution resistance. *J. Exp. Bot.* 50:1423–35.

Batty, L. E. 2005. The potential importance of mine sites for biodiversity. *Mine Water Environ.* 24:101–03.

Beltman, D. J., et al. 1999. Benthic invertebrate metals exposure, accumulation, and community-level effects downstream from a hard-rock mine site. *Environ. Toxicol. Chem.* 18:299–307.

Berckmoes V., et al. 2005. Effects of environmental pollution on microsatellite DNA diversity in wood mouse (*Apodemus sylvaticus*) populations. *Environ. Toxicol. Chem.* 24:2898–907.

Blanar, C. A., et al. 2009. Pollution and parasitism in aquatic animals: A meta-analysis of effect size. *Aquat. Toxicol.* 93:18–28.

Borja, A., J. Franco, and V. Perez. 2000. A marine biotic index to establish the ecological quality of soft-bottom benthos within European estuarine and coastal environments. *Mar. Pollut. Bull.* 40:1100–14.

Borja, A., I. Muxika, and J. Franco. 2003. The application of a marine biotic index to different impact sources affecting soft-bottom benthic communities along European coasts. *Mar. Pollut. Bull.* 46:835–45.

Borja, A., A. Ranasinghe, and S. B. Weisberg. 2009. Assessing ecological integrity in marine waters using multiple indices and ecosystem components: Challenges for the future. *Mar. Pollut. Bull.* 59:1–4.

Bradshaw, A. D., and T. McNeilly. 1981. *Evolution and pollution.* Institute of Biology Series 130. London: Edward Arnold.

Cain, D. J., S. N. Luoma, and W. G. Wallace. 2004. Linking metal bioaccumulation of aquatic insects to their distribution patterns in a mining-impacted river. *Environ. Toxicol. Chem.* 23:1463–73.

Canterbury, G. E., et al. 2000. Bird communities and habitat as ecological indicators of forest condition in regional monitoring. *Conserv. Biol.* 14:544–58.

Carlisle, D. M., and W.H Clements. 1999. Sensitivity and variability of metrics used in biological assessments of running waters. *Environ. Toxicol. Chem.* 18:285–91.

Chapman, P. 1996. Presentation and interpretation of sediment quality triad data. *Ecotoxicology* 5:327–39.

Chen, X. Y., et al. 2003. Genetic structure along a gaseous organic pollution gradient: A case study with *Poa annua* L. *Environ. Pollut.* 124:449–55.

Clements, W. H. 1999. Metal tolerance and predator-prey interactions in benthic macroinvertebrate stream communities. *Ecol. Appl.* 9:1073–84.

Clements, W. H. 2000. Integrating effects of contaminants across levels of biological organization: An overview. *J. Aquat. Ecosyst. Stress Recov.* 7:113–16.

Clements, W. H., and J. Rohr. 2009. Community responses to contaminants: Using basic ecological principles to predict ecotoxicological effects. *Environ. Toxicol. Chem.* 28:1789–800.

Coors, A., et al. 2009. Land use, genetic diversity, and tolerance of natural populations of *Daphnia magna. Aquat. Toxicol.* 95:71–79.

Deng, J., et al. 2007. The effects of heavy metal pollution on genetic diversity in zinc/cadmium hyperaccumulator *Sedum alfredii* populations. *Plant Soil* 297:83–92.

Eaton, L. 2001. Development and validation of biocriteria using benthic macroinvertebrates for North Carolina estuarine waters. *Mar. Pollut. Bull.* 42:23–30.

Engle, V., J. K. Summers, and G. R. Gaston. 1994. A benthic index of environmental condition of Gulf of Mexico estuaries. *Estuaries* 17:372–84.

Gallagher, F. 2008. The role of soil metal contamination in the vegetative assemblage development of an urban brownfield. PhD dissertation, Rutgers University, New Brunswick, NJ.

Ganasan, V., and R. M. Hughes. 1998. Application of an index of biological integrity (IBI) to fish assemblages of the rivers Khan and Kshipra (Madhya Pradesh), India. *Freshwater Biol.* 40:367–83.

Gaston, G. R., and J. C. Young. 1992. Effects of contaminants on macrobenthic communities in the upper Calcasieu estuary, Louisiana. *Bull. Environ. Contam. Toxicol.* 49:922–28.

Haap, T., and H. R. Kohler. 2009. Cadmium tolerance in seven *Daphnia magna* clones is associated with reduced hsp 70 baseline levels and induction. *Aquat. Toxicol.* 94:131–37.

Halmetoja, A., E. T. Valtonen, and E. Koskenniemi. 2000. Perch (*Perca fluviatilis* L.) parasites reflect ecosystem conditions: A comparison of a natural lake and two acidic reservoirs in Finland. *Int. J. Parasit.* 30:1437–44.

Harris, J. H., and R. Silveira. 1999. Large-scale assessments of river health using an index of biotic integrity with low-diversity fish communities. *Freshwater Biol.* 41:235–52.

Harrison, T. D., and A. K. Whitfield. 2004. A multi-metric fish index to assess the environmental condition of estuaries. *J. Fish Biol.* 65:683–710.

Hickey, C. W., and W. H. Clements. 1998. Effects of heavy metals on benthic macroinvertebrate communities in New Zealand streams. *Environ. Toxicol. Chem.* 7:2338–46.

Karr, J. R. 1991. Biological integrity: A long-neglected aspect of water resource management. *Ecol. Appl.* 1:66–84.

Kashian, D. R., et al. 2007. The cost of tolerance: Sensitivity of stream benthic communities to UV-B and metals. *Ecol. Appl.* 17:365–75.

Keane, B., M. Collier, and S. Rogstad. 2005. Pollution and genetic structure of North American populations of the common dandelion (*Taraxacum officinale*). *Environ. Monit. Assess.* 105:341–57.

Keane, B., et al. 1999. Genetic diversity of *Typha latifolia* (Typhaceae) and the impact of pollutants examined with tandem-repetitive DNA probes. *Am. J. Bot.* 86:1226–38.

Kim, S. J., et al. 2003. Emergent effects of heavy metal pollution at a population level: *Littorina brevicula* a study case. *Mar. Pollut. Bull.* 46:74–80.

Klemm, D. J., et al. 2003. Development and evaluation of a macroinvertebrate biotic integrity index (MBII) for regionally assessing mid-Atlantic streams. *Environ. Manage.* 31:656–69.

Kozyrenko, M., et al. 2007. Effect of fluoride pollution on genetic variability of *Larix gmelinii* (Pinaceae) in East Siberia. *J. Forest Res.* 12:388–92.

Lafferty, K. D. 1993. The marine snail *Cerithidea californica* matures at smaller sizes where parasitism is high. *Oikos* 68:3–11.

Lafferty, K. D. 1997. Environmental parasitology: What can parasites tell us about human impacts on the environment? *Parasit. Today* 13:251–55.

La Point, T. W., and J. F. Fairchild. 1992. Evaluation of sediment contaminant toxicity: The use of freshwater community structure. In *Sediment toxicity assessment*, ed. G. A. Burton, 87–110. Boca Raton, FL: Lewis Publishers.

Li, N., et al. 2004. Effects of traffic pollution on the genetic structure of *Poa annua* L. populations. *J. Environ. Sci.* 16:454–57.

Llanso, R. J., et al. 2002. An estuarine benthic index of biotic integrity for the mid-Atlantic region of the United States. 1. Classification of assemblages and habitat definition. *Estuaries* 25:1219–30.

Lopes, I., D. Baird, and R. Ribeiro. 2004. Genetic determination of tolerance to lethal and sublethal copper concentrations in field populations of *Daphnia longispina*. *Arch. Environ. Contam. Toxicol.* 46:43–51.

Lopes, I., et al. 2009. Genetic erosion and population resilience in *Daphnia longispina* O. F. Müller under simulated predation and metal pressures. *Environ. Toxicol. Chem.* 28:1912–19.

MacKenzie, K., et al. 1995. Parasites as indicators of water quality and the potential use of helminth transmission in marine pollution studies. *Adv. Parasit.* 35:85–144.

Marcogliese, D. J., and D. K. Cone. 1996. On the distribution and abundance of eel parasites in Nova Scotia: Influence of pH. *J. Parasit.* 82:389–99.

McGee, B. L., et al. 1995. Sediment contamination and biological effects in a Chesapeake Bay marina. *Ecotoxicology* 4:39–59.

Mebane, C. A., T. R. Maret, and R. M. Hughes. 2003. An index of biological integrity (IBI) for Pacific Northwest rivers. *T. Am. Fish. Soc.* 132:239–61.

Medina, M. H., J. A. Correa, and C. Barata 2007. Micro-evolution due to pollution: Possible consequences for ecosystem responses to toxic stress. *Chemosphere* 67:2105–14.

Miller, S. J., et al. 2006. A plant-based index of biological integrity (IBI) for headwater wetlands in central Pennsylvania. *Ecol. Indic.* 6:290–312.

Mousavi, S. K., R. Primicerio, and P. A. Amundsen. 2002. Diversity and structure of Chironomidae (Diptera) communities along a gradient of heavy metal contamination in a subarctic watercourse. *Sci. Tot. Environ.* 307:93–110.

Nevo, E., et al. 1986. Genetic diversity and resistance to marine pollution. *Biol. J. Linnean Soc.* 29:139–44.

Nowak, C., et al. 2009. Rapid genetic erosion in pollutant-exposed experimental chironomid populations. *Environ. Pollut.* 157:881–86.

Oberdorff, T., and R. M. Hughes. 1992. Modification of an index of biotic integrity based on fish assemblages to characterize rivers of the Seine basin, France. *Hydrobiologia* 228:117–30.

O'Connell, T. J., L. E. Jackson, and R. P. Brooks. 1998. *The bird community index: A tool for assessing biotic integrity in the mid-Atlantic highlands.* Report 98-4 of Penn State Cooperative Wetlands Center to U.S. Environmental Protection Agency, Region III.

Paul, J. F., et al. 2001. Developing and applying a benthic index of estuarine condition for the Virginian biogeographic province. *Ecol. Indic.*1:83–99.

Piola, R. F., and E. L. Johnston. 2008. Pollution reduces native diversity and increases invader dominance in marine hard-substrate communities. *Divers. Distrib.* 14:329–42.

Piola, R. F., and E. L. Johnston. 2009. Comparing differential tolerance of native and non-indigenous marine species to metal pollution using novel assay techniques. *Environ. Pollut.* 157:2853–64.

Prus-Glowacki, W., et al. 1999. Industrial pollutants tend to increase genetic diversity: Evidence from field-grown European Scots Pine populations. *Water Air Soil Poll.* 116:395–402.

Prus-Głowacki, W., et al. 2006. Effects of heavy metal pollution on genetic variation and cytological disturbances in the *Pinus sylvestris* L. population. *J. Appl. Genet.* 47:99–108.

Ranasinghe, J. A., et al. 2009. Calibration and evaluation of five indicators of benthic community condition in two California bay and estuary habitats. *Mar. Pollut. Bull.* 59:5–13.

Resh, V. H., and J. K. Jackson. 1993. Rapid assessment approaches to biomonitoring using benthic macroinvertebrates. In *Freshwater biomonitoring and benthic macroinvertebrates,* D. M. Rosenberg and V. H. Resh, 195–233. New York: Chapman & Hall.

Rodriguez, V., et al. 2007. A molecular fingerprint technique to detect pollution-related changes in river cyanobacterial diversity. *J. Environ. Qual.* 36:464–68.

Ross, K., et al. 2002. Genetic diversity and metal tolerance of two marine species: A comparison between populations from contaminated and reference sites. *Mar. Pollut. Bull.* 44:671–79.

Sandulli, R., et al. 2004. Meiobenthic biodiversity in areas of the Gulf of Taranto (Italy) exposed to high environmental impact. *Chem. Ecol.* 20:379–86.

Santiago Bass, C., and J. S. Weis. 2008. Increased abundance of snails and trematode parasites of *Fundulus heteroclitus* (L.) in restored New Jersey wetlands. *Wetl. Ecol. Manage.* 16:173–82.

Schmalz, W. F., A. Hernandez, and P. Weis. 2002. Hepatic histopathology in two populations of the mummichog, *Fundulus heteroclitus. Mar. Environ. Res.* 54:539–42.

Simon, T. P., and P. M. Stewart. 1998. Application of an index of biotic integrity for dunal palustrine wetlands: Emphasis on assessment of nonpoint source landfill effects on the Grand Calumet Lagoons. *Aquat. Ecosyst. Hlth. Manage.* 1:63–74.

Sures, B., H. Taraschewski, and M. Rydio. 1997. Intestinal fish parasites as heavy metal bioindicators: A comparison between *Acanthocephalus lucii* (Palaeacanthocephala) and the zebra mussel, *Dreissena polymorpha. Bull. Environ. Contam. Toxicol.* 59:14–21.

Tapio, E., E. Belskii, and B. Kuranov. 2006. Environmental pollution affects genetic diversity in wild bird populations. *Mutat. Res.* 608:8–15.

Theodorakis, C. W., et al. 2000. Relationship between genotoxicity, mutagenicity, and fish community structure in a contaminated stream. *J. Aquat. Ecosys. Stress Recov.* 7:131–43.

Toivonen, E. L., and M. V. Kozlov. 2008. Changes in species richness of vascular plants under the impact of air pollution: A global perspective. *Global Ecol. Biogeogr.* 17:305–19.

United Nations Convention on Biodiversity. 1992. UN Conference on Environment and Development, Rio de Janeiro. Article 2: Use of Terms. cbd.int/convention/articles.shtml?a=cbd-02

Van Dolah, R. F., et al. 1999. A benthic index of biological integrity for assessing habitat quality in estuaries of the southeastern USA. *Mar. Environ. Res.* 48:269–83.

Weis, J. S., J. Skurnick, and P. Weis. 2004. Studies of a contaminated brackish marsh in the Hackensack Meadowlands of Northeastern New Jersey: Benthic communities and metal contamination. *Mar. Pollut. Bull.* 49:1025–35.

Weis, J. S., and P. Weis.1989. Tolerance and stress in a polluted environment: The case of the mummichog. *BioScience* 39:89–96.

Xiao, L. M, D. L. Cowles, and R. L. Carter. 2000. Effect of pollution on genetic diversity in the bay mussel *Mytilus galloprovincialis* and the acorn barnacle *Balanus glandula. Mar. Environ. Res.* 50:559–63.

12 Cost of Tolerance

Catherine Mouneyrac, Priscilla T. Y. Leung,
and Kenneth M. Y. Leung

CONTENTS

12.1 INTRODUCTION

Living organisms, regularly submitted to fluctuations in the ambient environment, are able to regulate their inner environment to ensure its stability by homeostatic processes. Homeostatic processes may operate at different levels of biological organisation, and each level has the ability to absorb change without a reduction in fitness to survive (Lawrence and Hemingway 2007).

When organisms are exposed to chemical stress, a new physiological steady state appears in order to maintain homeostasis. If effects of chemical exposure outcompete homeostatic processes, then cellular damage may occur and repair mechanisms are likely to be activated in the organism. Seyle (1956) developed a stress concept describing the pattern of the physiological responses of an organism under stress, such as when exposed to a toxicant, as the general adaptation syndrome. It is characterised by three sequential phases. The first one is the stage of physiological alarm, in the course of which the effects of a chemical trigger initial physiological responses to compensate for specific effects according to the mode of action of the contaminant, but some nonspecific homeostatic processes (e.g., behaviour) are also involved in order to reestablish a homeostatic equilibrium. The second phase, the stage of resistance, occurs when physiological mechanisms set up against the contaminant stressor have been established. During this phase, counteracting the effects of the contaminant becomes part of the normal cost endured by the exposed animal. This stage may be associated with energy expenditure. The third phase, corresponding to the exhaustion stage, comes into effect if the stressor is of sufficient magnitude or is applied for a long time. Thus, an inability to maintain homeostasis may lead to diseases or death, a condition known as homeostatic imbalance.

For organisms exposed for a long time to sublethal levels of contaminants in their environment, their ability to tolerate these disturbances minimises the effects of damage to their Darwinian fitness and is essential for the sustainability of populations, communities, and ecosystems. As discussed in Chapter 2, an organism can acquire tolerance to pollutants through (1) physiological acclimation via existing cellular and enzymatic processes and (2) genetic adaptation via mutation and selection of genes that modify physiological functions (Klerks and Weis 1987; Posthuma and van Straalen 1993). Physiological processes at molecular and cellular levels involved to combat and ameliorate the undesirable effects of toxicants are well developed in Chapters 7 to 9. Acclimation is achieved merely by physiological adjustments without altering the genetic makeup, whereas genetic adaptation is primarily operated through the basic evolutionary mechanism, i.e., natural selection of favourable genetic variations and traits. Relatively, the genetic adaptation process requires a prolonged period of continuous exposure to the chemical contaminant(s) and often involves multigenerational genetic mutation and selection (May 1985).

Understanding the relation between toxicant effects at different levels of biological organisation is of crucial importance in ecotoxicological research (Maltby and Calow 1989). However, causal links between xenobiotic damage to individuals and populations are difficult to demonstrate, because of an increasing degree of complexity. Indeed, a secondary response can be considered as (partly) compensatory to the direct toxic effect. For instance, direct effects on organismal growth are not necessarily reflected in altered population dynamics, if the effect on somatic growth is indirectly compensated by an enhanced energy transfer to reproduction. For example, the total number of oocytes per female of the polychaete *Nereis diversicolor* and fecundity (total number of oocytes in mature females) were generally higher in worms from the contaminated Loire estuary compared to the reference site (Bay of Bourgneuf, France), suggesting an energy switch to reproductive effort in organisms under stress (Mouneyrac et al. 2010). Moreover, the same species (*N. diversicolor*), originating from the multicontaminated Seine estuary, despite being tolerant to at least one metal (i.e., zinc) and able to use a number of defence mechanisms, exhibited lower physiological and populational status than worms collected from the reference Authie estuary (Mouneyrac et al. 2009). However, such observations might have been confounded by other factors, such as local species compositions and interactions. Given that it is important to consider the whole ecosystem and community together with their ecological relationships, such a single species approach to ecotoxicology may no longer be adequate. Species interactions may produce surprising outcomes when several species are exposed to pollutants. Populations of opportunistic species, for example, may expand under pollution stress due to the loss of more sensitive competitors for a specific food resource (Leung et al. 2005).

Reproduction is one of the central processes on which chemical pollutants can impact with consequences for population and thus Darwinian fitness. As cellular and whole animal energy reserves are involved in combating pollution effects, egg size and fecundity may be reduced, leading to population decline and even extinction in severe cases. Within contaminated populations, over multiple generations, organisms may become tolerant to the toxicant and trade-off mechanisms among life history traits may appear (Kammenga et al. 2000; Xie and Klerks 2003, 2004). Although increased sensitivity to new stressors in tolerant organisms has often been reported in various animal species, the adaptive benefit of tolerance may have negative side effects on population fitness in addition to a potential energy cost of the tolerance. In a recent study on fitness cost and resistance development in the beet armyworm *Spodoptera exigua*, Jia et al. (2009) continuously exposed *S. exigua* to an insecticide, Tebufenozide, for sixty-one generations and observed that the insect developed a ninety-two-fold resistance to the chemical and had a reduced relative population fitness of 0.71 against the control (1.0).

Moreover, the development of increased tolerance may result in a reduction of the overall amount of genetic variation in the population, because a strong selective pressure may be exerted by chemical stressors (e.g., metals) in exposed populations (Newman and Unger 2003).

12.2 COST OF POLLUTANT TOLERANCE

The cost of pollutant tolerance can be categorised into two types: individual level and population level (Sibly and Calow 1989; Hoffmann and Parsons 1991). Development of tolerance can come at a physiological cost to individual organisms via fitness trade-offs. Fitness costs may be associated with (1) an increased energy allocation to develop and maintain pollutant tolerance that would otherwise be used for somatic growth and reproduction, (2) pollutant-mediated deleterious genetic mutations and alterations of protein functions, and perhaps (3) potentially reduced adaptability of the population to future environmental conditions (Calow 1991; Taylor and Feyereisen 1996; Meyer and Di Giulio 2003). In other words, populations with pollutant tolerance may have lower survival ability in the longer term. Fitness costs corresponding to laboratory selection experiments have been observed in many studies (Xie and Klerks 2004; Tsui and Wang 2007; Kwok et al. 2009; Pook et al. 2009), despite conflicting results reported in other studies (Medina et al. 2008; Brausch and Smith 2009). In this section, we take a closer look at the energy and fitness costs involved in the development of chemical tolerance through physiological acclimation and genetic adaptation, respectively.

12.2.1 COST OF PHYSIOLOGICAL ACCLIMATION

For physiological acclimation, there is often an additional cost required for the increased transcription and production (i.e., upregulation) of detoxifying proteins for

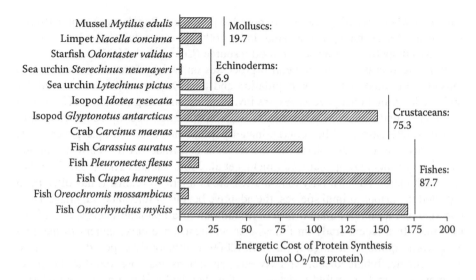

FIGURE 12.1   The average energy cost of protein synthesis in a range of aquatic organisms. Data are extracted from Bowgen et al. (2007). The results can be converted to kJ of energy per mg protein by using an energetic conversion factor of 484 kJ energy per mole of O_2.

combating toxic metals and xenobiotics, such as metallothioneins, heat shock proteins, glutathione S-transferases, cytochrome P450 monooxygenases, and carboxyl/cholinesterases in living organisms (Claudianos et al. 2006). The cost of protein synthesis can be measured by the correlation of oxygen consumption and absolute protein synthesis or the use of cycloheximide as an inhibitor of protein synthesis. The latter often gives a slightly lower estimation of the energy cost than the former method (Bowgen et al. 2007). Based on a meta-analysis of previous studies, the cost of protein synthesis varies significantly among different species and among taxonomic groups of aquatic animals (Figure 12.1). The average costs of protein synthesis in fishes and crustaceans are generally higher (87.7 and 75.3 μmol O_2 mg^{-1} protein) than those of molluscs (19.7 μmol O_2 mg^{-1} protein) and echinoderms (6.9 μmol O_2 mg^{-1} protein) (Figure 12.1). Regardless of species, the cost of synthesising detoxifying proteins, in terms of proportion of the overall metabolism cost per individual, would be considerable. For instance, Barber et al. (1990) used the correlation between oxygen consumption and protein synthesis and estimated that the cost of metallothionein production in Cd-challenged daphnids *Daphnia magna* was about 1.0 μL O_2 (i.e., ~0.04 μmol O_2) over twenty-four hours, and that was roughly equivalent to about 4% of total oxygen uptake (or total metabolism) in starved daphnids, or 2% for a feeding daphnid under chronic Cd exposure (5 μg.L^{-1}). It is therefore evident that combating the toxic effects of trace metals is metabolically costly. Yet, there is a lack of studies to detail the cost of production of detoxification proteins in an array of different aquatic organisms.

    Energy can be obtained from carbohydrates, lipids, and proteins in living organisms. In mammals, for example, the energy expenditures for adenosine triphosphatase (ATP) synthesis from glucose and lipid (tripalmitin) are estimated at 91.0 and 96.3

kJ/ATP, whereas for proteins, energy expenditures vary among different amino acids and range from 99.2 kJ/ATP (glutamate) to 153.2 kJ/ATP (cysteine) (Milgen 2002). It is therefore generally more costly to use proteins as the energy source. During chronic exposure to sublethal levels of copper or cadmium, the mussels *Perna viridis* mainly displayed significant reduction in lipid levels in their tissues, while both glycogen and protein contents remained fairly constant (unpublished data and personal communication, Jamius Yeung of the Swire Institute of Marine Science, Hong Kong). In the case of bivalve molluscs, lipids are possibly a more preferable energy source for combating the poisoning effects of trace metals.

It is also noteworthy that direct chemical toxicity can also cause a decline in mitochondrial capacity to synthesise ATP in cells, leading to a reduction of available energy for maintenance and reproduction, and thus lowering lifetime fitness. This has been shown in the oyster *Crassostrea virginica* exposed to elevated cadmium (Sokolova et al. 2005; Lannig et al. 2006a) and in mammals exposed to elevated zinc (Dineley et al. 2003). Besides, other physiological processes induced by chemical exposure, such as apoptosis and change in ion regulation, will also increase the overall metabolic energy demand (Rose et al. 2006).

12.2.2 Cost of Genetic Adaptation

Molecular mechanisms for genetic adaptation to pesticides and antibiotics have been extensively studied in bacteria, fungi, and insects over the past fifty years (Devonshire and Field 1991; Silver and Phung 1996; Sundin 1996; Taylor and Feyereisen 1996). The evolution of chemical tolerance in animals is usually associated with life history or population fitness costs (Carriere et al. 1994). Cousteau et al. (2000) reviewed the relevant literature and summarised them broadly into four classes:

1. Constitutive overproduction (i.e., upregulation) of tolerance genes and proteins
2. Constitutive underproduction (i.e., downregulation) of one gene product
3. Alternation of a target or receptor gene
4. An inducible change in gene regulation

Cousteau et al. (2000) also suggested three theoretical predictions: (1) genetic adaptation will likely involve at least a single major mutation leading to tolerance to xenobiotics; (2) such a mutation is probably associated with a significant physiological cost; and (3) this cost can potentially be corrected by epistasis with subsequent mutations in the absence of the selective pressure. The latter could also be accomplished via compensatory mutations that can reduce the cost of tolerance (Wijngaarden et al. 2005). The molecular mechanism (1) is relatively common in resistant insects and can be a consequence of the enhanced transcription of amplified genes or from a mutation leading to a constitutive overtranscription and overtranslation of the tolerance enzymes or other detoxification proteins (Devonshire and Field 1991; Taylor and Feyereisen 1996; Cousteau et al. 2000). Upregulation and overproduction of tolerance genes and proteins are energy demanding and hence pose a trade-off in energy allocation within an individual organism, leading to impairment of fitness. On the contrary, the other mechanisms are less energetically costly (Cousteau et al.

2000). Although there are many laboratory selection studies supporting the conclusion that the development of chemical tolerance can result in fitness reduction at individual and population levels (Table 12.1), only very few *in situ* field studies have been conducted to provide evidence to support this theory (Lenormand et al. 1998; Rubal et al. 2009).

12.2.3 FITNESS COST OF POLLUTANT TOLERANCE

In aquatic organisms, for example, fitness cost of chemical tolerance can be successfully identified through multigenerational acclimation studies and comparison of the intrinsic population growth rate (r) or reproductive output between chemically treated and untreated populations (i.e., in the presence and absence of the selective force). In several multigenerational acclimation experiments, significant trade-off costs have been identified when the animals established tolerance to the target toxicant (Table 12.1). In the example of the killifish *Heterandria formosa*, although the Cd-tolerant population presented threefold more tolerance to Cd than the control population, such a tolerance incurred a significant trade-off in fitness cost, as indirectly reflected by a significant reduction of brood size, postponement of reproduction, and shorter female life span (Xie and Klerks 2004). The reduction of fitness may be due to (1) reallocation of energy for defence, detoxification, and repair mechanisms that otherwise would be used for reproduction; (2) direct toxic effect of the chemical on reproduction-related physiological functions; and (3) direct fitness reduction associated with deleterious mutations (Raymond et al. 2004; Roux et al. 2005).

In contrast, other studies could not demonstrate fitness trade-offs resulting from the development of chemical tolerance. Ward and Robinson (2005) reported that there was no fitness trade-off cost for *Daphnia magna* in a Cd-adapted population, in terms of life span, reproduction output, and growth rate. Other kinds of costs, however, have been identified in the Cd-adapted *D. magna* population, which had a generally smaller body size, a lower genetic diversity, and was more sensitive to other toxicants, such as phenol (Ward and Robinson 2005). Such alternations could concomitantly reduce the probability of long-term survivorship in the Cd-adapted population. Brausch and Smith (2009) also demonstrated that there were no obvious fitness trade-offs, in terms of survival and reproduction output, in *D. magna*, which developed tolerance to cyfluthrin and naphthalene; albeit, there was an energy cost related to the upregulation of P450 enzymes. Marine copepods (*Tigriopus angulatus*) living in a site in northern Chile that is historically contaminated with an elevated level of Cu have developed a high tolerance to Cu (8.7 to 30.7 µg.L^{-1}), but do not show any decline in population fitness or any sign of negative effects on their development and fecundity (Medina et al. 2008). Perhaps, compensatory mutations might occur in this copepod population to match the reduced fitness cost due to Cu tolerance (Wijngaarden et al. 2005).

12.2.4 MATERNAL TRANSFER OF TOLERANCE

Tolerance to chemicals in aquatic animals such as crustaceans (Tsui and Wang 2005; Guan and Wang 2006) and fishes (Lin et al. 2000; Peake et al. 2004) can be

TABLE 12.1

Examples of Experiments Demonstrating the Various Fitness Trade-Offs Associated with the Development of Tolerance to Selected Chemicals

Species	Chemicals Adapted	Tolerance Level	Fitness Trade-Off	Descriptions	Other Cost(s)
Copepod *Tigriopus japonicus*[a]	Cu	100 µg.L⁻¹	Yes	↘ intrinsic population growth rate (r); dose-dependent ↘ in reproductive output	
Copepod *Tigriopus angulatus*[b]	Cu	8.7–30.7 µg.L⁻¹	No	r remained unaffected; absence of effects on development and fecundity	
Daphnid *Daphnia magna*[c]	Cd	61–180 µg.L⁻¹	Yes	Smaller body size, ↗ susceptibility to other chemicals; no effect on life span, growth rate, and reproductive output	↘ genetic diversity
Daphnid *Daphnia magna*[d]	Cd	3 µg.L⁻¹	Yes	In F5 and F6, ↘ growth and reproduction due to ↘ feeding rate and Cd tolerance	Change in Cd assimilation efficiencies
Daphnid *Daphnia magna*[e]	Cyfluthrin, naphthalene	LC50 values	No	No differences in survival, time to first brood, and total number of offspring	Lost chemical tolerance; ↗ physiological cost in detoxification (upregulation of cytochrome P450)
Killifish *Heterandria formosa*[f]	Cd	6 mg.L⁻¹, threefold more tolerance to Cd	Yes	↘ size offspring; ↘ brood size; ↗ time to first brood; ↘ female life span	

[a] Kwok et al. 2009.
[b] Medina et al. 2008.
[c] Ward and Robinson 2005.
[d] Guan and Wang 2006.
[e] Brausch and Smith 2009
[f] Xie and Klerks 2004.

established through multigenerational acclimation. For example, the marine cope-pod *Tigriopus japonicus* showed a gradual increase in copper (Cu) tolerance when they were continuously exposed to elevated Cu concentrations (10 or 100 μg.L^{-1}) for three generations (F0 to F2), and Cu tolerance of the copepod was increased even after one generation of acclimation to 100 μg.L^{-1} (Kwok et al. 2009). For a fixed exposure of 100 μg.L^{-1}, the intrinsic population growth rate (r) of the F0 generation was determined at 0.17, which gradually increased to $r = 0.22$ after one generation and then further to $r = 0.24$ after the second generation (Kwok et al. 2009). However, when the Cu-tolerant copepods were returned to clean seawater, their offspring would lose the Cu tolerance. The observed Cu tolerance in the copepod's offspring is probably developed through maternal transfer of detoxification mRNA, enzymes, or proteins.

Maternal transfer or maternal effect is an important mechanism for enhanc-ing tolerance in offspring. In this case, the cost of defence in the embryo and larvae mainly relies on the mother. For example, Peake et al. (2004) demonstrated that female fathead minnows *Pimephales promelas* exposed to Cu were able to maternally transfer Cu tolerance to their larval offspring. When challenged with Cu at high sublethal levels, their offspring showed significantly greater survivor-ship than larvae produced by the same females prior to Cu exposure. Peake et al. (2004) deduced three possible ways for maternal transfer of Cu tolerance. First, Cu ions may have been transferred from the mother to the embryo, perhaps associ-ated with binding of Cu to vitellogenin (Ghosh and Thomas 1995), and then the transferred Cu ions would induce detoxification processes in the embryo. It has been shown that pretreatment of eggs or larvae of fish and oyster with trace metals like cadmium, zinc, or mercury can make the larvae more tolerant to subsequent metal exposure, and such acclimation to metals after preexposure was attributable to stimulation and acceleration of the synthesis of metallothioneins, which form a nontoxic complex with the metals (Weis and Weis 1983; Roesijadi and Fellingham 1987; Roesijadi et al. 1997). In these processes, the energy cost of producing detox-ification proteins is drawn from the offspring themselves. Second, the mother can transfer mRNA of essential or detoxification proteins that are induced during the acclimation to the metal exposure. Lin et al. (2000) showed that female tilapias (*Oreochromis mossambicus*) exposed to Cd can maternally transfer mRNA of metallothioneins to enhance Cd tolerance of their offspring. Third, the mother may be able to directly transfer essential or detoxification proteins to the embryo; for example, maternal transfer of immunoglobulins has been reported for tilapia (Mor and Avtalion 1990; Takemura and Takano 1997). Apart from fishes, similar maternal effects have also been observed through multigenerational studies in the water flea *Daphnia magna* challenged with cyanobacterial toxins (Gustafsson et al. 2005) and in the marine copepod *Tigriopus japonicus* exposed to Cu (Rhodes et al. 2008; Kwok et al. 2009). Moreover, this phenomenon of maternal transfer is also common in insects. DDT-resistant *Drosophila melanogaster* can overtran-scribe the cytochrome P450 gene (Cyp6g1) and pass its mRNA and proteins to their eggs and larvae, which eventually enjoy such an inherited fitness benefit with-out any cost (McCart and Ffrench-Constant 2008). Nonetheless, more research is warranted to elucidate exactly what substances are being maternally transferred into embryos across various taxonomic groups of animals.

12.2.5 CHANGE OF GENETIC VARIATION IN A TOLERANT POPULATION

Pollutant-mediated genetic adaptation may also lead to a change of genetic variation in the tolerant population. Pollutants can act as strong selective forces for selecting tolerance genes and traits, so that the species can tolerate the chemical exposure and survive against the odds (Klerks and Weis 1987; Klerks and Levinton 1989; Posthuma and Van Straalen 1993; Van Straalen and Timmermans 2002). In fact, chemical pollution may indirectly change neutral genetic markers through genetic drift after a drastic reduction in population size (the bottleneck effect), and may directly change the genetic variability of life history traits through direct selection by the continuous selective force (i.e., chemical stressors; Cousyn et al. 2001; Van Straalen and Timmermans 2002). The cost associated with the indirect effect on neutral markers is generally accepted to be related to the decrease in genetic variability, and thus may lead to a decrease in adaptive potential toward other pollutants (Athrey et al. 2007). However, the results from different studies, which have searched for reductions of genetic variability during and after pollution events, are still inconclusive (Table 12.2).

In some studies, significant reduction in genetic diversity was reported in aquatic species associated with pollution tolerance (Table 12.2). For example, a decrease in genetic diversity was found in the crab *Pachygrapsus marmoratus* after exposure to trace metals in the field (Fratini et al. 2008), and in the mussel *Mytilus galloprovincialis* and the barnacle *Balanus glandula* historically exposed to various industrial pollutants (Ma et al. 2000). Some laboratory experiments further showed evidence that adaptation to environmental contaminants can result in loss of genetic variation. For instance, populations of the least killifish *Heterandria formosa*, which underwent eight generations of selection for an increased tolerance to Cd, showed a significant decrease in genetic variation (Athrey et al. 2007). On the contrary, Martins et al. (2009) revealed the genetic diversity in several populations of *Daphnia longispina* using amplified fragment length polymorphism (AFLP), and showed that there was no reduction of genetic diversity in metal-tolerant populations collected from a metal-contaminated site. Other studies also reached a similar conclusion that there was no apparent reduction in genetic variability caused by the development of chemical tolerance in the redbreast sunfish *Lepomis auritus* (Theodorakis et al. 2006) and in the midge *Chironomus riparius* (Vogt et al. 2007). Contaminants can affect genetic variation by altering migration patterns and mutation rate (Van Straalen and Timmermans 2002). For example, the population of the redbreast sunfish *Lepomis auritus* living in a contaminated site heavily affected by pulp mill effluent showed a higher genetic diversity than a population of the same species living in the reference site (Theodorakis et al. 2006). The increase in genetic diversity was thought to be related to the increase in mutation rate of the population in the polluted site and the alternation of migration patterns, i.e., a decrease in emigrants but an increase in immigrants at the polluted site.

TABLE 12.2

A Summary of Previous Studies on Genetic Diversity of Aquatic Species That Are Shown to Develop Tolerance to the Selected Pollutant(s)

Species	Source of Pollution (Study Type)	Molecular Markers*	Decrease in Genetic Diversity	Others Genetic Variations
Daphnid *Daphnia longispina*[a]	Acid mine drainage (lab and field)	AFLP	No	Genetic differentiation between contaminated and reference sites
Midge *Chironomus riparius*[b]	TBT (lab)	Microsatellite	No	Deviation from HWE in generations of treated group
Sunfish *Lepomis auritus*[c]	Pulp mill effluent (field)	RAPD	No	Level of genetic diversity was higher; increased mutation rate; alternation in migration patterns with fewer emigrants but an increase in immigrants
Killifish *Heterandria formosa*[d]	Cd (lab)	Microsatellite	Yes	Heterozygosity was lower, i.e., decrease in genetic diversity
Crab *Pachygrapsus marmoratus*[e]	Trace metals (field)	Microsatellite	Yes	
Mussel *Mytilus galloprovincialis*[f]	Industrial pollutant (field)	RAPD	Yes	
Barnacle *Balanus glandula*[f]	Industrial pollutant (field)	RAPD	Yes	

Note: AFLP = amplified fragment length polymorphism; RAPD = random amplification of polymorphic DNA.

[a] Martins et al. 2009.
[b] Vogt et al. 2007.
[c] Theodorakis et al. 2006.
[d] Athrey et al. 2007.
[e] Fratini et al. 2008.
[f] Ma et al. 2000.

12.3 THE METABOLIC COST CONCEPT AND THE REALLOCATION OF ENERGY

As far as animal life is concerned, an active organism has a certain amount of energy available to it through its feeding, which will be assimilated and stored in energy reserves. This assimilated energy is allocated toward two competitive needs (Figure 12.2): maintenance and production (Congdon et al. 1982; McNab 2002). Maintenance costs, essential for the continuity of life, are devoted to both standard (basal energy costs for living) and activity (e.g., energy for locomotion, foraging, digestion) maintenances. Energy allocated toward production is in favour of life history processes such as growth and reproduction (Congdon et al. 2001). For many organisms, assimilated energy is mainly devoted to maintenance costs and

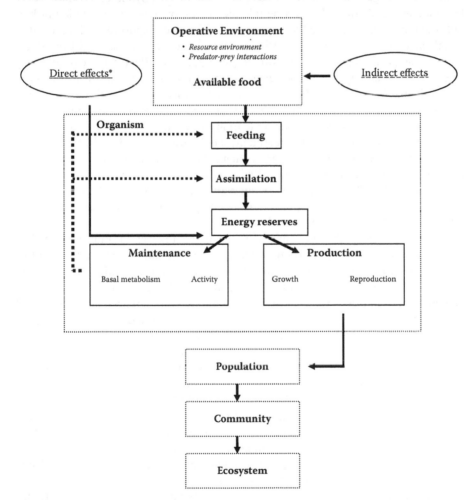

FIGURE 12.2 Direct and indirect effects of contaminants on energy reserve levels. Metabolic cost hypothesis: Under toxicant exposure, a shift in energy allocation will increase energy devoted to maintenance at the expense of production.

substantially less is allocated toward production (DuRant et al. 2007). For example, in reptiles, up to 80% of their energy budget is allocated toward maintenance costs, whereas production (for somatic growth, storage, and reproduction) accounts for the remaining 20% (Congdon et al. 1982). Thus, small increases in allocation to maintenance (e.g., those caused by contaminants) can result in proportionally larger reductions to production, which in turn can result in variation in life history phenotypes, such as offspring size or number, growth rate, size and age at maturity, and reproductive output. During times of limited available energy resources, energy is mainly allocated for living, whereas energy costs devoted to activity or production are often restricted (Lucas 1996; McNab 2002).

Exposure of organisms to chemical stressors is regarded as metabolically costly. This concept is generally referred to as the metabolic cost hypothesis (Calow 1991). The magnitude of stress that can be tolerated by an organism is a function of how well the organism is equipped to cope with the effects of exposure. To face an additional metabolic cost imposed by the tolerance to contaminants, a shift in allocation of energy would occur, for example, to repair the affected system, and may be exerted at the expense of other energy-demanding processes or by an enhanced energy intake (Beyers et al. 1999). For example, DuRant et al. (2007) examined in the lizard (*Sceloporus occidentalis*) changes in energy allocation after acute exposure to varying concentrations of pesticide (carbaryl). Results showed no differences in total energy expenditure among treatments, but lizards exposed to the highest dose of carbaryl allocated energy differently than animals in other groups. Compared to controls, these lizards allocated 16 to 30% more energy toward maintenance requirements, but 45 to 58% less energy toward additional energy expenditures. Thus, an increase in energy use for one portion of the energy budget would lead to an energy deficiency in others (i.e., energy trade-off).

Thus, considering that an understanding of energy reallocation and implied trade-offs is so important to explain effects of contaminants, one would expect that evidence for trade-offs has accumulated in ecotoxicological studies. According to Calow (1979), the energy costs of secreting mucus defences could be considerable. Indeed, in certain freshwater invertebrates, mucus secretion represents 13 to 32% of absorbed energy (Calow 1974; Calow and Woolhead 1977), and in the marine gastropod *Patella vulgata* 23% (Davies et al. 1990). Wicklum and Davies (1996) observed in the leech *Nephelopsis obscura* decreased ingestion and changes in energy allocation (e.g., increase in mucus secretion) of animals exposed in the laboratory to high Cd treatment. In the marine gastropod *Nucella lapillus* exposed to Cd, decreased oxygen consumption rate and glycogen level could be linked to both production of mucus and metallothionein synthesis (Leung et al. 2000). In oysters exposed to Ag, biomineralisation of this metal as Ag_2S in the basal membrane of the whole organs (Berthet et al. 1992) was accompanied by a strong depletion of glycogen, the concentrations of which were rapidly restored when oysters were transferred into clean water (Berthet et al. 1990). In *Ruditapes philippinarum*, the increased energy expenditure recorded in clams exposed in the laboratory to Cu was probably due to the increased cost of metal-binding protein synthesis (Munari and Mistri 2007). However, in some cases, metabolic costs involved in the induction of specific protective processes might be negligible compared to whole body energy expenditure.

In terms of cost per unit synthesis, proteins are the most energetically expensive molecules to synthesise (Jørgensen 1988; see also Section 12.3.1). Smith et al. (2001) investigated the tissue-specific energy cost of protein synthesis in the rainbow trout *Oncorhynchus mykiss* under Cu exposure. In cells (hepatocytes, gill cells) where protein synthesis rates are high, there is either no increase in synthesis cost following Cu exposure, or these cells or tissue types are able to avoid detrimental increases in the cost of synthesis when exposed to Cu. On the contrary, tissues (skin explants) with lower protein synthesis rates are susceptible to Cu since any further reduction in protein synthesis could result in an unacceptable increase in synthesis cost. Thus, the energy demand of protein synthesis may determine tissue sensitivity or susceptibility. Moreover, in isolation, does the induction of specific protective proteins have an appreciable effect on energy budgets relative to the overall expenditure of energy within organisms? For example, in metal-challenged daphnids, metallothionein synthesis represented less than 5% of total metabolism energy (Barber et al. 1990). In fish, after prolonged exposure to acid and aluminium, the increase in gill protein synthesis represents only approximately 2 to 3% of whole body protein synthesis, and thus the additional energy demand for metallothionein synthesis will probably be small, as long as metallothionein induction is the major energy-demanding process under sublethal metal exposure (Wilson 1996).

Another defence mechanism by which organisms cope with contaminant exposure is the biotransformation of xenobiotics through oxidation, reduction, hydrolysis, and conjugation reactions catalysed by multiple enzymes (Jimenez and Stegeman 1990). All data in the literature indicate that there is a quantifiable cost to organisms that detoxify xenobiotics. The biosynthetic energy costs of producing and managing a high level of detoxification enzymes might constitute a significant component of the energy budget of an organism smaller than a rodent (Brattsen 1988). For example, in ruffed grouse (*Bonasa umbellus*), the detoxification of coniferyl benzoate assimilated from the diet resulted in an estimated energy cost of 10 to 14% of metabolisable energy intake (Guglielmo et al. 1996). The induction of systems involved in xenobiotic metabolism is believed to be evidence that maintaining high levels of cytochrome P450 is metabolically costly (Brattsen et al. 1977). For example, in fish, a correlation has been observed between activities of biotransformation enzymes cytochrome P450 and energy expense (McMaster et al. 1991). But, the main difficulty is that the exact energy costs for biotransformation in organisms are nearly impossible to quantify due to the sum of all metabolic processes (Van Straalen and Hoffmann 2000; Morrow et al. 2004). Moreover, it must be duly noted that the levels of defence proteins such as heat shock proteins (HSPs) cannot increase unceasingly because the cost of their expression exceeds their benefits (Eckwert et al. 1997; Pyza et al. 1997).

Nearly twenty years ago, Calow (1991), reviewing empirical evidence for costs of combating toxicants that could explain trade-offs, concluded that there was some good, but not decisive, evidence for costs. For example, in the case of tolerance to metal toxicity, some authors report significant metabolic (and fitness) costs in invertebrates, and others do not (Barata et al. 2002; Arnold et al. 2008; Kwok et al. 2009). There are at least four possible explanations for an unchanged metabolic rate under stress:

1. Additional costs associated with repair mechanisms are masked by other toxicant effects, such as indirect effects of chemicals (see Section 12.8).
2. These energy demands are too small compared to whole metabolism costs to be evident, or tissue or organ level changes in metabolism may be masked by compensation in other tissues and organs, and consequently whole body metabolism will not be significantly impacted (Lyndon and Houlihan 1988).
3. Organisms have a susceptibility to reduce their metabolic rate as a scavenge response to toxic challenge (Hand and Hardewig 1996).
4. There are no additional costs due to chemical stress.

Moreover, the impact of toxicants on energy expenditure varies greatly among species and types of contaminants. More tolerant genotypes often have lower rates of metabolism and are therefore disconcerting.

12.4   METHODS FOR ASSESSMENT OF METABOLIC COST

Much of the interest in an energy approach to a discussion of the costs of tolerance is due to its reliance on the first principles of thermodynamics: that matter and energy are conserved. The rate of energy metabolism probably integrates more aspects of animal performance than any other single indicator of a physiological process (Bartholomew 1982; Giesy and Graney 1989). Moreover, energy is the common currency for translating biological effects to higher levels of biological organisation. It supports the means for quantifying the value of available resources as well as associated acquisition costs, and can account for the physiological effects of suboptimal environmental conditions that may deplete energy reserves (Beyers et al. 1999).

12.4.1   Standard Metabolic Rate: Oxygen Consumption

Usually, in a stressed organism, the reparation of homeostasis is associated with an increase in metabolic rate over the nonstressed condition (Beyers et al. 1999). Information about the metabolic rate of individuals can be obtained from the analysis of oxygen consumption. The measure of an organism's respiration rate provides a record of total metabolic activity because respiration is representative of the intensity of physiological processes at the time of measurement. With this methodological approach, effects of toxicants on the metabolic costs of the whole organism can be attributed to an increase in the production of defence and repair mechanisms, such as exclusion of chemical stressors, protective proteins (e.g., heat shock proteins, metallothioneins), changes in ion balance and hydration, and changes in metabolically active hormone levels. Increases in standard metabolic rate (SMR) estimated as oxygen consumption have been observed in different species (molluscs, crustaceans, amphibians, reptiles) exposed *in situ* or in the laboratory to metals (Rowe 1998; Hopkins et al. 1999; Rowe et al. 2001; Lannig et al. 2006b). In clams (*Macoma balthica*) experimentally exposed to polluted sediments (from the Bidassoa estuary in southern France), respiration was much higher than in clams from clean sediments (Hummel et al. 2000). On the contrary, other studies have reported a decrease in oxygen consumption in organisms, indicating a reduction in standard metabolic rate (Leung et al. 2000; Santos et al. 2000; Manyin and Rowe 2009). Such metabolic depression could be a strategy to minimise the uptake of the toxicant, and thereby reduce the exposure risk. The declines in respiration may also reflect the effects of toxicants at tissue, cellular, and biochemical levels. For example, gill necrosis would directly inhibit oxygen consumption (Nimmo et al. 1977; Soegianto et al. 1999). Cadmium has also been shown to reduce the number of mitochondria per unit cell volume in oysters (*Crassostrea virginica*), leading to a reduction of aerobic respiration (Cherkasov et al. 2006). At the cellular level, significant respiratory costs (up to 41%) were observed in trout hepatocytes during the metabolic processing of pyrene (Bains and Kennedy 2004). Rissanen et al. (2003) observed that the resin acid dehydroabietic acid (DHAA) increased trout hepatocyte oxygen consumption rates in a dose-dependent manner. Interpretation of results must also take into account that observed responses do not consider anaerobic energy production (Abele et al. 1998).

12.4.2 Energy Reserves

In animals, energy reserves are predominantly stored as glycogen and lipids. Glycogen is the form of carbohydrates stored by most animals and can be rapidly mobilised to produce high quantities of ATP. Lipids are comparatively more energy rich than glycogen (Mayer et al. 2002). Usually, lipids are mobilised to maintain homeostasis during toxicant exposure. The stimulation of lipid catabolism comprises formation of lipoproteins utilised for repair of damaged cell organelles, direct utilisation for energy requirements, and increased lipolysis. When these lipid energy reserves are depleted to a critical threshold, proteins can be used as energy sources (Le Gal et al. 1997). A reduction in total protein level shows that tissue protein may be submitted to proteolysis, resulting in the production of free amino acids used in the tricarboxylic acid cycle for energy production during stress conditions (Begum 2004). Thus, according to the metabolic cost hypothesis, energy reserves may be impacted by toxicant exposure. In general, a decrease of energy reserves is observed in different species following laboratory or field exposure to various types of contaminants: copper, zinc, mercury, cadmium, lead, organochlorines (lindane, endosulfan), organophosphates (parathion), polychlorinated biphenyls (PCBs), domestic or industrial effluents, or in polluted sediments (Table 12.3). In the long run, decreases in energy reserve levels may have consequences at higher levels of biological organisation, such as diminished growth, survival probability, and reproduction (Roex et al. 2003).

The mobilisation of one or another of the constituents of energy reserves is a function of the type of contaminant. For example, a mobilisation of glycogen and lipids is observed in the terrestrial isopod crustacean *Porcellio dilatatus* following laboratory dietary exposure to endosulfan (0.1 to 500 µg.g^{-1}) without modification of protein levels, while the latter also decrease further under parathion (0.1 in 500 µg.g^{-1}) exposure (Ribeiro et al. 2001). Physiological acclimation to toxicant exposure also affects energy reserve levels. For example, in *Daphnia magna*, individuals pre-exposed to zinc (and having acquired a tolerance toward this metal) do not mobilise

TABLE 12.3

A Summary of Previous Investigations on the Effects of Different Chemicals on Energy Reserves in Various Aquatic Organisms

Contaminant	Species	Exposure	Parameter	Response of Energy Reserves
Zn[a]	*Daphnia magna*	Doses: 0.1–1.0 μM	Proteins, sugars, lipids	Increase following preexposure
Zn, Cd[b]	*Poecilus cupreus*	Contaminated food	Caloric content	Decrease
Cd[c]	*Macoma balthica*	During 5 weeks Doses: 0, 10, 30, 100, 300 μg.L^{-1}	Glycogen	Decrease
Cd[d]	*Nucella lapillus* Digestive gland, gonad	During 80 days Doses: <0.01 and 400 μg.L^{-1}	Glycogen	No effect
Cd[e]	*Nereis virens* Posterior segments	From 2 to 80 days Doses: 0.5–40 mg.L^{-1}	Glycogen	Decrease
Cd[f]	*Anguilla anguilla* Muscle	During 1 month Doses: 0–5 μg.L^{-1}	Lipids	Decrease
Cu[g]	*Daphnia magna*	Doses: 0.5, 1, 5, 12, 35, 100 μg.L^{-1}	Energy reserves (proteins, glycogen, lipids)	Significant decrease of energy reserves in specimens of the first generation compared to those from the third
Pb[h]	*Porcellio scaber*	From 7 to 80 days Doses: 1–100 mg.L^{-1}	Protein, glycogen	Decrease
Metals[i]	*Nereis diversicolor*	*In situ*	Lipids, carbohydrates	Decrease
Hg or lindane[j]	*Daphnia magna*	During 96 hours Hg: 1.8, 3.2, 5.6, 10, 18, 24, 32 μg.L^{-1} Lindane: 0.056, 0.10, 0.18, 0.32, 0.56, 0.75 mg.L^{-1}	Polysaccharides, lipids, proteins	Decrease
Endosulfan[k]	*Anguilla anguilla* Muscle	During 12, 24, 48, 72, and 96 hours to 4.1–8.2 μg.L^{-1}	Glycogen	Decrease
Parathion[l]	*Porcellio dilatatus*	During 21 days to 0.1, 1, 10, 25, 50, 100, 250, 500 μg.g^{-1} of food	Glycogen, lipids, proteins	Decrease

Endosulfan[l]	*Porcellio dilatatus*		Glycogen, lipids, proteins	Decrease No effect
Fenhexamide[m]	*Tubifex tubifex*	During 2, 4, and 7 days 0.1, 1, 10 mg.L^{-1}	Glycogen, proteins	Significant decrease from 2 to 4 days of exposure
Chitosan[n]	*Tubifex tubifex*	During 2, 4, and 7 days 25, 62, 125 mg.L^{-1}	Proteins	Decrease
Tebuconazole[o]	*Daphnia magna*	During 5 days 0.41, 0.52, 0.71, 1.14 mg.L^{-1}	Lipids, carbohydrates, proteins	Decrease
PCBs[p]	*Crassostrea virginica* Adductor muscle, mantle, gonad	Laboratory during 8 weeks 0.35–3.5 µg.L^{-1}	Glycogen	Decrease No effect

Note: Laboratory experiments except for *Nereis diversicolor*. Waterborne contaminants except for *Poecilus cupreus* exposed to metals in contaminated food. Determination carried out in the whole organisms except when mentioned.

[a] Canli 2005.
[b] Maryanski et al. 2002.
[c] Duquesne et al. 2004.
[d] Leung and Furness 2001.
[e] Carr and Neff 1982.
[f] Pierron et al. 2007.
[g] Bossuyt and Janssen 2003.
[h] Knigge and Köhler 2000.
[i] Pook et al. 2009.
[j] De Coen and Janssen 1997.
[k] Gimeno et al. 1995.
[l] Ribeiro et al. 2001.
[m] Mosleh et al. 2005.
[n] Mosleh et al. 2007.
[o] Sancho et al. 2009.
[p] Encomio and Chu 2000.

their energy reserves further under laboratory exposure to dissolved zinc (0.1 and 1.0 µM) in contrast to those specimens from the control group without preexposure to zinc (Canli 2005).

12.4.3 Scope for Growth

Probably the most successful application of the metabolic cost hypothesis is the scope for growth (SfG) concept because it combines individual physiological responses, such as respiration, excretion rate, and assimilation efficiency, into a single, integrated bioassay (Widdows and Salkeld 1993). SfG measures the fraction of assimilated energy that is not consumed in basal maintenance and is available for growth and reproduction (Warren and Davis 1967; Widdows and Johnson 1988; Smolders et al. 2003; Crowe et al. 2004). Indeed, SfG is calculated by the difference between energy intake and metabolic output:

$$SfG = A - R - E \tag{12.1}$$

A corresponds to energy assimilated from food, R is the energy metabolised, and E is the energy lost in nitrogen excretion.

A positive SfG indicates that the organism is consuming less energy than it takes up, so energy is available for growth and reproduction, whereas a negative SfG shows that more energy is metabolised than is assimilated, and thus energy reserves are probably consumed. The SfG of freshwater and marine invertebrates has been successfully used as a bioassay to detect a range of chemical stresses, such as exposures to PAHs, TBT, DDT, PCBs, organochlorines, and metals. Table 12.4 illustrates some examples of the responses of SfG measures in organisms exposed *in situ* or the laboratory to various toxicants. Generally, negative correlations (Table 12.2) between SfG and chemicals have been observed. In some cases, reduction in SfG was the result of a decrease in energy intake rather than an increase in energy expenditure. For example, in the clam *Ruditapes philippinarum*, exposed in the laboratory to 4-nonylphenol, decreases in both respiration rate and clearance rate were associated with SfG reductions compared to controls (Matozzo et al. 2003). The positive correlation between clearance rate and respiration rate in 4-nonylphenol-treated clams was probably due to a decrease in feeding, which presumably led to a slower metabolism and lower energy demand (Sze and Lee 2000). The high correlation between SfG and reproduction in daphnids (*D. magna*) suggests that SfG is a good indicator of the general physiological condition of this species (Smolders et al. 2005). Even if measurement of SfG is labor intensive, the close coupling of this sensitive stress response and contaminant levels in the tissues of various organisms has provided a powerful and cost-effective method for assessing environmental pollution (Widdows 1998; Widdows et al. 2002).

12.4.4 Cellular Energy Allocation

The cellular energy allocation (CEA) biomarker was developed by De Coen and Janssen (1997) to provide measurements of energy allocation at the cellular level

TABLE 12.4
A Summary of Previous Investigations on the Responses of Scope for Growth (SfG) Measured in Aquatic Organisms Exposed *In Situ* or in the Laboratory to Various Toxicants; Decrease of SfG Observed in All Cases

Contaminant	Species	Exposure Conditions
Cadmium[a]	*Daphnia magna*	Laboratory for 2 weeks Doses: $0.1.10^{-8}$, 2.10^{-8}, 4.10^{-8} M
Chlorpyrifos[b]	*Neomysis integer*	Laboratory during 48, 96, and 168 hours Doses: 0.038, 0.056, 0.072, 0.100 µg.L^{-1}
Copper[c]	*Ruditapes decussatus*	Laboratory during 0, 2, 5, 9, 14, and 20 days Dose: 0.01 mg.L^{1}
Copper[d]	*Ruditapes philippinarum*	Laboratory during 0, 2, 5, 9, 14, and 20 days Dose: 10 µg.L^{-1}
Metals[e]	*Anadara trapezia*	Transplantation
Metals[f]	*Mytilus edulis*	Transplantation
Metals[g]	*Mytilus galloprovincialis*	*In situ* and laboratory during 4 weeks Doses: Low: mixture 20 mg Ni L^{-1}, 0.5 mg Cr L^{-1}, and 400 mg Fe L^{-1}; high: mixture 200 mg Ni L^{-1}, 5 mg Cr L^{-1}, and 4,000 mg Fe L^{-1}
Metals[h]	*Nereis diversicolor*	In situ
Metals, PCB, DDT, organochlorines[i]	*Mytilus galloprovincialis*	In situ
Biomonitoring (metals, organometals, organics)[j]	*Mytilus edulis*	In situ
4-Nonylphenol[k]	*Ruditapes philippinarum*	Laboratory during 7 days Doses: 0, 0.025, 0.05, 0.1, and 0.2 mg.L^{-1}

[a] Bailleul et al. 2005.
[b] Verslycke et al. 2004.
[c] Sobral and Widdows 1997.
[d] Munari and Mistri 2007.
[e] Burt et al. 2007.
[f] Wepener et al. 2008.
[g] Tsangaris et al. 2007.
[h] Pook et al. 2009.
[i] Widdows et al. 1997.
[j] Widdows et al. 2002.
[k] Matozzo et al. 2003.

(Verslycke and Janssen 2002; De Coen and Janssen 2003; Verslycke et al. 2004). Calorimetric measurements of available energy (protein, lipid, and carbohydrate contents) and energy consumed (electron transport system (ETS) activity as the cellular aspect of respiration) are combined into a value (Equation 12.2) expressing the net energy budget at the cellular level (De Coen and Janssen 1997).

$$CEA = \frac{\int_0^t Erdt - \int_0^t Ecdt}{t} \tag{12.2}$$

Er corresponds to energy reserves, *Ec* to the energy consumption, *t* to the duration of exposure, and *dt* to the time derivative.

CEA has been successfully applied in daphnids (De Coen and Janssen 1997; 2003; Muyssen et al. 2001), fish larvae (Nguyen 1997), and mysid crustaceans, *Neomysis integer* (Verslycke et al. 2003). Usually, the lipid budget was the key factor in describing the CEA, and *Ec* had only a minor contribution to the CEA (De Coen and Janssen 1997, 2003; Muyssen et al. 2001). The method has proven to be ecologically relevant as cellular effects have been linked to higher levels of biological organisation (De Coen and Janssen 2003; Smolders et al. 2004). For example, the biochemical responses of the CEA have been significantly correlated with effects at the population level in toxicant-exposed daphnids, such as the intrinsic rate of natural increase and the mean total offspring per female (De Coen and Janssen 1997). Vandenbrouck et al. (2009) examined toxic actions of waterborne nickel to *Daphnia magna* at both the gene transcription and suborganismal level (CEA and growth). Results indicated that several gene fragments (coding for haemoglobin, RNA terminal phosphate cyclase, a ribosomal protein), which are arguably linked with the energy budget, were covarying with the CEA measure, indicating a strong directly proportional relationship between both phenomena. These authors suggest that these genes could become molecular biomarkers with ecological relevance. CEA is also considered a sensitive biomarker, as changes in CEA have been shown to be an order of magnitude more sensitive in some species than, for example, the scope for growth assay, corroborating the theory of higher sensitivities of endpoints at a lower level of biological organisation (Verslycke et al. 2004). A disadvantage of the CEA response reported in the estuarine mysid *Neomysis integer* exposed in the laboratory to chlorpyrifos was the lack of a linear concentration-dependent response, limiting its value because it did not distinguish between low sublethal and high sublethal effects of this contaminant. Recently, CEA was tested as a biomarker to study the effects of oil-related compounds on Arctic benthic invertebrates (the amphipod crustaceans *Gammarus setosus* and *Onismus litoralis*) and the bivalve mollusc *Liocyma fluctuosa* (Olsen et al. 2007). Results showed significantly lower CEA values and higher ETS activity in *G. setosus* exposed to the water-accommodated fraction of crude oil, and higher CEA values and lower cellular respiration in *O. litoralis* subjected to drill cuttings, compared to controls. No differences were observed in the energy budget of *L. fluctuosa* between the treatments. These different responses to oil-related compounds according to the three species studied could be the result of differences in feeding and burrowing behaviour and species-specific sensitivity to petroleum-related compounds (Olsen et al. 2007).

12.4.5 Dynamic Energy Budget

Dynamic energy budgets (DEBs) reflect all physiological changes that exposed organisms have to make to survive during the entire exposure period. DEB models use differential equations to describe the rates at which individual organisms assimilate and utilise energy from food for maintenance, growth, reproduction, and development (Kooijman 1986; Kooijman and Bedaux 1996a). The greatest aim of the DEB modelling approach is to bring together the acquisition and expenditure of resources of all organisms in a single framework (Kooijman 2000; Kooijman et al. 2003). There are multiple DEB models. A basic bioenergetic budget for an organism can be expressed as

$$C = P + R + E \tag{12.3}$$

C is the energy consumed, P is the energy invested in production (e.g., growth, reproduction, energy storage, secretions), R is the respiratory dispense, and E is the energy loss to excretion and egestion.

Van der Meer (2006) reviewed applications of the DEB model to various species and discussed what sort of data sets are needed and have been used to estimate the various model parameters. The application of DEB theory to toxicants (DEBtox) corresponds to an extension of DEB theory by the integration of data from toxicity tests (Kooijman and Bedaux 1996a,b). In the DEBtox method, energy management at the organism level is assumed to be disturbed by a toxic compound when the internal concentration exceeds a threshold concentration called the no-observed effect concentration (NOEC). The DEBtox approach can take into account variations in exposure concentrations and time (Péry et al. 2001) and also provides information about the underlying mechanisms (Kooijman et al. 2003). DEBtox has been successfully applied in different invertebrate species (e.g., the midge *Chironomus riparius*) submitted to various chemicals, such as pesticides, PAHs, and metals (Ducrot et al. 2004; Alvarez et al. 2005; Péry et al. 2003, 2005). The development of DEBtox has been investigated further to analyse full life cycle tests (Ducrot et al. 2007) and mixtures of toxicants. One great interest of energy models is their relevance for the extrapolation of individual-level effects to population-level effects (Lopes et al. 2005). The introduction of DEBtox models into a matrix population model allows the observation of the changes of population growth rate continuously against chemical exposure concentrations, and compares the consequences at the population level of choosing one or another assumption about the mode of action of the contaminant at the individual level (Billoir et al. 2007).

12.5 CONFOUNDING FACTORS: INDIRECT CONTAMINANT EFFECTS

The most commonly examined effects of contaminants are those related to direct toxicity. In laboratory experiments, standardisation of conditions (e.g., fixed food availability and temperature) is required to distinguish direct toxic effects from those due to nontoxicological variables. Generally, chronic toxic bioassays are conducted with excessive food resources, but under ecologically realistic conditions, such resources are often less abundant and limited (i.e., density dependent). In the field, establishing definitive causal relationships between toxic stress and observed effects (e.g., energy reserves) is difficult due, in part, to the interference of many natural factors (environmental: temperature, season, salinity; biological: sex, weight, reproductive status). For example, besides variations induced by toxicants, energy reserve levels are greatly influenced by biotic and abiotic factors, particularly those linked seasonally to reproduction processes (Pellerin-Massicotte et al. 1994; LeGal et al. 1997; Gauthier-Clerc et al. 2002; Mouneyrac et al. 2006; Durou and Mouneyrac 2007). Moreover, seasonal and life stage changes in feeding patterns can also influence the energy reserves and susceptibility of exposed organisms (Chapman et al. 2003).

Feeding behaviour, food, and its availability are crucial factors in how an animal responds metabolically to a toxicant. Decreased resource quality or quantity can lead to deficiencies, altering the condition of studied organisms, potentially influencing their sensitivity to contaminants (Ankley and Blazer 1988; Bridges et al. 1997). Several studies have explored the responses of invertebrates to contaminants under conditions of limited resources (Rao and Sarma 1986; Chandini 1988, 1989; Klüttgen and Ratte 1994; Bridges et al. 1997). For example, Hopkins et al. (2002) investigated the interaction between resource level and sediment toxicity in a fish, the lake chubsucker (*Erimizon sucetta*). Results showed higher standard metabolic rates in exposed fish with reduced rations compared to control fish fed similar rations, which may be explained by the apparent inability of the fish with reduced rations to maintain a positive energy balance. This study showed that exposed fish with reduced rations failed to exhibit energy-saving decreases in metabolism, suggesting that these animals may face increased energy requirements due to costs associated with detoxification or pollutant-induced tissue damage. Moreover, reduced food ingestion frequently accompanies toxicant exposure, thereby negatively impacting an individual's energy budget through decreased acquisition (Carr 2002). A decrease in the feeding rate of the ragworm *N. diversicolor* originating from a contaminated estuary (Loire, France), in addition to disturbed assimilation (due to decreased digestive enzyme activities), could explain that the general condition of these worms was diminished compared to those from a reference site (Kalman et al. 2009). Barber et al. (1990) did not find an increase in O_2 consumption in *Daphnia magna* exposed to cadmium and dichloranaline as expected because of evidence for the deployment of energetically costly mechanisms of stress tolerance after exposure to these contaminants. However, the authors (Barber et al. 1990) noted a reduction in food intake and inferred that an increase in basal metabolism most likely did occur but was offset by a decrease in metabolic costs associated with consuming and digesting food. Lizards (*Sceloporus occidentalis*) exposed to the highest dietary (250 µg.g^{-1}) dose of carbaryl would still experience negative consequences compared to controls due to a 30 to 34% reduction in food consumption (DuRant et al. 2007). On the contrary, the high energy status of daphnids (*Daphnia magna*) from the provision of high food concentrations at the start of a laboratory exposure to cadmium did not provide an increased capacity to cope with additional stress (Smolders et al. 2005).

Effects of toxicants on the organism's environment, called indirect effects, are often much less obvious but may be equally or more important. These indirect effects correspond to changes in habitat, food availability, predator-prey interactions, and competition under chemical stress (see review by Fleeger et al. 2003). The direct effects of toxicants on predators can lead to cascading indirect effects on tolerant species at other trophic levels. Indirect effects of toxicants may lead to increase (via loss of competitors) or decrease (via loss of prey) in the available resources of an organism's environment. For example, herbicides can reduce plant cover and food availability while altering levels of risk associated with exposure to predators or changes in predator densities. Otherwise, pesticides can reduce fish populations indirectly by reducing their crustacean prey populations (Congdon et al. 2001). According to Chapman (2004), "in the case of food-web mediated effects, the following sequence exposure can and does apply: chronic contaminant exposure — reduced food abundance of certain dietary components — increased energy costs of feeding — reduced growth efficiency," with consequences at higher levels of biological organisation.

12.6 CONCLUSIONS

A priori, there are many reasons why there might be energy costs associated with tolerance to stress, so in this chapter we have tried to answer the question: Do previous ecotoxicological studies confirm this postulation? The impact of chemical stress on energy expenditure varies greatly among species, type of stressors, and patterns of energy expenditure. Therefore, allocation and acquisition under stressful conditions can be complex. Simultaneous monitoring of several parameters related to energy use can provide insight into the ultimate energy consequences of a stressor to an individual. The magnitude of stress that can be tolerated by an organism is a function of how well the animal is equipped to compensate for the effects of exposure. The compensatory ability of an organism in a given environment is dependent on its evolutionary history and the energy reserves that can be allocated to offset the effects of the stressor. If less energy is available, food demands may increase and make an animal more vulnerable to predation by increasing foraging time. Increased investment in the tolerance mechanism may initiate a trade-off between the benefits of adaptation and the costs.

In conclusion, clearly, much environmental stress (natural or chemical) can substantially reduce the total assimilated energy that is available for maintenance and production of an organism. Even if it is generally recognised that contaminants elicit their toxic effects by interaction with biological systems at the cellular and subcellular levels of organisation, linking these cellular and biological responses to the individual and population or community levels is necessary in ecotoxicological studies. Using energy as a common ecological currency, chemically mediated effects at a cellular level will reduce the energy availability for growth and maintenance for homeostasis, and thus ultimately result in undesirable impairment of the reproductive and developmental processes (Kooijman and Bedaux 1996b). While many laboratory studies support the fitness trade-off of chemical tolerance associated with both physiological acclimation and genetic adaptation, the results of only very few *in situ* studies actually agree with this hypothesis. Therefore, parallel laboratory and field studies should be carried out to decouple the possible conflicting results and pinpoint the uncertainty of extrapolating laboratory results to the field situation. As a result of the complexity of field settings, more concerted efforts should be devoted to study the evolution of chemical tolerance in natural populations and communities with consideration of other environmental factors, such as seasonal migration in relation to the gene flow, population dynamics (i.e., density dependency), and concomitant effects of chemical mixtures (Relyea and Hoverman 2006).

REFERENCES

Abele, D., et al. 1998. Exposure to elevated temperatures and hydrogen peroxide elicits oxidative stress and antioxidant response in the Antarctic intertidal limpet *Nacella concinna*. *Comp. Biochem. Physiol.* B120:425–35.

Alvarez, O., et al. 2005. Responses to stress of *Caenorhabditis elegans* populations with different reproductive strategies. *Funct. Ecol.* 19:656–64.

Ankley, G. T., and V. S. Blazer. 1988. Effects of diet on PCB-induced changes in xenobiotic metabolism in the liver of channel catfish (*Ictalurus punctatus*). *Can. J.Fish. Aquat. Sci.* 45:132–37.

Arnold, R. E., M. E. Hodson, and C. J. Langdon. 2008. A copper tolerant population of the earthworm *Dendrodrilus rubidus* (Savigny, 1862) at Coniston Copper Mines, Cumbria, UK. *Environ. Pollut.* 152:713–22.

Athrey, N. R. G., P. L. Leberg, and P. L. Klerks. 2007. Laboratory culturing and selection for increased resistance to cadmium reduce genetic variation in the least killifish, *Heterandria formosa. Environ. Toxicol. Chem.* 26:1916–21.

Bailleul, M., R. Smolders, and R. Blust. 2005. The effect of environmental stress on absolute and mass-specific scope for growth in *Daphnia magna* stress. *Comp. Biochem. Physiol.* C140:364–73.

Bains, O. S., and C. J. Kennedy. 2004. Energetic costs of pyrene metabolism in isolated hepatocytes of rainbow trout, *Oncorhynchus mykiss. Aquat. Toxicol.* 67:217–26.

Barata, C., et al. 2002. Among- and within-population variability in tolerance to cadmium stress in natural populations of *Daphnia magna*: Implications for ecological risk assessment. *Environ. Toxicol. Chem.* 21:1058–64.

Barber, I., D. J. Baird, and P. Calow. 1990. Clonal variation in general responses of *Daphnia magna* Straus to toxic stress. 2. Physiological effects. *Funct. Ecol.* 4:409–14.

Bartholomew, G. A. 1982. Energy metabolism. In *Animal physiology: Principles and adaptations*, ed. M. S. Gordon et al., 46–93. 4th ed. New York: MacMillan Publishing Co.

Begum, G. 2004. Carbofuran insecticide induced biochemical alterations in the liver and the muscle tissues of the fish *Clarias batrachus* (Linn) and recovery response. *Aquat. Toxicol.* 66:83–92.

Berthet, B., C. Amiard-Triquet, and R. Martoja. 1990. Chemical and histological effects of depuration in *Crassostrea gigas* Thunberg exposed previously to silver. *Water Air Soil Pollut.* 50:355–69.

Berthet, B., et al. 1992. Bioaccumulation, toxicity and physico-chemical speciation of silver in bivalve molluscs: Ecotoxicological and health consequences. *Sci. Total Environ.* 125:97–122.

Beyers, D. W., et al. 1999. Estimating physiological cost of chemical exposure: Integrating energetics and stress to quantify toxic effects in fish. *Can. J. Fish. Aquat. Sci.* 56:814–22.

Billoir, E., A. R. R. Péry, and S. Charles. 2007. Integrating the lethal and sublethal effects of toxic compounds into the population dynamics of *Daphnia magna*: A combination of the DEBtox and matrix population models. *Ecol. Model.* 203(3–4):204–21.

Bossuyt, B. T. A., and C. R. Janssen. 2003. Acclimation of *Daphnia magna* to environmentally realistic copper concentrations. *Comp. Biochem. Physiol.* C136:253–64.

Bowgen, A. D., et al. 2007. Energetic cost of synthesizing proteins in Antarctic limpet, *Nacella concinna* (Strebel, 1908), is not temperature dependent. *Am. J. Physiol. Regul. Integr. Comp. Physiol.* 292:R2266–74.

Brattsen, L. B. 1988. Enzymic adaptations in leaf-feeding insects to host–plant allelochemicals. *J. Chem. Ecol.* 14:1919–40.

Brattsen, L. B., C. F. Wilkinson, and T. Eisner. 1977. Herbivore–plant interactions: Mixed-function oxidases and secondary plant substances. *Science* 196:1349–52.

Brausch, J. M., and P. N. Smith. 2009. Development of resistance to cyfluthrin and naphthalene among *Daphnia magna*. *Ecotoxicology* 18:600–9.

Bridges, T. S., J. D. Farrar, and B. M. Duke. 1997. The influence of food ration on sediment toxicity in *Neanthes arenaceodentata* (Annelida: Polychaeta). *Environ. Toxicol. Chem.* 16:1659–65.

Burt, A., et al. 2007. The accumulation of Zn, Se, Cd and Pb and physiological condition of *Anadara trapezia* transplanted to a contamination gradient in Lake Macquarie, New South Wales, Australia. *Mar. Environ. Res.* 64:54–78.

Calow, P. 1974. Some observations on locomotory strategies and their metabolic effects in two species of freshwater gastropods, *Ancylus fluviatilis* Müll., and *Planorbis contortus* Linn. *Oecologia* 16:149–61.

Calow, P. 1979. Why some metazoan mucus secretions are more susceptible to microbial attack than others. *Am. Nat.* 114:149–51.

Calow, P. 1991. Physiological costs of combating chemical toxicants: Ecological implications. *Comp. Biochem. Physiol.* C100:3–6.

Calow, P., and A. S. Woolhead. 1977. Locomotory strategies in freshwater triclads and their effects on the energetics of degrowth. *Oecologia* 27:353–62.

Canli, M. 2005. Dietary and water-borne Zn exposures affect energy reserves and subsequent Zn tolerance of *Daphnia magna*. *Comp. Biochem. Physiol.* C141:110–16.

Carr, J. A. 2002. Stress, neuropeptides, and feeding behavior: A comparative perspective. *Integr. Comp. Biol.* 42:582–90.

Carr, R. S., and J. M. Neff. 1982. Biochemical indices of stress in the sandworm *Neanthes virens* (Sars). II. Sublethal responses to cadmium. *Aquat. Toxicol.* 2:319–33.

Carriere, Y., et al. 1994. Life-history costs associated with the evolution of insecticide resistance. *Proc. R. Soc. Lond.* B258:35–40.

Chandini, T. 1988. Effects of different food (*Chlorella*) concentrations on the chronic toxicity of cadmium to survivorship, growth and reproduction of *Echinisca triserialis* (Crustacea:Cladocera). *Environ. Pollut.* 54:139–54.

Chandini, T. 1989. Survival, growth, and reproduction of *Daphnia carinata* (Crustacea: Cladocera) exposed to chronic cadmium stress at different food (*Chlorella*) levels. *Environ. Pollut.* 60:29–45.

Chapman, P. M. 2004. Indirect effects of contaminants. *Mar. Pollut. Bull.* 48:411–12.

Chapman, P. M., et al. 2003. Conducting ecological risk assessments of inorganic metals and metalloids: Current status. *Hum. Ecol. Risk. Assess.* 9:641–97.

Cherkasov, A. S., A. H. Ringwood, and I. M. Sokolova. 2006. Combined effects of temperature acclimation and cadmium exposure on mitochondrial function in eastern oysters *Crassostrea virginica* Gmelin (Bivalva: Ostreidae). *Environ. Toxicol. Chem.* 25:2461–69.

Claudianos, C., et al. 2006. A deficit of detoxification enzymes: Pesticide sensitivity and environmental response in the honeybee. *Insect Mol. Biol.* 15:615–36.

Congdon, J. D., A. E. Dunham, and D. W. Tinkle. 1982. Energy budgets and life histories of reptiles. In *Biology of the reptilia*, ed. C. Gans and F. H. Pough, 155–99. Vol. 13. London: Academic Press.

Congdon, J. D., et al. 2001. Resource allocation-based life histories: A conceptual basis for studies of ecological toxicology. *Environ. Toxicol. Chem.* 20:1698–703.

Cousteau, C., C. Chevillon, and R. ffrench-Constant. 2000. Resistance to xenobiotics and parasites: Can we count the cost? *Trends Ecol. Evol.* 15:378–83.

Cousyn, C., et al. 2001. Rapid, local adaptation of zooplankton behavior to changes in predation pressure in the absence of neutral genetic changes. *Proc. Natl. Acad. Sci. USA* 98:6256–60.

Crowe, T. P., et al. 2004. Measurements of sublethal effects on individual organisms indicate community-level impacts of pollution. *J. Appl. Ecol.* 41:114–23.

Davies, M. S., S. J. Hawkins, and H. D. Jones. 1990. Mucus production and physiological energetics in *Patella vulgata* L. 1990. *J. Mollusc. Stud.* 56:499–503.

De Coen, W. M., and C. R. Janssen. 1997. The use of biomarkers in *Daphnia magna* toxicity testing. IV. Cellular energy allocation: A new methodology to assess the energy budget of toxicant-stressed *Daphnia* populations. *J. Aquat. Stress Recov.* 6:43–55.

De Coen, W. M., and C. R. Janssen. 2003. The missing biomarker link: Relationship between effects on the cellular energy allocation biomarker of toxicant-stressed *Daphnia magna* and corresponding population characteristics. *Environ. Toxicol. Chem.* 22:1632–41.

Devonshire, A. L., and L. M. Field. 1991. Gene amplification and insecticide resistance. *Annu. Rev. Entomol.* 36:1–23.

Dineley, K. E., T. V. Votyakova, and I. J. Reynolds. 2003. Zinc inhibition of cellular energy production: Implications for mitochondria and neurodegeneration. *J. Neurochem.* 85:563–70.

Ducrot, V., et al. 2004. Energy based modelling as a basis for the analysis of reproduction data with the midge *Chironomus riparius*. *Environ. Toxicol. Chem.* 23:225–31.

Ducrot, V., et al. 2007. Rearing and estimation of life-cycle parameters of the tubicifid worm *Branchiura sowerbyi*: Application to ecotoxicity testing. *Sci. Total Environ.* 384:252–63.

Duquesne, S., M. Liess, and D. J. Bird. 2004. Sub-lethal effects of metal exposure: Physiological and behavioural responses of the estuarine bivalve *Macoma balthica*. *Mar. Environ. Res.* 58:245–50.

DuRant, S. E., W. A. Hopkins, and L. G. Talent. 2007. Energy acquisition and allocation in an ectothermic predator exposed to a common environmental stressor. *Comp. Biochem. Physiol.* C145:442–48.

Durou, C., and C. Mouneyrac. 2007. Linking steroid hormone levels to sexual maturity index and energy reserves in *Nereis diversicolor* from clean and polluted estuaries. *Gen. Comp. Endocrinol.* 150:106–13.

Eckwert, H., G. Alberti, and H. R. Köhler. 1997. The induction of stress proteins (HSP) in *Oniscus asellus* (Isopoda) as a molecular marker of multiple heavy metal exposure. I. Principles and toxicological assessment. *Ecotoxicology* 6:249–62.

Encomio, V., and F.-L. Chu. 2000. The effect of PCBs on glycogen reserves in the eastern oyster *Crassostrea virginica*. *Mar. Environ. Res.* 50:45–49.

Fleeger, J. W., K. R. Carman, and R. M. Nisbet. 2003. Indirect effects of contaminants in aquatic ecosystems. *Sci. Total Environ.* 317:207–33.

Fratini, S., et al. 2008. Relationship between heavy metal accumulation and genetic variability decrease in the intertidal crab *Pachygrapsus marmoratus* (Decapoda; Grapsidae). *Estuar. Coast. Shelf Sci.* 79:679–86.

Gauthier-Clerc, S., et al. 2002. Delayed gametogenesis of *Mya arenaria* in the Saguenay fjord (Canada): A consequence of endocrine disruptors? *Comp. Biochem. Physiol.* C131:457–67.

Ghosh, P., and P. Thomas. 1995. Binding of metals to red drum vitellogenin and incorporation into oocytes. *Mar. Environ. Res.* 39:165–68.

Giesy, J. P., and R. L. Graney. 1989. Recent developments in and intercomparisons of acute and chronic bioassays and bioindicators. *Hydrobiologia* 188/189:21–60.

Gimeno, L., et al. 1995. Pesticide effects on eel metabolism pesticide. *Ecotoxicol. Environ. Saf.* 31:153–57.

Guan, R., and W. X. Wang. 2006. Multigenerational cadmium acclimation and biokinetics in *Daphnia magna*. *Environ. Pollut.* 141:343–52.

Guglielmo, G. C., W. H. Karasov, and W. J. Jakubas. 1996. Nutritional costs of plant secondary metabolite explain selective foraging by ruffed grouse. *Ecology* 77:1103–15.

Gustafsson, S., K. Rengefors, and L. A. Hansson. 2005. Increased consumer fitness following transfer of toxin tolerance to offspring via maternal effects. *Ecology* 86:2561–67.

Hand, S. C., and I. Hardewig. 1996. Downregulation of cellular metabolism during environmental stress: Mechanisms and implications. *Annu. Rev. Physiol.* 58:539–63.

Hoffmann, A. A., and P. A. Parsons. 1991. *Evolutionary genetics and environmental stress*. Oxford: Oxford Science Publishers.

Hopkins, W. A., C. L. Rowe, and J. D. Congdon. 1999. Elevated trace element concentrations and standard metabolic rate in banded water snakes (*Nerodia fasciata*) exposed to coal combustion wastes. *Environ. Toxicol. Chem.* 18:1258–63.

Hopkins, W. A., et al. 2002. Effects of food ration on survival and sublethal responses of lake chubsuckers (*Erimyzon sucetta*) exposed to coal combustion wastes. *Aquat. Toxicol.* 57:191–202.

Hummel, H., et al. 2000. The respiratory performance and survival of the bivalve *Macoma balthica* at the southern limit of its distribution area: A translocation experiment. *J. Exp. Mar. Biol. Ecol.* 251:85–102.

Jia, B. T., et al. 2009. Inheritance, fitness cost and mechanism of resistance to tebufenozide in *Spodoptera exigua* (Hubner) (Lepidoptera: Noctuidae). *Pest. Manage. Sci.* 65:996–1002.

Jimenez, B. D., and J. J. Stegeman. 1990. Detoxification enzymes as indicators of environmental stress on fish. *Am. Fish. Soc. Symp.* 8:145–66.

Jørgensen, C. B. 1988. Metabolic costs of growth and maintenance in the toad, *Bufo bufo*. *J. Exp. Biol.* 138:319–31.

Kalman, J., et al. 2009. Assessment of the health status of populations of the ragworm *Nereis diversicolor* using biomarkers at different levels of biological organisation. *Mar. Ecol. Prog. Ser.* 393:55–67.

Kammenga, J., M. J. Arts, and A. Doroszuk. 2000. Multi-generation effects at the population level: Fitness maximisation and optimal allocation in a nematode. In *Demography in ecotoxicology*, ed. J. Kammenga and R. Laskowski, 163–78. Chichester, UK: John Wiley & Sons Ltd.

Klerks, P. L., and J. S. Levinton. 1989. Rapid evolution of metal resistance in a benthic oligochaete inhabiting a metal-polluted site. *Biol. Bull.* 176:135–41.

Klerks, P. L., and J. S. Weis. 1987. Genetic adaptation to heavy metals in aquatic organisms— A review. *Environ. Pollut.* 45:173–205.

Klüttgen, B., and H. T. Ratte. 1994. Effects of different food doses on cadmium toxicity to *Daphnia magna*. *Environ. Toxicol. Chem.* 13(10):1619–27.

Knigge, T., and H.-R. Köhler. 2000. Lead impact on nutrition, energy reserves, respiration and stress protein (hsp 70) level in *Porcellio scaber* (Isopoda) populations differently preconditioned in their habitats. *Environ. Pollut.* 108:209–17.

Kooijman, S. A. L. M. 1986. Energy budgets can explain body size relations. *J. Theor. Biol.* 121:269–82.

Kooijman, S. A. L. M. 2000. *Dynamic energy and mass budgets in biological systems*. Cambridge: Cambridge University Press.

Kooijman, S. A. L. M., and J. J. M. Bedaux. 1996a. *The analysis of aquatic toxicity data*. Amsterdam: VU University Press.

Kooijman, S. A. L. M., and J. J. M. Bedaux. 1996b. Analysis of toxicity tests on *Daphnia* survival and reproduction. *Water Res.* 30:1711–23.

Kooijman, S. A. L. M., et al. 2003. *Water quality—Guidance document on the statistical analysis of ecotoxicity data*, chap. 7. ISO/TC 147/SC 5 N 18, ISO/WD 1. Geneva, Switzerland: International Organization for Standardization.

Kwok, K. W. H., E. P. M. Grist, and K. M. Y. Leung. 2009. Acclimation effect and fitness cost of copper resistance in the marine copepod *Tigriopus japonicus. Ecotoxicol. Environ. Saf.* 72:358–64.

Lannig, G., A. S. Cherkasov, and I. M. Sokolova. 2006a. Temperature-dependent effects of cadmium on mitochondrial and whole-organism bioenergetics of oysters (*Crassostrea virginica*). *Mar. Environ. Res.* 62:S79–82.

Lannig, G., J. F. Flores, and I. M. Sokolova. 2006b. Temperature-dependent stress response in oysters, *Crassostrea virginica*: Pollution reduces temperature tolerance in oysters. *Aquat. Toxicol.* 79:278–87.

Lawrence, A., and K. Hemingway. 2007. *Effects of pollution on fish: Molecular effects and population responses.* Oxford, UK: Blackwell.

Le Gal, Y., et al. 1997. Charge énergétique en adénylates (CEA) et autres biomarqueurs associés au métabolisme énergétique. In *Biomarqueurs en Écotoxicologie: Aspects Fondamentaux*, ed. L. Lagadic, T. Caquet, J. C. Amiard, and F. Ramade, 241–85. Paris: Masson.

Lenormand, T., et al. 1998. Appearance and sweep of a gene amplification: Adaptive response and potential for new functions in the mosquito *Culex pipiens. Evolution* 52:1705–12.

Leung, K. M. Y., and R. W. Furness. 2001. Survival, growth, metallothionein and glycogen levels of *Nucella lapillus* (L.) exposed to sub-chronic cadmium stress: The influence of nutritional state and prey type. *Mar. Environ. Res.* 52:173–94.

Leung, K. M. Y., A. C. Taylor, and R. W. Furness. 2000. Temperature-dependant physiological responses of the dogwhelk *Nucella lapillus* to cadmium exposure. *J. Mar. Biol. Assoc. U.K.* 80:647–60.

Leung, K. M. Y., et al. 2005. Deriving sediment quality guidelines using field-based species sensitivity distributions. *Environ. Sci. Technol.* 39:5148–56.

Lin, H. C., S. C. Hsu, and P. P. Hwang. 2000. Maternal transfer of cadmium tolerance in larval *Oreochromis mossambicus. J. Fish Biol.* 57:239–49.

Lopes, C., et al. 2005. Ecotoxicology and population dynamics: Using DEBtox models in a Leslie modeling approach. *Ecol. Model.* 188:30–40.

Lucas, A. 1996. *Bioenergetics of aquatic animals.* London: Taylor and Francis Ltd.

Lyndon, A. R., and D. F. Houlihan. 1998. Gill protein turnover: Costs of adaptation. *Comp. Biochem. Physiol.* A119:27–34.

Ma, X. L., D. L. Cowles, and R. L. Carter. 2000. Effect of pollution on genetic diversity in the bay mussel *Mytilus galloprovincialis* and the acorn barnacle *Balanus glandula. Mar. Environ. Res.* 50:559–63.

Maltby, L., and P. Calow. 1989. The application of bioassays in the resolution of environmental problems; past, present and future. *Hydrobiologia* 188/189:65–76.

Manyin, T., and C. L. Rowe. 2009. Bioenergetic effects of aqueous copper and cadmium on the grass shrimp, *Palaemonetes pugio. Comp. Biochem. Physiol.* C150:65–71.

Martins, N., et al. 2009. Effects of acid mine drainage on the genetic diversity and structure of a natural population of *Daphnia longispina. Aquat. Toxicol.* 92:104–12.

Maryanski, M., et al. 2002. Decreased energetic reserves, morphological changes and accumulation of metals in carabid beetles (*Poecilus cupreus* L.). *Ecotoxicology* 11:127–39.

Matozzo, V., et al. 2003. Evaluation of 4-nonylphenol toxicity in the clam *Tapes philippinarum. Environ. Res.* 91(3):179–85.

May, R. M. 1985. Evolution of pesticide resistance. *Nature* 315:12–13.

Mayer, F. L., et al. 2002. Physiological and nonspecific biomarkers. In *Biomarkers: Biochemical, physiological and histological markers of anthropogenic stress*, ed. R. J. Huggett, R. A. Kimerle, P. M. J. Mehrle, and H. L. Bergman, 5–85. SETAC Special Publication Series. Boca Raton, FL: Lewis Publishers.

McCart, C., and R. H. ffrench-Constant. 2008. Dissecting the insecticide-resistance-associated cytochrome P450 gene Cyp6g1. *Pest. Manage. Sci.* 64:639–45.

McMaster, M. E., et al. 1991. Changes in hepatic mixed function oxygenase (MFO) activity, plasma steroid-levels and age at maturity of a white sucker (*Catostomus commersoni*) population exposed to bleached kraft pulp mill effluent. *Aquat. Toxicol.* 21:199–237.

McNab, B. K. 2002. *The physiological ecology of vertebrates: A view from energetics.* Ithaca, NY: Cornell University Press.

Medina, M. H., B. Morandi, and J. A. Correa. 2008. Copper effects in the copepod *Tigriopus angulatus* Lang, 1933: Natural broad tolerance allows maintenance of food webs in copper-enriched coastal areas. *Mar. Freshwater Res.* 59:1061–66.

Meyer, J. N., and R. T. Di Giulio. 2003. Heritable adaptation and fitness costs in killifish (*Fundulus heteroclitus*) inhabiting a polluted estuary. *Ecol. Appl.* 13:490–503.

Milgen, J. V. 2002. Modelling biochemical aspects of energy metabolism in mammals. *J. Nutr.* 132:3195–202.

Mor, A., and R. R. Avtalion. 1990. Transfer of antibody activity from immunized mother to embryo in tilapias. *J. Fish Biol.* 37:249–55.

Morrow, M. D., D. Higgs, and C. J. Kennedy. 2004. The effects of diet composition and ration on biotransformation enzymes and stress parameters in rainbow trout, *Oncorhynchus mykiss. Comp. Biochem. Physiol.* C137:143–154.

Mosleh, Y. Y., et al. 2005. Metallothionein induction, antioxidative responses, glycogen and growth changes in *Tubifex tubifex* (Oligochaete) exposed to the fungicide, fenhexamid. *Environ. Pollut.* 135:73–82.

Mosleh, Y. Y., et al. 2007. Effects of chitosan on oxidative stress and metallothioneins in aquatic worm *Tubifex tubifex* (Oligochaeta, Tubificidae). *Chemosphere* 67:167–75.

Mouneyrac, C., H. Perrein-Ettajani, and C. Amiard-Triquet. 2010. The use of fitness, reproduction and burrowing behaviour of the polychaete *Nereis diversicolor* in the assessment of estuarine sediment quality. *Environ. Pollut.* 158:121–28.

Mouneyrac, C., et al. 2009. Linking energy metabolism, reproduction abundance and structure of *Nereis diversicolor* populations. In *Environmental assessment of estuarine ecosystems: A case study*, ed. C. Amiard-Triquet and P. S. Rainbow, 155–78. Boca Raton, FL: CRC Press, Taylor and Francis.

Mouneyrac, C., et al. 2006. *In situ* relationship between energy reserves and steroid hormone levels in *Nereis diversicolor* (O. F. Müller) from clean and contaminated sites. *Ecotoxicol. Environ. Saf.* 65:181–187.

Munari, C., and M. Mistri. 2007. Effect of copper on the scope for growth of clams (*Tapes philippinarum*) from a farming area in the Northern Adriatic Sea. *Mar. Environ. Res.* 64:347–57.

Muyssen, B. T. A., C. R. Janssen, and B. T. A. Bossuyt. 2001. Tolerance and acclimation to zinc of field-collected *Daphnia magna* populations. *Aquat. Toxicol.* 56:69–79.

Newman, M. C., and M. A. Unger, eds. 2003. *Fundamentals of ecotoxicology.* 2nd ed. Boca Raton, FL: Lewis Publishers/CRC Press.

Nguyen, T. H. L. 1997. *Potential and Limitations of Early Life Stage Toxicity Tests with Fish.* PhD thesis, University of Ghent, Ghent, Belgium.

Nimmo, D. W. R., D. V. Lightner, and L. H. Bahner. 1977. Effects of cadmium on shrimps *Penaeus duodarum, Palaemonetes pugio* and *Palaemonetes vulgaris*. In *Physiological responses of marine biota to pollutants*, ed. F. J. Vernberg, A. Calabrese, F. P. Thurberg, and W. B. Vernberg, 131–84. New York: Academic Press.

Olsen, G. H., et al. 2007. Alterations in the energy budget of Arctic benthic species exposed to oil-related compounds. *Aquat. Toxicol.* 83:85–92.

Peake, E. B., et al. 2004. Copper tolerance in fathead minnows. II. Maternal transfer. *Environ. Toxicol. Chem.* 23:208–11.

Pellerin-Massicotte, J., et al. 1994. Seasonal variability in biochemical composition of the polychaete *Nereis virens* (Sars) in two tidal flats with different geographic orientations. *Comp. Biochem. Physiol.* A107:509–16.

Péry, A. R. R., et al. 2001. Analysis of bioassays with time-varying concentrations. *Water Res.* 35:3825–32.

Péry, A. R. R., et al. 2003. Modelling toxicity and mode of action of chemicals to analyse growth and emergence tests with the midge *Chironomus riparius. Aquat. Toxicol.* 65:281–92.

Péry, A. R. R., et al. 2005. Body residues: A key variable to analyse toxicity tests with *Chironomus riparius* exposed to copper-spiked sediments. *Ecotoxicol. Environ. Saf.* 61:160–67.

Pierron, F., et al. 2007. Impairment of lipid storage by cadmium in the European eel (*Anguilla anguilla*). *Aquat. Toxicol.* 81:304–11.

Pook, C., C. Lewis, and T. Galloway. 2009. The metabolic and fitness costs associated with metal resistance in *Nereis diversicolor. Mar. Pollut. Bull.* 58:1063–71.

Posthuma, L., and N. M. Van Straalen. 1993. Heavy-metal adaptation in terrestrial invertebrates: A review of occurrence, genetics, physiology and ecological consequences. *Comp. Biochem. Physiol.* C106:11–38.

Pyza, E., et al. 1997. Heat shock proteins (HSP70) as biomarkers in ecotoxicological studies. *Ecotoxicol. Environ. Saf.* 38:244–51.

Rao, T. R., and S. S. S. Sarma. 1986. Demographic parameters of *Brachionus patulus* Müller (Rotifera) exposed to sublethal DDT concentrations at low and high food levels. *Hydrobiologia* 139:193–200.

Raymond, M., et al. 2004. Insecticide resistance in the mosquito *Culex pipiens*: What have we learned about adaptation? *Genetica* 112–113:287–96.

Relyea, R. A., and J. T. Hoverman. 2006. Assessing the ecology in ecotoxicology: A review and synthesis in freshwater systems. *Ecol. Lett.* 9:1157–71.

Rhodes, J. R., et al. 2008. A Bayesian mixture model for estimating inter-generation chronic toxicity. *Environ. Sci. Technol.* 42:8108–14.

Ribeiro, S., et al. 2001. Effect of endosulfan and parathion on energy reserves and physiological parameters of the terrestrial isopod *Porcellio dilatatus. Ecotoxicol. Environ. Saf.* 49:131–38.

Rissanen, E., G. Krumschnabel, and M. Nikinmaa. 2003. Dehydroabietic acid, a major component of wood industry effluents, interferes with cellular energetics in rainbow trout hepatocytes. *Aquat. Toxicol.* 62:45–53.

Roesijadi, G., and G. W. Fellingham. 1987. Influence of Cu, Cd and Zn preexposure on Hg toxicity in the mussel *Mytilus edulis. Can. J. Fish. Aquat. Sci.* 44:680–84.

Roesijadi, G., K. M. Hansen, and M. E. Unger. 1997. Metallothionein mRNA accumulation in early development stages of *Crassostrea virginica* following pre-exposure and challenge with cadmium. *Aquat. Toxicol.* 39:185–94.

Roex, E. W. M., R. Keijzers, and C. A. M. VanGestel. 2003. Acetylcholinesterase and increased food consumption rate in the zebrafish, *Danio rerio*, after chronic exposure to parathion. *Aquat. Toxicol.* 64:451–60.

Rose, W. L., et al. 2006. Using an integrated approach to link biomarker responses and physiological stress to growth impairment of cadmium-exposed larval topsmelt. *Aquat. Toxicol.* 80:298–308.

Roux, F., et al. 2005. Epistatic interactions among herbicide resistances in *Arabidopsis thaliana*: The fitness cost of multi-resistance. *Genetics* 171:1277–88.

Rowe, C. L. 1998. Elevated standard metabolic rate in a freshwater shrimp (*Palaemonetes paludosus*) exposed to trace element-rich coal combustion waste. *Comp. Biochem. Physiol.* A121:299–304.

Rowe, C. L., et al. 2001. Metabolic costs incurred by crayfish (*Procambarus acutus*) in a trace element-polluted habitat: Further evidence of similar responses among diverse taxonomic groups. *Comp. Biochem. Physiol.* C129:275–83.

Rubal, M., L. M. Guilhermino, and M. H. Medina. 2009. Individual, population and community level effects of subtle anthropogenic contamination in estuarine meiobenthos. *Environ. Pollut.* 157:2751–58.

Sancho, E., et al. 2009. Disturbances in energy metabolism of *Daphnia magna* after exposure to tebuconazole. *Chemosphere* 74:1171–78.

Santos, M. H. S., N. Troca da Cunha, and A. Bianchini. 2000. Effects of copper and zinc on growth, feeding and oxygen consumption of *Farfantepenaeus paulensis* postlarvae (Decapoda: Penaeidae). *J. Exp. Mar. Biol. Ecol.* 247:233–42.

Seyle, H. 1956. *The stress of life*. New York: McGraw-Hill Book Company.

Sibly, R. M., and P. Calow. 1989. A life-cycle theory of responses to stress. *Biol. J. Linn. Soc.* 37:101–16.

Silver, S., and L. T. Phung. 1996. Bacterial heavy metal resistance: New surprises. *Annu. Rev. Microbiol.* 50:753–89.

Smith, R. W., et al. 2001. Protein synthesis costs could account for the tissue-specific effects of sub-lethal copper on protein synthesis in rainbow trout (*Oncorhynchus mykiss*). *Aquat. Toxicol.* 53:265–77.

Smolders, R., M. Baillieul, and R. Blust. 2005. Relationship between the energy status of *Daphnia magna* and its sensitivity to environmental stress. *Aquat. Toxicol.* 7:155–70.

Smolders, R., et al. 2003. A conceptual framework for using mussels as biomonitors in whole effluent toxicity. *Hum. Ecol. Risk. Assess.* 9:741–60.

Smolders, R., et al. 2004. Cellular energy allocation in zebra mussels exposed along a pollution gradient: Linking cellular effects to higher levels of biological organization. *Environ. Pollut.* 129:99–112.

Sobral, P., and J. Widdows. 1997. Effects of copper exposure on the scope for growth of the clam *Ruditapes decussatus* from southern Portugal. *Mar. Pollut. Bull.* 12:992–1000.

Soegianto, A., et al. 1999. Impact of cadmium on the structure of gills and epipodites of the shrimp *Penaeus japonicus* (Crustacea: Decapoda). *Aquat. Liv. Resour.* 12:57–70.

Sokolova, I. M., E. P. Sokolow, and K. M. Ponnappa. 2005. Cadmium exposure affects mitochondrial bioenergetics and gene expression of key mitochondrial proteins in the eastern oyster *Crassostrea virginica* Gmelin (Bivalvia: Ostreidae). *Aquat. Toxicol.* 73:242–55.

Sundin, G. W. 1996. Evolution and selection of antibiotics and pesticide resistance: A molecular genetic perspective. *Mol. Genet. Evol. Pest. Resist.* 64:97–105.

Sze, P. W. C., and S. Y. Lee. 2000. Effects of chronic copper exposure on the green mussel *Perna viridis*. *Mar. Biol.* 137:379–92.

Takemura, A., and K. Takano. 1997. Transfer of maternally-derived immunoglobulin (IgM) to larvae in tilapia, *Oreochromis mossambicus*. *Fish Shellfish Immunol.* 7:355–63.

Taylor, M., and R. Feyereisen. 1996. Molecular biology and evolution of resistance to toxicants. *Mol. Biol. Evol.* 13:719–34.

Theodorakis, C. W., et al. 2006. Evidence of altered gene flow, mutation rate, and genetic diversity in redbreast sunfish from a pulp-mill-contaminated river. *Environ. Sci. Technol.* 40:377–86.

Tsangaris, C., E. Papathanasiou, and E. Cotou. 2007. Assessment of the impact of heavy metal pollution from a ferro-nickel smelting plant using biomarkers. *Ecotoxicol. Environ. Saf.* 66:232–43.

Tsui, M. T. K., and W. X. Wang. 2005. Influences of maternal exposure on the tolerance and physiological performance of *Daphnia magna* under mercury stress. *Environ. Toxicol. Chem.* 24:1228–34.

Tsui, M. T. K., and W. X. Wang. 2007. Biokinetics and tolerance development of toxic metals in *Daphnia magna*. *Environ. Toxicol. Chem.* 26:1023–32.

Van der Meer, J. 2006. An introduction to dynamic energy budget (DEB) models with special emphasis on parameter estimation. *J. Sea Res.* 56:85–102.

Van Straalen, N. M., and A. A. Hoffmann. 2000. Review of experimental evidence for physiological costs of tolerance to toxicants. In *Demography in ecotoxicology*, ed. J. Kammenga and R. Laskowski, 115–24. Chichester, UK: John Wiley & Sons Ltd.

Van Straalen, N. M., and M. Timmermans. 2002. Genetic variation in toxicant-stressed populations: An evaluation of the "genetic erosion" hypothesis. *Hum. Ecol. Risk Assess.* 8:983–1002.

Vandenbrouck, T., et al. 2009. Nickel and binary metal mixture responses in *Daphnia magna*: Molecular fingerprints and (sub)organismal effects. *Aquat. Toxicol.* 92:18–29.

Verslycke, T., and C. R. Janssen. 2002. Effects of a changing abiotic environment on the energy metabolism in the estuarine mysid shrimp *Neomysis integer* (Crustacea: Mysidacea). *J. Exp. Mar. Biol. Ecol.* 279:61–72.

Verslycke, T., et al. 2004. Cellular energy allocation and scope for growth in the estuarine mysid *Neomysis integer* (Crustacea: Mysidacea) following chlorpyrifos exposure: A method comparison. *J. Exp. Mar. Biol. Ecol.* 306:1–16.

Verslycke, T., et al. 2003. Cellular energy allocation in the estuarine mysid shrimp *Neomysis integer* (Crustacea: Mysidacea) following tributyltin exposure. *J. Exp. Mar. Biol. Ecol.* 288:167–79.

Vogt, C., et al. 2007. Multi-generation studies with *Chironomus riparius*—Effects of low tributyltin concentrations on life history parameters and genetic diversity. *Chemosphere* 67:2192–200.

Ward, T. J., and W. E. Robinson. 2005. Evolution of cadmium resistance in *Daphnia magna*. *Environ. Toxicol. Chem.* 24:2341–49.

Warren, C. E., and G. E. Davis. 1967. Laboratory studies on the feeding, bioenergetics and growth of fish. In *The biological basis for freshwater fish production*, ed. S. D. Gerking, 175–214. Oxford: Blackwell Scientific.

Weis, P., and J. S. Weis. 1983. Effects of embryonic pre-exposure to methylmercury and Hg^{2+} on larval tolerance in *Fundulus heteroclitus*. *Bull. Environ. Contam. Toxicol.* 31:530–34.

Wepener, V., et al. 2008. Metal exposure and biological responses in resident and transplanted mussels (*Mytilus edulis*) from the Scheldt estuary. *Mar. Pollut. Bull.* 57:624–31.

Wicklum, D., and R. W. Davies. 1996. The effects of chronic cadmium stress on energy acquisition and allocation in a freshwater benthic invertebrate predator. *Aquat. Toxicol.* 35:237–52.

Widdows, J. 1998. Marine and estuarine invertebrate toxicity tests. In *Handbook of ecotoxicology*, ed. P. Calow, 145–66. Oxford: Blackwell Science.

Widdows, J., and D. Johnson. 1988. Physiological energetics of *Mytilus edulis*: Scope for growth. *Mar. Ecol. Prog. Ser.* 46:113–21.

Widdows, J., and P. N. Salkeld. 1993. *Practical procedures for the measurement of scope for growth*, 147–72. Vol. 71, MAP Technical Reports Series. Athens, Greece: United Nations Environment Programme.

Widdows, J., et al. 2002. Measurement of stress effects (scope for growth) and contaminant levels in mussels (*Mytilus edulis*) collected from the Irish Sea. *Mar. Environ. Res.* 53:327–56.

Widdows, J., C. Nasci, and V. U. Fossato. 1997. Effects of pollution on the scope for growth of mussels (*Mytilus galloprovincialis*) from the Venice lagoon. *Mar. Environ. Res.* 43:69–79.

Wijngaarden, P. J., et al. 2005. Adaptation to the cost of resistance: A model of compensation, recombination and selection in a haploid organism. *Proc. R. Soc. Lond.* B272:85–89.

Wilson, R. W. 1996. Physiological and metabolic costs of acclimation to chronic sub-lethal acid and aluminium exposure in rainbow trout. In *Toxicology of aquatic pollution. Physiological, molecular and cellular approaches*, ed. E. W. Taylor, 143–67. Seminar Series. Cambridge: Cambridge University Press, Society for Experimental Biology.

Xie, L. T., and P. L. Klerks. 2003. Responses to selection for cadmium resistance in the least killifish, *Heterandria formosa. Environ. Toxicol. Chem.* 22:313–20.

Xie, L. T., and P. L. Klerks. 2004. Fitness cost of resistance to cadmium in the least killifish (*Heterandria formosa*). *Environ. Toxicol. Chem.* 23:1499–503.

13 Tolerance and the Trophic Transfer of Contaminants

Claude Amiard-Triquet and Philip S. Rainbow

CONTENTS

13.1 INTRODUCTION

All organisms have strategies allowing them to cope with chemical stress, including exposure to toxic contaminants, and all are inevitably members of food webs, consisting of several interrelating food chains. Furthermore, as has been well exemplified in other chapters in this volume, many organisms can develop tolerance to contaminant exposure, whether by acclimation or by genetic adaptation. This chapter addresses the question of whether the acquisition of tolerance to a contaminant by an organism has any implications for the trophic transfer of that contaminant to animals higher up any food chain.

According to the nature of the contaminant and the associated physiological processes responsible for the development of tolerance to the contaminant, any ecotoxicological risk associated with the transfer of the contaminant between successive links of any food chain may be more or less critical. As shown in Figure 13.1, the biomass of the different constituents of a theoretical food chain may be represented by a pyramid, a given biomass of primary producers being needed to produce a

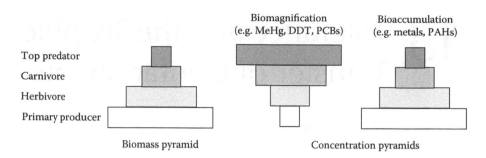

FIGURE 13.1 Biomagnification versus bioaccumulation in aquatic food chains. (After Amiard-Triquet, C., and J. C. Dauvin, in *Environmental Assessment of Estuarine Ecosystems. A Case Study*, ed. C. Amiard-Triquet and P. S. Rainbow, CRC Press, Boca Raton, FL, 2009, pp. 1–17. With permission.)

lower biomass of herbivores, and so on. For inorganic contaminants, the concentration pyramid is typically orientated like the biomass pyramid, as is the case for most metals along most aquatic food chains, except for those like mercury, which are at least partly in organometallic form in the environment and in prey organisms. The opposite orientation of the pyramid is the typical case for many organic contaminants, which undergo biomagnification along food chains. Due to their lipophilic character, organic contaminants have a high potential for biomagnification, but the pattern is very contrasted between those which are very stable (such as polychlorinated biphenyls (PCBs)) and those which are easily biodegraded (such as polyaromatic hydrocarbons (PAHs) in vertebrates) with consequent loss of definition of the inverted pyramid.

Briefly, tolerance to a contaminant may result from (1) the limitation of uptake, (2) increased elimination, or (3) storage in nontoxic form. In the two first cases, given that the details are rarely so simple, the body burden of a contaminant in a prey organism remains low, and so any transfer to a predator is clearly limited. In the last case, this tolerance mechanism may be responsible for the presence of a food species containing a very high concentration of an environmental contaminant, with ecotoxicological implications for predators higher up the food chain. However, detoxified contaminants in plants, invertebrates, and vertebrates may be present in very diverse physicochemical forms, and the characteristics of these in turn appear to control to some degree their subsequent transfer to the next trophic level.

On the other hand, avoidance reactions by feeding animals to prey with high contaminant concentrations are well documented in both invertebrates and vertebrates. The best known examples have been published for herbivorous species in which food selection is an important component of defence against toxic molecules, either natural toxins such as plant secondary metabolites or anthropogenic contaminants such as metals accumulated in tolerant plants. Thus, animals feeding on metal-contaminated food may reduce feeding rates, although reduced feeding may result from either behavioural avoidance of the contamination or poisoning of the digestive tract. The amphipod crustacean *Orchestia gammarellus*, for example, had decreased feeding rates when fed a diet of decaying kelp *Laminaria digitata* enriched with Cu or

Zn (Weeks 1993). The ingestion rate of the copepod crustacean *Acartia spinicauda* feeding on the planktonic dinoflagellate *Prorocentrum minimum* was reduced by increased concentrations of Cd and Se (but not Zn) in the food particles (Xu et al. 2001). Woodward et al. (1995) found a reduction in feeding rate of trout fed frozen insect larvae collected from a metal-contaminated river.

The relative importance of the development of tolerance to contaminants in assessing the environmental consequences of food web contamination by environmental pollutants will depend on the nature of the physiological mechanisms of acclimation or genetic adaptation of organisms to the relevant pollutant, and the degree to which it is possible to generalise about the taxonomic distribution of different tolerance mechanisms. Thus, in this chapter we will review the literature data indicating the extent to which tolerance to different classes of contaminants exists in different taxa, and the physiological nature of the tolerance mechanisms concerned.

13.2   TOLERANCE TO TOXIC METALS

Some metals, such as Cu, Fe, and Zn, are oligo-nutrients and consequently are essential in small quantities for most organisms. However, even these essential elements can be toxic when they are present at very high availabilities. Controlling metal homeostasis thus appears to be a necessity for all organisms in order to adjust to changing levels of bioavailable trace metals in their medium. On the other hand, elements such as Cd, Hg, and Pb are generally considered potentially toxic at all bioavailabilties, whereas the essentiality of others (As, Ni) is still a topic of discussion.

Metal tolerance is particularly well documented in plants (see Chapter 15), but many examples of tolerance in vertebrates and invertebrates have been reported (see Tables 2.1 and 13.2). Mechanisms of tolerance to metals have been reviewed in aquatic organisms (Mason and Jenkins 1995) and in plants (Clemens 2006). In Chapter 6 it was concluded that genetically adapted metal tolerance in invertebrates often involves increased rates of storage detoxification of metals. Many different strategies of metal detoxification exist between and within different taxa, governing the total metal concentrations accumulated, the importance of sequestration of metals in detoxified pools, and the nature of metal binding with different ligands. Both the total amount of metals accumulated in a food species and their physicochemical forms of storage have the potential to affect their subsequent transfer to predators as well as to human consumers.

Aquatic food chains typically do not show biomagnification of trace metals. The accumulated concentration of a metal in an organism does not depend on the trophic level occupied, but on the metal accumulation mechanism shown by an organism irrespective of its position in a food chain (Wang 2002; Luoma and Rainbow 2008). Thus, all different combinations of the distributions of relative concentrations between trophic levels are possible according to the organisms making up that food chain. The highest or lowest accumulated metal concentration may, therefore, be found at the bottom, top, or middle of a food chain. There are necessarily some examples of biomagnification, such as in the cases of radiocesium (^{137}Cs) or of mercury as methyl mercury, arguably for the latter as a result of its organic lipophilicity. For aquatic invertebrates and fish, the importance of metal trophic transfer is still a topic of discussion (Fisher and Reinfelder 1995; Wang 2002), whereas prey are the main source of metals for marine mammals, as they are for terrestrial consumers.

13.2.1 TERRESTRIAL PLANTS

To avoid metal toxicity and fulfil their physiological needs, all plants have evolved basic tolerance mechanisms. Two categories have been described in the literature: nonaccumulating plants, which—even when growing on contaminated soils—never reach very high metal concentrations in their tissues, and hyperaccumulators, which do attain high bioaccumulated metal concentrations (Table 13.1). Among the latter, almost all taxa are metal tolerant and endemic to metal-rich soils. In certain areas, metal concentrations in soils may be orders of magnitude higher than background levels. Metal tolerance and hyperaccumulation are thought to have evolved through adaptations of metal homeostasis processes, including metal uptake, chelation, trafficking, and storage (Courbot et al. 2007 and literature quoted therein).

While tolerance and the potential of hyperaccumulators for phytoremediation of contaminated soils have been documented for a long time (Baker and Brooks 1989; Chaney et al. 1997), progress into the biochemical and molecular understanding of plant tolerance has occurred mainly in the last decade. The review by Clemens (2006) was mainly based on cadmium data, as cadmium is the most widely studied of the nonessential metals. Briefly, transition metal ions such as Cd are bound in the cell by low molecular weight (LMW) complexes, mainly phytochelatin (PC). The second step in the physiological processing of Cd consists of the transport of PC-Cd complexes into vacuoles where high molecular weight (HMW) complexes are formed. It is generally assumed that the major sites of metal sequestration are the vacuoles of the root cells. Complementary to this PC pathway, one of the most important detoxification mechanisms is the sequestration of metals in specific subcellular compartments. For instance, Mazen (2004) has shown Ca oxalate-crystal formation sequestering Al in the nalta jute *Corchorus olitorius*, whereas Choi et al. (2001) and Sarret et al. (2006) described the production of Ca/Cd and Ca/Zn

---

TABLE 13.1

Metal Concentrations (µg.g⁻¹ dry weight) in Soils and Different Categories of Plants

	Zn	Pb	Cd
French standard for contaminated soil	>300	>100	>2
Nonaccumulating plants on uncontaminated soil	<100	<10	<1
Nonaccumulating plants on contaminated soil	<1,000	<100	<10
Threshold of hyperaccumulation	>10,000	>1,000	>100

Source: After Bert, V., et al., *New Phytol.*, 155, 47–57, 2002.

containing grains in the trichomes (leaf hairs) of tobacco, leading to a substantial elimination of these metals.

Accumulation of metals in tolerant edible plants is considered the main source of metals in food for humans and animals. Because all plants possess a minimal tolerance to some metals/metalloids (Cd, Co, Mo, Se), their concentrations in food plants can reach a level high enough to pose a potential toxic risk for human health while no phytotoxicity occurs. On the other hand, elements like Cu, Mn, Ni, or Zn are phytotoxic at concentrations in the plant that do not pose toxic risks for consumers (Clemens 2006 and literature quoted therein).

Itai-itai disease is a dramatic illustration of health problems induced by the consumption of contaminated food plants. Itai-itai disease is considered to be a type of acquired Fanconi syndrome characterised by renal tubular dysfunction and osteomalacia. Itai-itai disease was found among middle-aged and elderly women after relatively frequent pregnancies, living in the Jinzu River basin (Japan), a region where Cd analyses revealed a severe pollution of soils in the paddy fields. In the most heavily polluted zones, the prevalence of itai-itai disease in the women over fifty years of age was $\geq 20\%$ (Kanazawa Medical University, http://www.kanazawa-med. ac.jp/~pubhealt/cadmium2/itaiitai-e/itai01.html).

13.2.2 PROKARYOTIC MICROORGANISMS

As testified by the review of Silver and Phung (1996), bacterial resistance to a large range of toxic metal ions, including Ag^+, AsO_2^-, AsO_4^{3-}, Cd^{2+}, Co^{2+}, CrO_4^{2-}, Cu^{2+}, Hg^{2+}, Ni^{2+}, Pb^{2+}, Sb^{3+}, TeO_3^{2-}, Tl^+, and Zn^{2+}, is well established. More recent studies have allowed a better understanding of the molecular genetics and environmental roles of these resistances (see Chapter 14). The largest group of metal resistance systems functions by energy-dependent efflux of toxic ions. For instance, the *cadA* gene codes for a type P ATPase pump responsible for Cd^{2+} efflux. Bacterial resistance to mercury is one of the numerous examples of the genetic and physiological adaptation of microbial communities exposed to high available concentrations of contaminants in their environment. The mechanism of resistance typically involves the *merA* gene, which codes for an enzyme ensuring the reduction of Hg^{2+} to Hg^0, which is non-bioavailable. In such cases, it seems that the risk of transfer between microbes and their consumers is probably limited. The situation is not so clear when the defence mechanism is based on bacterial metallothionein, sequestering and detoxifying ions as demonstrated for Zn in many bacteria (Blindauer et al. 2002). In addition, chelation and surface adsorption constitute an important response of microorganisms in metal-polluted environments, thus leading to significant metal accumulation.

Thus, it is not surprising that reports exist on the role of bacteria in passing metals up food chains. Patrick and Loutit (1976) have demonstrated the passage of six metals (Cr, Cu, Mn, Fe, Pb, and Zn) from bacteria to tubificid worms. The role of cadmium-resistant or sensitive marine bacteria in the accumulation of this metal by mussels *Mytilus edulis* has been investigated (Flatau and Gauthier 1983). Cadmium accumulation after ingestion of bacteria previously contaminated by cadmium was greater when the metal was bound to the resistant strain. A bacterial copper-resistant strain identified as *Vibrio* sp. (the lowest concentration of Cu^{2+} required for its complete inhibition was 50 μg.ml^{-1}) was found to accumulate copper (Miranda and Rojas 2006). When scallop *Argopecten purpuratus* larvae were exposed to Cu-enriched *Vibrio*, they accumulated high levels of Cu.

13.2.3 PROTISTANS INCLUDING PHYTOPLANKTON

Some species and ecotypes of unicellular eukaryotic algae can live in the presence of toxic metal concentrations that are lethal for other species or populations (Table 13.2). With chronic exposure to sublethal concentrations, phytoplankters that exhibit metal tolerance are then potentially toxic for grazing species at higher trophic levels. After extended periods of culture in sublethal concentrations of Cd and Cu, estuarine unicellular algae developed tolerances to metal concentrations that were inhibitory upon initial exposure (Wikfors and Ukeles 1982). When such tolerant strains of *Isochrysis galbana* grown in a medium with 15.3 μg.ml^{-1} Cd or 47.3 μg.ml^{-1} Cu were fed to veliger larvae of the oyster *Crassostrea virginica*, poor growth and high mortalities were observed in grazing larvae.

As in higher plants, both micro- and macroalgae have developed detoxification mechanisms to avoid metal poisoning. Metal uptake may be controlled by reducing the bioavailability of the metal in the surrounding environment through the production of extracellular metal-complexing ligands, which is well documented, particularly for Cu, in phytoplankton species, marine and estuarine fungi, and marine cyanobacteria (Mason and Jenkins 1995). This process ensures the protection not only of microorganisms but also of other organisms living in the same area. Other detoxificatory physiological processes are not so favourable to consumers, such as the complexation of metals at the surface of the organisms (Mason and Jenkins 1995; Ng et al. 2005) or their intracellular sequestration by means of phytochelatins (Perales-Vela et al. 2006; Pawlik-Skowronska et al. 2007). In the green alga *Chlamydomonas reinhardtii*, up to 70% of the total Cd found in extracts of Cd-treated cells are sequestered in such peptide-metal complexes (Howe and Merchant 1992).

Metal accumulation by photosynthetic protistans is at least partly governed by ambient P concentrations, probably as a consequence of increasing metal binding capacity associated with formation of polyphosphate bodies. It has been proposed that intracellular polyphosphates are important in sequestering metals and serve as protection from metal toxicity by binding incoming metals in a detoxified form (Wang and Dei 2006 and literature quoted therein). In the phytoplanktonic alga *Tetraselmis suecica*, the relationship between storage of metals (Ag, Cd, Cu) in polyphosphate bodies and toxicity has been investigated in detail (Ballan-Dufrançais et al. 1991; Nassiri et al. 1996). Among cells exposed to Cu, 40% became severely disorganised ultrastructurally compared to controls (Figure 13.2), even if starch was still abundant and osmiophilic vesicles were identical to those of controls. On the other hand, 60% of Cu-exposed cells displayed well-preserved structures such as mitochondria, dictyosomes, and flagellar apparatus. In these cells, starch was strongly depleted and the aspect of osmiophilic vesicles was clearly modified with a dilated membrane around a circular electron dense core, with a peripheric light space (Figure 13.2). Metals—Cu as well as others present at background level in culture medium—were stored in these osmiophilic P-rich vesicles, which correspond

TABLE 13.2
Metal Tolerance in Organisms Chronically Exposed to Metal Pollution in the Field or Preexposed in the Laboratory

Taxon	Species	Element	Reference
Ciliate	*Uronema nigricans*[1]	Hg	Berk et al. 1978
Ciliate	Soil ciliate species[+]	Cd, Cu, Zn	Diaz et al. 2006
Microalgae	Many different species[1]	Cd, Cu, Zn	Takamura et al. 1989
Microalga	*Scenedesmus acutus*[1]	Cr, Cd, Cu	Twiss et al. 1993; Corradi et al. 1995; Torricelli et al. 2004; Gorbi et al. 2006
Microalga	*Scenedesmus* sp.[1]	Hg	Capolino et al. 1997
Microalga	*Chlorella* sp.	Cd	Kaplan et al. 1995
Microalga	*Gomphonema parvulum*[1]	Zn	Ivorra et al. 2002
Microalga	*Pseudokirchneriella subcapitata*[1]	Cu	Bossuyt and Janssen 2004a
Microalga	*Amphidinium caterii*[2]	Fluorure	Antia and Klut 1981, in Cosper et al. 1987
Macroalga	*Stigeoclonium tenue*[1]	Zn	Pawlik Skowrońska 2003
Macroalgae	*Ectocarpus silicosus*[2] *Fucus vesiculosus*[2]	Cu	Review by Bryan 1984
Nematodes	Estuarine communities[2]	Cu	Millward and Grant 1995
Annelid	*Limnodrilus hoffmeisteri*[1]	Cd, Ni	Klerks and Levinton 1989
Annelid	*Sparganophilus pearsei*	Hg	Vidal and Horne 2003
Annelid	*Eisenia fetida*[+]	Cd	Reinecke et al. 1999
Annelid	*Dendrodrilus rubidus*[+]	Cu	Arnold et al. 2008
Annelid	*Nereis diversicolor*[2]	Cd, Cu, Zn	Ait Alla et al. 2006 and literature quoted therein; Burlinson and Lawrence 2007
Bivalve	*Macoma balthica*[2]	Cu	Luoma et al. 1983
Bivalve	*Scrobicularia plana*[2]	Zn	Amiard 1991
Bivalve	*Ostrea edulis*[2]	Cu, Zn	Bryan et al. 1987
Bivalve	*Crassostrea gigas* (larvae)[2]	Cu	Damiens et al. 2006
Bivalve	*Mytilus edulis*[2]	Hg	Roesijadi et al. 1982
Bivalve	*Mytilus edulis* (embryos)[2]	Cu	Hoare et al. 1995
Crustacean	*Daphnia magna*[1]	Cd, Cu, Hg, Ni, Zn	Bossuyt and Janssen 2004b and literature quoted therein; Tsui and Wang 2005
Crustacean	*Acartia clausi*[2]	Cd, Cu	Moraitou-Apostolopoulou et al. 1979; Luoma et al. 1983
Crustacean	*Tisbe holothuriae*[2]	Cd	Review by Bryan 1984
Crustacean	*Artemia salina*[2]	Cu	Review by Bryan 1984
Crustacean	*Gammarus duebeni*[2]	Zn	Jones and Johnson 1992
Crustacean	*Gammarus pulex*[1]	Cd, Zn	Naylor et al. 1990; Stuhlbacher and Maltby 1992
Crustacean	*Asellus aquaticus*[1]	Zn	Naylor et al. 1990
Crustacean	*Asellus meridianus*[1]	Cu, Pb	Review by Bryan 1984

(continued)

TABLE 13.2

Metal Tolerance in Organisms Chronically Exposed to Metal Pollution in the Field or Preexposed in the Laboratory (Continued)

Taxon	Species	Element	Reference
Crustacean	*Platynympha longicaudata*[2]	Cd, Cu, Mn, Pb, Zn	Ross et al. 2002
Crustacean	*Palaemonetes pugio*[2]	Hg	Kraus et al. 1988
Crustacean	*Carcinus maenas*[2]	Zn	Review by Bryan 1984
Crustacean	*Eriocheir sinensis*[1]	Cd	Silvestre et al. 2006 and literature quoted therein
Insect	*Chironomus tentans* (larvae)[1]	Mixture (Cd, Cr, Zn)	Wentsel et al. 1978
		Cd	Postma and Davids 1995
Insect	*Chironomus riparius* (larvae)[1]	Zn	Miller and Hendricks 1996
Insects	*Hydropsyche* spp.[1] *Baetis* spp.[1]	Cu	Cain et al. 2004
Fish	*Fundulus heteroclitus*[+]	Methyl mercury	Burnett et al. 2007
Fish	*Heterandria formosa*[1]	Cd	Xie and Klerks 2004
Fish	*Catostomus commersoni*[1]	Cd, Cu	Duncan and Klaverkamp 1983; Munkittrick and Dixon 1988
Fish	*Salmo gairdneri*[1]	Zn	Bradley et al. 1985
Fish	*Oncorhynchus mykiss*[1]	Cd	Hollis et al. 1999, in McGeer et al. 2000; Chowdhury et al. 2004
		Cu	Taylor et al. 2000, in McGeer et al. 2000
		Zn	Aslop et al. 1999, in McGeer et al. 2000

Note: [+], terrestrial; [1], freshwater; [2], estuarine and marine.

to polyphosphate bodies (Ballan-Dufrançais et al. 1991). The same bioaccumulation pattern was described in unicellular algae exposed to Ag and Cd. Osmiophilic vesicles can be excreted during cell division, thus contributing to metal detoxification processes (Nassiri et al. 1996).

The total amount of metals accumulated by phytoplankton will affect the quantity of trace metals transferred trophically to their consumers, but the physico-chemical form of metals in the unicellular algae also has the potential to influence this transfer. Intuitively, cytosolic metals, associated with ligands easily degradable during digestion at acid pH and in the presence of digestive enzymes, appear to be the best candidates for trophic transfer. In agreement with this hypothesis, Reinfelder and Fisher (1991) demonstrated a linear 1:1 relationship between the metal assimilated by marine copepods from a diet of diatoms and the metal partitioned in the cytoplasm of the ingested diatoms. Similarly, oysters *Crassostrea gigas* fed with contaminated diatoms *Haslea ostrearia* retained 93% of Cu, corresponding to the potentially available fraction (cytosolic fraction + exchangeable

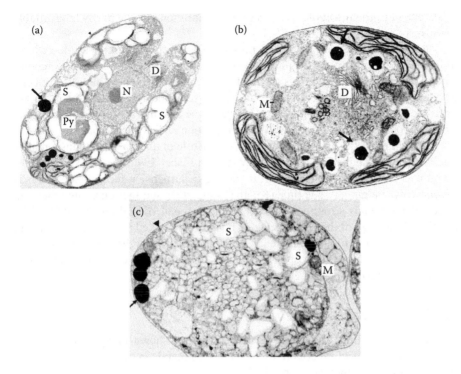

FIGURE 13.2 Distribution of and tolerance to Cu in the microalga *Tetraselmis suecica*: (a) control, (b and c) exposed to 50 µg Cu L^{-1} (b, tolerant cell; c, nontolerant cell). N = nucleus, M = mitochondria, S = starch, D = dictyosome, Py = pyrenoid. Arrows: osmiophilic vesicles. (After Amiard-Triquet, C., C. Cossu-Leguille, and C. Mouneyrac, in *Les biomarqueurs dans l'évaluation de l'état écologique des milieux aquatiques*, ed. J. C. Amiard and C. Amiard-Triquet, Lavoisier, Paris, 2008, pp. 55–94. With permission.)

Cu loosely linked onto the cell wall) in the microalgae (90%). However, only half (21%) of this potentially available Cu in the prasinophycean *T. suecica* (42%) was readily assimilated in oysters (Amiard-Triquet et al. 2006). Great variation of Cd, Se, and Zn assimilation efficiency (AE) has been observed in oysters feeding on different types of particles. In the estuarine oyster *Crassostrea rivularis*, Cd and Se from *Tetraselmis levis* were assimilated at a much lower efficiency than Cd and Se from the diatom *Thalassiosira pseudonana*, but no consistent differences related to the food source were observed for Zn. For all three metals, the nature of the food did not influence AE in the coastal oyster *Saccostrea glomerata* (Ke and Wang 2001). In a comprehensive study, Ng et al. (2005) attempted to examine the relationship between the AE of Ag, Cd, and Zn in three marine filter feeders (the green mussel *Perna viridis*, the clam *Ruditapes philippinarum*, and the barnacle *Balanus amphitrite*) and metal fractionation in seven phytoplankton species. There was no support, however, for a generalised conclusion that any of the labile, cytosolic, or insoluble fractions isolated represents the sole form of phytoplankton-accumulated metal that is bioavailable for trophic transfer to an herbivore. Even trace metals bound to the insoluble fraction in phytoplankton may be partly bioavailable to herbivores.

Bioavailability determined from metal speciation in phytoplankton as food allows a relevant prediction of the trophic transfer in some cases, but caution is recommended in generalising from this mode of assessment, since many other factors can intervene. For example, the influence of the nature of the cell walls—the rigid cell wall of a crystalline nature of green unicellular algae contains sporopollenin, a highly refractory compound resistant to enzymatic digestion and physical breakdown, and is very different from the undigestible siliceous frustule of diatoms—remains a matter of debate (Fisher and Reinfelder 1995; Wang and Fisher 1996; Ke and Wang 2001). In a number of species, the AE of certain metals in suspension-feeding herbivores is influenced by the gut passage time (Ke and Wang 2001 and literature cited therein), a factor that depends strongly on the available suspended food (Bayne 1998). The digestive physiology of the consumer may be also be invoked as having a significant effect on relative assimilation efficiencies as demonstrated by the interspecies differences observed in the AE of suspension feeders under otherwise identical conditions (Ke and Wang 2001; Ng et al. 2005).

13.2.4 Invertebrates

A review by Posthuma and Van Straalen (1993) has shown many cases of metal adaptation in terrestrial invertebrates. Tolerance to metals in other invertebrate species is also well documented (Table 13.2).

In invertebrates, many different systems contribute to metal detoxification. Limitation of uptake has been cited as a strategy in preventing metal toxicity in aquatic organisms provided with carapaces, integument, tests, shells, and scales, even if contamination from the food sources and physiological changes associated with molting limit the efficiency of impermeability as a mechanism of defence (Mason and Jenkins 1995). Mucus secretion—a normal mechanism involved in different aspects of invertebrate biology—is enhanced in many different cases of invertebrates under stress, including exposure to increased doses of metals. In the leech *Nephelopsis obscura*, exposure to Cd (but also chlorine, salinity, or hyperoxia) increases mucus production (Wicklum and Davies 1996 and literature quoted therein). In the common ragworm *Nereis diversicolor*, a field population adapted to a metal-rich area produced more mucus when exposed to metals than specimens from a clean site (Figure 13.3) (Mouneyrac et al. 2003). In the neogastropod *Nucella lapillus*, mucus production was observed in specimens exposed to Cd (Leung et al. 2000). In mussels (*Mytilus edulis*, *Perna viridis*, *Septifer virgatus*), mucus secretion is a significant response to heightened trace metal exposure (Hietanen et al. 1988; Sze and Lee 1995). In *P. viridis* exposed to acute Cu exposure, the metal content of the mucus was about eighteen times that in the control and six times that in the soft tissues, suggesting that mucus may be an effective agent for Cu depuration (Sze and Lee 1995). However, according to Mason and Jenkins (1995), the story is not so simple since the species that produce mucus can then engulf the mucus in feeding or by processes of endo- or pinocytosis. In addition, mucus secreted by one organism can be consumed by another.

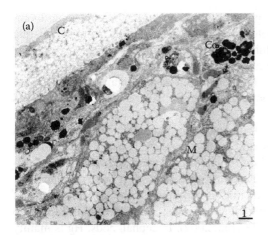 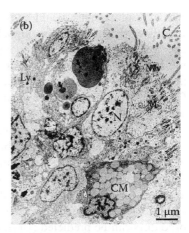

FIGURE 13.3 Production of mucus in invertebrates exposed to environmental contaminants. (a) Ultrastructural aspect of mucus cells in the epidermis of the polychaete *Nereis diversicolor* exposed to zinc: M, mucus; C, cuticle; Co, mineral concretions (for more details, see Mouneyrac et al. 2003). (b) Gill epithelium of oyster *Crassostrea gigas* exposed to bromadiolone: CM, cell full of mucus grains; Ly, lysosome; N, nucleus; M, mitochondria; Mv, microvilli. (After Jeantet, A.-Y., C. Ballan-Dufrançais, and A. Anglo, in *Biomarqueurs en écotoxicologie. Aspects fondamentaux*, ed. L. Lagadic et al., Masson, Paris, 1997, pp. 315–53. With permission.)

Another way to keep the body burden low is increased elimination. Invertebrates have developed a variety of routes and modes for eliminating metals (Mason and Jenkins 1995). In some cases, the process is highly specific, as exemplified in tolerant oligochaetes *Sparganophilus pearsei* exposed to mercury-contaminated sediments (Vidal and Horne 2003). These organisms accumulated Hg in their caudal segments, which were then jettisoned via the process of autotomy. This process was interpreted as a contribution to detoxification but may represent a novel exposure route for other organisms via feeding on discarded tails.

However, in general, mechanisms that limit uptake or enhance elimination are not as important as metal storage under nontoxic physicochemical forms, such as metal binding with metallothioneins (MTs) or biomineralisation in very stable compounds (see Chapter 6). Both of these strategies were recently reviewed in aquatic invertebrates (Marigomez et al. 2002; Amiard et al. 2006). Metallothioneins are ubiquitous in the plant (as phytochelatins or class III MTs) and animal kingdoms. MTs were found first in equine renal cortex, but a number of invertebrate species (mussels, oysters, crabs, lobsters, etc.) have class I MTs, which are very similar to mammalian MTs. Metallothioneins are cytosolic nonenzymatic proteins with a high cysteine content, the thiol groups (–SH) of which enable them to bind a number of metals, essential (Cu, Zn) and nonessential (Ag, Cd, Hg). Depending on the species and the level of exposure, the pool of metal bound to MTs can represent the major store of metals in a given species. As cytosolic proteins, it is highly probable that they are degraded in the conditions prevailing in the digestive tract of any predator, including a low pH and the presence of digestive enzymes. Thus, it has been hypothesised that cytosolic metals would be more prone to trophic transfer than metals in biomineralised structures. In the oligochaete worm *Limnodrilus hoffmeisteri*, nontolerant specimens store Cd mainly under soluble form, whereas tolerant specimens store Cd mainly as detoxification granules. In agreement with the previous hypothesis, decapod crustaceans *Palaemonetes pugio* fed nontolerant worms retained about 75% of Cd present in the prey, whereas, in contrast, shrimps fed tolerant specimens retained only 21% (Wallace et al. 1998).

Cephalopods constitute a primary food source for many marine predators, such as fish, marine mammals, or seabirds, and are also extensively fished and consumed by humans. The high levels of cadmium reported for cephalopod species from different areas of the world suggest that these organisms have developed efficient storage detoxification processes for this metal. As cadmium in the digestive gland of cephalopods is mainly associated with soluble ligands, a high potential transfer to predators can be predicted (Bustamante et al. 2002). In cephalopods, Hg is mainly present in organic form, highly bioavailable, so these species should be considered a significant source of Hg for consumers and predators (Bustamante et al. 2006a). On the other hand, Ag is held mainly in insoluble form in the digestive gland of the cuttlefish *Sepia officinalis* (Bustamante et al. 2006b), as it is in many bivalves, being stored as silver sulphide (Ag_2S), a very stable compound, less prone to trophic transfer to predators/consumers (Berthet et al. 1992).

Certain insoluble forms of storage are indeed so stable that they pass through the gut of the predator without apparent change in elemental composition. Phosphate granules in prey species (the scallop *Chlamys opercularis*, the gastropod *Littorina littorea*, the barnacle *Semibalanus* (as *Balanus*) *balanoides*)) appear to be unavailable to carnivores (the neogastropods *Nassarius reticulatus* and *Nucella lapillus*) since granules with their constituent metals were recovered in faecal pellets apparently unchanged according to electron microscopy and x-ray microanalysis techniques (Nott and Nicolaidou 1990). The same authors used the digestive gland of marine gastropods as a food source for hermit crabs as predators. Significant proportions of detoxified Mn, Ni, Cu, Zn, and Ag (bound electrostatically to phosphate or covalently to sulphur within membrane-bound intracellular compartments) were again shown to be unavailable to the predator. On the contrary, Cd, which was bound to cytosolic sulphur-rich proteins in the digestive glands of the gastropods, was transferred to the hermit crabs (Nott and Nicolaidou 1994). Rainbow et al. (2004) examined the trophic transfer of metals between tolerant worms *Nereis diversicolor* from a metal-rich area and carnivorous worms *Nereis virens*. *N. virens* showed net accumulation of Cd, Cu, Pb, and Zn from the prey, accumulation by the predator increasing with increasing prey concentration. In the prey species, 68% of accumulated Cu was bound in metal-rich granules and was not trophically available. The remainder, either in soluble form or bound to other cellular material, was trophically available and responsible for high body Cu concentrations in the predator. From a study of trophic transfer of Cd and Zn between bivalves (*Macoma balthica* and *Potamocorbula amurensis*) and a decapod crustacean (*Palaemon macrodactylus*), Wallace and Luoma (2003) defined a trophically available metal (TAM) fraction as including metal bound to total proteins in the soluble fraction (MT-like protein (MTLP) and other constitutive proteins) plus trace metals bound to cell organelles.

TABLE 13.3

Assimilation Efficiencies (AE, mean percentage ± SD) of Metals of the Neogastropod *Nassarius festivus* Fed Gelatin-Packeted Isolated Subcellular Fractions of Accumulated Metal in the Digestive Glands of Bivalves Radiolabelled with ^{110m}Ag, ^{109}Cd, and ^{65}Zn from Water (W) or Phytoplankton (P)

	Ag	Cd	Zn
Scallop *Chlamys nobilis*			
MTLP (W)	80 ± 7	97 ± 9	89 ± 11
MTLP (P)		81 ± 13	82 ± 9
Organelles (W)	82 ± 6	78 ± 4	
Organelles (P)		84 ± 6	
MRG (W)	58 ± 14		
Oyster *Saccostrea cucullata*			
MRG (W)	74 ± 6	75 ± 14	

Source: After Rainbow, P. S., et al., *Mar. Ecol. Prog. Ser.*, 348, 125–38, 2007.

Note: MTLP: metallothionein-like proteins; MRG: metal-rich granules.

The subcellular compartmentalisation proposed by these authors was tested considering other food chains, including a decapod crustacean (*Palaemonetes varians*) fed tolerant and nontolerant worms *N. diversicolor* (Rainbow et al. 2006a) or a carnivorous neogastropod (*Nassarius festivus*) fed selected tissues of four bivalve prey species (Rainbow et al. 2007). Depending on the elements (Ag, Cd, Zn) and the species, assimilation occurs from metal bound in subcellular fractions right across the insoluble and soluble spectrum. The more striking result is shown in Table 13.3. Assimilation efficiencies of Cd, Ag, and Zn were determined in specimens of *N. festivus* fed subcellular fractions of metals accumulated in the digestive gland of their prey. As expected, for all of the three metals, AE from the MTLP fraction was high (≥80%), as was assimilation from organelles (≥78%), but surprisingly, AE was significant in *N. festivus* receiving MRG (metal-rich granules), reaching 58 to 74% for Ag and 75% for Cd. The neogastropod *N. festivus* feeds via an eversible proboscis with very strong external digestive abilities, and these atypically strong digestive powers may make strongly bound metals in MRG assimilable by this predator, while they are not trophically available to most invertebrate predators (Rainbow et al. 2007). The same ability of a neogastropod predator (in this case *Thais clavigera*) to assimilate ingested trace metals bound in MRG had been shown by Cheung and Wang (2005) feeding on a range of invertebrate prey, including zinc in barnacle prey. The high concentrations of zinc accumulated by barnacles are detoxified in the form of zinc pyrophosphate granules (Rainbow 2002; Luoma and Rainbow 2008), and so it appears that even this zinc can be trophically available to the right predator. The neogastropods as a taxon appear to have atypically strong digestive powers of metals accumulated in prey. The concept of metal detoxification along the food chain proposed by Nott and Nicolaidou (1990) is, therefore, only partly verified and cannot be considered to be of universal applicability.

Thus, differences in trophic availabilities of metals affected by the chemical form of the metal accumulated in the prey are further affected by the relative strength of the digestion processes of the predator, rendering impossible the generalised definition of any TAM (Rainbow et al. 2007).

13.2.5 VERTEBRATES

In vertebrates, metal tolerance has been established in a number of species among fish (Table 13.2). In mammals, very high levels of toxic metals have been determined in some species, such as Cd and Hg in marine mammals. In a significant fraction of the pilot whale *Globicephala melas* population sampled in the vicinity of the Faroe Islands, blood and urine Cd concentrations are higher than minimum adverse-effect levels established for human beings. However, the frequency-at-age data have not shown anomalies likely to reveal a major toxic problem, suggesting a remarkable tolerance of this species to trace metals (Caurant et al. 1994; Caurant and Amiard-Triquet 1995).

As in invertebrates, complexation of metals at the surface of organisms has been proposed as a mechanism of defence. In the rainbow trout *Oncorhynchus mykiss*, a substantial hypertrophy of gill mucus cells coincided with the first appearance of increased resistance (Wilson et al. 1994). In the brook trout *Salvelinus fontinalis*, changes in either mucus cell production or secretion or changes in mucus chemistry contribute, in part, to acclimation to Al (Mueller et al. 1991). In several fish species, protection appears to be conferred by sequestration of potentially toxic metals with a mucoid matrix in the gut lumen (Mason and Jenkins 1995). Because the sites where mucus is produced and metals are detoxified are usually not consumed—at least by humans—the protective role of mucus for fish species may also be efficient for consumers.

In mammalian species, MT plays a major role in metal detoxification, a fact supported by extensive evidence from both *in vivo* and *in vitro* studies (Miles et al. 2000). In vertebrates in general, metallothioneins are usually the main detoxificatory ligands for toxic metals, allowing survival to metal exposure in their environment, including their food. As mentioned above for invertebrates, the originally defined TAM (total proteins including MTs + organelles) of Wallace and Luoma (2003) may represent an approximate minimum for the trophic availability of metals. If so, this preferential method of metal detoxification in vertebrates can favour trophic transfer of the metal to their consumers. However, as considered above, biomagnification of metals in the food chain is a far from frequent phenomenon.

Two radionuclides constitute exceptions, ^{137}Cs in aquatic food chains as established very early in a review by Amiard and Amiard-Triquet (1980), and both ^{137}Cs and [85]Sr in terrestrial food chains (Ramade 2007). The ecotoxicological risk is higher for elements such as mercury, which have organometallic forms, the biomagnification potential of which was clearly demonstrated following the Minamata disaster

(Harada 1995). This phenomenon leads to increasing Hg concentrations at successively higher trophic levels. Despite Hg having a high affinity for MT, in a number of biological matrices only a minor fraction of this metal is stored in association with MT. Mercury in fish occurs mainly in organic form, as in muscles of marine mammals. On the contrary, in the liver, the proportion of organic mercury is much lower (Caurant et al. 1996). This is likely because methyl mercury is detoxified by a chemical mechanism involving selenium (Law 1996; Seppänen et al. 1998), as suggested by the strong correlation between these two elements in the liver and kidney of marine mammals (Caurant et al. 1994 and literature quoted therein). In Greenland marine animals, a 1:1 molar ratio between Se and Hg was found in tissues of marine mammals with high mercury concentrations (above approximately 10 nmol.g^{-1}). This was most clearly demonstrated for liver and kidney tissues of the polar bear *Ursus maritimus* and for the ringed seal *Pusa hispida* with high mercury concentrations in the liver (Dietz et al. 2000). In marine mammals and cormorants (*Phalacrocorax carbo*), Nigro and Leonzio (1996) have shown that the ratio between Hg and Se levels was close to equimolarity, whereas most of the European porpoises (*Phocoena phocoena*) studied by Lahaye et al. (2007) exhibited a Hg:Se ratio lower than 1:1, indicating an excess of Se compared to Hg, and thus also availability of Se for other metabolic functions (e.g., as a constituent of glutathione peroxidase). A large excess of selenium in relation to mercury was observed in fish and moreover in shellfish (Dietz et al. 2000; Plessi et al. 2001). Tiemannite (HgSe) has been identified as the probable product of the demethylation of mercury by selenium in cetaceans, sea lions, and cormorants, but not in fish (Nigro and Leonzio 1996 and literature quoted therein). As a very stable compound, these crystals of mercury selenide are responsible for the mineralisation of mercury in a presumed nontoxic nonbioavailable form in the trophic web.

In terrestrial food chains, biomagnification of mercury has also been demonstrated, particularly in bird food chains, as documented by Berg et al. 1966 (in Ramade 2007). The depletion of terrestrial raptors in Sweden in the 1950s and 1960s was probably attributable to the biomagnification of mercury in their food chains (Ramade 2007, cited opinion).

Virtually all humans who consume fish and shellfish have at least trace quantities of methyl mercury (MeHg) present in their tissues. At high doses, methyl mercury has been well established as a neurotoxic agent for humans and many species of animals, including nonhuman primates. Three major outbreaks of severe methyl mercury poisoning occurred in Japan (Tsubaki and Irukayama 1977; Harada 1995) during the 1950s and early 1960s (in Minamata, Niigata, and Kumamoto). Within the past decade there have been clinically obvious cases in the Songhua River region of China (Chun et al. 2001). Broad contamination of the Amazon River Basin with inorganic mercury, subsequently methylated and incorporated into fish, has produced severe exposures and neurological symptoms among people routinely consuming fish from these waters. Many reviews describe the nature and extent of this environmental pollution problem (e.g., Malm 1998; Maurice-Bourgoin et al. 2000). However, in the general population it is necessary to consider the benefit-risk balance of the consumption of seafood products as sources of MeHg and omega-3 fatty acids that improve cardiovascular health and provide benefits for *in utero* development (Mahaffey et al. 2008). For instance, salmon followed by shrimp are principal sources of omega-3 fatty acids and are lesser sources of MeHg, in contrast with tuna, which provides omega-3 fatty acids, but considerably higher levels of MeHg.

The intake of mercury and cadmium by human consumption of marine mammals, especially pilot whales, caught off the Faroe Islands and described as highly tolerant (Caurant and Amiard-Triquet 1995), has been a source of concern, and was carefully assessed in countries where marine mammal flesh is a traditional food item (Andersen et al. 1987; Endo et al. 2004). As the level of Cd contamination in the odontocetes caught off the coast of Japan is markedly lower than the levels of total mercury (T-Hg) and MeHg contamination, Cd intoxication due to the consumption of whale products may not be as serious a problem as T-Hg and MeHg intoxication. These data have been used to advise populations about the risk associated with the consumption of the most contaminated marine mammal organs, namely, liver and kidney.

13.3 TOLERANCE TO ORGANIC CONTAMINANTS

13.3.1 Is Tolerance to Organic Contaminants a Usual Phenomenon?

Pesticide resistance is ranked as one of the top environmental problems due to organic contaminants (Chapter 16). The extensive use of pesticides has often resulted in the development and evolution of pesticide resistance in insects, plant pathogens, and weeds, as well as in nontarget species, as exemplified in Table 13.4 (see also Table 2.1). About 520 insect and mite species, a total of nearly 150 plant pathogen species, and about 273 weed species are now resistant to pesticides (Stuart 2003, in Pimentel 2005). It has been clearly shown that a number of microbial strains are tolerant to structurally unrelated antimicrobial agents, such as antibiotics, organic solvents, and biocides (see mini-reviews by Sardessai and Bhosle 2002a, and Fernandes et al. 2003). Field studies have revealed the presence of organic solvent-tolerant bacteria in the mangrove ecosystem (Sardessai and Bhosle 2002b). Tolerance to different classes of organic contaminants has been demonstrated in several taxa, including microorganisms, invertebrates, and vertebrates (Table 13.4). Even if counterexamples exist, tolerance to organic contaminants is clearly a frequent phenomenon, and thus it is necessary to investigate the potential consequences for trophic transfer.

13.3.2 Biological Mechanisms Involved in Tolerance to Organic Compounds

A number of biological mechanisms have been described that can contribute to the acquisition of tolerance to organic contaminants and may interfere with food chain transfer.

Mucus production was observed in the leech *Nephelopsis obscura* exposed to chlorpyrifos, an organophosphate pesticide (Singhal et al. 1989). In oysters exposed to bromadiolone, a rodenticide containing hydroxyl-4-coumarine, hypersecretion of mucus was also observed, associated with an increase of the number of mucus cells compared to controls (Figure 13.3). The dense core of mucus grains contained bromine, whereas in controls, the rare mucus cells were bromine-free (Jeantet et al.

TABLE 13.4
Tolerance in Organisms Chronically Exposed to Organic Chemicals in the Field or Preexposed in the Laboratory

Taxon	Species	Contaminant Class	Molecule	Reference
Microalga	*Asterionella japonica*[2]	PCB		Cosper et al. 1984
Microalga	*Ditylum brightwellii*[2]	PCB		Cosper et al. 1984
Microalgae	*Asterionella glacialis*[2]	PCB		Cosper et al. 1988
	Thalassiosira nordenskioldii[2]	PCB		
Phytoplankton	Microplankton, nanoplankton[2]		Tri-butyl-tin	Petersen and Gustavson 1998
Microalgae	Phytoplankton communities[2]	Biocide in antifouling paint	4,5-Dichloro-2-n-octyl-isothiazoline-3-one	Larsen et al. 2003
Microalgae	Microphytobenthos[2]	Herbicide	Isoproturon	Schmitt-Jansen and Altenburger 2005a
	Periphyton[1]	Herbicides	Atrazine, prometryn, isoproturon	Schmitt-Jansen and Altenburger 2005b
Microalgae	Assemblage of soil microalgae[+]	Herbicide	Atrazine	García-Villada and Reboud 2007
Microalgae	Phytoplankton community[1]	Herbicide	Atrazine	Seguin et al. 2002
Microalga	*Ditysphaerium pulchellum*[1]	Herbicide	Monuron	Bernarz 1981, in Cosper et al. 1987
Microalga	*Chlorella protothecoides*[1]	OP insecticide	Methyl parathion	Saroja-Subbaraj and Bose 1983, in Cosper et al. 1987
Cyanophyceae	*Microcystis aeruginosa*[1]	Pesticide	Dinitrophenol	Genoni et al. 2001
Cyanophyceae	*Anabaena variabilis*[1]		Hydroxylamine	Jain et al. 1967, in Cosper et al. 1988
Nematodes		PAHs		Carman et al. 1995
Annelids	*Nereis virens*[2]	PAHs		Lewis and Galloway 2008
Annelids	*Monopylephorus rubroniveus*[2]	PAHs	Fluoranthene	Weinstein et al. 2003
Copepods, Ostracods			Diesel	Carman et al. 2000
Crustaceans	*Daphnia magna*[1]	OP insecticide	Ethyl parathion	Barata et al. 2001
Crustaceans	*Daphnia magna*[1]	Herbicide	Molinate	Sanchez et al. 2004

(continued)

TABLE 13.4

Tolerance in Organisms Chronically Exposed to Organic Chemicals in the Field or Preexposed in the Laboratory (Continued)

Taxon	Species	Contaminant Class	Molecule	Reference
Crustaceans	*Daphnia magna*[1]	Pesticide	Toxaphene	Kashian 2004
Crustaceans	*Daphnia magna*[1]	Pharmaceuticals	17α-Ethinylestradiol faslodex	Clubbs and Brooks 2007
Crustaceans	*Hyalella curvispina*[1]	OP insecticide	Azinphosmethyl	Anguiano et al. 2008
Fish	Several species of minnows[1]	Residual chlorine		Lotts and Stewart 1995
Fish	*Fundulus heteroclitus*[2]	TCDD, PCBs, PAHs		Burnett et al. 2007
Fish	*Microgadus tomcod*[2]	PAHs	B[a]P	Sorrentino et al. 2004
Fish	*Microgadus tomcod*[2]	PCB, PCDD		Yuan et al. 2006
Amphibians	Toad embryos[1]	Pesticides		Anguiano et al. 2001

Note: [+], terrestrial; [1], freshwater; [2], estuarine and marine.

1997). In fish, the intestinal lumen-villus interface is composed of replicating villus epithelial cells, covered by a mucus surface coat that provides functions ranging from lubrication of food passage to modulation of nutrient uptake. A feeding experiment with catfish (*Ictalurus punctatus*) has shown the association of the polycyclic aromatic hydrocarbon (PAH) benzo[a]pyrene (B[a]P) with the mucus surface coat of the colon in correlation to high levels of B[a]P in villi of those regions (Kleinow et al. 1996). The authors suggest that the surface coat either is acting as a reservoir of B[a]P for the villi or is in some manner restricting clearance of B[a]P from the regional environment of the colon. Thus, instead of being a protective effect of mucus, this prolonged contact could favour B[a]P and metabolite transfer through the intestine barrier.

When a xenobiotic enters the cell, the induction of a multixenobiotic defence mechanism can prevent cellular accumulation in exposed biota thanks to a transmembrane P-glycoprotein (P-gp) acting as an energy-dependent efflux pump (Chapter 10). This was recognised first in certain tumor cells resistant to cytotoxic drugs used in chemotherapy, but P-gp-like proteins have been described in many nonmammalian organisms, including invertebrates. This multixenobiotic resistance (MXR) acts as a protective system against environmental contaminants at the individual level, but by limiting bioaccumulation, it may represent a transfer of detoxification to predator species.

Despite the role of MXR, most xenobiotics are able to accumulate in biota to different degrees, depending on both the properties of the chemical and the characteristics of each species (Abarnou 2009). Norström (2002; Figure 13.4) has described the relationships between the chemical properties of persistant organic pollutants (POPs)

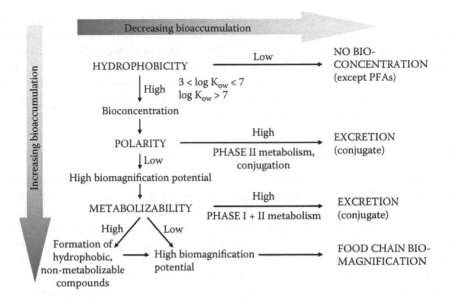

FIGURE 13.4 Chemical properties that strongly influence biomagnification potential. (After Norström, R. J., *Plenary Lecture, Understanding Bioaccumulation of POPs in Food Webs: Chemical, Biological and Environmental Considerations, Dioxin—2002*, Barcelona, Spain, August 12–15, 2002. With permission.)

and the cascade of events leading or not to bioconcentration and biomagnification. The hydrophobicity of a chemical may be assessed by using the octanol-water partition coefficient, K_{ow}. In many cases, the bioconcentration factor (BCF; which is the ratio between the concentration of a chemical in biota and its concentration in ambient water) depends on hydrophobicity, a relationship that may be modelled as follows:

$$Log \; BCF = a \; log \; (K_{ow}) + b$$

Compounds with low K_{ow} are poorly bioconcentrated with the exception of polyfluorinated alkyl acids (PFAs). According to Martin et al. (2003), hydrophobicity is not the sole predictor of bioaccumulation or bioconcentration potential for PFAs, and one must consider the effects of the acid function. Moderately hydrophobic compounds ($3 < K_{ow} < 7$) in an aquatic environment partition onto organic surfaces with subsequent bioconcentration. A number of molecules that have very high K_{ow} ($K_{ow} > 7$), high molecular masses, and large molecular dimensions are prone to be adsorbed onto solid particles, thus limiting bioaccumulation. Compounds that possess polar functional groups may be enzymatically conjugated and excreted, whereas compounds with low polarity have a high biomagnification potential. Once incorporated at higher trophic levels, biotransformation reactions are controlled by structural characteristics of the compounds (size, stereochemistry, coplanarity). Low metabolisation favours food chain biomagnification, whereas high metabolisation can lead to the excretion of an important proportion of the contaminants entering the organism. However, in certain cases, the molecules resulting from biotransformation are hydrophobic, nonmetabolisable compounds, and these have a high bioaccumulation potential. Well-known examples are hydroxylated and methyl-sulphonated metabolites of polychlorinated biphenyls (PCBs) or dichlorodiphenyldichloroethylene (DDE), a compound resulting from dichlorodiphenyltrichloroethane (DDT) metabolisation. Several derivatives of PBDEs (polybrominated diphenyl ethers) fall into this category of bioaccumulated metabolites. For additional details, refer to the reviews by Norström and Letcher (1997) and Abarnou (2009).

13.3.3 TRANSFER OF PAHs ALONG THE FOOD CHAIN

It is probable that PAH compounds do not accumulate along food chains as a result of their rapid biotransformation at higher trophic levels, as illustrated in the food web of a multipolluted area of the Northwest coast of France, the Baie de Seine (Figure 13.5). Similarly, Wan et al. (2007) observed a "trophic dilution of PAHs in a marine food web" from Bohai Bay, North China. PAHs tend to undergo biotransformation processes in vertebrates like fish. As a result of biotransformation, more polar metabolites are formed. For example, hydroxylated metabolites have been reported from fish bile and faeces. It is generally considered that invertebrates are not as efficient as vertebrates in biotransforming organic compounds (see Chapter 8). Because annelid (oligochaete and polychaete) worms are particularly tolerant to PAHs and are important constituents at the bottom of aquatic food webs, they deserve special attention.

The oligochaete *Monopylephorus rubroniveus* has a high toxicological tolerance to the PAH fluoranthene and the ability to bioaccumulate large amounts of sediment-associated fluoranthene (Weinstein et al. 2003). As a consequence of the relatively low efficiency in metabolising and eliminating PAHs, high levels of contaminants bioaccumulated in oligochaetes can be passed on to primary consumers, as demonstrated in the case of the grass shrimp *Palaemonetes pugio* (Filipowicz et al. 2007). However, the efficiency of mechanisms of biotransformation and elimination of PAHs has been recognised in certain invertebrates, such as some marine polychaetes, as reviewed by Jorgensen et al. (2008). In a feeding experiment with two *Capitella* species, the carnivorous worm *Nereis virens* fed the biotransforming species, *Capitella* sp. 1, accumulated significantly more fluorenthene equivalents than worms fed *Capitella* sp. S, which has a very limited biotransformation ability (Palmqvist et al. 2006).

It must be kept in mind that in efficient biotransformers, it is precisely this process that can lead to carcinogenic metabolites responsible for most of biological damage, even in the absence of biomagnification (see Chapter 8). Polychaetes (*Armandia brevis*) exposed for four weeks to field or spiked sediments (PAHs, chlorinated compounds) were then fed to juvenile English sole (*Pleuronectes vetulus*). Fish growth was generally lower in groups fed contaminant-exposed worms than in reference groups. Exposure to contaminated food induced increased expression of CYP1A, but hepatic DNA adducts were observed only in fish exposed to B[a]P-contaminated worms (Rice et al. 2000). Thus, polychaetes may play a key role in the transfer of contaminants and their toxicity from sediment to biota, particularly as food sources for bottom-feeding fish and wading birds. PAHs have been reported in the stomach

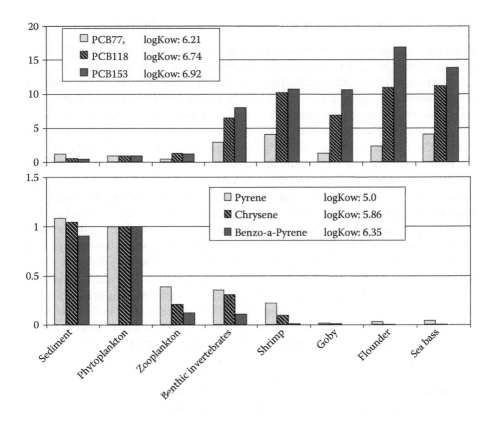

FIGURE 13.5 Apparent biomagnification factors in the sea bass (*Dicentrarchus labrax*) and flounder (*Platichthys flesus*) food webs from the Seine estuary. BF = concentration in organisms/concentration in phytoplankton for a few PCBs and PAHs. The species are benthic invertebrates including molluscs (*Abra alba*, *Ensis ensis*), ophiuroids (*Acrocnida brachiata*), polychaete worms (*Owenia fusiformis*, *Pectinaria koreni*), natantian crustaceans (shrimp: *Crangon crangon*), and fish (goby *Pomatoschistus microps*). (After Abarnou, A., in *Environmental Assessment of Estuarine Ecosystems. A Case Study*, ed. C. Amiard-Triquet and P. S. Rainbow, CRC Press, Boca Raton, FL, 2009, pp. 107–34. With permission.)

contents of bottom-feeding fish, and they have been correlated with a higher tumour incidence in such fishes (reported in Chandler et al. 1997). Similarly, a transfer of toxicity (genotoxicity) was shown in mammals fed mussels contaminated with hydrocarbons released into the field after the oil spill of the tanker *Erika* (Lemière et al., 2005).

In the Baie de Seine, concentrations of the PAH anthracene were determined in different size classes of plankton. An inverse relationship was depicted, the smaller size group—mainly consisting of phytoplankton—having the highest concentrations, whereas the lowest concentrations were determined in the bigger zooplankton. These results support the hypothesis that PAHs are very rapidly biotransformed even by the smallest zooplanktonic species (Abarnou 2009).

13.3.4 Transfer of Persistent Organic Pollutants (POPs)

It seems that biomagnification was observed first by Hunt and Bischoff (1960, quoted by Ramade 2007) studying the effects of the application of the insecticide 1,1-dichloro-2,2-bis(*p*-chlorophenyl)ethane (DDD), very similar in structure to dichlorodiphenyltrichloroethane (DDT), in the food web of Clear Lake in California. Since this date, many other examples have arisen in many parts of the world, and the dangers of DDT have become widely known and appreciated, in large part due to Rachel Carson's famous book *Silent Spring*. An emblematic example is the story of the bald eagle, which was officially declared an endangered species in 1967, under a law that preceded the Endangered Species Act of 1973, in all areas of the United States south of the fortieth parallel. In addition to hunting and habitat loss, DDT biomagnification in the food chain (water → aquatic plants → small animals → fish → bald eagles) interfered with the bald eagle's ability to develop strong shells for its eggs. As a result, bald eagles and many other bird species began laying eggs with shells so thin that they often broke during incubation or otherwise failed to hatch, leading to reproduction disruption. DDT use in the United States peaked in the early 1960s and was prohibited in 1972, playing a major role in the ongoing recovery of the bald eagle. In agroecosystems, until the 1980s, the transfer of DDT in the food chain (soil → grass → cattle (milk, meat) → women → breast milk) led to concentrations in breast milk exceeding the maximum acceptable concentration in human food (0.14 mg DDT kg^{-1} milk lipids), with concentrations as high as 19.8 mg DDT.kg^{-1} milk lipids registered in India (quoted by Ramade 2007).

Whereas environmental concentrations of organochlorine (OC) pesticides have generally declined during the past thirty years, concerns nevertheless remain, because these substances persist in the environment (Van Metre and Mahler 2005) and accumulate in the food chain. There continue to be fish consumption advisories based on unacceptable levels of OC pesticides in sport and commercial fish from the Great Lakes (Kannan et al. 2006). Similarly, although PCB concentrations have decreased consistently in the environment after being prohibited for decades, their detection in soils, aquatic sediment, and biota remains very common (Matthiessen and Law 2002; Van Metre and Mahler 2005; Munschy et al. 2008), leading still to restrictions in human food consumption of local biological products in certain areas, for instance, in both Europe and North America.

PCB153, as an indicator of total PCBs, was analysed in different size classes of plankton in the multipolluted Baie de Seine. Contrary to the ranking observed for PAHs, a positive relationship was observed between accumulated contaminant concentration and zooplankton size, suggesting a poor biotransformation of PCBs in the zooplankton (Abarnou 2009). In the laboratory, crustaceans exposed to PCBs in water and food accumulated more contaminant (319 versus 22 mg.kg^{-1}) when eating the tolerant unicellular alga *Nannochloropsis oculata*, accumulating more PCBs (257 mg.kg^{-1}) than those fed the sensitive unicellular alga *Isochrysis galbana* (64 mg.kg^{-1}) (Wang et al. 1998). Contrary to the decreasing PAH concentrations observed in the Seine food web with increasing trophic level, a clear biomagnification trend was depicted for three PCBs, biomagnification being more and more noticeable with increasing lipophilicity (log K$_{ow}$) of the PCB (Figure 13.5). Brominated flame retardants such as PBDEs share a number of chemical similarities with PCBs and also share their ability to biomagnify in fish food chains (Bragigand et al. 2006).

It is not only the organisms at the bottom of the phylogenetic tree that are unable to eliminate PCBs, for most biological systems metabolise them poorly. This is one of the reasons why PCBs biomagnify along food chains, in addition to their great stability and persistence. From this point of view, the long food chains of the Great Far North represent a crucial model that has been studied in many important recent programs (e.g., Hoekstra et al. 2003; Buckman et al. 2004; Vorkamp et al. 2004; Muir et al. 2006; Bentzen et al. 2008). Human beings as top predators of both aquatic and terrestrial food chains are particularly at risk, and the presence of PCBs, PBDEs, dioxins, and polychlorodibenzofurans in breast milk is clearly a topic of concern.

Because tolerance is so poorly documented in the case of dioxins and polychlorodibenzofurans, their distribution in the food chain will not be discussed further here. As far as we know, no literature exists on tolerance to perfluorooctane sulphonate (PFOS), whereas biomonitoring data give evidence of the significant bioaccumulation potential of PFOS, since high trophic level predators tend to have the highest concentrations (Giesy and Kannan 2001). From a study carried out with selected species from the Barents Sea food web (amphipod *Gammarus wilkitzkii*, cod *Boreogadus saida*, guillemot *Cepphus grylle*, and gull *Larus hyperboreus*), per- and polyfluorinated alkyl substances (PFASs) appear to accumulate through the food web to the same extent as the lipid-soluble POPs (Haukås et al. 2007).

13.4 CONCLUSIONS

The literature provides a number of reports that do not show acclimation or adaptation in organisms chronically exposed to environmental contaminants. Because publication of negative results is usually less easily achieved than publication of positive results, it may be suspected that tolerance is not as general as suggested, for instance, in Tables 13.3 and 13.4. However, tolerance to chemicals, by genetic adaptation or by acclimation as a response to living in contaminated environments, has been frequently observed in organisms belonging to many different taxa exposed to a large range of different contaminants. Thus, it is necessary to evaluate the consequences of this phenomenon.

The present review underlines the fact that the mechanisms that allow organisms to cope with environmental contaminants are generally those that are normally used as responses to natural stress (e.g., production of mucus) or to handle natural compounds involved in normal metabolism, such as the storage and homeostasis of essential metals and the biotransformation of both endogenous and exogenous substrates.

If the metal tolerance mechanism of an invertebrate involves increased storage detoxification, there is a real risk of increased trophic transfer of the metal up a local food chain. This transfer may be ecotoxicologically significant. Thus, decapod crustaceans *Palaemonetes varians* fed on metal-rich Restronguet Creek *Nereis diversicolor* showed significant mortality (Rainbow et al. 2006b). Furthermore, zebrafish *Danio rerio* fed on Restronguet Creek *N. diversicolor* in the laboratory showed reduced reproductive outputs, attributed by the authors to the trophic transfer of arsenic from these worms (Boyle et al. 2008).

From the different studies presented above, it may be concluded that what is trophically available to one predator (feeding on one prey type) is not necessarily trophically available to another (taxonomically separated) predator, even if feeding on the same prey, given the variability between animal digestive systems. TAM, as defined by Wallace and Luoma (2003), may represent an approximate minimum accumulated component for trophically available metals in prey, but is not a fully accurate predictor applicable to a large range of prey and predatory animals. Thus, with a view to risk assessment, it has been necessary to develop alternative methodologies based on *in vitro* digestion with the objective of modelling directly the fraction of metals bioaccumulated in the prey that can be solubilised under the conditions prevailing in the digestive tract of the predator.

In the case of deposit feeders, important quantities of sediment—a major sink for most environmental contaminants—are ingested as food, while predators typically select prey. Several authors have used gut juice extracted from the deposit feeder or predator to predict the bioavailability of metals, radionuclides, PAHs, PCBs, etc., in the food (Weston and Mayer 1998; Shull and Mayer 2002; Weston and Maruya 2002; Zhong and Wang 2006, 2008). Other authors have used solutions mimicking the pH of the digestive juice and consisting of a mixture of enzymes that are known to intervene in the gut juice of different predators (Amiard et al. 1995; Ettajani et al. 1996). Our own species is a consumer of many food products that are contaminant or toxin accumulators, among which are many shellfish species, and the applicability of an *in vitro* model of human digestion has been verified in assessing the bioaccessibility of mycotoxins and metals from food (Versantvoort et al. 2005; Amiard et al. 2008). A similar method has been applied to assess the health risk associated with soilborne contaminants, an important route of exposure in children because of their hand-to-mouth behaviour (Minhas et al. 2006).

The outcome of this review emphasises the necessity to include the aspect of secondary poisoning in hazard assessment schemes for environmental pollutants. The process for assessing secondary poisoning has been described by ECETOC (1995) and is in line with the EU risk assessment procedure as described in the EU Technical Guidance Document (TGD 2003).

As depicted by Norström (2002) (Figure 13.4), the best candidate contaminants for biomagnification are organic contaminants with a log K_{ow} between 3 and 7, as well as low polarity and low metabolisability. Among those priority substances (listed in regulatory bodies such as USEPA, EU directives) discharged into the environment as a result of human activities and considered hazardous to sensitive species of natural resources, many share such physicochemical characteristics. It is particularly dramatic in the case of substances known for, or suspected of, carcinogenicity, mutagenicity, and teratogenicity, as reviewed recently by Eisler (2007). Tolerance is able to protect organisms again direct toxic effects on the short term, but toxicity may be postponed, leading to cancer in the exposed individual itself, as shown in tolerant fish for the North American estuary (Chapter 8), or to deleterious effects on progeny.

REFERENCES

Abarnou, A. 2009. Organic contaminants in coastal and estuarine food webs. In *Environmental assessment of estuarine ecosystems. A case study*, ed. C. Amiard-Triquet and P. S. Rainbow, 107–34. Boca Raton, FL: CRC Press.

Ait Alla, A., et al. 2006. Tolerance and biomarkers as useful tools for assessing environmental quality in the Oued Souss estuary (Bay of Agadir, Morocco). *Comp. Biochem. Physiol.* 143C:23–29.

Amiard, J. C. 1991. Réponses des organismes marins aux pollutions métalliques. In *Réactions des êtres vivants aux changements de l'environnement*, 197–205. Paris: Actes des Journée de l'Environnement du CNRS-PIREN.

Amiard, J. C., and C. Amiard-Triquet. 1980. Le transfert des polluants radioactifs dans les chaines alimentaires aquatiques. *Année Biol.* 19:117–46.

Amiard, J. C., et al. 1995. Bioavailability and toxicity of sediment-bound lead to a filter-feeder bivalve *Crassostrea gigas* (Thunberg). *Biometals* 8:280–89.

Amiard, J. C., et al. 2006. Metallothioneins in aquatic invertebrates: Their role in metal detoxification and their use as biomarkers. *Aquat. Toxicol.* 76:160–202.

Amiard, J. C., et al. 2008. Bioaccessibility of essential and non-essential metals in commercial shellfish from Western Europe and Asia. *Food Chem. Toxicol.* 46:2010–22.

Amiard-Triquet, C., C. Cossu-Leguille, and C. Mouneyrac. 2008. Les biomarqueurs de défense, la tolérance et ses conséquences écologiques. In *Les biomarqueurs dans l'évaluation de l'état écologique des milieux aquatiques*, ed. J. C. Amiard and C. Amiard-Triquet, 55–94. Paris: Lavoisier.

Amiard-Triquet, C., and J. C. Dauvin. 2009. Introduction. In *Environmental assessment of estuarine ecosystems. A case study*, ed. C. Amiard-Triquet and P. S. Rainbow, 1–17. Boca Raton, FL: CRC Press.

Amiard-Triquet, C., et al. 2006. Significance of physico-chemical forms of storage in microalgae in predicting copper transfer to filter-feeders (*Crassostrea gigas*). *Environ. Toxicol.* 21:1–7.

Andersen, A., et al. 1987. Trace elements intake in the Faroe Islands. II. Intake of mercury and other elements by consumption of pilot whales (*Globicephalus meleanus*). *Sci. Total Environ.* 65:63–68.

Anguiano, O. L., A. Caballero de Castro, and A. M. Pechen de D'Angelo. 2001. The role of glutathion conjugation in the regulation of early toad embryos' tolerance to pesticides. *Comp. Biochem. Physiol.* 128C:35–43.

Anguiano, O. L., et al. 2008. Enhanced esterase activity and resistance to azinphosmethyl in target and nontarget organisms. *Environ. Toxicol. Chem.* 27:2117–23.

Arnold, R. E., M. E. Hodson, and C. J. Langdon. 2008. A Cu tolerant population of the earthworm *Dendrodrilus rubidus* (Savigny, 1862) at Coniston Copper Mines, Cumbria, UK. *Environ. Pollut.* 152:713–22.

Baker, A. J. M., and R. R. Brooks. 1989. Terrestrial higher plants which hyperaccumulate metallic elements. A review of their distribution, ecology and phytochemistry. *Biorecovery* 1:81–126.

Ballan-Dufrançais, C., C. Marcaillou, and C. Amiard-Triquet. 1991. Response of the phytoplanktonic alga *Tetraselmis suecica* to copper and silver exposure: Vesicular metal bioaccumulation and lack of starch bodies. *Biol. Cell.* 72:103–12.

Barata, C., et al. 2001. Biochemical factors contributing to response variation among resistant and sensitive clones of *Daphnia magna* Straus exposed to ethyl parathion. *Ecotoxicol. Environ. Saf.* 49:155–63.

Bayne, B. L. 1998. The physiology of suspension feeding by bivalve molluscs: An introduction to the Plymouth "TROPHEE" workshop. *J. Exp. Mar. Biol. Ecol.* 219:1–19.

Bentzen, T. W., et al. 2008. Dietary biomagnification of organochlorine contaminants in Alaskan polar bears. *Can. J. Zool.* 86:177–91.

Berk, S. G., et al. 1978. Effects of ingesting mercury-containing bacteria on mercury tolerance and growth rates of ciliates. *Microb. Ecol.* 4:319–30.

Bert, V., et al. 2002. Do *Arabidopsis halleri* from nonmetallicolous populations accumulate zinc and cadmium more effectively than those from metallicolous populations? *New Phytol.* 155:47–57.

Berthet, B., et al. 1992. Bioaccumulation, toxicity and physico-chemical speciation of silver in bivalve molluscs: Ecotoxicological and health consequences. *Sci. Total Environ.* 125:97–122.

Blindauer, C. A., et al. 2002. Multiple bacteria encode metallothioneins and SmtA-like zinc fingers. *Mol. Microbiol.* 45:1421–32.

Bossuyt, B. T. A., and C. R. Janssen. 2004a. Long-term acclimation of *Pseudokirchneriella subcapitata* (Korshikov) Hindak to different copper concentrations: Changes in tolerance and physiology. *Aquat. Toxicol.* 68:61–74.

Bossuyt, B. T. A., and C. R. Janssen. 2004b. Influence of multigeneration acclimation to copper on tolerance, energy reserves, and homeostasis of *Daphnia magna* Straus. *Environ. Toxicol. Chem.* 23:2029–37.

Boyle, D., et al. 2008. Natural arsenic contaminated diets perturb reproduction in fish. *Environ. Sci. Technol.* 42:5354–60.

Bradley, R. W., C. Duquesnay, and J. B. Sprague. 1985. Acclimation of rainbow trout, *Salmo gairdneri* Richardson, to zinc: Kinetics and mechanism of enhanced tolerance induction. *J. Fish. Biol.* 27:367–79.

Bragigand, V., et al. 2006. Influence of biological and ecological factors on the bioaccumulation of polybrominated diphenyl ethers in aquatic food webs from French estuaries. *Sci. Total Environ.* 368:615–26.

Bryan, G. W. 1984. Pollution due to heavy metals and their compounds. In *Marine ecology*, ed. O. Kinne, 1289–431. New York: John Wiley & Sons Ltd.

Bryan, G. W., et al. 1987. Copper, zinc and organotin as long-term factors governing the distribution of organisms in the Fal Estuary in Southwest England. *Estuaries* 10:208–19.

Buckman, A. H., et al. 2004. Organochlorine contaminants in seven species of Arctic seabirds from northern Baffin Bay. *Environ. Pollut.* 128:327–38.

Burlinson, F. C., and A. J. Lawrence. 2007. Development and validation of a behavioural assay to measure the tolerance of *Hediste diversicolor* to copper. *Environ. Pollut.* 145:274–78.

Burnett, K. G., et al. 2007. *Fundulus* as the premier teleost model in environmental biology: Opportunities for new insights using genomics. *Comp. Biochem. Physiol.* 2D:257–86.

Bustamante, P., et al. 2002. Cadmium detoxification processes in the digestive gland of cephalopods in relation to accumulated cadmium concentrations. *Mar. Environ. Res.* 53:227–41.

Bustamante, P., et al. 2006a. Total and organic Hg concentrations in cephalopods from the North Eastern Atlantic waters: Influence of geographical origin and feeding ecology. *Sci. Total Environ.* 368:585–96.

Bustamante, P., et al. 2006b. Subcellular distribution of Ag, Cd, Co, Cu, Fe, Mn, Pb, and Zn in the digestive gland of the common cuttlefish *Sepia officinalis*. *J. Shellfish. Res.* 25:987–93.

Cain, D. J., S. N. Luoma, and W. G. Wallace. 2004. Linking metal bioaccumulation of aquatic insects to their distribution patterns in a mining-impacted river. *Environ. Toxicol. Chem.* 23:1463–73.

Capolino, E., et al. 1997. Tolerance to mercury chloride in *Scenedesmus* strains. *Biometals* 10:85–94.

Carman, K. R., J. W. Fleeger and S. M. Pomarico. 2000. Does historical exposure to hydrocarbon contamination alter the response of benthic communities to diesel contamination? *Mar. Environ. Res.* 49:255–78.

Carman, K. R., et al. 1995. Experimental investigation of the effects of polynuclear aromatic hydrocarbons on an estuarine sediment food web. *Mar. Environ. Res.* 40:289–318.

Caurant, F., and C. Amiard-Triquet. 1995. Cadmium contamination in pilot whales *Globicephala melas*: Source and potential hazard to the species. *Mar. Pollut. Bull.* 30:207–10.

Caurant, F., M. Navarro, and J. C. Amiard. 1996. Mercury in pilot whales: Possible limits to the detoxification process. *Sci. Total Environ.* 186:95–104.

Caurant, F., et al. 1994. Ecological and biological factors controlling the concentrations of trace elements (As, Cd, Cu, Hg, Se, Zn) in delphinids *Globicephala melas* from the North Atlantic Ocean. *Mar. Ecol. Prog. Ser.* 103:207–19.

Chandler, G. T., M. R. Shipp, and T. L. Donelan. 1997. Bioaccumulation, growth and larval settlement effects of sediment-associated polynuclear aromatic hydrocarbons on the estuarine polychaete, *Streblospio benedicti* (Webster). *J. Exp. Mar. Biol. Ecol.* 213:95–110.

Chaney, R. L., et al. 1997. Phytoremediation of soil metals. *Curr. Opin. Biotechnol.* 8:279–84.

Cheung, M. S., and W. X. Wang. 2005. Influence of subcellular compartmentalization in different prey on the transfer of metals to a predatory gastropod from different prey. *Mar. Ecol. Prog. Ser.* 286:155–66.

Choi, Y. E., et al. 2001. Detoxification of cadmium in tobacco plants: Formation and active excretion of crystals containing cadmium and calcium through trichomes. *Planta* 213:45–50.

Chowdhury, M. J., E. F. Pane, and C. M. Wood. 2004. Physiological effects of dietary cadmium acclimation and waterborne cadmium challenge in rainbow trout: Respiratory, ionoregulatory, and stress parameters. *Comp. Biochem. Physiol.* 139C:163–73.

Chun, F., et al. 2001. Survey on the neuropsychic symptoms and signs of the residents living in the mercury-polluted area along the Songhua River. In *Proceedings of the Sixth International Conference on Mercury as a Global Pollutant*, Minamata, Japan, Abstract HE-70.

Clemens, S. 2006. Toxic metal accumulation, responses to exposure and mechanisms of tolerance in plants. *Biochimie* 88:1707–19.

Clubbs, R. L., and B. W. Brooks. 2007. *Daphnia magna* responses to a vertebrate estrogen receptor agonist and an antagonist: A multigenerational study. *Ecotoxicol. Environ. Saf.* 67:385–98.

Corradi, M. G., et al. 1995. Chromium-induced sexual reproduction gives rise to a Cr-tolerant progeny in *Scenedesmus acutus*. *Ecotoxicol. Environ. Saf.* 32:12–18.

Cosper, E. M., C. F. Wurster, and M. F. Bautista. 1988. PCB-resistant diatoms in the Hudson River estuary. *Estuar. Coast. Shelf Sci.* 26:215–26.

Cosper, E. M., C. F. Wurster, and R. G. Rowland. 1984. PCB resistance within phytoplankton populations in polluted and unpolluted marine environments. *Mar. Environ. Res.* 12:209–23.

Cosper, E. M., et al. 1987. Induced resistance to polychlorinated biphenyls confers cross-resistance and altered environmental fitness in a marine diatom. *Mar. Environ. Res.* 23:207–22.

Courbot, M., et al. 2007. A major quantitative trait locus for cadmium tolerance in *Arabidopsis halleri* colocalizes with *HMA4*, a gene encoding a heavy metal ATPase. *Plant Physiol.* 144:1052–65.

Damiens, G., et al. 2006. Metal bioaccumulation and metallothionein concentrations in larvae of *Crassostrea gigas*. *Environ. Pollut.* 140:492–99.

Díaz, S., A. Martín-González, and J. Carlos Gutiérrez. 2006. Evaluation of heavy metal acute toxicity and bioaccumulation in soil ciliated protozoa. *Environ. Int.* 32:711–17.

Dietz, R., F. Riget, and E. W. Born. 2000. An assessment of selenium to mercury in Greenland marine animals. *Sci. Total Environ.* 245:15–24.

Duncan, D. A. and J. F. Klaverkamp. 1983. Tolerance and resistance to cadmium in white suckers (*Catostomus commersoni*) previously exposed to cadmium, mercury, zinc, or selenium. *Can. J. Fish. Aquat. Sci.* 40:128–38.

ECETOC. 1995. *The role of bioaccumulation in environmental risk assessment: The aquatic environment and related food webs.* Technical Report 67. Brussels: ECETOC.

Eisler, R. 2007. *Eisler's encyclopedia of environmentally hazardous priority chemicals.* Amsterdam: Elsevier.

Endo, T., et al. 2004. Contamination by mercury and cadmium in the cetacean products from Japanese market. *Chemosphere* 54:1653–62.

Ettajani, H., et al. 1996. Fate and effects of soluble or sediment-bound arsenic in oysters (*Crassostrea gigas* Thun). *Arch. Environ. Contam. Toxicol.* 31:38–46.

Fernandes, P., B. S. Ferreira, and J. M. S. Cabral. 2003. Solvent tolerance in bacteria: Role of efflux pumps and cross-resistance with antibiotics. *Int. J. Antimicrob. Agents* 22:211–16.

Filipowicz, A. B., J. E. Weinstein, and D. M. Sanger. 2007. Dietary transfer of fluoranthene from an estuarine oligochaete (*Monopylephorus rubroniveus*) to grass shrimp (*Palaemonetes pugio*): Influence of piperonyl butoxide. *Mar. Environ. Res.* 63:132–45.

Fisher, N. S., and J. R. Reinfelder. 1995. The trophic transfer of metals in marine systems. In *Metal speciation and bioavailability in aquatic systems*, ed. A. Tessier and D. R. Turner, 363–406. Chichester, UK: Wiley.

Flatau, G., and M. J. Gauthier. 1983. Accumulation du cadmium par *Mytilus edulis* en présence de souches bactériennes sensibles ou résistantes à ce métal. *Can. J. Microbiol.* 29:210–17.

García-Villada, L., and X. Reboud. 2007. Induction of atrazine tolerance in a natural soil assemblage of microalgae reared in the laboratory. *Ecotoxicol. Environ. Saf.* 66:102–6.

Genoni, G. P., et al. 2001. Complex dynamics of adaptation in a nonaxenic *Microcystis* culture. 1. Effects of dinitrophenol on population growth. *Ecotoxicol. Environ. Saf.* 48:235–40.

Giesy, J. P., and K. Kannan. 2001. Global distribution of perfluorooctane sulfonate in wildlife. *Environ. Sci. Technol.* 35:1339–42.

Gorbi, G., et al. 2006. Differential responses to Cr(VI)-induced oxidative stress between Cr-tolerant and wild-type strains of *Scenedesmus acutus* (Chlorophyceae). *Aquat. Toxicol.* 79:132–39.

Harada, M. 1995. Minamata disease: Methylmercury poisoning in Japan caused by environmental pollution. *Crit. Rev. Toxicol.* 25:1–24.

Haukås, M., et al. 2007. Bioaccumulation of per- and polyfluorinated alkyl substances (PFAS) in selected species from the Barents Sea food web. *Environ. Pollut.* 148:360–71.

Hietanen, B., I. Sunila, and R. Kristoffersson. 1988. Toxic effects of zinc on the common mussel *Mytilus edulis* L. (Bivalvia) in brackish water. I. Physiological and histopathological studies. *Ann. Zool. Fenn.* 25:341–47.

Hoare, K., A. R. Beaumont, and J. Davenport. 1995. Variation among populations in the resistance of *Mytilus edulis* embryos to copper: Adaptation to pollution? *Mar. Ecol. Prog. Ser.* 120:155–61.

Hoekstra, P. F., et al. 2003. Trophic transfer of persistent organochlorine contaminants (OCs) within an Arctic marine food web from the southern Beaufort-Chukchi Seas. *Environ. Pollut.* 124:509–22.

Howe, G., and S. Merchant. 1992. Heavy metal-activated synthesis of peptides in *Chlamydomonas reinhardtii*. *Plant Physiol.* 98:127–36.

Ivorra, N., et al. 2002. Metal-induced tolerance in the freshwater microbenthic diatom *Gomphonema parvulum*. *Environ. Pollut.* 116:147–57.

Jeantet, A. Y., C. Ballan-Dufrançais, and A. Anglo. 1997. Pollution par les métaux et atteintes cytologiques chez les bivalves marins. In *Biomarqueurs en écotoxicologie. Aspects fondamentaux*, ed. L. Lagadic et al., 315–53. Paris: Masson.

Jones, M. B., and I. Johnson. 1992. Responses of the brackish-water amphipod *Gammarus duebeni* (Crustacea) to saline sewage. *Neth. J. Sea Res.* 30:141–47.

Jorgensen, A., et al. 2008. Biotransformation of polycyclic aromatic hydrocarbons in marine polychaetes. *Mar. Environ. Res.* 65:171–86.

Kannan, K., J. Ridal, and J. Struger. 2006. Pesticides in the Great Lakes. In *Handbook of environmental chemistry*, ed. R. A. Hites, 151–99. Berlin: Springer-Verlag.

Kaplan, D., et al. 1995. Cadmium toxicity and resistance in *Chlorella* sp. *Plant Sci.* 109:129–37.

Kashian, D. R. 2004. Toxaphene detoxification and acclimation in *Daphnia magna*: Do cytochrome P-450 enzymes play a role? *Comp. Biochem. Physiol.* 137C:53–63.

Ke, C., and W. X. Wang. 2001. Bioaccumulation of Cd, Se and Zn in an estuarine oyster (*Crassostrea rivularis*) and a coastal oyster (*Saccostrea glomerata*). *Aquat. Toxicol.* 56:33–51.

Kleinow, K. M., et al. 1996. Role of the mucous surface coat, dietary bulk and mucosal cell turnover in the intestinal disposition of benzo(a)pyrene (BaP). *Mar. Environ. Res.* 42:65–73.

Klerks, P. L., and J. S. Levinton. 1989. Rapid evolution of metal resistance in a benthic oligochaete inhabiting a metal-polluted site. *Biol. Bull.* 176:135–41.

Kraus, M. L., J. S. Weis, and P. Weis. 1988. Effects of mercury on larval and adult grass shrimp (*Palaemonetes pugio*). *Arch. Environ. Contam. Toxicol.* 17:355–63.

Lahaye, V., et al. 2007. Biological and ecological factors related to trace element levels in harbour porpoises (*Phocoena phocoena*) from European waters. *Mar. Environ. Res.* 64:247–66.

Larsen, D. K., et al. 2003. Long-term effect of Sea-Nine on natural coastal phytoplankton communities assessed by pollution induced community tolerance. *Aquat. Toxicol.* 62:35–44.

Law, R. J. 1996. Metals in marine mammals. In *Environmental contaminants in wildlife: Interpreting tissue concentrations*, ed. W. N. Beyer, G. H. Heinz, and A. W. Redmond-Norwood, 357–76. Boca Raton, FL: CRC Press.

Lemière, S., et al. 2005. DNA damage measured by the single-cell gel electrophoresis (Comet) assay in mammals fed with mussels contaminated by the "Erika" oil-spill. *Mutat. Res.-Fund. Mol. M.* 581:11–21.

Leung, K. M. Y., A. C. Taylor, and R. W. Furness. 2000. Temperature-dependent physiological responses of the dogwhelk *Nucella lapillus* to cadmium exposure. *J. Mar. Biol. Assoc. U.K.* 80:647–60.

Lewis, C., and T. Galloway. 2008. Genotoxic damage in polychaetes: A study of species and cell-type sensitivities. *Mutation Res./Genetic Toxicol. Environ. Mutagenesis* 654:69–75.

Lotts, J. W., and A. J. Stewart. 1995. Minnows can acclimate to total residual chlorine. *Environ. Toxicol. Chem.* 14:1365–74.

Luoma, S. N., and P. S. Rainbow, 2008. *Metal contamination in aquatic environment. Science and lateral management.* Cambridge: Cambridge University Press.

Luoma, S. N., et al. 1983. Variable tolerance to copper in two species from San Francisco bay. *Mar. Environ. Res.* 10:209–22.

Mahaffey, K. R., R. P. Clickner, and R. A. Jeffries. 2008. Methylmercury and omega-3 fatty acids: Co-occurrence of dietary sources with emphasis on fish and shellfish. *Environ. Res.* 107:20–29.

Malm, O. 1998. Gold mining as a source of mercury exposure in the Brazilian Amazon. *Environ. Res.* 77A:73–78.

Marigomez, I., et al. 2002. Cellular and subcellular distribution of metals in molluscs. *Microscopy Res. Technique* 56:358–92.

Martin, J. W., et al. 2003. Dietary accumulation of perfluorinated acids in juvenile rainbow trout (*Oncorhynchus mykiss*). *Environ. Toxicol. Chem.* 22:189–95.

Mason, A. Z., and K. D. Jenkins. 1995. Metal detoxication in aquatic organisms. In *Metal speciation and bioavailability in aquatic systems*, ed. A. Tessier and D. R. Turner, 479–608. Chichester, UK: John Wiley & Sons.

Matthiessen, P., and R. J. Law. 2002. Contaminants and their effects on estuarine and coastal organisms in the United Kingdom in the late twentieth century. *Environ. Pollut.* 120:739–57.

Maurice-Bourgoin, L., et al. 2000. Mercury distribution in waters and fishes of the upper Madeira rivers and mercury exposure in riparian Amazonian populations. *Sci. Total Environ.* 260:73–86.

Mazen, A. M. A. 2004. Calcium oxalate deposits in leaves of *Corchorus olitorius* as related to accumulation to toxic metals. *Russ. J. Plant Physiol.* 51:281–85.

McGeer, J. C., et al. 2000. Effects of chronic sublethal exposure to waterborne Cu, Cd or Zn in rainbow trout. 1. Iono-regulatory disturbance and metabolic costs. *Aquat. Toxicol.* 50:231–43.

Miles, A. T., et al. 2000. Induction, regulation, degradation, and biological significance of mammalian metallothioneins. *Crit. Rev. Biochem. Mol. Biol.* 35:35–70.

Millward, R. N., and A. Grant. 1995. Assessing the impact of copper on nematode communities from a chronically metal-enriched estuary using pollution-induced community tolerance. *Mar. Pollut. Bull.* 30:701–6.

Minhas, J. K., et al. 2006. Mobilization of chrysene from soil in a model digestive system. *Environ. Toxicol. Chem.* 25:1729–37.

Miranda, C. D., and R. Rojas. 2006. Copper accumulation by bacteria and transfer to scallop larvae. *Mar. Pollut. Bull.* 52:293–300.

Miller, M. P., and A. C. Hendricks. 1996. Zinc resistance in *Chironomus riparius*: Evidence for physiological and genetic components. *J. N. Am. Benthol. Soc.* 15:106–16.

Moraitou-Apostolopoulou, M., and G. Verriopoulos. 1979. Some effects of sub-lethal concentrations of copper on a marine copepod. *Mar. Pollut. Bull.* 10:88–92.

Moraitou-Apostolopoulou, M., G. Verriopoulos, and P. Palla. 1979. Temperature and adaptation to pollution as factors influencing the acute toxicity of Cd to the planktonic copepod *Acartia clausi. Tethys* 9:97–101.

Mouneyrac, C., et al. 2003. Trace-metal detoxification and tolerance of the estuarine worm *Hediste diversicolor* chronically exposed in their environment. *Mar. Biol.* 143:31–44.

Mueller, M. E., et al. 1991. Nature and time course of acclimation to aluminium in juvenile brook trout (*Salvelinus fontinalis*). II. Gill histology. *Can. J. Fish. Aquat. Sci.* 48:2016–27.

Muir, D. C. G., et al. 2006. Brominated flame retardants in polar bears (*Ursus maritimus*) from Alaska, the Canadian Arctic, East Greenland, and Svalbard. *Environ. Sci. Technol.* 40:449–55.

Munkittrick, K. R., and D. G. Dixon. 1988. Evidence for a maternal yolk factor associated with increased tolerance and resistance of feral white sucker (*Catostomus commersoni*) to waterborne copper. *Ecotoxicol. Environ. Saf.* 15:7–20.

Munschy, C., et al. 2008. Polychlorinated dibenzo-p-dioxins and dibenzofurans (PCDD/Fs) in marine mussels from French coasts: Levels, patterns and temporal trends from 1981 to 2005. *Chemosphere* 73:945–53.

Nassiri, Y., et al. 1996. Effects of heavy metals on *Tetraselmis suecica*: Ultrastructural and energy-dispersive x-ray spectroscopic studies. *Biol. Cell.* 86:151–60.

Naylor, C., L. Pindar, and P. Calow. 1990. Inter- and intraspecific variation in sensitivity to toxins: The effects of activity of acidity and zinc on the freshwater crustaceans *Asellus aquaticus* (L.) and *Gammarus pulex* (L.). *Water Res.* 24:757–62.

Ng, T. Y. T., et al. 2005. Physico-chemical form of trace metals accumulated by phytoplankton and their assimilation by filter-feeding invertebrates. *Mar. Ecol. Prog. Ser.* 299:179–91.

Nigro, M., and C. Leonzio. 1996. Intracellular storage of mercury and selenium in different marine vertebrates. *Mar. Ecol. Prog. Ser.* 135:137–43.

Norström, R. J. 2002. *Plenary lecture, understanding bioaccumulation of POPs in food webs: Chemical, biological and environmental considerations, dioxin—2002*, Barcelona, Spain, August 12–15.

Norström, R. J., and R. J. Letcher. 1997. Role of biotransformation in bioconcentration and bioaccumulation, annex 1. In *Biotransformation in environmental risk assessment*, ed. D. Sijn et al., 103–13. Bruxelles: Society of Environmental Toxicology and Chemistry (SETAC).

Nott, J. A., and A. Nicolaidou. 1990. Transfer of metal detoxification along marine food chains. *J. Mar. Biol. Assoc. U.K.* 70:905–12.

Nott, J. A., and A. Nicolaidou. 1994. Variable transfer of detoxified metals from snails to hermit crabs in marine food chains. *Mar. Biol.* 120:369–77.

Palmqvist, A., L. J. Rasmussen, and V. E. Forbes. 2006. Influence of biotransformation on trophic transfer of the PAH, fluoranthene. *Aquat. Toxicol.* 80:309–19.

Patrick, F. M., and M. Loutit. 1976. Passage of metals in effluents, through bacteria to higher organisms. *Water Res.* 10:333–35.

Pawlik-Skowrońska, B. 2003. When adapted to high zinc concentrations the periphytic green alga *Stigeoclonium tenue* produces high amounts of novel phytochelatin-related peptides. *Aquat. Toxicol.* 62:155–63.

Pawlik-Skowrońska, B., J. Pirszel, and M. T. Brown. 2007. Concentrations of phytochelatins and glutathione found in natural assemblages of seaweeds depend on species and metal concentrations of the habitat. *Aquat. Toxicol.* 83:190–99.

Perales-Vela, H. V., J. M. Pena-Castro, and R. O. Canizares-Villanueva. 2006. Heavy metal detoxification in eukaryotic microalgae. *Chemosphere* 64:1–10.

Petersen, S., and K. Gustavson. 1998. Toxic effects of tri-butyl-tin (TBT) on autotrophic pico-, nano-, and microplankton assessed by a size fractionated pollution-induced community tolerance (SF-PICT) concept. *Aquat. Toxicol.* 40: 253–64.

Pimentel, D. 2005. Environmental and economic costs of the application of pesticides primarily in the United States. *Environ. Dev. Sustain.* 7:229–52.

Plessi, M., D. Bertelli and A. Monzani. 2001. Mercury and selenium content in selected seafood. *J. Food Comp. Anal.* 14:461–67.

Posthuma, L., and N. M. Van Straalen. 1993. Heavy-metal adaptation in terrestrial invertebrates: A review of occurrence, genetics, physiology and ecological consequences. *Comp. Biochem. Physiol.* 106C:11–38.

Postma, J. F., and C. Davids. 1995. Tolerance induction and life cycle changes in cadmium-exposed *Chironomus riparius* (Diptera) during consecutive generations. *Ecotoxicol. Environ. Saf.* 30:195–202.

Rainbow, P. S. 2002. Trace metal concentrations in aquatic invertebrates: Why and so what? *Environ. Pollut.* 120:497–507.

Rainbow, P. S., et al. 2004. Enhanced food chain transfer of copper from a diet of copper-tolerant estuarine worms. *Mar. Ecol. Prog. Ser.* 271:183–91.

Rainbow, P. S., et al. 2006a. Trophic transfer of trace metals: Subcellular compartmentalization in a polychaete and assimilation by a decapod crustacean. *Mar. Ecol. Prog. Ser.* 308:91–100.

Rainbow, P. S., et al. 2006b. Trophic transfer of trace metals from the polychaete worm *Nereis diversicolor* to the polychaete *N. virens* and the decapod crustacean *Palaemonetes varians*. *Mar. Ecol. Prog. Ser.* 321:167–81.

Rainbow, P. S., et al. 2007. Trophic transfer of trace metals: Subcellular compartmentalization in bivalve by prey, assimilation by a gastropod predator and *in vitro* digestion simulations. *Mar. Ecol. Prog. Ser.* 348:125–38.

Ramade, F. 2007. *Introduction à l'écotoxicologie. Fondements et applications.* Paris: Lavoisier.

Reinecke, S. A., M. W. Prinsloo, and A. J. Reinecke. 1999. Resistance of *Eisenia fetida* (Oligochaeta) to cadmium after long-term exposure. *Ecotoxicol. Environ. Saf.* 42:75–80.

Reinfelder, J. R., and N. S. Fisher. 1991. The assimilation of elements ingested by marine copepods. *Science* 251:794–96.

Rice, C. A., et al. 2000. From sediment bioassay to fish biomarker—Connecting the dots using simple trophic relationships. *Mar. Environ. Res.* 50:527–33.

Roesijadi, G., et al. 1982. Enhanced mercury tolerance in marine mussels and relationship to low molecular weight, mercury-binding proteins. *Mar. Pollut. Bull.* 13:250–53.

Ross, K., et al. 2002. Genetic diversity and metal tolerance of two marine species: A comparison between populations from contaminated and reference sites. *Mar. Pollut. Bull.* 44:671–79.

Sánchez, M., E. Andreu-Moliner, and M. D. Ferrando. 2004. Laboratory investigation into the development of resistance of *Daphnia magna* to the herbicide molinate. *Ecotoxicol. Environ. Saf.* 59:316–23.

Sardessai, Y., and S. Bhosle. 2002a. Tolerance of bacteria to organic solvents. *Res. Microbiol.* 153:263–68.

Sardessai, Y. N., and S. Bhosle. 2002b. Organic solvent-tolerant bacteria in mangrove ecosystem. *Curr. Sci.* 82:622–23.

Sarret, G., et al. 2006. Trichomes of tobacco excrete zinc as zinc-substituted calcium carbonate and other zinc-containing compounds. *Plant Physiol.* 141:1021–34.

Schmitt-Jansen, M., and R. Altenburger. 2005a. Toxic effects of isoproturon on periphyton communities—A microcosm study. *Estuar. Coast. Shelf Sci.* 62:539–45.

Schmitt-Jansen, M., and R. Altenburger. 2005b. Predicting and observing responses of algal communities to photosystem II. Herbicide exposure using pollution-induced community tolerance and species-sensitivity distributions. *Environ. Toxicol. Chem.* 24:304–12.

Seguin, F., et al. 2002. A risk assessment of pollution: Induction of atrazine tolerance in phytoplankton communities in freshwater outdoor mesocosms, using chlorophyll fluorescence as an endpoint. *Water Res.* 36:3227–36.

Seppänen, K., et al. 1998. Mercury-binding capacity of organic and inorganic selenium in rat blood and liver. *Biol. Trace Elem. Res.* 65:197–210.

Shull, D. H., and L., M. Mayer. 1998. Dissolution of particle-reactive radionuclides in deposit-feeder digestive fluids. *Limnol. Oceanogr.* 47:1530–36.

Silver, S., and L. T. Phung. 1996. Bacterial heavy metal resistance: New surprises. *Annu. Rev. Microbiol.* 50:753–89.

Silvestre, F., et al. 2006. Differential protein expression profiles in anterior gills of *Eriocheir sinensis* during acclimation to cadmium. *Aquat. Toxicol.* 76:46–58.

Singhal, R. N., H. B. Sarnat, and R. W. Davies. 1989. Unimpaired RNA synthesis in neurons and epithelial cells in a freshwater leech exposed to the organophosphate insecticide chlorpyrifos. *Sci. Total Environ.* 83:195–202.

Sorrentino, C., et al. 2004. B[a]P-DNA binding in early life-stages of Atlantic tomcod: Population differences and chromium modulation. *Mar. Environ. Res.* 58:383–88.

Stuhlbacher, A., and L. Maltby. 1992. Cadmium resistance in *Gammarus pulex* (L.). *Arch. Environ. Contam. Toxicol.* 22:319–24.

Sze, P. W. C., and S. Y. Lee. 1995. The potential role of mucus in the depuration of copper from the mussels *Perna viridis* (L.) and *Septifer virgatus* (Wiegmann). *Mar. Pollut. Bull.* 31:390–93.

Takamura, N., F. Kasai, and M. M. Watanabe. 1989. Effects of Cu, Cd and Zn on photosynthesis of freshwater benthic algae. *J. Appl. Phycol.* 1:39–52.

TGD. 2003. Technical Guidance Document on Risk Assessment in support of Commission Directive 93/67/EEC on Risk Assessment for new notified substances, Commission Regulation (EC) n° 1488/94 on Risk Assessment for existing substances and Directive 98/8/EC of the European Parliament and of the Council concerning the placing of biocidal products on the market. EUR 20418 EN/2. European Commission, Joint Research Centre.

Torricelli, E., et al. 2004. Cadmium tolerance, cysteine and thiol peptide levels in wild type and chromium-tolerant strains of *Scenedesmus acutus* (Chlorophyceae). *Aquat. Toxicol.* 68:315–23.

Tsubaki, T., and K. Irukayama. 1977. *Minamata disease*. Amsterdam: Elsevier.

Tsui, M. T. K., and W.-X. Wang. 2005. Influences of maternal exposure on the tolerance and physiological performance of *Daphnia magna* under mercury stress. *Environ. Toxicol. Chem.* 24:1228–34.

Twiss, M. R., P. M. Welbourn, and E. Schwärtzel. 1993. Laboratory selection for copper tolerance in *Scenedesmus acutus* (Chlorophyceae). *Can. J. Bot.* 71:333–38.

Van Metre, P. C., and B. J. Mahler. 2005. Trends in hydrophobic organic contaminants in urban and reference lake sediments across the United States, 1970–2001. *Environ. Sci. Technol.* 39:5567–74.

Versantvoort, C. H. M., et al. 2005. Applicability of an *in vitro* digestion model in assessing the bioaccessibility of mycotoxins from food. *Food Chem. Toxicol.* 43:31–40.

Vidal, D. E., and A. J. Horne. 2003. Mercury toxicity in the aquatic oligochaete *Sparganophilus pearsei*. II. Autotomy as a novel form of protection. *Arch. Environ. Contam. Toxicol.* 45:462–67.

Vorkamp, K., et al. 2004. Chlorobenzenes, chlorinated pesticides, coplanar chlorobiphenyls and other organochlorine compounds in Greenland biota. *Sci. Total Environ.* 331:157–75.

Wallace, W. G., G. R. Lopez, and J. S. Levinton. 1998. Cadmium resistance in an oligochaete and its effect on cadmium trophic transfer to an omnivorous shrimp. *Mar. Ecol. Prog. Ser.* 172:225–37.

Wallace, W. G., and S. N. Luoma. 2003. Subcellular compartmentalization of Cd and Zn in two bivalves. II. The significance of trophically available metal (TAM). *Mar. Ecol. Prog. Ser.* 257:125–37.

Wan, Y., et al. 2007. Trophic dilution of polycyclic aromatic hydrocarbons (PAHs) in a marine food web from Bohai Bay, North China. *Environ. Sci. Technol.* 41:3109–14.

Wang, J. S., et al. 1998. Uptake and transfer of high PCB concentrations from phytoplankton to aquatic biota. *Chemosphere* 36:1201–10.

Wang, W. X. 2002. Interactions of trace metals and different marine food chains. *Mar. Ecol. Prog. Ser.* 243:295–309.

Wang, W. X., and R. C. H. Dei. 2006. Metal stoichiometry in predicting Cd and Cu toxicity to a freshwater green alga *Chlamydomonas reinhardtii*. *Environ. Pollut.* 142:303–12.

Wang, W. X., and N. S. Fisher. 1996. Assimilation of trace elements by the mussel *Mytilus edulis*: Effects of diatom chemical composition. *Mar. Biol.* 125:715–24.

Weeks, J. M. 1993. Effects of dietary copper and zinc concentrations on feeding rates of two species of talitrid amphipods (Crustacea). *Bull. Environ. Contam. Toxicol.* 50:883–90.

Weinstein, J. E., D. M. Sanger and A. F. Holland. 2003. Bioaccumulation and toxicity of fluoranthene in the estuarine oligochaete *Monopylephorus rubroniveus*. *Ecotoxicol. Environ. Saf.* 55:278–86.

Wentsel, R., A. McIntosh, and G. Atchison. 1978. Evidence of resistance to metals in larvae of the midge *Chironomus tentans* in a metal contaminated lake. *Bull. Environ. Contam. Toxicol.* 20:451–55.

Weston, D. P., and K. A. Maruya. 2002. Predicting bioavailability and bioaccumulation with *in vitro* digestive fluid extraction. *Environ. Toxicol. Chem.* 21:962–71.

Weston, D. P., and L. M. Mayer. 1998. *In vitro* digestive fluid extraction as a measure of the bioavailability of sediment-associated polycyclic aromatic hydrocarbons: Sources of variations for partitioning models. *Environ. Toxicol. Chem.* 17:820–29.

Wicklum, D., and R. W. Davies. 1996. The effects of chronic cadmium stress on energy acquisition and allocation in a freshwater benthic invertebrate predator. *Aquat. Toxicol.* 35:237–52.

Wikfors, G. H., and R. Ukeles. 1982. Growth and adaptation of estuarine unicellular algae in media with excess of metal-contaminated algal food on *Crassostrea virginica* larvae. *Mar. Ecol. Prog. Ser.* 7:191–206.

Wilson, R. W., H. L. Bergman, and C. M. Wood. 1994. Metabolic costs and physiologial consequences of acclimation to aluminium in juvenile rainbow trout (*Oncorhynchus mykiss*). 2. Gill morphology, swinning performance, and aerobic scope. *Can. J. Fish. Aquat. Sci.* 51:536–44.

Woodward, D. F., et al. 1995. Metals-contaminated benthic invertebrates in the Clark Fork River, Montana: Effects on age-0 brown trout and rainbow trout. *Can. J. Fish. Aquat. Sci.* 52:1994–2004.

Xie, L., and P. L. Klerks. 2004. Fitness cost of resistance to cadmium in the least killifish (*Heterandria formosa*). *Environ. Toxicol. Chem.* 23:1499–503.

Xu, Y., W. X. Wang, and D. P. H Hsieh. 2001. Influences of metal concentration in phytoplankton and seawater on metal assimilation and elimination in marine copepods. *Environ. Toxicol. Chem.* 20:1067–77.

Yuan, Z., et al. 2006. Evidence of spatially extensive resistance to PCBs in an anadromous fish of the Hudson River. *Environ. Health Perspect.* 114:77–84.

Zhong, H., and W. X. Wang. 2006. Influences of aging on the bioavailability of sediment-bound Cd and Zn to deposit-feeding sipunculans and soldier crabs. *Environ. Toxicol. Chem.* 25:2775–80.

Zhong, H., and W. X. Wang. 2008. Methylmercury extraction from artificial sediments by the gut juice of the sipunculan, *Sipunculus nudus. Environ. Toxicol. Chem.* 27:138–45.

Section

Case Studies

14 Bacterial Tolerance in Contaminated Soils

Potential of the PICT Approach in Microbial Ecology

Gwenaël Imfeld, Françoise Bringel, and Stéphane Vuilleumier

CONTENTS

14.1 TOLERANCE VERSUS RESISTANCE

The pristine soil microbiome constitutes one of the largest and most diverse microbial communities on earth (Curtis et al. 2002). It remains substantially unexplored in terms of functional diversity (Crawford et al. 2005). The structure and the physiology of the soil microbial compartment are shaped by exposure to a wide range of environmental factors, including chemicals originating from microbial, plant, animal, and human activities (Dumbrell et al. 2009). The fate of chemical compounds released into the environment as a consequence of human activities and lifestyle has become a serious concern, mainly due to their impact on health (Williams et al. 2009). The soil microbial compartment resorts to complementary, nonexclusive strategies to deal with pollution: development of resistance, by way of neutralisation or degradation of the toxicant, and development of tolerance, defined as the ability of both strains and communities to withstand toxic insults inflicted by pollutants on the ecosystem. Sequencing of the soil community metagenome has revealed the unexpected density of resistance genes that constitute the natural antibiotic "resistome," which originates from the fact that over 80% of antibiotics in clinical use originated either directly as soil natural products or as their semisynthetic derivatives (D'Costa et al. 2007; Martinez et al. 2009). Similarly, the need for rehabilitation of environmental contamination caused by manufactured chemicals, which do not occur in significant levels in nature, has contributed to the study of microorganisms, genes, and enzymes that provide efficient degradation pathways for organic chemicals (Copley 2009), or contribute to the dissemination of adaptation to toxic metal contaminants, including radionuclides (Sobecky and Coombs 2009). Characterising the resistance to contaminants and the capacity to degrade them has been a mainstay for research in microbial ecology. This particularly pertains to the isolation and detailed characterisation of numerous "superbugs" highly resistant to high concentrations of trace metals and organic pollutants (Singer et al. 2005; Pandey et al. 2009) and, for the latter class of compounds, the mineralisation of them. In comparison, the subject of tolerance of bacterial strains or bacterial communities to environmental contamination has not yet been given the same level of attention. As a result of growing awareness of environmental issues, and of the need to develop tools for the management of polluted sites and risk evaluation, the focus of investigations has shifted to the measurement, understanding, and quantification of the extent and modes of development of overall tolerance in microbial communities, without necessarily considering the presence or the expression of degradation or resistance genes to tolerate exposure to toxic contaminants. One major challenge remains—to disentangle changes caused by an anthropogenic disturbance, such as industrial or diffuse contamination, from stochastic variability within the microbial compartment as a consequence of normal fluctuations of environmental conditions, in order to establish cause-effect relationships between the presence of toxicants in a given environment and the overall structure and function of the microbial compartment.

The topic of this chapter is to discuss application of the concept of pollution-induced community tolerance (PICT) (Rutgers and Breure 1999; Siciliano and Roy 1999; Boivin et al. 2002; Millward and Klerks 2002; van Beelen 2003) for the characterisation of soil ecosystem tolerance, with particular emphasis on contaminant-induced changes and tolerance at the bacterial level (Section 14.2). A battery of experimental tools (Section 14.3) may help to disentangle the different environmental factors contributing to bacterial tolerance, such as the intrinsic properties of the soil matrix and contaminant bioavailability (Section 14.4). The growing literature on the use of bacteria as proxies to assess the magnitude and toxicity of contamination in complex ecosystems applying the PICT concept, for both inorganic and organic contaminants, is reviewed in Section 14.5. This is a relatively new field of enquiry, and we identify a series of challenges (Section 14.6) that will need to be addressed to evaluate environmental contamination using bacterial-based indicators in the context of the PICT approach, and to develop a general risk assessment framework that is widely accepted.

14.2 THE PICT CONCEPT FROM A BACTERIAL POINT OF VIEW

Ideally, the characterisation of shifts in the structure and function of soil microbial communities following contaminant exposure should allow us to address, in a global and integrated manner, soil functioning, the food web structure, and to make inferences on associated aboveground ecosystems. This approach is warranted by the critical role of the microbial compartment in a wide range of soil processes, including cycling of nutrients, breakdown of organic matter, interaction with plant roots, and formation of soil structure (Khan and Scullion 2000). Soil microorganisms are sensitive to environmental perturbations, so that changes in bacterial community structures and functions can provide early signs of anthropogenic disturbances and even predict the outcome of the envisaged restoration process. To be complete, microbial analysis of the effect of contaminant in soil ecosystems should necessarily encompass the fungal/mycorrhizal compartment (Liu et al. 2006). For reasons of limited space, however, this chapter specifically focuses on bacteria.

Originally introduced as an ecotoxicological tool applied to algal communities, the PICT concept aims at providing a quantitative assessment of pollutant-induced contaminant tolerance in various ecosystems. The principle, application, and limits of the PICT concept are the focus of Chapter 4. The magnitude of increase of tolerance of a microbial community, investigated experimentally by PICT-type approaches (Blanck and Wangberg 1988; Blanck 2002; Boivin et al. 2002), can be used as a quantitative measure of the degree of ecological disturbance. The degree of PICT is generally correlated to the concentration of the toxicant (Blanck and Wangberg 1988; Baath 1992). In soil ecosystems, PICT relying on microorganisms usually develops rapidly, and an increase in PICT in soil within a few days of exposure of soil to toxicants has been observed (e.g., Muller et al. 2001; Schmitt et al. 2005, 2006; Demoling and Baath 2008a). PICT-type investigations, using approaches described in Section 14.3 and summarised in Table 14.1, offer the advantage of being more sensitive that the more classical measures of enzymatic functions related to nutrient cycles. This is because the abundance of specific microbial groups can be strongly affected, even when little impact on overall ecosystem functioning is detected, as a result of extensive functional redundancy in the microbial compartment (Siciliano and Roy 1999; Griffiths et al. 2001). In addition, enzyme activity in soil is not necessarily correlated with viable cell populations, due to persistence of functional enzymes released from dead cells (Wallenstein and Weintraub 2008).

Microorganisms react quickly and sensitively to contamination, as their high surface-to-volume ratio renders them more exposed to their environment than organisms with a small surface-to-volume ratio and stronger homeostasis, such as birds and mammals (Boivin et al. 2002). In addition, motility of microorganisms can be regarded as negligible in the face of contamination exposure. Consequently, the observed distribution of microorganisms can be closely linked to the small-scale distribution of contaminants in soils, providing the advantage of a straightforward correlation between the extent of estimated contamination and soil ecosystem status. It is thus not surprising that an increasing body of evidence indicates that microorganisms, globally, are far more sensitive to contaminant stress than soil animals and plants (Giller et al. 1998), and that, therefore, bacteria represent organisms of choice to assess the effects of environmental contamination, in particular when contaminant effects are expected to be small. An additional key advantage of using bacteria for characterising the effects of contaminants using the PICT approach is the ease in sample collection, handling, and analysis that, at least in part, can be automated. Hence, assessing the occurrence of both inorganic and organic contaminant effects using microorganisms (see Table 14.1) requires relatively little effort.

14.3 MEASURING PICT IN BACTERIAL SYSTEMS

The toolbox of methods and techniques used to detect and define PICT in bacteria may often involve techniques specific to microbiology, and offers the opportunity to perform complementary experiments on those used for other living organisms to define PICT in the environment of interest.

14.3.1 PRINCIPLES OF TOLERANCE DEVELOPMENT AND MONITORING IN BACTERIAL PICT EXPERIMENTS

A preferred experimental approach to assess tolerance of the bacterial compartment is to add contaminants to uncontaminated soil, either in the laboratory or in the field, as endogenous soil microbes are likely to be well adapted to the particular matrix to be investigated (e.g., McLaughlin and Smolders 2001). In the selection phase of bacterial PICT experiments (see Section 4.3 for details), a PICT response will be defined when the ratio between sensitive and tolerant microorganisms has changed due to the elimination of sensitive phylotypes or phenotypes. Second, the subsequent detection phase will involve the evaluation of the evolution of tolerance and resistance levels in the studied bacterial communities following contaminant exposure. Because the observed changes both among and within phylotypes in the global bacterial community can be due to loss, replacement, or changes of phylotypes, detection should preferably involve a functional indicator matched to the mode of action of the toxicant, or a superior higher-level indicator able to integrate the effects arising from all modes of action of the contaminant under study (Siciliano and Roy 1999; Boivin et al. 2006a).

TABLE 14.1
Use of the PICT Concept in Microbial Studies of Soil Systems

Contaminant	Experimental Setup	Findings	References
	Inorganic Contaminants		
Zn	Microcalorimetric analysis, cell enumeration	Soil microbial communities and activities can adapt to a certain extent: Zn pollution higher than 6,000 $\mu g.g^{-1}$ inhibit fungus growth	Zhou et al. 2009
Al, Ni, U	Functional gene array	Diversity and community structure vary in relation to contamination	Waldron et al. 2009
Zn, Pb, Cu	Biolog ECO plates, CLPP	PICT was increased in polluted meadow soils when compared to control meadow soils	Stefanowicz et al. 2009
Cu	IC50 determination for Cu concentrations, T-RFLP	Microbial community IC50 monitoring was recommended as a sensitive and consistent indicator of metal pollution on estuarine sediment	Ogilvie and Grant 2008
Cu	IC50 determination for Cu concentrations	The long-term use of Cu in vineyards has a toxic effect on the soil bacterial community, resulting in an increased tolerance	Diaz-Ravina et al. 2007
Cu, Zn	Biolog ECO plates, microbial biomass (C-mic), basal soil respiration (BAS), CLPP	Biolog assay was the most effective method; association of multiple stresses decreases microbial PICT	Niklinska et al. 2006
Cu, Zn, Pb	Biolog ECO plates	Enhanced tolerance of the microbial community to Cu stress in the polluted site; Biolog assay provided sensitive detection of microbial PICT changes	Kamitani et al. 2006
Cu	DGGE, CLPP	Resilience monitoring of copper-contaminated water	Boivin et al. 2006b

(continued)

TABLE 14.1

Use of the PICT Concept in Microbial Studies of Soil Systems (Continued)

Contaminant	Experimental Setup	Findings	References
Pb, Cu	[^{14}C]-leucine and [^{3}H]-thymidine incorporation rates, CO_2 evolution, biomass indicators, and N mineralisation rates	Shifts in CLPP and DGGE patterns were not correlated to metal contamination in grassland soil; no filtration of the potential metal effects from the total variation by multivariate analyses has been performed	Boivin et al. 2006a
Zn	Biolog ECO plates	Functional diversity of microbial communities decreased with increasing zinc concentrations in pore water	Lock and Janssen 2005
Zn, Cd	Sequential dispersion/ density gradient centrifugation to separate free and loosely attached cells from strongly attached cells, [^{3}H]-thymidine incorporation labelling	Zn ion activity was close to toxicity level, and could have caused the observed PICT Zn in free and loosely attached cells; temporal/spatial fluctuations of soil pH played a critical role in the observed PICT effect	Almas et al. 2005
Cr, Cu, As	Modified potential ammonium oxidation assay	Culturable microbial community and the ammonium oxidising bacterial community both responded to increased Cr concentrations in soil	Gong et al. 2002
As	PFLA, [^{3}H]-thymidine incorporation labelling	The advantages and disadvantages of PFLA and [^{3}H]-thymidine incorporation labelling tools are discussed and exemplified by field data and laboratory experiments from different soils	Baath et al. 1998

Organic Contaminants

Contaminant	Experimental Setup	Findings	References
Sulphamethoxazole	Leucine incorporation labelling, Biolog ECO plate, CLPP, PFLA	Different methods, different thresholds of sulphamethoxazole; initial decrease in bacterial growth rates of sensitive species, which eventually transformed into more tolerant species, alter the community composition	Demoling et al. 2009

TABLE 14.1
Use of the PICT Concept in Microbial Studies of Soil Systems (Continued)

Contaminant	Experimental Setup	Findings	References
Sulphadiazine	[³H]-leucine incorporation labelling	Carbon substrate amendment increased the microbial PICT response; rhizosphere and manure-soil interfaces may comprise key sites for proliferation of bacteria that are resistant or tolerant to antibiotics	Brandt et al. 2009
Sulphonamide Sulphachloropyridazine	Biolog ECO plates	For the detection of antibiotic effects on soil microbial communities, it is recommended to use nutrient amendments (possibly fresh pig slurry), and to extend the period of Biolog plate measurement beyond seven days	Schmitt et al. 2005
Sulphonamide Sulphachloropyridazine	CLPP	PICT effect was accompanied by small changes in CLPP profiles	Schmitt et al. 2004
Trinitrotoluene	DGGE, CLPP, cell enumeration	PICT was observed as TNT concentrations increased; the transformation to amino derivatives decreased at high concentrations of TNT, indicative of inhibition of microbial TNT transformation	Travis et al. 2008
2,3,6-Trinitrotoluene	DGGE, Biolog ECO plates, seed germination assays	Effective assay to rapidly detect TNT exposure and toxicity in soil microbial communities	Siciliano et al. 2000
Diuron and other agricultural contaminants	DGGE, pesticide dosage	Measurable effects on small river biofilm communities depend on pollutant exposure procedures	Tlili et al. 2008
2,4-Dichlorophenoxy-acetic acid (2,4-D)	DGGE, methane oxidation capacity, soil supplementation of increasing amounts of (2,4-D)	Effective assays to distinguish a long-term herbicide-treated soil from an untreated soil	Seghers et al. 2003

(continued)

TABLE 14.1

Use of the PICT Concept in Microbial Studies of Soil Systems (Continued)

Contaminant	Experimental Setup	Findings	References
Various phenols	[³H]-leucine incorporation labelling, PFLA	The development of PICT in phenol- and 2-chlorophenol-polluted soils was shown	Demoling and Baath 2008a,c
Tylosin	[³H]-]leucine incorporation labelling	PICT will return to prepollution levels when the selective pressure of the toxicant is removed, and will thus be a useful technique for monitoring remediation measures	Demoling and Baath 2008c

Note: IC50, half maximal inhibitory concentration; CLPP, community-level physiological profiling; 16S rDNA genes analysis using denaturing gradient gel electrophoresis (DGGE) or terminal restriction fragment length polymorphism (T-RFLP).

The duration of exposure to contaminants is crucial for the dynamics of the selection phase. Chronically contaminated ecosystems are characterised by long- or middle-term exposure of endogenous bacterial communities, resulting in a gradual change in bacterial composition over extended periods of time, and in which natural selection and genetic exchange both contribute to the reshaping of the bacterial community to improve its adequacy to the prevailing conditions (Joynt et al. 2006). Adaptation seems more likely in the case of plots or field gradients where pollutants have been added progressively, and have equilibrated with the soil colloidal matrix over several years (Giller et al. 2009). When the contamination is only transient or subject to rapid variations, alternative subsistence strategies to avoid episodic contamination events may be favoured. In this case, initial decrease in biomass and activity as a tolerant community develops in soils is often only temporary (e.g., Ranjard et al. 2006).

14.3.2 THE MICROBIOLOGICAL TOOLBOX

Any observable effect of contamination relevant to microbiology, including relative phylotype abundance, overall bacterial community composition, structure, and function, as assessed by cultivation-dependent and -independent approaches (Rutgers et al. 1998; Siciliano and Roy 1999; Blanck 2002; Seghers et al. 2003; van Beelen 2003; Davis et al. 2004; Boivin et al. 2006a; Niklinska et al. 2006; Diaz-Ravina et al. 2007; Travis et al. 2008; Stefanowicz et al. 2009), can be the experimental target for investigation using the PICT approach.

Classically, soil bacterial characteristics have been evaluated by traditional counting and quantification methods, such as the number of phylotypes in bacterial communities (Gremion et al. 2004), the magnitude of biochemical functions including respiration and mineralisation (Iyyemperumal et al. 2007), carbon dioxide evolution (Kahkonen et al. 2007), and the quantity of microbial biomass (Anderson and Domsch 1978; Brookes et al. 1985). Nevertheless, several other, generally more recently developed, methods can also be used, some of which are presented below and summarised in Table 14.1.

14.3.2.1 Growth and Substrate Use Analysis

Although the amount of microbial biomass may be a sensitive indicator of contaminant stress, its suitability as an indicator of soil pollution is limited because of its spatial variability (Broos et al. 2007) and technical shortcomings in its measurement (Dalal 1998). In particular, growth rate cannot easily be calculated from increase in biomass or cell numbers, because increases are generally balanced by losses. Nevertheless, several other growth-related short-term tests can be used for PICT detection, among them the rate of metabolism of simple organic substrates. This method has been highly systematised by the development of Biolog® plates, containing thirty-one different organic substrates, which allow for the detection of multiple microbial metabolic activities (e.g., Rutgers et al. 1998; Schmitt et al. 2004, 2005; Stefanowicz et al. 2009). Growth and metabolisation of a substrate can be further analysed colorimetrically, using the pH-dependent tetrazolium dye, which gives the cellular redox state.

Community-level population profiling (CLPP) is another useful approach for the analysis of bacterial community structures that is amenable to analysis in microtiter format (Garland et al. 2003; Vinther et al. 2003; Niklinska et al. 2006; Ramsey et al. 2006; Yang et al. 2006). The pattern of utilisation of different organic substrates by the investigated community defines its so-called physiological fingerprint, which can then be compared between samples using multivariate analysis and ordination methods. Both Biolog and CLPP techniques can also be combined to reveal changes in community tolerance accompanied by changes in potential physiological capacity (Dahllof et al. 2001; Barranguet et al. 2003; Schmitt et al. 2004).

14.3.2.2 Incorporation of Labelled Leucine and Thymidine

Bacterial growth and reproduction are very sensitive to contamination, and thus represent useful indicators of stress (Baath 1992). Hypothetically, in contaminated environments, the growth rate of stressed microorganisms is expected to decrease because energy is diverted from synthesis of new biomass to cell maintenance (i.e., detoxification).

Growth assessment methods based on thymidine and leucine incorporation, that measure incorporation of labelled precursors into DNA and protein, are commonly used methods in the context of soil contamination (e.g., Diaz-Ravina et al. 1994; Demoling and Baath 2008a; Brandt et al. 2009; Demoling et al. 2009). This represents a less direct, but highly sensitive and economical method that can replace or at least complement the classical, more insensitive soil respiration tests that are extensively used for laboratory testing of chemicals in soil. In particular, the Leu-PICT protocol, based on short-term [^{3}H]-leucine incorporation into growing cells, is now a well-established method. Using [^{3}H]-thymidine and [^{14}C]-leucine in a dual-label approach, both parameters being measured in the same assay (Blanck 2002; Boivin et al. 2006a), provides information on biomass production, as well as on the number of active cells.

14.3.2.3   Culture-Independent Methods

Culture-dependent methods are highly biased when analysing environmental samples (Amann et al. 1995), as only a restricted subset of the staggering bacterial diversity of soil can routinely be cultivated under laboratory conditions (Torsvik et al. 1998; Ogram 2000; Gans et al. 2005; Schloss and Handelsman 2006; Roesch et al. 2007; Beg 2008). Culture-independent methods have been developed in the last twenty years to characterise the structure and, increasingly in recent years, the functions of bacterial communities, and their dynamics under changing environmental conditions (Nocker et al. 2007; Wilmes et al. 2009). Many of these methods are based on the taxonomic value of the comparison of 16S rRNA gene sequences amplified using universal primers in polymerase chain reactions (PCRs). In particular, DNA fingerprinting patterns obtained after the separation of the amplified segments by amplified ribosomal DNA restriction analysis (ARISA) (Kumari et al. 2009), terminal-restriction fragment length polymorphism (T-RFLP) (Schutte et al. 2008), or denaturing gradient gel electrophoresis (DGGE) (Muyzer et al. 1993; Muyzer and Smalla 1998) have been used to compare bacterial communities in contaminated soils (MacNaughton et al. 1999; Kozdroj and van Elsas 2001; Feris et al. 2004; Gremion et al. 2004; Boivin et al. 2006a,b; Joynt et al. 2006), because multiple samples can be simultaneously and quickly analysed to discriminate different communities. This growing molecular biological toolbox allows microbial ecologists to monitor and quantify the *in situ* response of specific microbial functions to pollution with increasing precision, most often by using PCR-based protocols.

The advent of efficient and economical deep-sequencing approaches (MacLean et al. 2009) may very soon find a regular application in PICT, as they promise to yield a global picture of the composition and dynamics of both structural and functional genes in bacterial communities, at the DNA but also at the mRNA level (Christen 2008). Beyond technical issues of nucleic acid isolation, sequencing accuracy, or data management, our present incapacity to sample a significant and representative part of the microbial diversity of soils, and of pristine soils in particular (Vogel et al. 2009), is probably the main remaining bottleneck preventing the full exploitation of the power of metagenomic approaches for PICT analysis.

Of course, molecules of biochemical importance other than nucleic acids may also serve as valuable markers in PICT. Most prominently, membrane fatty acid patterns of bacterial communities may also vary as a function of pollutant toxicity and environmental stress (Kandeler et al. 1992; Heipieper and Debont 1994; Heipieper et al. 1996; Joynt et al. 2006; Demoling and Baath 2008b; Demoling et al. 2009). Lipid biomarker-based techniques (Hedrick et al. 1991; MacNaughton et al. 1999) thus provide culture-independent insights into several important characteristics of microbial communities, such as viable biomass, community structure, nutritional status, or physiological stress response of bacteria.

Finally, it should also be emphasised that benchmark comparisons of different methods for detection and estimation of PICT described above are still rare at the present stage. Niklinska et al. (2005) highlighted the higher sensitivity of a PICT approach based on the Biolog ECO plate assay compared to microbial indices such as microbial biomass, basal soil respiration, or community-level physiological profile (CLPP) in studying the effect of Cu and Zn on soil microbial communities. Some other recent pioneering studies also suggest that the PICT response in soil largely varies according to the approach used (Demoling and Baath 2008a; Brandt et al. 2009). Thus, further studies are required to assess in the context of PICT the relevance and sensitivities of the above-described methods under various environmental conditions, and how these methods can be combined in an integrative approach.

14.4 CONTRIBUTING FACTORS TO PICT IN BACTERIA

Environmental systems are complex, with many factors, not all of them readily measurable experimentally, contributing to the adaptation of bacterial communities to toxicant exposure. Prominent among these, soil characteristics and contaminant bioavailability are briefly discussed below.

14.4.1 PHYSICOCHEMICAL CHARACTERISTICS OF THE SOIL

Physicochemical characteristics of soil codetermine the response of microbial communities to contaminants. For example, anoxic conditions in soil resulting from abundant rainfall or flooding may mask effects of metals prevailing under oxic conditions, as observed in the amendment cover of Demmerikse polder (the Netherlands) (van der Welle et al. 2007; Rutgers 2008). Contaminant mobility is also affected by amendments and differs between amendment-soil mixtures, and between different contaminants within a given amendment-soil mixture. Importantly, carbon substrate load has been observed to correlate well with the potential for contaminant impact on overall microbial resistance in soil. In a recently reported study, Brandt and coauthors (2009) compared the impact of the frequently used sulphonamide antibiotic sulphadiazine on the PICT response in bulk soil to soil hot spots amended with nutrients and displaying elevated bacterial activity, and noted an increased PICT response of the soil bacterial community as a result of increased carbon substrate amendment per se. Similarly, Joynt et al. (2006) investigated the relationship between the presence of metal and aromatic contaminants, changes in bacterial community structure, and biomass levels as a function of total organic carbon (TOC) and nitrogen. In fact, several other studies had previously indicated that soil organic carbon content is a more significant predictor of bacterial biomass than contaminant content itself (Landmeyer et al. 1993; Kunito et al. 1999), although this also raises the possibility that contaminants in the investigated soils were poorly available. Moreover, amendment of soils with metal salts is often accompanied by a significant decrease of soil pH (Stevens et al. 2003; Stuczynski et al. 2003), representing a driving factor controlling metal bioavailability (Del Val et al. 1999; Oorts et al. 2006), mobility (Smolders et al. 2004; Ginocchio et al. 2006; Oorts et al. 2006), speciation (Ginocchio et al. 2006), and toxicity, and thereby affecting bacterial community structure, activity, biomass, and substrate utilisation (Niklinska et al. 2005).

14.4.2 BACTERIAL BIOFILMS

The fraction of contaminant that causes an effect on microorganisms, or bioavailable fraction (refer also to Section 4.3.1.2), is difficult to evaluate in the field, due to soil heterogeneity and the associated differential capacity of bacterial populations of variable physiological status to sense, transform, and degrade contaminants in micro-environments with different physicochemical characteristics. As a complicating factor, microorganisms in soil live mostly as biofilms (i.e., heterogeneous biotic assemblages of hydrated exopolymers, microorganisms, and cellular debris) on mineral and organic surfaces (Davey and O'Toole 2000; Young et al. 2008), on microhabitats on the surface of clays or soil organic matter, or partly trapped within microaggregates (Almas et al. 2005). The heterogeneous concentrations and transport conditions of nutrients and contaminants in these structures, in particular at the micro-scale, depend in part on bacterial density and metabolic activity, and will in turn impact on the chemodynamics of contaminants in a way that will be very different to that experienced by planktonic cells in the water column. The fate of contaminants in biofilms may be considered to include three aspects: (1) diffusion into the biofilm, (2) binding to the biofilm polymeric matrix, and (3) contact with the organisms in the biofilm. Hence, either the time required for the pollutant to enter into the biofilm may be very long or, if contamination is transient, the pollutant may not enter the biofilm at all, with important implications for exposure of bacterial cells to contaminants over both spatial and temporal scales, and in different types of environmental contamination (Buffle et al. 2009). Nevertheless, despite these difficulties in assessing PICT in heterogeneous and dynamic environmental systems where physicochemical and microbial components tightly interact over variable spatial and temporal scales, the PICT concept relying on microorganisms has been increasingly applied in recent studies, a selection of which are briefly presented below.

14.5 EXPLORATION AND VALIDATION OF THE PICT CONCEPT USING BACTERIA

A number of studies have documented how PICT develops in soil bacterial communities in the presence of a toxic substance, be it inorganic or organic (Table 14.1). Encouragingly, a quantitative correlation of PICT to classical parameters associated with structural and functional community shifts is becoming apparent.

14.5.1 PICT STUDIES OF BACTERIAL TOLERANCE TO INORGANIC CONTAMINANTS

Metal and metalloid concentrations in soil exert an enormous influence on the diversity, composition, and activity of soil microorganisms. At low concentrations, microorganisms compete for essential trace elements that are required to support growth, and thus also affect plant health through the production of metal chelators. At high concentrations, the toxic effects of metals result in reduced bacterial diversity and altered rates of key biological processes that underlie ecosystem function (Giller et al. 2009). This has become of great concern as large land areas across the globe are now contaminated with metals from both land application and atmospheric deposition of wastes.

The sensitivity of microbial communities to metal contamination has been studied in both field-contaminated and laboratory-spiked soils. High levels of soil metal contamination have been correlated with decreased bacterial activity (Pennanen et al. 1996; Smolders et al. 2004) and biomass (Khan and Scullion 2000), as well as impaired enzyme function and nutrient cycling (Kandeler et al. 1996; Del Val et al. 1999). Metal contamination has been shown to exert selective pressure on soil bacterial communities, leading to shifts in community structures (Diaz-Ravina et al. 1994; Pennanen et al. 1996; Joynt et al. 2006; Ashraf and Ali, 2007) and a decrease in overall diversity (Joynt et al. 2006; Stefanowicz et al. 2008). Examples of immediate effects of metal stress and toxicity that have been observed include a decrease in bacterial biomass (e.g., inhibition of bacterial growth, Knight et al. 1997), in phylotype diversity (Joynt et al. 2006), or loss of biochemical activities essential for soil function (Kandeler et al. 1996).

Changes in microbial communities have been used as evidence for long-term metal contamination, and have suggested a positive correlation between environmental metal concentration and increased microbial community tolerance (Pennanen et al. 1996; Diaz-Ravina and Baath 2001; Lock and Janssen 2005; Diaz-Ravina et al. 2007). This increased tolerance correlated with changes in community structures using molecular fingerprinting techniques, such as phospholipid fatty acid (PFLA) patterns (Pennanen et al. 1996; Baath et al. 2005; Hinojosa et al. 2005), DGGE, or T-RFLP (Clement et al. 1998; Gremion et al. 2004; Gillan et al. 2005; Grandlic et al. 2006). More recently, increasingly powerful tools have allowed dissection of the PICT response at the bacterial level at an unprecedented level of detail. For example, the effects of different concentrations of Zn on soil microbial communities and activities were analysed by applying different doses to the soil of a wheat culture using a combination of cultivation-dependent and -independent methods (Zhou et al. 2009). The bacterial metabolic patterns revealed a clearly bimodal pattern at higher Zn concentrations (>1,920 μg.g^{-1}), suggesting differential sensitivity of the microbial communities to zinc poisoning. Fungal and bacterial growth experiments performed in parallel showed that soil microbial communities and activities could adapt to Zn pollution to a certain extent. In a related study, the effect of soil organic status on copper impact was investigated in microcosm experiments (250 μg Cu.g^{-1}, incubation of 35 days) (Lejon et al. 2008), showing that organic matter controlled copper distribution, speciation, and bioavailability, resulting in differential copper toxicity and impact on microorganisms. Importantly, biomass carbon and fungal ARISA measurements, unlike bacterial ARISA, did not correlate with copper speciation and bioavailability, suggesting that the specific composition of indigenous soil communities predetermines its level of sensitivity to copper toxicity.

Several open questions remain regarding the capacity of metals to shape microbial communities in soils and the uniformity of microbial community responses to metal stress across soil types. For instance, metals are less easy to remove from a contaminated site than organic pollutants, and for this reason are more likely to cause long-term tolerance with lasting shifts in microbial community structures. Indeed, it is possible that there is no generally applicable threshold for metal toxicity, and that it may represent a site-specific property (Bunemann et al. 2006). To obtain more reliable predictions of contaminant toxicity for microorganisms, integration of microbial micro-location relative to hot spots of both nutrient sources and contaminants, if it can be determined, may turn out to be highly advantageous in the future (Giller et al. 2009).

14.5.2 PICT STUDIES OF BACTERIAL TOLERANCE TO ORGANIC CONTAMINANTS

The most striking observation apparent from a survey of the literature on PICT in microbial systems today is that studies dedicated to organic pollutants are much rarer than for inorganic pollutants. This may well reflect the fact that, in contrast to the situation with inorganic pollutants, the focus in this case has been on the search for degradation potential of endogenous microbes to degrade and remove such contaminants, rather than on the use of microbial communities to characterise the response of ecosystems.

Early bioremediation studies in soils or microcosms have often been limited to an assessment of contamination decrease. As a consequence, the microbial compartment has often just represented a black box (Frostegard et al. 1996), despite repeated observations demonstrating that organic toxicants can strongly affect the microbial community structures (e.g., Long et al. 1995; White et al. 1999; Travis et al. 2008). Two opposed effects of organic pollutants on the soil bacterial community can be defined. On the one hand, organic pollutants will be used by some microorganisms as the carbon source for growth, and this will tend to increase the microbial diversity of a given site (Feris et al. 2004). On the other hand, organic pollutants will pose a toxic threat to many microorganisms and, consequently, tend to detrimentally affect abundance or composition of the microbial community (Bachoon et al. 2001). A recent study on the community shifts taking place in natural microbial communities in soils heavily contaminated with jet fuel at different stages of a bioremediation air sparging treatment nicely illustrates both types of effects (Kabelitz et al. 2009). Lipid biomarker analysis, potentially reporting on the physiological status of the community, indicated little differentiation among the bacterial phylotypes present on site, but DNA-based methods revealed significant taxonomic shifts in bacterial community structure following contaminant treatment. As lower concentrations of aromatic compounds became prevalent, increased biomass content, and taxon complexity of bacterial assemblages were both observed. The authors suggested that reduction in the concentration of toxicants, apart from increasing the population of potential degraders, also allowed the development of sensitive bacterial types not necessarily able to degrade aromatic compounds, but possibly cross-feeding on the ensemble of metabolites secreted from the degraders. In a related field study to determine the long-term effect of fertilizer addition on the degradation rate and the toxicity of oil residues in sub-Antarctic conditions, Delille et al. (2007) observed that the toxicity of contaminated soil was maintained despite apparent significant degradation of alkanes and aromatics, suggesting that the less toxic material was degraded first in this case, but that a large proportion of toxic residues remained even after four years.

Finally, it is important to emphasise that only a few studies so far have addressed mixed-waste contaminated soils, although these are clearly very common. In one well-designed study, the impact on the bacterial community of long-term environmental exposure to lead, chromium, and organic chemicals was investigated along a contaminant gradient (Joynt et al. 2006). This study showed that soils contaminated with both metals and hydrocarbons for several decades underwent significant changes in community composition, but still contained a phylogenetically diverse group of bacteria. Observations of this type would clearly warrant further study in the context of investigations on bacterial community tolerance to contaminants.

14.5.3 Validity of the PICT Concept Using Bacteria

As described above, the existence of a qualitative or quantitative link between the degrees of PICT, the concentrations of toxicants, and other ecological parameters is now increasingly documented in the literature. Nevertheless, coming up with generally applicable predictions within a defined risk assessment framework based on PICT measurements still appears beyond reach at present due to several limits that still challenge the validity of the PICT concept using microorganisms.

First, the occurrence of cotolerance patterns at the microbial community level complicates the interpretation of the PICT signal arising from a cocktail of contaminants. Indeed, cotolerance to several contaminants may occur in some instances (Blanck and Wangberg 1991), although measurement of PICT has the potential of being rather specific for the toxicant being tested (van Beelen et al. 2004). Cotolerance may arise due to the fact that genes for resistance to, or transformation of, different contaminants are found on the same mobile genetic element, such as a plasmid or a transposon, thus eliciting cotolerance to contaminants that are *a priori* unrelated structurally or functionally (Top and Springael 2003; Wright et al. 2008). As a consequence, a microbial community may develop tolerance for a compound to which it was never previously exposed. Also, in the present state of knowledge, compounds that are related in terms of their effect on microorganisms are best considered collectively as a group when using the PICT approach (Blanck 2002; Schmitt et al. 2006).

Second, discrimination at the phylotype level when characterising the PICT response may not be informative, and the strains really contributing to community changes may be difficult to define with precision. Indeed, the species concept is currently hotly debated in the case of prokaryotes (Doolittle and Zhaxybayeva 2009), as functional properties of bacterial taxa do not necessarily correlate with phylogenetic pedigree, and may rapidly change following exposure to contaminants, due to the ability of microbes for lateral exchange of genetic material. This is a notable confounding factor with respect to ecotoxicological approaches, such as the species sensitivity approach (Escher and Hermens 2002), which assume that the organisms tested are representative of the ecosystem under study, and that counts of all populations by random sampling will yield a representative view of the prevailing situation.

Finally, some important questions remain on the correlation of observed PICT responses and measured residual contamination in follow-up studies of treated contaminated sites. How long will PICT be detected after a toxicant has disappeared from the ecosystem? How quickly will the community return to the tolerance levels observed before the toxic event? Obtaining detailed information on such "recovery phases" seems essential in the context of the environmental rehabilitation of contaminated sites and associated safety issues. So far, it has been shown that ecosystem resilience and long-term recovery after toxicant removal may be reflected by changes in the microbial community structures (Matthews et al. 1996; Curtis et al. 2002). Indeed, several studies suggest that community recovery follows contaminant removal, as indicated by recolonisation of the site of interest by more sensitive populations, for both inorganic (Diaz-Ravina and Baath 2001) and organic (Demoling and Baath 2008b) contaminants.

14.6 THE BACTERIAL FUTURE FOR PICT

Soil ecosystems often represent ultimate sinks for many metals and organic contaminants, and the associated bacterial diversity has emerged as a key indicator with significant untapped potential for use as a proxy to assess the effect of contamination. Ideally, further development of microbiology-based indicators of toxicant exposure (Table 14.2) will allow the integration of the ecodynamic of contaminants in much the same way as macro-biological indicators can achieve, and the PICT approach holds great promise in this context (Blanck and Wangberg 1988; Boivin et al. 2002), if the challenges already mentioned in part above, as summarised briefly below, can be addressed and resolved satisfactorily.

14.6.1 THE ELUSIVE RELATIONSHIP BETWEEN CONTAMINATION AND BACTERIAL COMMUNITY STRUCTURE AND FUNCTION

Knowledge of bacterial community composition and dynamics in soil, and of its role for ecosystem processes in the context of exposure to contaminants, is currently inadequate. In part, this can be attributed to the high phylotype richness, functional redundancy, and complexity of soil ecosystems (Wall and Virginia 1999), the major part of which is currently noncultivated, and which next-generation sequencing approaches has revealed to be even more staggering than had been hitherto estimated (e.g., Roesch et al. 2007; Vogel et al. 2009).

Clearly, a better understanding of the relationship between contaminant exposure and the structure and function of microbial communities will need to await the establishment of reliable, generally accepted global strategies and concepts for the analysis of the enormous amounts of information now afforded by deep-sequencing approaches (Christen 2008; MacLean et al. 2009).

These developments may largely contribute to answer key issues on the ecological functioning of contaminated sites. For example, it is often assumed that an increase in community tolerance is accompanied by a decrease in ecosystem function and/ or phylotype richness, so that a shift of communities toward greater tolerance coincides with the apparent extinction of sensitive phylotypes. There is some evidence that higher bacterial diversity confers stability and resilience in ecosystem function (Giller et al. 1997; Fitter et al. 2005; Brussaard et al. 2007), but more research is needed on the relationship between diversity and stability of ecosystem functions

TABLE 14.2
Bacterial-Based Ecological Indicators Described for Contaminated Soil Systems

Bioindicator	Applications and Comments	References
[^{3}H]-leucine incorporation	Integrated response of the community; only applicable to toxicants affecting metabolism	Baath et al. 1998; Diaz-Ravina and Baath 2001; Shi et al. 2002a,b; Brandt et al. 2004, 2009; Petersen et al. 2004; Rajapaksha et al. 2004; Baath et al. 2005; Boivin et al. 2006a; Tobor-Kaplon et al. 2006; Diaz-Ravina et al. 2007; Ipsilantis and Coyne 2007; Jia et al. 2007; Demoling and Baath 2008a, b; Demoling et al. 2009; Ricart et al. 2009
Biomass	Limited utility due to spatial heterogeneity and technical shortcomings in its measurement	Rutgers and Breure 1999; Thompson et al. 1999; Kandeler et al. 2000; Shi et al. 2002a; Jiang et al. 2003; Stuczynski et al. 2003; Wilke et al. 2004; Baath et al. 2005; Hinojosa et al. 2005; Niklinska et al. 2005, 2006; Shi et al. 2005; Boivin et al. 2006a; Joynt et al. 2006; Pereira et al. 2006; Ranjard et al. 2006; Broos et al. 2007; Ipsilantis and Coyne 2007; Jia et al. 2007; Demoling and Baath 2008a; Lejon et al. 2008; Stefanowicz et al. 2008, 2009; Teng et al. 2008; Anderson et al. 2009; Demoling et al. 2009; Hui et al. 2009; Kabelitz et al. 2009; Mathe-Gaspar et al. 2009; Wang et al. 2009; Zhou et al. 2009
Microcalorimety	Integrative measurement of the metabolic activity of soil bacteria and fungi	Yao et al. 2007; Zhou et al. 2009
Community-level physiological profiling (CLPP), Biolog profiling	Readout of bacterial community structure physiological profiling; detection of multiple microbial metabolic activities using the pH-dependent tetrazolium dye, which reports on the cellular redox state; compatible with microtiter plate format; fingerprints can be compared between samples using multivariate analysis and ordination methods	Knight et al. 1997; Rutgers et al. 1998; Garland and Mills 1999; Ellis et al. 2001; Gremion et al. 2004; Niklinska et al. 2004, 2005, 2006; Schmitt et al. 2005, 2006; Ramsey et al. 2006; Yang et al. 2006; Travis et al. 2008; Demoling et al. 2009; Mathe-Gaspar et al. 2009; Stefanowicz et al. 2009; Wang et al. 2009

(continued)

TABLE 14.2

Bacterial-Based Ecological Indicators Described for Contaminated Soil Systems (Continued)

Bioindicator	Applications and Comments	References
Taxonomic community structure	PCR- and DNA-based profiling (DGGE, ARISA, T-RFLP); patterns can be compared between samples using multivariate analysis and ordination methods	Muyzer and Smalla 1998; Torsvik et al. 1998; Stephen et al. 1999; White et al. 1999; Kandeler et al. 2000; Klopfenstein et al. 2001; Kozdroj and van Elsas 2001; Seghers et al. 2003; Feris et al. 2004; Gillan 2004; Gremion et al. 2004; van Bleijswijk and Muyzer 2004; Boivin et al. 2006a,b; Kourtev et al. 2006; Ramsey et al. 2006; Takahata et al. 2006; Bordenave et al. 2007; Naslund et al. 2008; Ogilvie and Grant 2008; Tlili et al. 2008; Travis et al. 2008; Moreno et al. 2009
Lipid composition	Provides culture-independent insights into important characteristics, such as viable biomass, community structure, nutritional status, and physiological stress response of the strains composing the community; patterns can be compared between samples using multivariate analysis and ordination methods	Pennanen et al. 1998a,b; MacNaughton et al. 1999; Stephen et al. 1999; White et al. 1999; Huertas et al. 2000; Kandeler et al. 2000; Khan and Scullion 2000; Elsgaard et al. 2001; Grcman et al. 2001; Priha et al. 2001; Shi et al. 2002a; Petersen et al. 2004; Bundy et al. 2004; Rajapaksha et al. 2004; Wilke et al. 2004, 2005; Baath et al. 2005; Hinojosa et al. 2005; Ramsey et al. 2005; Joynt et al. 2006; Ramsey et al. 2006; Antizar-Ladislao et al. 2007; Lazaroaie 2007; Demoling and Baath 2008a; Ringelberg et al. 2008; Demoling et al. 2009; Kabelitz et al. 2009

before a predictive understanding of ecosystem response to contaminant stress may be developed.

14.6.2 DEFINING APPROPRIATE BIOINDICATORS

Selecting appropriate microbial indicators for contaminant exposure is difficult because several other environmental variables also modulate the response of microbial functions, and thus contribute to confound effects that are specifically due to contaminant exposure. Accordingly, such indicators should be both spatially and temporally defined, and also ideally account for heterogeneity at a given spatial and temporal scale (Conroy et al. 1996). Nevertheless, the majority of environmental soil monitoring programmes today (reviewed in Nielsen and Winding 2002) still use data sets with only minimal biological components to describe soil quality (Winder 2003).

Baseline, classically used microbial bioindicators have the advantage to potentially yield relatively inexpensive and easy-to-measure integrative estimates of the functional status of the overall microbial community, and can be further included in statistical models. However, the usually measured types of such measurements have been increasingly criticised as being inconsistent and displaying high coefficients of variation, e.g., in response to seasonal climatic changes (Gil-Sotres et al. 2005).

Alternative PICT-type approaches, potentially including both baseline and multiparametric assays, and a combination of culture-based and cultivation-independent approaches (Table 14.1), may thus prove a more suitable integrative approach for the comprehensive assessment of soil ecosystems in the future (e.g., Winding et al. 2005; Weiss and Cozzarelli 2008; Imfeld et al. 2009).

14.6.3 ADDRESSING DIFFUSE CONTAMINATIONS

The analysis of diffuse contamination, where contaminant-independent inputs are often likely to be comparable in magnitude to contaminant-specific effects (Boivin et al. 2002), will differ from environmental problems emanating from point sources, where adverse effects are dominant and immediately obvious. In line with the PICT concept, factors potentially masking contaminant effects may be efficiently dealt with using multiendpoint observation techniques coupled with multivariate analyses (Rutgers 2008). Such approaches will need to be more systematically developed in the future, in particular for aspects regarding the bacterial compartment.

14.6.4 ADDRESSING UNCERTAINTIES IN RISK ASSESSMENT PROCEDURES

Clearly, each endpoint of a microbially based indicator, including the PICT approach, introduces conceptual uncertainty: To which extent does the indicator effectively report on the expected target? Uncertainty originates from our imperfect knowledge of functioning of ecosystems and may only be reduced through method development and by using alternative methods and improved models (Winding et al. 2005). A clear-cut interpretation to establish a causative relationship between the detection of community tolerance and the presence of a localised specific toxic effect for a given contaminant is therefore often challenging at present (Boivin et al. 2002, 2006a). Being able to define in experimental terms the normal operating range of a system will also reduce probabilistic uncertainty (Beer 2006). A major drawback in the interpretation of experimental data is that the microbial response measured in a particular assay, or the mechanism behind it, often remains unknown. In this sense, risk assessment methodology still remains an essentially operational, nonexplanatory approach, a situation that applies to bacterial-based PICT-type approaches as well. Some ways forward are beginning to be explored in this context, such as the weight of evidence (WOE) approach, which relies on data from independent scientific disciplines to increase the precision of the answer (Long and Chapman 1985; Suter et al. 2000; Iannuzzi et al. 2008). The development of WOE approaches in risk assessment, including multivariate statistics as filtering tools, fits well with current concerns in the field of stress ecology (Batley et al. 2002; Burton et al. 2002), and in our view could be extended with profit within the PICT framework (Faber 2006).

14.6.5   THE SCALE ISSUE

Another remaining, long-standing challenge will be to master the transition from laboratory scale experiments to the field. The strength of laboratory-based ecotoxicology, which already includes PICT-type approaches, is well established. However, its relevance to effects occurring in natural systems is often weak, and this has become a major focal point for developments. The emerging evidence suggests that bacterial responses observed in short-term assays (and for the main usually involving acute toxicity) bear little resemblance to those observed in long-term field studies (which integrate chronic toxicity and stress) (e.g., Renella et al. 2002a,b; Yao et al. 2007). Clearly, concerted, integrative, and complementary investigations on both laboratory and field scales have the potential to strengthen the usefulness of the PICT approach for investigations of contaminated soil ecosystems, as well as our understanding of processes driving microbial tolerance.

In conclusion, it can be hoped that, through judicious selection of model ecosystems with reduced complexity at the laboratory, microcosm, or field scales, novel, yet hidden, aspects of microbial mechanisms of tolerance and adaptation to contaminants will be revealed, which eventually will also apply to more complex contaminated ecosystems. The yet unheard of quantities of biological information that are being collected in ongoing high-throughput, large-scale sequencing efforts (Vogel et al. 2009) may also eventually contribute to a better understanding of ecosystem functioning. In any event, current interest in the PICT approach is expected to boost the much needed development of models to describe and predict the response of soil ecosystems to environmental changes and disturbances (Freckman and Virginia 1997), and fundamental research into the mechanisms by which contaminants affect individual microorganisms and microbial communities and their activities in soils. Such developments can further reach a point where biologically informed integrated strategies for risk assessment may be implemented to manage soil bodies and services, and prevent their degradation in the future.

ACKNOWLEDGMENTS

G.I., F.B., and S.V. are members of REALISE, the Network of Laboratories in Engineering and Sciences for the Environment of Région Alsace (France) (http://realise.u-strasbg.fr), from which support is gratefully acknowledged.

REFERENCES

Almas, A. R., J. Mulder, and L. R. Bakken. 2005. Trace metal exposure of soil bacteria depends on their position in the soil matrix. *Environ. Sci. Technol.* 39:5927–32.

Amann, R. I., W. Ludwig, and K. H. Schleifer. 1995. Phylogenetic identification and *in-situ* detection of individual microbial-cells without cultivation. *Microbiol. Rev.* 59:143–69.

Anderson, J. A. H., et al. 2009. Molecular and functional assessment of bacterial community convergence in metal-amended soils. *Microbial Ecol.* 58:10–22.

Anderson, J. P. E., and K. H. Domsch. 1978. Physiological method for quantitative measurement of microbial biomass in soils. *Soil Biol. Biochem.* 10:215–21.

Antizar-Ladislao, B., et al. 2007. The influence of different temperature programmes on the bioremediation of polycyclic aromatic hydrocarbons (PAHs) in a coal-tar contaminated soil by in-vessel composting. *J. Hazard. Mater.* 144:340–47.

Ashraf, R., and T. A. Ali. 2007. Effect of heavy metals on soil microbial community and mung beans seed germination. *Pakistan J. Bot.* 39:629–36.

Baath, E. 1992. Measurement of heavy-metal tolerance of soil bacteria using thymidine incorporation into bacteria extracted after homogenization centrifugation. *Soil Biol. Biochem.* 24:1167–72.

Baath, E., M. Diaz-Ravina, and L. R. Bakken. 2005. Microbial biomass, community structure and metal tolerance of a naturally Pb-enriched forest soil. *Microbial Ecol.* 50:496–505.

Baath, E., et al. 1998. Microbial community-based measurements to estimate heavy metal effects in soil: The use of phospholipid fatty acid patterns and bacterial community tolerance. *Ambio* 27:58–61.

Bachoon, D. S., et al. 2001. Microbial community assessment in oil-impacted salt marsh sediment microcosms by traditional and nucleic acid-based indices. *J. Microbiol. Meth.* 46:37–49.

Barranguet, C., et al. 2003. Copper-induced modifications of the trophic relations in riverine algal-bacterial biofilms. *Environ. Toxicol. Chem.* 22:1340–49.

Batley, G. E., et al. 2002. Uncertainties in sediment quality weight-of-evidence (WOE) assessments. *Hum. Ecol. Risk Assess.* 8:1517–47.

Beer, T. 2006. Ecological risk assessment and quantitative consequence analysis. *Hum. Ecol. Risk Assess.* 12:51–65.

Beg, Q. K. 2008. *Accessing uncultivated microorganisms from the environment to organisms and genomes and back*, ed. K. Zengler. Washington, DC: American Society for Microbiology.

Blanck, H. 2002. A critical review of procedures and approaches used for assessing pollution-induced community tolerance (PICT) in biotic communities. *Hum. Ecol. Risk Assess.* 8:1003–34.

Blanck, H., and S. A. Wangberg. 1988. Pollution-induced community tolerance—A new ecotoxicological tool. In *Functional testing of aquatic biota for estimating hazards of chemicals*, ed. J. Cairns Jr. and J. R. Pratt, 219–30. West Conshohocken, PA: American Society for Testing and Materials.

Blanck, H., and S. A. Wangberg. 1991. Pattern of cotolerance in marine periphyton communities established under arsenate stress. *Aquat. Toxicol.* 21:1–14.

Boivin, M. E. Y., et al. 2002. Determination of field effects of contaminants—Significance of pollution-induced community tolerance. *Hum. Ecol. Risk Assess.* 8:1035–55.

Boivin, M. E. Y., et al. 2006a. Discriminating between effects of metals and natural variables in terrestrial bacterial communities. *Appl. Soil Ecol.* 34:103–13.

Boivin, M. E. Y., et al. 2006b. Functional recovery of biofilm bacterial communities after copper exposure. *Environ. Pollut.* 140:239–46.

Bordenave, S., et al. 2007. Effects of heavy fuel oil on the bacterial community structure of a pristine microbial mat. *Appl. Environ. Microbiol.* 73:6089–97.

Brandt, K. K., et al. 2004. Microbial community-level toxicity testing of linear alkylbenzene sulfonates in aquatic microcosms. *FEMS Microbiol. Ecol.* 49:229–41.

Brandt, K. K., et al. 2009. Increased pollution-induced bacterial community tolerance to sulfadiazine in soil hotspots amended with artificial root exudates. *Environ. Sci. Technol.* 43:2963–68.

Brookes, P. C., et al. 1985. Chloroform fumigation and the release of soil-nitrogen—A rapid direct extraction method to measure microbial biomass nitrogen in soil. *Soil Biol. Biochem.* 17:837–42.

Broos, K., et al. 2007. Limitations of soil microbial biomass carbon as an indicator of soil pollution in the field. *Soil Biol. Biochem.* 39:2693–95.

Brussaard, L., P. C. de Ruiter, and G. G. Brown. 2007. Soil biodiversity for agricultural sustainability. *Agric. Ecosyst. Environ.* 121:233–44.

Buffle, J., K. J. Wilkinson, and H. P. van Leeuwen. 2009. Chemodynamics and bioavailability in natural waters. *Environ. Sci. Technol.* 43:7170–74.

Bundy, J. G., G. I. Paton, and C. D. Campbell. 2004. Combined microbial community level and single species biosensor responses to monitor recovery of oil polluted soil. *Soil Biol. Biochem.* 36:1149–59.

Bunemann, E. K., G. D. Schwenke, and L. Van Zwieten. 2006. Impact of agricultural inputs on soil organisms—A review. *Aust. J. Soil Res.* 44:379–406.

Burton, G. A., et al. 2002. A weight-of-evidence framework for assessing sediment (or other) contamination: Improving certainty in the decision-making process. *Hum. Ecol. Risk Assess.* 8:1675–96.

Christen, R. 2008. Global sequencing: A review of current molecular data and new methods available to assess microbial diversity. *Microbes Environ.* 23:253–68.

Clement, B. G., et al. 1998. Terminal restriction fragment patterns (TRFPs), a rapid, PCR-based method for the comparison of complex bacterial communities. *J. Microbiol. Meth.* 31:135–42.

Conroy, M. J., et al. 1996. Statistical inference on patch-specific survival and movement rates from marked animals. *Environ. Ecol. Stat.* 3:99–116.

Copley, S. D. 2009. Evolution of efficient pathways for degradation of anthropogenic chemicals. *Nat. Chem. Biol.* 5:560–67.

Crawford, J. W., et al. 2005. Towards an evolutionary ecology of life in soil. *Trends Ecol. Evol.* 20:81–87.

Curtis, T. P., W. T. Sloan, and J. W. Scannell. 2002. Estimating prokaryotic diversity and its limits. *Proc. Natl. Acad. Sci. USA* 99:10494–99.

Dahllof, I., et al. 2001. The effect of TBT on the structure of a marine sediment community—A boxcosm study. *Mar. Pollut. Bull.* 42:689–95.

Dalal, R. C. 1998. Soil microbial biomass—What do the numbers really mean? *Aust. J. Exp. Agric.* 38:649–65.

Davey, M. E., and G. A. O'Toole. 2000. Microbial biofilms: From ecology to molecular genetics. *Microbiol. Mol. Biol. Rev.* 64:847–67.

Davis, M. R. H., F. J. Zhao, and S. P. McGrath. 2004. Pollution-induced community tolerance of soil microbes in response to a zinc gradient. *Environ. Toxicol. Chem.* 23:2665–72.

D'Costa, V. M., E. Griffiths, and G. D. Wright. 2007. Expanding the soil antibiotic resistome: Exploring environmental diversity. *Curr. Opin. Microbiol.* 10:481–89.

Delille, D., F. Coulon, and E. Pelletier. 2007. Long-term changes of bacterial abundance, hydrocarbon concentration and toxicity during a biostimulation treatment of oil-amended organic and mineral sub-Antarctic soils. *Polar Biol.* 30:925–33.

Del Val, C., J. M. Barea, and C. Azon-Aguilar. 1999. Assessing the tolerance to heavy metals of arbuscular mycorrhizal fungi isolated from sewage sludge-contaminated soils. *Appl. Soil Ecol.* 11:261–69.

Demoling, L. A., and E. Baath. 2008a. Use of pollution-induced community tolerance of the bacterial community to detect phenol toxicity in soil. *Environ. Toxicol. Chem.* 27:334–40.

Demoling, L. A., and E. Baath. 2008b. No long-term persistence of bacterial pollution-induced community tolerance in tylosin-polluted soil. *Environ. Sci. Technol.* 42:6917–21.

Demoling, L. A., and E. Baath. 2008c. The use of leucine incorporation to determine the toxicity of phenols to bacterial communities extracted from soil. *Appl. Soil Ecol.* 38:34–41.

Demoling, L. A., et al. 2009. Effects of sulfamethoxazole on soil microbial communities after adding substrate. *Soil Biol. Biochem.* 41:840–48.

Diaz-Ravina, M., and E. Baath. 2001. Response of soil bacterial communities pre-exposed to different metals and reinoculated in an unpolluted soil. *Soil Biol. Biochem.* 33:241–48.

Diaz-Ravina, M., R. C. de Anta, and E. Baath. 2007. Tolerance (PICT) of the bacterial communities to copper in vineyards soils from Spain. *J. Environ. Qual.* 36:1760–64.

Diaz-Ravina, M., et al. 1994. Multiple heavy-metal tolerance of soil bacterial communities and its measurement by a thymidine incorporation technique. *Appl. Environ. Microbiol.* 60:2238–47.

Doolittle, W. F., and O. Zhaxybayeva. 2009. On the origin of prokaryotic species. *Genome Res.* 19:744–56.

Dumbrell, A. J., et al. 2009. Relative roles of niche and neutral processes in structuring a soil microbial community. *ISME J.* 4:337–45.

Ellis, R. J., et al. 2001. Comparison of microbial and meiofaunal community analyses for determining impact of heavy metal contamination. *J. Microbiol. Meth.* 45:171–85.

Elsgaard, L., S. O. Petersen, and K. Debosz. 2001. Effects and risk assessment of linear alkylbenzene sulfonates in agricultural soil. 1. Short-term effects on soil microbiology. *Environ. Toxicol. Chem.* 20:1656–63.

Escher, B. I., and J. L. M. Hermens. 2002. Modes of action in ecotoxicology: Their role in body burdens, species sensitivity, QSARs, and mixture effects. *Environ. Sci. Technol.* 36:4201–17.

Faber, J. H. 2006. European experience on application of site-specific ecological risk assessment in terrestrial ecosystems. *Hum. Ecol. Risk Assess.* 12:39–50.

Feris, K. P., et al. 2004. A shallow BTEX and MTBE contaminated aquifer supports a diverse microbial community. *Microbial Ecol.* 48:589–600.

Fitter, A. H., et al. 2005. Biodiversity and ecosystem function in soil. *Funct. Ecol.* 19:369–77.

Freckman, D. W., and R. A. Virginia. 1997. Low-diversity Antarctic soil nematode communities: Distribution and response to disturbance. *Ecology* 78:363–69.

Frostegard, A., A. Tunlid, and E. Baath. 1996. Changes in microbial community structure during long term incubation in two soils experimentally contaminated with metals. *Soil Biol. Biochem.* 28:55–63.

Gans, J., M. Wolinsky, and J. Dunbar. 2005. Computational improvements reveal great bacterial diversity and high metal toxicity in soil. *Science* 309:1387–1390.

Garland, J. L., and A. L. Mills. 1999. Importance of pattern analysis in community-level physiological profiles (CLPP): A reply to the letter from P. J. A. Howard. *Soil Biol. Biochem.* 31:1201–2.

Garland, J. L., et al. 2003. Community-level physiological profiling performed with an oxygen-sensitive fluorophore in a microtiter plate. *Appl. Environ. Microbiol.* 69:2994–98.

Gillan, D. C. 2004. The effect of an acute copper exposure on the diversity of a microbial community in North Sea sediments as revealed by DGGE analysis—The importance of the protocol. *Mar. Pollut. Bull.* 49:504–13.

Gillan, D. C., et al. 2005. Structure of sediment-associated microbial communities along a heavy-metal contamination gradient in the marine environment. *Appl. Environ. Microbiol.* 71:679–90.

Giller, K. E., E. Witter, and P. McGrath. 1998. Toxicity of heavy metals to microorganisms and microbial processes in agricultural soils: A review. *Soil Biol. Biochem.* 30:1389–414.

Giller, K. E., E. Witter, and S. P. McGrath. 2009. Heavy metals and soil microbes. *Soil Biol. Biochem.* 41:2031–37.

Giller, K. E., et al. 1997. Agricultural intensification, soil biodiversity and agroecosystem function. *Appl. Soil Ecol.* 6:3–16.

Gil-Sotres, F., et al. 2005. Different approaches to evaluating soil quality using biochemical properties. *Soil Biol. Biochem.* 37:877–87.

Ginocchio, R., et al. 2006. Agricultural soils spiked with copper mine wastes and copper concentrate: Implications for copper bioavailability and bioaccumulation. *Environ. Toxicol. Chem.* 25:712–18.

Gong, P., et al. 2002. Assessment of pollution-induced microbial community tolerance to heavy metals in soil using ammonia-oxidizing bacteria and biolog assay. *Hum. Ecol. Risk Assess.* 8:1067–81.

Grandlic, C. J., et al. 2006. Lead pollution in a large, prairie-pothole lake (Rush Lake, WI, USA): Effects on abundance and community structure of indigenous sediment bacteria. *Environ. Pollut.* 144:119–26.

Grcman, H., et al. 2001. EDTA enhanced heavy metal phytoextraction: Metal accumulation, leaching and toxicity. *Plant Soil* 235:105–14.

Gremion, F., et al. 2004. Impacts of heavy metal contamination and phytoremediation on a microbial community during a twelve-month microcosm experiment. *FEMS Microbiol. Ecol.* 48:273–83.

Griffiths, B. S., et al. 2001. An examination of the biodiversity-ecosystem function relationship in arable soil microbial communities. *Soil Biol. Biochem.* 33:1713–22.

Hedrick, D. B., et al. 1991. Disturbance, starvation, and overfeeding stresses detected by microbial lipid biomarkers in high-solids high-yield methanogenic reactors. *J. Ind. Microbiol.* 8:91–98.

Heipieper, H. J., and J. A. M. Debont. 1994. Adaptation of *Pseudomonas putida* S12 to ethanol and toluene at the level of fatty-acid composition of membranes. *Appl. Environ. Microbiol.* 60:4440–44.

Heipieper, H. J., et al. 1996. Effect of environmental factors on the trans/cis ratio of unsaturated fatty acids in *Pseudomonas putida* S12. *Appl. Environ. Microbiol.* 62:2773–77.

Hinojosa, M. B., et al. 2005. Microbial response to heavy metal-polluted soils: Community analysis from phospholipid-linked fatty acids and ester-linked fatty acids extracts. *J. Environ. Qual.* 34:1789–800.

Huertas, M. J., et al. 2000. Tolerance to sudden organic solvent shocks by soil bacteria and characterization of *Pseudomonas putida* strains isolated from toluene polluted sites. *Environ. Sci. Technol.* 34:3395–400.

Hui, N., et al. 2009. Influence of lead on organisms within the detritus food web of a contaminated pine forest soil. *Boreal Environ. Res.* 14:70–85.

Iannuzzi, T. J., et al. 2008. Sediment quality triad assessment of an industrialized estuary of the northeastern USA. *Environ. Monit. Assess.* 139:257–75.

Imfeld, G., et al. 2009. Monitoring and assessing processes of organic chemicals removal in constructed wetlands. *Chemosphere* 74:349–62.

Ipsilantis, I., and M. S. Coyne. 2007. Soil microbial community response to hexavalent chromium in planted and unplanted soil. *J. Environ. Qual.* 36:638–45.

Iyyemperumal, K., D. W. Israel, and W. Shi. 2007. Soil microbial biomass, activity and potential nitrogen mineralization in a pasture: Impact of stock camping activity. *Soil Biol. Biochem.* 39:149–57.

Jia, Y., et al. 2007. Kinetics of microbial growth and degradation of organic substrates in subsoil as affected by an inhibitor, benzotriazole: Model based analyses of experimental results. *Soil Biol. Biochem.* 39:1597–608.

Jiang, X. J., et al. 2003. Changes in soil microbial biomass and Zn extractability over time following Zn addition to a paddy soil. *Chemosphere* 50:855–61.

Joynt, J., et al. 2006. Microbial community analysis of soils contaminated with lead, chromium and petroleum hydrocarbons. *Microbial Ecol.* 51:209–19.

Kabelitz, N., et al. 2009. Enhancement of the microbial community biomass and diversity during air sparging bioremediation of a soil highly contaminated with kerosene and BTEX. *Appl. Microbiol. Biot.* 82:565–77.

Kahkonen, M. A., M. Tuomela, and A. Hatakka. 2007. Microbial activities in soils of a former sawmill area. *Chemosphere* 67:521–26.

Kamitani, T., H. Oba, and N. Kaneko. 2006. Microbial biomass and tolerance of microbial community on an aged heavy metal polluted floodplain in Japan. *Water Air Soil Pollut.* 172:185–200.

Kandeler, E., C. Kampichler, and O. Horak. 1996. Influence of heavy metals on the functional diversity of soil microbial communities. *Biol. Fert. Soils* 23:299–306.

Kandeler, E., G. Luftenegger, and S. Schwarz. 1992. Soil microbial processes and testacea (Protozoa) as indicators of heavy-metal pollution. *Z. Pflanz. Bodenkunde* 155:319–22.

Kandeler, E., et al. 2000. Structure and function of the soil microbial community in microhabitats of a heavy metal polluted soil. *Biol. Fert. Soils* 32:390–400.

Khan, M., and J. Scullion. 2000. Effect of soil on microbial responses to metal contamination. *Environ Pollut.* 110:115–25.

Klopfenstein, N. B., et al. 2001. Molecular genetic approaches to risk assessment in forest ecosystems. In *Proceedings of the Society of American Foresters*, 2000 (Nov.) National Convention, Washington, DC, pp. 108–21.

Knight, B. P., S. P. McGrath, and A. M. Chaudri. 1997. Biomass carbon measurements and substrate utilization patterns of microbial populations from soils amended with cadmium, copper, or zinc. *Appl. Environ. Microbiol.* 63:39–43.

Kozdroj, J., and J. D. van Elsas. 2001. Structural diversity of microbial communities in arable soils of a heavily industrialised area determined by PCR-DGGE fingerprinting and FAME profiling. *Appl. Soil Ecol.* 17:31–42.

Kumari, N., et al. 2009. Molecular approaches towards assessment of cyanobacterial biodiversity. *Afr. J. Biotechnol.* 8:4284–98.

Kunito, T., et al. 1999. Influences of copper forms on the toxicity to microorganisms in soils. *Ecotoxicol. Environ. Saf.* 44:174–81.

Kourtev, P. S., C. H. Nakatsu, and A. Konopka. 2006. Responses of the anaerobic bacterial community to addition of organic C in chromium(VI)- and iron(III)-amended microcosms. *Appl. Environ. Microbiol.* 72:628–37.

Landmeyer, J. E., P. M. Bradley, and F. H. Chapelle. 1993. Influence of Pb on microbial activity in Pb-contaminated soils. *Soil Biol. Biochem.* 25:1465–66.

Lazaroaie, M. 2007. The tolerance of gram-negative bacterial strain to saturated and aromatic hydrocarbons. *Rom. Biotechnol. Lett.* 12:3329–38.

Lejon, D. P. H., et al. 2008. Copper dynamics and impact on microbial communities in soils of variable organic status. *Environ. Sci. Technol.* 42:2819–25.

Liu, B. R., et al. 2006. A review of methods for studying microbial diversity in soils. *Pedosphere* 16:18–24.

Lock, K., and C. R. Janssen. 2005. Influence of soil zinc concentrations on zinc sensitivity and functional diversity of microbial communities. *Environ. Pollut.* 136:275–81.

Long, E. R., and P. M. Chapman. 1985. A sediment quality triad—Measures of sediment contamination, toxicity and infaunal community composition in Puget Sound. *Mar. Pollut. Bull.* 16:405–15.

Long, S. C., et al. 1995. A comparison of microbial community characteristics among petroleum-contaminated and uncontaminated subsurface soil samples. *Microbial Ecol.* 30:297–307.

MacLean, D., J. D. G. Jones, and D. J. Studholme. 2009. Application of 'next-generation' sequencing technologies to microbial genetics. *Nat. Rev. Microbiol.* 7:287–96.

MacNaughton, S. J., et al. 1999. Microbial population changes during bioremediation of an experimental oil spill. *Appl. Environ. Microbiol.* 65:3566–74.

Martinez, J. L., et al. 2009. Functional role of bacterial multidrug efflux pumps in microbial natural ecosystems. *FEMS Microbiol. Rev.* 33:430–49.

Mathe-Gaspar, G., et al. 2009. Environmental impact of soil pollution with toxic elements from the lead and zinc mine at Gyongyosoroszi (Hungary). *Commun. Soil Sci. Plant Anal.* 40:294–302.

Matthews, R. A., W. G. Landis, and G. B. Matthews. 1996. The community conditioning hypothesis and its application to environmental toxicology. *Environ. Toxicol. Chem.* 15:597–603.

McLaughlin, M. J., and E. Smolders. 2001. Background zinc concentrations in soil affect the zinc sensitivity of soil microbial processes—A rationale for a metalloregion approach to risk assessments. *Environ. Toxicol. Chem.* 20:2639–43.

Millward, R. N., and P. L. Klerks. 2002. Contaminant-adaptation and community tolerance in ecological risk assessment: Introduction. *Hum. Ecol. Risk Assess.* 8:921–32.

Moreno, B., et al. 2009. Restoring biochemical activity and bacterial diversity in a trichloroethylene-contaminated soil: The reclamation effect of vermicomposted olive wastes. *Environ. Sci. Pollut. Res. Int.* 16:253–64.

Müller, A. K., L. D. Rasmussen, and S. J. Sorensen. 2001. Adaptation of the bacterial community to mercury contamination. *FEMS Microbiol. Lett.* 204:49–53.

Muyzer, G., E. C. Dewaal, and A. G. Uitterlinden. 1993. Profiling of complex microbial populations by denaturing gradient gel-electrophoresis analysis of polymerase chain reaction-amplified genes coding for 16s ribosomal-RNA. *Appl. Environ. Microbiol.* 59:695–700.

Muyzer, G., and K. Smalla. 1998. Application of denaturing gradient gel electrophoresis (DGGE) and temperature gradient gel electrophoresis (TGGE) in microbial ecology. *Anton. Leeuw. Int. J. G.* 73:127–41.

Naslund, J., J. E. Hedman, and C. Agestrand. 2008. Effects of the antibiotic ciprofloxacin on the bacterial community structure and degradation of pyrene in marine sediment. *Aquat. Toxicol.* 90:223–27.

Nielsen, M. N., and A. Winding. 2002. *Microorganisms as indicators of soil health.* Technical Report N° 388. Denmark: National Environmental Research Institute.

Niklinska, M., M. Chodak, and R. Laskowski. 2006. Pollution-induced community tolerance of microorganisms from forest soil organic layers polluted with Zn or Cu. *Appl. Soil Ecol.* 32:265–72.

Niklinska, M., M. Chodak, and A. Stefanowicz. 2004. Community level physiological profiles of microbial communities from forest humus polluted with different amounts of Zn, Pb, and Cd-preliminary study with BIOLOG ecoplates. *Soil Sci. Plant Nutr.* 50:941–44.

Niklinska, M., et al. 2005. Characterization of the forest humus microbial community in a heavy metal polluted area. *Soil Biol. Biochem.* 37:2185–94.

Nocker, A., M. Burr, and A. K. Camper. 2007. Genotypic microbial community profiling: A critical technical review. *Microbial Ecol.* 54:276–89.

Ogilvie, L. A., and A. Grant. 2008. Linking pollution induced community tolerance (PICT) and microbial community structure in chronically metal polluted estuarine sediments. *Mar. Environ. Res.* 65:187–98.

Ogram, A. 2000. Soil molecular microbial ecology at age 20: Methodological challenges for the future. *Soil Biol. Biochem.* 32:1499–504.

Oorts, K., et al. 2006. Soil properties affecting the toxicity of $CuCl_2$ and $NiCl_2$ for soil microbial processes in freshly spiked soils. *Environ. Toxicol. Chem.* 25:836–44.

Pandey, J., A. Chauhan, and R. K. Jain. 2009. Integrative approaches for assessing the ecological sustainability of *in situ* bioremediation. *FEMS Microbiol. Rev.* 33:324–75.

Pennanen, T., et al. 1996. Phospholipid fatty acid composition and heavy metal tolerance of soil microbial communities along two heavy metal-polluted gradients in coniferous forests. *Appl. Environ. Microbiol.* 62:420–28.

Pennanen, T., et al. 1998a. Structure of a microbial community in soil after prolonged addition of low levels of simulated acid rain. *Appl. Environ. Microbiol.* 64:2173–80.

Pennanen, T., et al. 1998b. Prolonged, simulated acid rain and heavy metal deposition: Separated and combined effects on forest soil microbial community structure. *FEMS Microbiol. Ecol.* 27:291–300.

Pereira, S. I. A., A. I. G. Lima, and E. M. D. P. Figueira. 2006. Heavy metal toxicity in *Rhizobium leguminosarum* biovar viciae isolated from soils subjected to different sources of heavy-metal contamination: Effects on protein expression. *Appl. Soil Ecol.* 33:286–93.

Petersen, D. G., I. Dahllof, and L. P. Nielsen. 2004. Effects of zinc pyrithione and copper pyrithione on microbial community function and structure in sediments. *Environ. Toxicol. Chem.* 23:921–28.

Priha, O., et al. 2001. Microbial community structure and characteristics of the organic matter in soils under *Pinus sylvestris*, *Picea abies* and *Betula pendula* at two forest sites. *Biol. Fert. Soils* 33:17–24.

Rajapaksha, R. M. C. P., M. A. Tobor-Kaplon, and E. Baath. 2004. Metal toxicity affects fungal and bacterial activities in soil differently. *Appl. Environ. Microbiol.* 70:2966–73.

Ramsey, P. W., et al. 2005. Relationship between communities and processes; new insights from a field study of a contaminated ecosystem. *Ecol. Lett.* 8:1201–10.

Ramsey, P. W., et al. 2006. Choice of methods for soil microbial community analysis: PLFA maximizes power compared to CLPP and PCR-based approaches. *Pedobiologia* 50:275–80.

Ranjard, L., et al. 2006. Field and microcosm experiments to evaluate the effects of agricultural Cu treatment on the density and genetic structure of microbial communities in two different soils. *FEMS Microbiol. Ecol.* 58:303–15.

Renella, G., P. C. Brookes, and P. Nannipieri. 2002a. Cadmium and zinc toxicity to soil microbial biomass and activity. *Soil Miner. Organ. Matter Microorg. Interactions Ecosyst. Health* 28b:267–73.

Renella, G., A. M. Chaudri, and P. C. Brookes. 2002b. Fresh additions of heavy metals do not model long-term effects on microbial biomass and activity. *Soil Biol. Biochem.* 34:121–24.

Ricart, M., et al. 2009. Effects of low concentrations of the phenylurea herbicide diuron on biofilm algae and bacteria. *Chemosphere* 76:1392–401.

Ringelberg, D., et al. 2008. Utility of lipid biomarkers in support of bioremediation efforts at army sites. *J. Microbiol. Meth.* 74:17–25.

Roesch, L. F., et al. 2007. Pyrosequencing enumerates and contrasts soil microbial diversity. *ISME J.* 1:283–90.

Rutgers, M. 2008. Field effects of pollutants at the community level—Experimental challenges and significance of community shifts for ecosystem functioning. *Sci. Total Environ.* 406:469–78.

Rutgers, M., and A. M. Breure. 1999. Risk assessment, microbial communities, and pollution-induced community tolerance. *Hum. Ecol. Risk Assess.* 5:661–70.

Rutgers, M., et al. 1998. Rapid method for assessing pollution-induced community tolerance in contaminated soil. *Environ. Toxicol. Chem.* 17:2210–13.

Schloss, P. D., and J. Handelsman. 2006. Toward a census of bacteria in soil. *Plos Comput. Biol.* 2:786–93.

Schmitt, H., H. Haapakangas, and P. van Beelen. 2005. Effects of antibiotics on soil microorganisms: Time and nutrients influence pollution-induced community tolerance. *Soil Biol. Biochem.* 37:1882–92.

Schmitt, H., et al. 2004. Pollution-induced community tolerance of soil microbial communities caused by the antibiotic sulfachloropyridazine. *Environ. Sci. Technol.* 38:1148–53.

Schmitt, H., et al. 2006. On the limits of toxicant-induced tolerance testing: Cotolerance and response variation of antibiotic effects. *Environ. Toxicol. Chem.* 25:1961–68.

Schutte, U. M. E., et al. 2008. Advances in the use of terminal restriction fragment length polymorphism (T-RFLP) analysis of 16S rRNA genes to characterize microbial communities. *Appl. Microbiol. Biot.* 80:365–80.

Seghers, D., et al. 2003. Pollution induced community tolerance (PICT) and analysis of 16S rRNA genes to evaluate the long-term effects of herbicides on methanotrophic communities in soil. *Eur. J. Soil Sci.* 54:679–84.

Shi, W., et al. 2002a. Association of microbial community composition and activity with lead, chromium, and hydrocarbon contamination. *Appl. Environ. Microbiol.* 68:3859–66.

Shi, W., et al. 2002b. Long-term effects of chromium and lead upon the activity of soil microbial communities. *Appl. Soil Ecol.* 21:169–77.

Shi, W., et al. 2005. Microbial catabolic diversity in soils contaminated with hydrocarbons and heavy metals. *Environ. Sci. Technol.* 39:1974–79.

Siciliano, S. D., and R. Roy. 1999. The role of soil microbial tests in ecological risk assessment: Differentiating between exposure and effects. *Hum. Ecol. Risk Assess.* 5:671–82.

Siciliano, S. D., et al. 2000. Assessment of 2,4,6-trinitrotoluene toxicity in field soils by pollution-induced community tolerance, denaturing gradient gel electrophoresis, and seed germination assay. *Environ. Toxicol. Chem.* 19:2154–60.

Singer, A. C., C. J. van der Gast, and I. P. Thompson. 2005. Perspectives and vision for strain selection in bioaugmentation. *Trends Biotechnol.* 23:74–77.

Smolders, E., et al. 2004. Soil properties affecting toxicity of zinc to soil microbial properties in laboratory-spiked and field-contaminated soils. *Environ. Toxicol. Chem.* 23:2633–40.

Sobecky, P. A., and J. M. Coombs. 2009. Horizontal gene transfer in metal and radionuclide contaminated soils. *JM Methods Mol. Biol.* 532:455–72.

Stefanowicz, A. M., M. Niklinska, and R. Laskowski. 2008. Metals affect soil bacterial and fungal functional diversity differently. *Environ. Toxicol. Chem.* 27:591–98.

Stefanowicz, A. M., M. Niklinska, and R. Laskowski. 2009. Pollution-induced tolerance of soil bacterial communities in meadow and forest ecosystems polluted with heavy metals. *Eur. J. Soil Biol.* 45:363–69.

Stephen, J. R., et al. 1999. Microbial characterization of a JP-4 fuel-contaminated site using a combined lipid biomarker/polymerase chain reaction-denaturing gradient gel electrophoresis (PCR-DGGE)-based approach. *Environ. Microbiol.* 1:231–41.

Stevens, D. P., M. J. McLaughlin, and T. Heinrich. 2003. Determining toxicity of lead and zinc runoff in soils: Salinity effects on metal partitioning and on phytotoxicity. *Environ. Toxicol. Chem.* 22:3017–24.

Stuczynski, T. I., G. W. McCarty, and G. Siebielec. 2003. Response of soil microbiological activities to cadmium, lead, and zinc salt amendments. *J. Environ. Qual.* 32:1346–55.

Suter, G. W., et al. 2000. *Ecological risk assessment for contaminated sites.* Boca Raton FL: CRC Press.

Takahata, Y., et al. 2006. Rapid intrinsic biodegradation of benzene, toluene, and xylenes at the boundary of a gasoline-contaminated plume under natural attenuation. *Appl. Microbiol. Biot.* 73:713–22.

Teng, Y., et al. 2008. Tolerance of grasses to heavy metals and microbial functional diversity in soils contaminated with copper mine tailings. *Pedosphere* 18:363–70.

Thompson, I. P., et al. 1999. Concentration effects of 1,2-dichlorobenzene on soil microbiology. *Environ. Toxicol. Chem.* 18:1891–98.

Tlili, A., et al. 2008. Responses of chronically contaminated biofilms to short pulses of diuron—An experimental study simulating flooding events in a small river. *Aquat. Toxicol.* 87:252–63.

Tobor-Kaplon, M. A., J. Bloem, and P. C. De Ruiter. 2006. Functional stability of microbial communities in contaminated soils near a zinc smelter (Budel, The Netherlands). *Ecotoxicology* 15:187–97.

Top, E., and D. Springael. 2003. The role of mobile genetic elements in bacterial adaptation to xenobiotic organic compounds *Curr. Opin. Biotechnol.* 14:262–69.

Torsvik, V., et al. 1998. Novel techniques for analysing microbial diversity in natural and perturbed environments. *J. Biotechnol.* 64:53–62.

Travis, E. R., N. C. Bruce, and S. J. Rosser. 2008. Short term exposure to elevated trinitro-toluene concentrations induced structural and functional changes in the soil bacterial community. *Environ. Pollut.* 153:432–39.

van Beelen, P. 2003. A review on the application of microbial toxicity tests for deriving sediment quality guidelines. *Chemosphere* 53:795–808.

van Beelen, P., et al. 2004. Location-specific ecotoxicological risk assessment of metal-polluted soils. *Environ. Toxicol. Chem.* 23:2769–79.

van Bleijswijk, J., and G. Muyzer. 2004. Genetic diversity of oxygenic phototrophs in microbial mats exposed to different levels of oil pollution. *Ophelia* 58:157–64.

van der Welle, M. E. W., et al. 2007. Predicting metal uptake by wetland plants under aerobic and anaerobic conditions. *Environ. Toxicol. Chem.* 26:686–94.

Vinther, F. P., G. K. Mortensen, and L. Elsgaard. 2003. Effects of linear alkylbenzene sulfonates on functional diversity of microbial communities in soil. *Environ. Toxicol. Chem.* 22:35–39.

Vogel, T. M., et al. 2009. TerraGenome: A consortium for the sequencing of a soil metagenome. *Nat. Rev. Microbiol.* 7:252.

Waldron, P. J., et al. 2009. Functional gene array-based analysis of microbial community structure in groundwaters with a gradient of contaminant levels. *Environ. Sci. Technol.* 43:3529–34.

Wall, D. H., and R. A. Virginia. 1999. Controls on soil biodiversity: Insights from extreme environments. *Appl. Soil Ecol.* 13:137–50.

Wallenstein, M. D., and M. N. Weintraub. 2008. Emerging tools for measuring and modeling the *in situ* activity of soil extracellular enzymes. *Soil Biol. Biochem.* 40:2098–106.

Wang, Q. Y., et al. 2009. Indication of soil heavy metal pollution with earthworms and soil microbial biomass carbon in the vicinity of an abandoned copper mine in Eastern Nanjing, China. *Eur. J. Soil Biol.* 45:229–34.

Weiss, J. V., and I. M. Cozzarelli. 2008. Biodegradation in contaminated aquifers: Incorporating microbial/molecular methods. *Ground Water* 46:305–22.

White, D. C., et al. 1999. Molecular characterization of microbial communities in a JP-4 fuel contaminated soil. In *In situ bioremediation of petroleum hydrocarbon and other organic compounds*, ed. B. C. Allerman, 345–50. Columbus, OH: Battelle Press.

Wilke, B. M., et al. 2004. Phospholipid fatty acid composition of a 2,4,6-trinitrotolune contaminated soil and an uncontaminated soil as affected by a humification remediation process. *Soil Biol. Biochem.* 36:725–29.

Wilke, B. M., et al. 2005. Effects of fresh and aged copper contaminations on soil microorganisms. *J. Plant Nutr. Soil Sci.* 168:668–75.

Williams, E. S., J. Panko, and D. J. Paustenbach. 2009. The European Union's REACH regulation: A review of its history and requirements. *Crit. Rev. Toxicol.* 39:553–75.

Wilmes, P., et al. 2009. The dynamic genetic repertoire of microbial communities. *FEMS Microbiol. Rev.* 33:109–32.

Winder, J. 2003. Soil quality monitoring programs: A literature review. Report for *Alberta Environmentally Sustainable Agriculture (AESA) Soil Quality Monitoring Program*, 7000–13. No. 206. Edmonton, AB: Alberta Agriculture, Food and Rural Development, Conservation Branch.

Winding, A., K. Hund-Rinke, and M. Rutgers. 2005. The use of microorganisms in ecological soil classification and assessment concepts. *Ecotoxicol. Environ. Saf.* 62:230–48.

Wright, M. S., et al. 2008. Influence of industrial contamination on mobile genetic elements: Class 1 integron abundance and gene cassette structure in aquatic bacterial communities. *ISME J.* 2:417–28.

Yang, Y. G., et al. 2006. Microbial indicators of heavy metal contamination in urban and rural soils. *Chemosphere* 63:1942–52.

Yao, J., et al. 2007. An *in vitro* microcalorimetric method for studying the toxic effect of cadmium on microbial activity of an agricultural soil. *Ecotoxicology* 16:503–9.

Young, I. M., et al. 2008. Microbial distribution in soils: Physics and scaling. *Adv. Agron.* 100:81–121.

Zhou, Y., et al. 2009. A combination method to study microbial communities and activities in zinc contaminated soil. *J. Hazard. Mater.* 169:875–81.

15 Adaptation to Metals in Higher Plants
The Case of Arabidopsis halleri (Brassicaceae)

Hélène Frérot, Patrick de Laguérie, Anne Créach,
Claire-Lise Meyer, Maxime Pauwels,
and Pierre Saumitou-Laprade

CONTENTS

15.1 INTRODUCTION

Soils polluted by metals occur either naturally on metal outcrops or as a result of anthropogenic industrial activities. These environments host plant communities consisting of a few species that exhibit adaptations to such restrictive conditions. These species are good models to study the evolutionary mechanisms of adaptation to extreme environments (Bradshaw 1970; Antonovics et al. 1971; Baker 1987) and may eventually provide biological tools for phytoremediation, which consists of the use of green plants either to remove the pollutants from contaminated soils through phytoextraction or to render them harmless through phytostabilisation (Salt et al. 1995, 1998).

Metal tolerance was initially seen as the result of the ability of normally non-tolerant species to evolve tolerant ecotypes on contaminated soil through natural selection acting on the appropriate genetic variability (Antonovics et al. 1971). Correspondingly, populations of the same species growing on noncontaminated soils should not be tolerant. Prat (1934), in one of the first studies of metal tolerance, observed that seeds of *Melandrium silvestre* (later referred to as *Silene vulgaris*) harvested from a copper mine site grew far better in soils with high copper concentrations than seeds of the same species from noncontaminated origins. However, some plant species have been reported to exhibit metal tolerance throughout all their geographic range. Tolerance in such species is thus defined as constitutive (Baker and Proctor 1990; Meharg 1994; Boyd and Martens 1998). McNaughton et al. (1974), Taylor and Crowder (1984), and Ye et al. (1997) compared *Typha latifolia* populations from noncontaminated origin and contaminated origin in standard conditions and found no differences in tolerance to zinc (Zn), lead (Pb), and cadmium (Cd). Constitutive tolerance to Zn was also demonstrated in *Thlaspi caerulescens* (Ingrouille and Smirnoff 1986; Lloyd-Thomas 1995; Meerts and van Isacker 1997). While tolerance in species where it is restricted to metallicolous populations can be directly related to the occurrence of metal contamination, an understanding of the evolution of constitutive metal tolerance remains elusive.

Tolerance to metal-contaminated soil can be achieved either by exclusion or by accumulation in a nontoxic form (Baker 1981). Metal accumulation can be defined as the capacity of a plant to accumulate metal in its shoots to concentrations greater than those found in the soil or the nutritive medium in which the plant grows (Baker et al. 2000). Hyperaccumulators are able to accumulate exceptional concentrations of metals in their aboveground biomass, when growing in their natural habitat, compared with nonaccumulator plants growing in the same habitat (Brooks 1998a). Thresholds of metal contents defining hyperaccumulators are approximately ten times higher than the metal concentrations usually found in nonaccumulating plants growing in the same conditions. Metal hyperaccumulation is not a common feature in higher plants. Approximately four hundred taxa have so far been identified as hyperaccumulators in a wide range of families (Baker et al. 2000). Interestingly, *Brassicaceae* are particularly well represented in the hyperaccumulators: for instance, *Thlaspi caerulescens* and *Arabidopsis halleri* are well-known Zn and Cd hyperaccumulators (Lloyd-Thomas 1995; Meerts and van Isacker 1997).

Arabidopsis halleri (L.) O'Kane and Al-Shehbaz is found in Northern France exclusively near metallurgical factories, on soils heavily polluted by Zn and Cd. It has been included by phytosociologists in metalliferous plant communities, e.g., *Armerietum halleri* (Ernst 1976) or *Holco-Cardaminopsidetum halleri* (Hulbüsh 1981). Moreover, Zn concentrations in leaves are extremely high, giving *A. halleri* a status of Zn hyperaccumulator (Brooks 1998b) (the threshold for Zn hyperaccumulation is 10,000 $\mu g.g^{-1}$ dry weight). Throughout its geographical range (roughly Central Europe, from Northern Italy and Switzerland to Ukraine and Romania; Clapham and Akeroyd 1993; O'Kane and Al-Shehbaz 1997), *A. halleri* is also found in areas exhibiting significantly high levels of Zn, Pb, and Cd (in France, Germany, Poland, and Italy), resulting from human activities such as metal industry, mining, refuse sites, and metal emissions. But it is mainly a mountain species growing on noncontaminated sites at elevations below the timberline, preferably on fresh sandy and oligotrophic soil (Moravec 1965; Clapham and Akeroyd 1993), a member of the *Melandrio-Trisetetum* or the *Cardaminopsidi (halleri)-Agrostitetum* communities (Moravec 1965; Ellenberg 1987).

The occurrence of both metallicolous (M) and nonmetallicolous (NM) populations, its high level of tolerance and accumulation, and its phylogenetic proximity with *A. thaliana*, providing numerous molecular tools for genetic studies, make *A. halleri* an excellent model species for investigating the evolutionary origin, intraspecific variability, and genetic architecture of metal tolerance and hyperaccumulation. Therefore, the first part of this chapter deals with the screening of phenotypic variability of metal tolerance and hyperaccumulation (particularly Zn) and questions the evolutionary origin of this variability. The second part presents how these results can be used in order to investigate the genetic architecture of tolerance to and hyperaccumulation of Zn.

15.2 AN INTEGRATIVE APPROACH TO INVESTIGATE INTRASPECIFIC VARIABILITY OF METAL TOLERANCE AND HYPERACCUMULATION

15.2.1 PHENOTYPIC SCREENING OF ZINC TOLERANCE AND HYPERACCUMULATION

Phenotypic screening of Zn tolerance in *Arabidopsis halleri* has been performed over a large European scale. Among the thirty-one European M and NM populations tested (Pauwels et al. 2006), most populations were scattered throughout the European mountain ranges at middle to high altitudes in noncontaminated environments. They were located in the Tatras and High Tatras mountains (Slovakia), in the Apuseni Mountains (Romania), in the Bohemian Forest (Germany and Czech Republic), and on northern and southern slopes of the Alps. The other populations were sampled at low to moderate altitudes in polluted environments, i.e., in Northern France, Silesia (Poland), and Harz (Germany). Intraspecific variability of Zn tolerance has been investigated using a test established by Schat and Ten Bookum (1992) that is developed on short periods in hydroponic culture (Bert et al. 2000; Bert et al. 2003; Pauwels et al. 2006). It provides a measure of tolerance by sequentially transferring plants into increasing concentrations of a metal, and by determining for each individual the lowest concentration at which no new root is produced (EC100, lowest concentration for 100% growth inhibition). At the beginning of each sequence, roots of plants were blackened with activated charcoal and immersed in the nutritive solution with the corresponding metal concentration. After one week, the occurrence of new (white) roots was examined. At each experimental metal concentration, root growth was encoded for each plant as a binary variable and interpreted as individual survival or mortality. Finally, population survival proportions at each metal concentration were used to draw survival curves. Figure 15.1 shows the results of this phenotypic screening. Zn tolerance was proven to be constitutive, as all populations showed tolerance compared to nontolerant control species. However, a continuous

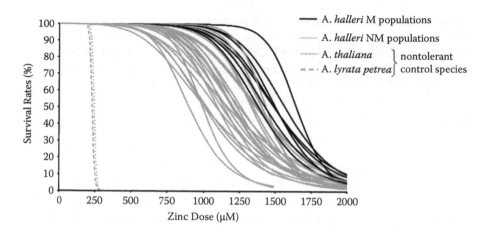

FIGURE 15.1 Population survival curves obtained after a phenotypic screening of metallicolous (M) and nonmetallicolous (NM) populations of *Arabidopsis halleri* and nontolerant control species (*A. lyrata petrea* and *A. thaliana*) in hydroponic culture. (Adapted from Pauwels, M., et al., *J. Evol. Biol.*, 19, 1838–50, 2006.)

range of variation in tolerance has been observed from NM to M populations. Indeed, M populations were shown to be significantly more Zn tolerant than NM ones.

Zn hyperaccumulation has been described in a few *A. halleri* M and NM populations (Macnair et al. 1999; Bert et al. 2000, 2002). As all populations tested exhibited high Zn concentrations in shoots, it has been assumed to be also constitutive in the species, although a tendency toward higher accumulation was shown in NM populations (Bert et al. 2000). This has been recently confirmed by Kostecka (2009) on several Polish populations. Although an exhaustive population screening of hyperaccumulation has never been performed, the physiology of the trait is better documented than for tolerance. The major Zn storage compartment in *A. halleri* is the mesophyll tissue (Küpper et al. 2000; Zhao et al. 2000; Sarret et al. 2002, 2009). The cellular localisation is still unclear, but Zn is probably sequestered in the vacuolar compartment (Küpper et al. 2000), as has been shown in *Thlaspi caerulescens* (Küpper et al. 1999; Frey et al. 2000). Trichomes were also found to exhibit high metal concentrations at their base (Küpper et al. 2000; Zhao et al. 2000; Sarret et al. 2002, 2009), but their possible role in metal detoxification remains uncertain. The chemical forms of Zn have been recently studied with spectroscopic tools (Sarret et al. 2002, 2009). The major form is Zn malate, but Zn could be complexed in leaves with other different organic acids or be free in solution (Sarret et al. 2009). Interestingly, different Zn partitioning has been observed between the veins and the leaf tissue. The vein/tissue ratio was negatively correlated with Zn accumulation (Sarret et al. 2009). These results suggested that in *A. halleri*, Zn is translocated very efficiently toward the leaves where it is stored in the mesophyll, probably sequestered in the vacuolar compartment without serious perturbations of the cellular metabolism. It implies the development of complex handling of the metal in plants from root uptake, upon xylem loading, translocation to shoots, and finally xylem unloading to leaf cells with the involvement of many membrane transporters and chelating molecules.

A. halleri is not only Zn but also Cd tolerant and a Cd hyperaccumulator. Cd tolerance and hyperaccumulation appear to vary among, as well as within, populations from polluted and nonpolluted sites (Bert et al. 2000, 2002, 2003). However, current knowledge is insufficient to discuss rigorously both qualitative and quantitative variation in Cd tolerance and hyperaccumulation among populations. Clearly, further surveys of natural variation are required.

Considering the level of quantitative polymorphism for Zn tolerance and hyperaccumulation abilities (in particular translocation toward aerial parts) revealed in *A. halleri* populations, Zn tolerance has recently been investigated using a new test. Indeed, the sequential test described above (Schat and Ten Bookum 1992) was considered to potentially lead to an underestimation of tolerance variability: first, because it considered tolerance as a binary trait (presence or absence of root growth), whereas tolerance actually shows quantitative variation; and second, because it focused on root growth, whereas shoot performance may be a more relevant trait in hyperaccumulator species. The translocation of most Zn toward the leaves likely enhances the metal toxicity in aerial parts. In leaves, Zn toxicity could reduce shoot growth and chlorophyll synthesis (through the induction of Fe deficiency or chloroplast degradation), and could also interfere with P, Mg, and Mn uptake (Carroll and Loneragan 1968; Boawn and Rasmussen 1971; Foy et al. 1978). Hence, the new test includes measures of morphological and physiological traits in both roots and shoots, namely, the length and biomass of roots, the width and biomass of leaves, the effective quantum yield of Photosystem II (Φ_{PSII}) (Genty et al. 1989), and the content of chlorophyll of young leaves. For each trait, a tolerance index was calculated as the ratio of the trait values measured from plants cultivated in contaminated (2,000 µM Zn) and control (10 µM Zn) conditions. Experiments were performed using geographically close M and NM populations, sampled along the Polish-Slovak border (Kostecka 2009; Meyer et al. 2010). Measures on shoots and roots have shown congruent trends toward higher mean values of Zn tolerance for M populations. In addition, the response of the shoot, mainly through Photosystem II yield, seemed most informative to capture the variability of tolerance, in particular among populations. Interestingly, for all traits used to estimate Zn tolerance, a high phenotypic variability was observed within populations of both edaphic types.

15.2.2 Understanding the Spatial Distribution of Intraspecific Polymorphism

Once the distribution of polymorphism of metal-related traits in populations is well characterised, understanding this distribution requires an ability to identify evolutionary factors shaping the geographic distribution of adaptive traits.

The main factor generally invoked to explain a differentiation of populations related to environmental variables is selection. As observed above, in metallophytes, the spatial distribution of metal-related traits is generally nonrandom and distinguishes M populations from NM ones. It is thus related to metal concentration in soils. This is assumed to result from the evolution that occurred after the colonisation of polluted sites, by selection of genotypes tolerating a higher toxic metal concentration in soils. In such a scenario, selection and adaptive differentiation of populations are usually expected to have occurred several times independently, each time the species colonises a new metalliferous site. However, consider a previously nonmetallophyte species that first colonised and adapted to a polluted site. It simultaneously expanded its ecological range and acquired the ability to colonise other polluted sites for which it is now preadapted. Further colonisations of polluted sites could occur by migration from this first M population, without any further selection event.

Obviously, both scenarios are not exclusive. Within a single metallophyte species, colonisations of polluted sites could have occurred from NM populations or preexisting M ones when they existed. But these distinct scenarios result in strikingly different evolutionary and genetic situations. In the first one, evolution toward higher fitness in metalliferous conditions occurred each time a NM population founded an M one. Thus, independent selection events could have involved various genetic resources and favoured distinct genetic mechanisms for adaptation. In the second one, adaptation occurred only once, during the first colonisation event, and genetic mechanisms conferring higher fitness in distinct M populations are expected to be similar.

In this context, assessing the relative contribution of past demographic events and selective processes in shaping the current distribution of metal-related traits could greatly help in understanding their genetics and evolution. One strategy consists in confronting the spatial distribution of adaptive traits to historical relationships of populations inferred from neutral genetic data. This can be performed using the theoretical framework of phylogeography (Box 15.1). Phylogeographers seek to deduce historical processes in population demographics from the pattern of congruence or noncongruence between genealogical relationships among genetic variants of neutral genes and their present-day geographic distribution (Box 15.1a–e). In practice, phylogeographic studies classically deploy a two-step argument: (1) nongenetic evidence, including geological, palaeoenvironmental, and ecological data, is used to infer past demographic scenarios and historical relationships among populations (Willis 1996; Tribsch and Schonswetter 2003; Willis and van Andel 2004); and (2) biogeographic assumptions are tested by searching the footprints of hypothetical past demographic processes in population genetic data (Avise et al. 1987; Avise 2000).

In *A. halleri*, M populations are discontinuously distributed, peripheral to the main species range, and occur on anthropogenic polluted sites. As a result, enhanced tolerance is generally assumed to have evolved recently, in response to local ecological pressures, independently in distant geographic areas. However, the existence of historical and genetic relationships among faraway M populations cannot be excluded: human migrations among industrial sites might have been frequent and could have favoured seed dispersal; intentional introduction of *A. halleri* in highly disturbed polluted sites from preexisting M populations is suspected, for example, in northern France. To test both historical hypotheses, the genetic relationships among *A. halleri* populations were investigated using the phylogeographic toolbox. Sixty-five populations scattered in the European species range were sampled (Box 15.1f–g), including M populations from northern France, Harz in Germany, and Silesia in Poland. Individuals were genotyped using neutral molecular markers from both the chloroplast (cpDNA) and the nuclear (nDNA) genomes. CpDNA and nDNA data were congruent and revealed

BOX 15.1   FUNDAMENTAL CONCEPTS OF PHYLOGEOGRAPHY (a–e) AND THEIR APPLICATION TO THE EVOLUTION OF METAL TOLERANCE IN *ARABIDOPSIS HALLERI* (f, g)

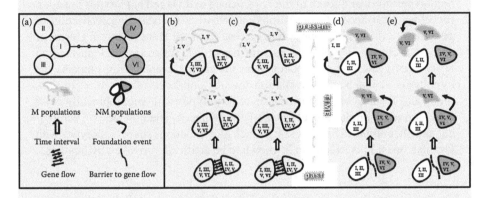

(a) Genealogical relationships between hypothetical haplotypes. Extant and missing (i.e. extinct or unsampled) haplotypes are represented by large open and small dark circles, respectively. Each branch in the tree represents a single mutational step. A discontinuity in the tree allows distinguishing two lineages (I-III and IV-VI). (b – e) Geographic distribution of haplotypes according to different population histories. Two scenarii for the origin of M populations are illustrated: in (b, d), M populations have been founded separately from distinct NM ones; in (c, e) M populations derived from a single NM one. In (b, c) ancestral NM populations were composed of both haplotypes lineages, suggesting gene flow. Oppositely, in (d, e), the genetic discontinuity among haplotypes lineages was geographically oriented in founder populations, suggesting long term genetic isolation. In (d, e), ancestral populations represented clear phylogeographic units. It appears that the occurrence of a phylogeographic structure in founder populations (d, e) is required to distinguish among both scenarios for the origin of M populations.

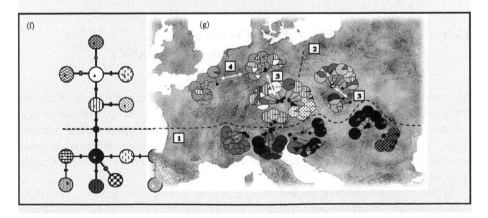

Genealogical relationships (f) and geographic distribution (g) of haplotypes in
Arabidopsis halleri. For each population in (g), a pie chart indicates the proportion of haplotypes with corresponding artwork in the haplotype tree (f). 1:
haplotypes north and south of the Alps belong to diverged haplotypes lineages,
revealing vicariance of populations caused by a longstanding restriction of
gene flow. 2: M populations (white circles) from the Harz region in Germany
and from Silesia in Poland share no haplotype, suggesting a longstanding
absence of gene flow and independent foundation events. This supports the
hypothesis of independent evolution towards enhanced tolerance. 3: in Germany and Poland, M populations share haplotypes with geographically close
NM ones (dark circles), revealing a maintenance of gene flow despite ecological and phenotypic differences. 4: French M populations share haplotypes with
German ones. This exception is known to be artificial, provoked by humans.

a phylogeographic pattern of distribution of neutral genetic diversity: phylogenetic
relationships among populations were better explained by geographic closeness than
by ecological type (M or NM). Therefore, M populations from distinct geographic
areas (e.g., Harz and Silesia) were highly genetically differentiated, whereas neighbouring M and NM populations were, in comparison, more similar. This pattern
highly supported the assumption that the colonisation of polluted sites by A. halleri
occurred several times independently at isolated polluted sites.

The logical implication of the phylogeographic study performed on A. halleri
is that any genetic study of Zn tolerance or hyperaccumulation should consider
that genetic mechanisms responsible for the higher level of tolerance of a given M
accession (e.g., from Harz) could be specific for this accession and could have differed if a different accession (e.g., Silesia) had been investigated. This also suggests
that several genetic studies should be performed in parallel to seek if similar M
phenotypes in independently evolved M populations are based on identical genetic
mechanisms, or if they could have been reached through distinct adaptive walks/
genetic pathways.

15.3   AN INTEGRATIVE APPROACH TO SEEK
GENETIC ARCHITECTURE OF ZINC TOLERANCE
AND HYPERACCUMULATION

As previously mentioned, to investigate the origin and potential multiplicity of
mechanisms underlying an adaptive trait, it is necessary to know the phenotype distribution in populations as well as the genetic relationships between these populations. In parallel, physiological or molecular observations give information about
possible genetic pathways. The question thus remaining is about the genes actually
responsible for the trait.

In the case of A. halleri, the following questions can be specifically asked: How
many genes are involved in the evolution of metal tolerance and hyperaccumulation?

Are they involved in a pleiotropic effect (i.e., response to different metals)? Do ecologically similar environments select for the same genes or different genetic mechanisms? In answering these questions, one first needs to identify the genes that influence metal tolerance and hyperaccumulation. This goal could be achieved using integrative approaches based on either experimental populations (e.g., quantitative traits loci (QTL) mapping) or natural populations (e.g., genome scan, linkage disequilibrium (LD) mapping) (Figure 15.2a).

15.3.1 INVESTIGATING GENETICS OF BETWEEN-SPECIES DIFFERENCES USING EXPERIMENTAL POPULATIONS

QTL mapping is a method that aims to identify and localise genomic regions potentially including the major gene(s) responsible for a given quantitative trait. It uses an experimental population created from a controlled cross and segregating for the trait under study (Figures 15.2a and Figure 15.3). Polymorphic molecular markers with Mendelian segregation are used to genotype parent and progeny and construct a genetic map (Box 15.2) by translating estimated recombination rates between each pair of markers into genetic distances in centimorgans. The resulting genetic map represents the whole genome of the species implied in the crossbreeding. If a quantitative trait locus is linked to a given marker, then individuals with different marker genotypes will have different mean values of the quantitative trait (Lander and Botstein 1989) (Figures 15.2a and Figure 15.3). Several close markers in linkage disequilibrium (Box 15.2) can be significantly linked to the trait so that a genomic region can correspond to a QTL region and may contain one or more genes responsible for the adaptive trait.

QTL mapping was used to indentify QTLs for tolerance in *A. halleri* (Figure 15.2b). Because Zn tolerance was demonstrated to be constitutive in this species, trait segregation was not ensured by making intraspecific crosses. Hence, interspecific crosses were performed between *A. halleri* and *A. lyrata petraea*, a nontolerant and nonaccumulator relative. First, a back-cross progeny (Figure 15.3) was produced, segregating for Zn and Cd tolerance but not for Zn hyperaccumulation. A robust genetic map of *A. halleri* was thus obtained, and the genetic architecture for Zn and Cd tolerance was investigated (Courbot et al. 2007; Willems et al. 2007). Three major QTLs were revealed for Zn tolerance, and among them, one was also linked to Cd tolerance. Therefore, the subsequent question was: Is it possible to find in these major QTL regions of *A. halleri* some genes that could be actually involved in Zn or Cd tolerance? Roosens et al. (2008) developed a genomic tool based on the synteny (Box 15.2) between *A. halleri* and *A. thaliana* genomes. Indeed, these two species are close relatives, and they show *circa* 94% nucleotide identity within coding regions (Becher et al. 2004). The advantage of *A. thaliana* is to be a model species whose genome is completely sequenced and whose gene functions are mostly described. Thus, by using molecular markers positioned on both a genetic map of *A. halleri* and a physical map (Box 15.2) of *A. thaliana*, two markers flanking a QTL region also delimit a genomic *A. thaliana* region that can contain interesting genes, namely, genes whose function is consistent with

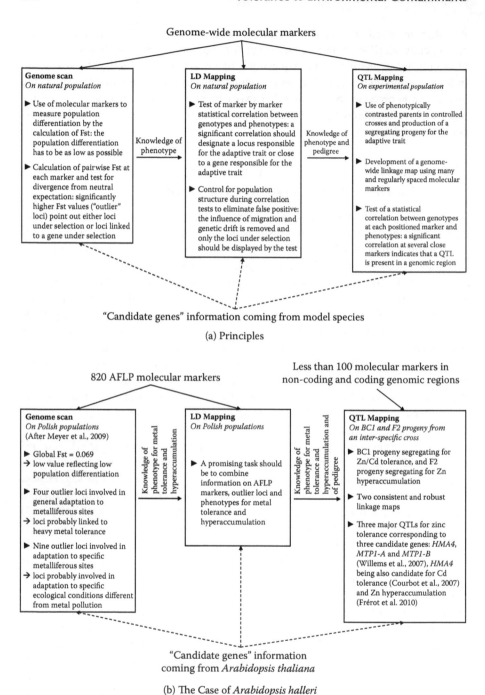

FIGURE 15.2a,b Integrative genomic approaches using genome-wide molecular markers aiming to identify the major genes responsible for an adaptive trait. (According to Stinchcombe, J. R., and Hoekstra, H. E., *Heredity*, 100, 158–70, 2008.)

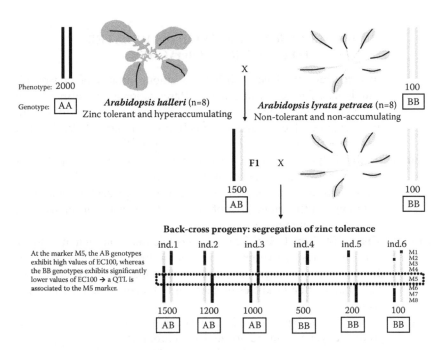

FIGURE 15.3 Production of a back-cross progeny segregating for zinc tolerance and principle of QTL identification for this trait. Two bars are used to represent two homologue chromosomes. The phenotypic values of zinc tolerance (EC100 values, see details in Section 15.2.1 of the text) are indicated below chromosomes. Genotypes at the gene controlling zinc tolerance are specified by two alleles A and B.

physiological knowledge of metal tolerance or hyperaccumulation. Such genes are currently called candidate genes (Figure 15.2a). Then, it is possible to list a set of candidate *A. thaliana* genes that may have corresponding homologues in *A. halleri*. Nevertheless, QTL regions are generally so large that hundreds of genes are potential candidate genes. A sorting can be achieved by taking into account transcriptomic (Box 15.2) studies and assuming reasonable expectations about gene functions. Therefore, for metal tolerance, transcriptomic studies (Becher et al. 2004; Weber et al. 2004; Talke et al. 2006) show that some genes involved in metal homeostasis are overexpressed in *A. halleri* in comparison to *A. thaliana* and are good candidate genes. Using such candidate gene information, it was thus demonstrated that the major QTL region common to Zn and Cd tolerance colocalises with *HMA4*, a gene coding for a protein from the metal transporting ATPase family, and responsible for the root-to-shoot translocation of metals. The importance of this gene was recently confirmed by functional studies (Hanikenne et al. 2008). Indeed, *A. halleri* plants with reduced expression of *HMA4* translocated less Zn from roots to shoots, but were also more sensitive to Zn and Cd toxicity. The two other QTL regions involved in Zn tolerance each colocalised with a copy of *MTP1*, a gene coding for a protein from the metal transporter protein family, and mediating the detoxification of Zn in the cell vacuole (Dräger et al. 2004).

BOX 15.2   SOME DEFINITIONS RELATED
TO GENETIC APPROACHES

Genetic differentiation: Genetic difference among groups of conspecific individuals. The degree of genetic differentiation is determined by the interplay among evolutionary factors like natural selection, genetic drift (changes due to random sampling of parental alleles in finite populations), mutation, and migration (or gene flow).

Genetic map: Or linkage map, i.e., a map based on linkage disequilibrium analysis between molecular markers. Markers in high linkage disequilibrium generate linkage groups, which represent chromosomes assuming a sufficient number of markers. On a genetic map, distances are given in centimorgans.

Linkage disequilibrium (LD): Statistical correlation (nonrandom association) of alleles at two or more polymorphic loci. This correlation declines with increasing distance between loci due to crossing over. The combination of alleles at genes in LD tends to be inherited as a group.

Physical map: A map based on DNA sequencing. In particular, genes and predicted genes can be located on chromosomes. On a physical map, distances are given in number of DNA base pairs (kilobases or megabases).

Synteny: Colinearity between genomes. Using molecular markers, synteny corresponds to the conservation of marker order between two genomes.

Transcriptomics: Transcriptomics refers to the study of the transcriptome. The genes are transcribed to premessenger RNA, further processed to messenger RNA (transcripts), and finally translated into protein. The transcriptome constitutes the complete set of different transcripts that is synthesised in the lifetime of a cell or a tissue.

Furthermore, metal tolerance and hyperaccumulation are both characteristic traits of metallophyte species and also key traits to be simultaneously improved for phytoremediation. Hence, the genetic architecture of metal hyperaccumulation and its genetic relationships with metal tolerance have to be investigated. In this context, a F2 progeny from a cross between *A. halleri* and *A. lyrata petraea* was also produced, segregating for Zn hyperaccumulation. A QTL region common with Zn or Cd tolerance was revealed colocalizing with HMA4 (Frérot et al. 2010). As such a common QTL region between zinc hyperaccumulation and Zn/Cd tolerance appeared, the corresponding candidate gene will provide a good basis for genetic programmes concerning phytoremediation.

15.3.2 INVESTIGATING GENETICS OF BETWEEN-POPULATION DIFFERENCES USING NATURAL POPULATIONS

To identify genes involved in adaptive traits, an alternative approach consists of using natural populations instead of laboratory crosses (Figure 15.2a). Typically, natural populations are used in two different ways: to detect natural selection at the molecular level (see genome scan, Figure 15.2a) or to find statistical associations between polymorphism and adaptive traits (LD mapping, Figure 15.2a). These approaches take advantage of hundreds to thousands of recombination events accumulated over time in these populations (compared to a few generations in QTL mapping). Hence, the resolution is high and it is possible to identify small-sized genomic regions or directly the genes involved in adaptive traits. The main bias when using these methods is to confound demographic events (such as migration, reduction, or expansion of population) with natural selection because both phenomena could have similar effects on population genetic differentiation (Box 15.2) (Storz 2005; Zhao et al. 2007; Excoffier et al. 2009). This bias can be limited by combining analyses on natural populations and on laboratory crosses. Furthermore, combining genetics methods seems to be the most likely approach to be successful in connecting adaptive processes to specific genes and polymorphisms.

The most common method used on natural populations to investigate the genetics of between-population differences in nonmodel species is genome scan (for review, see Nosil et al. 2009). This method aims to identify loci underlying local adaptation using populations with distinct ecologies and without *a priori* assumptions about the traits. Its basic principle is to scan the genome for polymorphism at many molecular markers on many individuals from several ecologically differentiated populations in order to identify markers corresponding or linked to the genes causing adaptation to local environmental conditions. The underlying idea is that these markers (called outliers) are expected to show higher population differentiation than the genomic background because of divergent selection between ecotypes (Luikart et al. 2003). Because this method does not focus on traits, it could be a very efficient way to investigate the genetic basis of adaptation to complex habitats such as metalliferous sites. Indeed, many abiotic constraints (high concentration of metals, particular soil structure, low water content, and low nutrient concentration) could be present in these sites. To maximise the statistical power of genome scan, samples are usually chosen according to two criteria: first, populations have to show weak neutral genetic differentiation to minimise the background level of population differentiation, and second, populations should have a relative common history to minimise the detection of false positives due to demographic events.

A genome scan approach was applied on *A. halleri* (see genome scan, Figure 15.2b) to investigate the genetic basis of differences between M and NM populations (Meyer et al. 2009). Four Slovak and Polish populations that had previously been shown to be weakly differentiated at neutral genetic loci (Pauwels et al. 2005) were analysed. As in most of the genomic surveys on nonmodel organisms, the genome of *A. halleri* was scanned using anonymous amplified fragment length polymorphism (AFLP) markers (820 loci, i.e., 1 marker/0.3 megabase). In

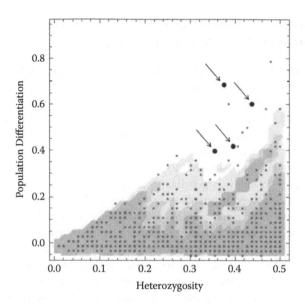

FIGURE 15.4 Example of metallicolous vs. nonmetallicolous pairwise comparison per-formed with Dfdist software: plot of heterozygosity (H_e) and population differentiation (F_{ST}) for the population pair M1-NM2 (modified from Meyer et al. 2009). Each dot represents at least one AFLP marker. The 90%, 95% and 99% confidence regions of the neutral distribu-tion are shown from darker to lighter grey, respectively. Outlier loci could then be identified by identifying loci with FST value that exceed the confidence region. Loci that are outliers in several pairwise comparisons involving M1 were pointed out by arrows.

order to identify outlier loci showing higher differentiation than the genomic back-ground, pairwise (i.e., between pairs of populations) genetic differentiation at the AFLPs was compared to neutral expected differentiation (according to Beaumont and Nichols 1996 and Vitalis et al. 2001). Loci diverging from neutral expectations (see example in Figure 15.4) in several M versus NM population comparisons were considered to be potentially involved in adaptation to metalliferous sites. Taking all the M populations together in the comparisons allowed the identification of four loci likely to be systematically involved in adaptation to polluted sites. In addi-tion, by taking each M population separately in the comparisons, nine loci were revealed and were therefore assumed to be implicated in adaptation to specific metalliferous sites. The fact that the latter loci differed between M populations could be explained by possible ecological differences among the metalliferous sites sampled (soil structure, nutrient concentration, or water content). Alternatively, this result could also reflect convergent evolution in these populations, i.e., selec-tion of different genetic solutions in response to ecologically similar environments. Indeed, recent empirical studies of the genetics of adaptation have shown that dif-ferent populations within a species may evolve the same phenotype with different genetic bases (Arendt and Reznick 2007).

Two subsequent questions can be raised by genome scan approaches: (1) What are the genes responsible for behaviour of outlier loci, and thus what are the targets of divergent selection? (2) Are outlier loci diverging from neutrality associated with differentiation for quantitative traits? An answer to the first question is possible by investigating the genomic regions surrounding outliers that diverge from neutrality. AFLP loci could be very efficiently isolated and sequenced following the method of Brugmans et al. (2003), or directly by using a second-generation genome sequencer (see Van Orsouw et al. 2007). Possible homologies between outlier sequences and model species DNA sequences taken from Genebank (www.ncbi.nlm.nih.gov) may then lead to the identification of candidate genes probably involved in adaptive processes. Nevertheless, the success of this method depends on the genetic proximity between the species studied and the model. In the case of *A. halleri*, the fact that it is one of the closest relatives of the model species *A. thaliana* (Al-Shehbaz and O'Kane 2002) opens up the opportunity to characterise quite easily the outlier loci involved in adaptation to metalliferous sites. Furthermore, as for QTL mapping (see above), it is possible to use the high synteny between *A. thaliana* and *A. halleri* maps to access the candidate genes localised in *A. halleri* regions of interest.

To answer the second question, links between outlier loci involved in local adaptation and quantitative traits could be investigated by following up the genome scan with LD mapping. LD mapping (also known as genome-wide association (GWA) mapping, association genetics) is a powerful method to investigate the genetic basis of quantitative traits in natural populations. Molecular markers whose allelic frequencies correlate with the phenotypes are then identified. In the case of *A. halleri*, a promising task should be to combine information on AFLP markers, outlier loci, and phenotypes for metal tolerance and hyperaccumulation to uncover the link between these quantitative traits, the molecular mechanisms, and the evolutionary processes.

15.4 PERSPECTIVES

The study of plant biology includes both fundamental issues (How do plants evolve in a changing environment?) and environmental or agronomic considerations (How will plant communities react to the local or global changes provoked by human activities? How shall plants be used to feed expanding human populations?). In this context, the study of plant adaptation to metal-polluted environments not only allows the investigation of plant evolution in extreme conditions, but also helps to provide ways either to manage metal pollution (phytoremediation; see Pilon-Smits 2005; Zhao and McGrath 2009) or to improve the nutritional content of vegetable food, and therefore reduce malnutrition of the most vulnerable people (biofortification; see Hirschi 2008; Zhao and McGrath 2009). Both purposes clearly imply the identification of gene(s) underlying metal homeostasis, tolerance, or hyperaccumulation.

The fundamental study of plant adaptation might be significantly enhanced by the availability of preexisting physiological, molecular, and genomic data and usually requires model species. In the last decade, the metallophyte close relative of *A. thaliana*, *A. halleri*, has proven to be pertinent for genetic analysis of metal homeostasis, tolerance, and hyperaccumulation. In particular, integrating molecular ecology approaches has revealed that although M populations differ from NM ones for several metal-related traits, both population types occur in distinct phylogeographic units. This suggests that the local adaptation of M populations to metalliferous conditions might have occurred several times independently and might have involved distinct genetic mechanisms. Furthermore, the QTL mapping approach allowed the addressing of questions about the number and type of genes responsible for these genetic mechanisms. Finally, the high synteny among *A. halleri* and *A. thaliana* genetic maps made the exploitation of the genomic and physiological knowledge on *A. thaliana* possible in order to identify candidate genes for Zn and Cd tolerance and hyperaccumulation.

Unfortunately, such model species are rarely relevant for direct applications. Indeed, the low biomass production of hyperaccumulators prevents their use in phytoremediation of metal-polluted soils. Likewise, although they are *Brassicaceae* species, neither *A. thaliana* nor *A. halleri* can serve as crops, and therefore cannot directly increase the content of bioavailable nutrients in food. Phytoremediation and biofortification strategies will probably require more relevant species. However, they will also require an ability to monitor the uptake of metals from different kinds of soil, the distribution of metals within plant, and their storage (or sequestration) in edible parts in trophically available or unavailable forms.

Definitely, the knowledge gathered on model species will be of great interest. Indeed, candidate genes identified in model species as responsible for enhanced metal tolerance or hyperaccumulation could represent target genes, alleles, or promoter regions further contributing to transgenic strategies or marker-assisted selection for the improvement of crops (Hirschi 2008; Palmgren et al. 2008; Zhao and McGrath 2009).

REFERENCES

Al-Shehbaz, I. A., and S. L. J. O'Kane. 2002. Taxonomy and phylogeny of *Arabidopsis* (*Brassicaceae*). In *The arabidopsis book*, ed. C. R. Somerville and E. M. Meyerowitz. Rockville, ND: American Society of Plant Biologist. doi: /10.1199/tab.0001, http://www.aspb.org/publications arabidopsis.

Antonovics, J., A. D. Bradshaw, and R. G. Turner. 1971. Heavy metal tolerance in plants. *Adv. Ecol. Res.* 7:1–85.

Arendt, J., and D. Reznick. 2007. Convergence and parallelism reconsidered: What have we learned about the genetics of adaptation? *Science* 23:26–32.

Avise, J. C. 2000. *Phylogeography: The history and formation of species*. Cambridge, MA: Harvard University Press.

Avise, J. C., et al. 1987. Intraspecific phylogeography: The mitochondrial DNA bridge between population genetics and systematics. *Annu. Rev. Ecol. Syst.* 7:489–522.

Baker, A. J. M. 1981. Accumulators and excluders—Strategies in the response of plants to heavy metals. *J. Plant Nutr.* 3:643–54.

Baker, A. J. M. 1987. Metal tolerance. *New Phytol.* 106(Suppl.):93–111.

Baker, A. J. M., and J. Proctor. 1990. The influence of cadmium, copper, lead, and zinc on the distribution and evolution of metallophytes in the British Isles. *Plant Syst. Evol.* 173:91–108.

Baker, A. J. M., et al. 2000. Metal hyperaccumulator plants: A review of the ecology and physiology of a biological resource for phytoremediation of metal-polluted soils. In *Phytoremediation of contaminated soils and water*, ed. N. Terry and G. Banuelos, 171–88. Boca Raton, FL: CRC Press.

Beaumont, M. A., and R. A. Nichols. 1996. Evaluating loci for use in the genetic analysis of population structure. *Proc. R. Soc. B* 263:1619–26.

Becher, M., et al. 2004. Cross-species microarray transcript profiling reveals high constitutive expression of metal homeostasis genes in shoots of the zinc hyperaccumulator *Arabidopsis halleri*. *Plant J.* 37:251–68.

Bert, V., et al. 2000. Zinc tolerance and accumulation in metallicolous and non-metallicolous populations of *Arabidopsis halleri* (Brassicaceae). *New Phytol.* 146:225–33.

Bert, V., et al. 2002. Do *Arabidopsis halleri* from nonmetallicolous populations accumulate zinc and cadmium more effectively than those from metallicolous populations? *New Phytol.* 155:47–57.

Bert, V., et al. 2003. Genetic basis of Cd tolerance and hyperaccumulation in *Arabidopsis halleri*. *Plant Soil* 249:9–18.

Boawn, L., and P. Rasmussen. 1971. Crop response to excessive fertilization of alkaline soil. *Agron. J.* 63:874–76.

Boyd, R. S., and S. N. Martens. 1998. The significance of metal hyperaccumulation for biotic interactions. *Chemoecology* 8:1–7.

Bradshaw, A. D. 1970. Plants and industrial waste. *Trans. Bot. Soc. Edinb.* 41:71–84.

Brooks, R. R. 1998a. General introduction. In *Plants that hyperaccumulate heavy metals*, ed. R. R. Brooks, 1–14. Oxford: CAB International.

Brooks, R. R. 1998b. Their role in phytoremediation, microbiology, mineral exploration and phytomining. In *Plants that hyperaccumulate heavy metals*, ed. R. R. Brooks, 327–356. Oxford: CAB International.

Brugmans, B., et al. 2003. A new and versatile method for the successful conversion of AFLP markers into single locus markers. *Nucleic Acids Res.* 31:e55.

Carroll, M. D., and J. Loneragan. 1968. Response of plant species to concentrations of zinc in solution. *Aust. J. Agric. Res.* 19:859–68.

Clapham, A. R., and J. R. Akeroyd. 1993. *Cardaminopsis*. In: *Flora Europea*, ed. T. G. Tutin, V. H. Heywood, N. A. Burges, et al., 290. Cambridge: Cambridge University Press.

Courbot, M., et al. 2007. A major quantitative trait locus for cadmium tolerance in *Arabidopsis halleri* colocalizes with HMA4, a gene encoding a heavy metal ATPase. *Plant Physiol.* 144:1052–65.

Dräger, D. B., et al. 2004. Two genes encoding *Arabidopsis halleri* MTP1 metal transport proteins co-segregate with zinc tolerance and account for high *MTP1* transcript levels. *Plant J.* 39:425–39.

Ellenberg, H. 1987. *Vegetation ecology of Central Europe*. Cambridge: Cambridge University Press.

Ernst, W. H. O. 1976. *Violetea calaminariae*. In *Of the European Prodrome plant communities*, ed. R. J. Tüxen, 1–25. Vaduz, Germany: J. Cramer.

Excoffier, L., T. Hofer, and M. Foll. 2009. Detecting loci under selection in a hierarchically structured population. *Heredity* 103:285–98.

Foy, C., R. Chanel, and M. White. 1978. The physiology of metal toxicity in plants. *Annu. Rev. Plant Physiol.* 29:511–66.

Frérot, H., et al. 2010. Genetic architecture of zinc hyperaccumulation in *Arabidopsis halleri*: The essential role of QTL × environment interactions. *New Phytol* 187:355–367.

Frey, B., et al. 2000. Distribution of Zn in functionally different leaf epidermal cells of the hyperaccumulator *Thlaspi caerulescens*. *Plant Cell Environ.* 23:675–87.

Genty, B., J. Briantais, and N. Baker. 1989. The relationship between the quantum yield of photosynthetic electron-transport and quenching of chlorophyll fluorescence. *Biochim. Biophys. Acta* 990:87–92.

Hanikenne, M., et al. 2008. Evolution of metal hyperaccumulation required cis-regulatory changes and triplication of HMA4. *Nature* 453:391–95.

Hirschi, K. 2008. Nutritional improvements in plants: Time to bite on biofortified foods. *Trends Plant Sci.* 13:459–63.

Hulbüsh, K. 1981. *Cardaminopsis halleri*-Gesellschaften im Harz. In *Syntaxonomie*, ed. H. Dierschke, 343–61. Vaduz, Germany: J. Cramer.

Ingrouille, M. J., and N. Smirnoff. 1986. *Thlaspi caerulescens* J. & C. presl. (*T. alpestre* L.) in Britain. *New Phytol.* 102:219–33.

Kostecka, A. A. 2009. *Adaptations of Arabidopsis halleri to habitats rich in heavy metals in southern Poland.* Warsaw: Polish Academy of Sciences.

Küpper, H., F. J. Zhao, and S. P. McGrath. 1999. Cellular compartimentation of zinc in the leaves of the hyperaccumulator *Thlaspi caerulescens. Plant Physiol.* 212:75–84.

Küpper, H., et al. 2000. Cellular compartimentation of cadmium and zinc in relation to other elements in the hyperaccumulator *Arabidopsis halleri. Planta* 212:75–84.

Lander, E., and D. Botstein. 1989. Mapping mendelian factors underlying quantitative traits using RFLP linkage maps. *Genetics* 121:185–99.

Lloyd-Thomas, D. H. 1995. *Heavy metal hyperaccumulation by Thlaspi caerulescens.* PhD thesis. Sheffield, UK: University of Sheffield.

Luikart, G., et al. 2003. The power and promise of population genomics: From genotyping to genome typing. *Nat. Rev. Genet.* 4:981–94.

Macnair, M. R., et al. 1999. Zinc tolerance and hyperaccumulation are genetically independent characters. *Proc. R. Soc. Lond. B* 266:2175–79.

McNaughton, S. J., et al. 1974. Heavy metal tolerance in *Typha latifolia* without the evolution of tolerant races. *Ecol. Lett.* 55:1163–65.

Meerts, P., and N. van Isacker. 1997. Heavy metal tolerance and accumulation in metallicolous and nonmetallicolous populations of *Thlaspi caerulescens* from continental Europe. *Plant Ecol.* 133:221–31.

Meharg, A. A. 1994. Integrated tolerance mechanisms: Constitutive and adaptive plant responses to elevated metal concentrations in the environment. *Plant Cell Environ.* 17:989–93.

Meyer, C. L., et al. 2009. Genomic pattern of adaptive divergence in *Arabidopsis halleri*, a model species for tolerance to heavy metal. *Mol. Ecol.* 18:2050–62.

Meyer, C. L., et al. 2010. Variability of zinc tolerance among and within populations of the pseudometallophyte species *Arabidopsis halleri* and possible role of directional selection. *New Phytol.* 185:130–42.

Moravec, J. 1965. Wiesen im mittleren Teil des Böhmerwaldes (Sumava). In *Vegetace CSSR A1 Synökologische Studien über Röhrichte, Wiesen und Auenwälder*, ed. R. Neuhäusl, J. Moravec, and Z. Neuhäuslova-Novotna, 202–52. Prague, Czech Republic: Verlag der Tschechoslowakischen Akademie der Wissenscaften.

Nosil, P., D. J. Funk, and D. Ortiz-Barrientos. 2009. Divergent selection and heterogeneous genomic divergence. *Mol. Ecol.* 18:375–402.

O'Kane, S. L., and I. A. Al-Shehbaz. 1997. A synopsis of *Arabidopsis (Brassicaceae). Novon* 7:323–27.

Palmgren, M. G., et al. 2008. Zinc biofortification of cereals: Problems and solutions. *Trends Plant Sci.* 13:464–73.

Pauwels, M., et al. 2005. Multiple origin of metallicolous populations of the pseudometallophyte *Arabidopsis halleri (Brassicaceae)* in Central Europe: The cpDNA testimony. *Mol. Ecol.* 14:4403–14.

Pauwels, M., et al. 2006. A broad-scale analysis of population differentiation for Zn tolerance in an emerging model species for tolerance study: *Arabidopsis halleri (Brassicaceae). J. Evol. Biol.* 19:1838–50.

Pilon-Smits, E. 2005. Phytoremediation. *Annu. Rev. Plant Biol.* 56:15–39.

Prat, S. 1934. Die Erblichkeit der Resistenz gegen Kupfer. *Berichte Deutschen Botanischen Gesellschaft* 52:65–78.

Roosens, N., et al. 2008. The use of comparative genome analysis and syntenic relationships allows extrapolating the position of Zn tolerance QTL regions from *Arabidopsis halleri* into *Arabidopsis thaliana*. *Plant Soil* 306:105–16.

Salt, D. E., R. D. Smith, and I. Raskin. 1998. Phytoremediation. *Annu. Rev. Plant Physiol. Plant Mol. Biol.* 49:643–68.

Salt, D. E., et al. 1995. Phytoremediation: A novel strategy for the removal of toxic metals from the environment using plants. *Biotechnology* 13:468–74.

Sarret, G., et al. 2002. Forms of zinc accumulated in the hyperaccumulator *Arabidopsis halleri*. *Plant Physiol.* 130:1815–26.

Sarret, G., et al. 2009. Zn distribution and speciation in *Arabidopsis halleri* x *Arabidopsis lyrata* progenies presenting various Zn accumulation capacities. *New Phytol.* 184:581–95.

Schat, H., and W. M. Ten Bookum. 1992. Genetic control of copper tolerance in *Silene vulgaris*. *Heredity* 68:219–29.

Stinchcombe, J. R., and H. E. Hoekstra. 2008. Combining population genomics and quantitative genetics: Finding the genes underlying ecologically important traits. *Heredity* 100:158–70.

Storz, J. F. 2005. Using genome scans of DNA polymorphism to infer adaptive population divergence. *Mol. Ecol.* 14:671–88.

Talke, I. N., M. Hanikenne, and U. Krämer. 2006. Zinc-dependent global transcriptional control, transcriptional deregulation, and higher gene copy number for genes in metal homeostasis of the hyperaccumulator *Arabidopsis halleri*. *Plant Physiol.* 142:148–67.

Taylor, G. J., and A. A. Crowder. 1984. Copper and nickel tolerance in *Typha latifolia* clones from contaminated and uncontaminated environments. *Can. J. Botany* 62:1304–8.

Tribsch, A., and P. Schonswetter. 2003. Patterns of endemism and comparative phylogeography confirm palaeoenvironmental evidence for Pleistocene refugia in the Eastern Alps. *Taxon* 52:477–97.

Van Orsouw, N. J., et al. 2007. Complexity reduction of polymorphic sequences (CRoPS): A novel approach for large-scale polymorphism discovery in complex genomes. *PLoS ONE* 2(11):e1172.

Vitalis, R., K. Dawson, and P. Boursot. 2001. Interpretation of variation across marker loci as evidence of selection. *Genetics* 158:1811–23.

Weber, M., et al. 2004. Comparative microarray analysis of *Arabidopsis thaliana* and *Arabidopsis halleri* roots identifies nicotianamine synthase, a ZIP transporter and other genes as potential metal hyperaccumulation factors. *Plant J.* 37:269–81.

Willems, G., et al. 2007. The genetic basis of zinc tolerance in the metallophyte *Arabidopsis halleri* ssp. *halleri* (Brassicaceae): An analysis of quantitative trait loci. *Genetics* 176:659–74.

Willis, K. J. 1996. Where did all the flowers go? The fate of temperate European flora during glacial periods. *Endeavour* 20:110–14.

Willis, K. J., and T. H. van Andel. 2004. Trees or no trees? The environments of central and eastern Europe during the Last Glaciation. *Quaternary Sci. Rev.* 23:2369–87.

Ye, Z. H., et al. 1997. Copper and nickel uptake, accumulation and tolerance in *Typha latifolia* with and without iron plaque on the root surface. *New Phytol.* 136:481–88.

Zhao, F., et al. 2000. Zinc hyperaccumulation and cellular distribution in *Arabidopsis halleri*. *Plant Cell Environ.* 23:507–14.

Zhao, F., and S. McGrath. 2009. Biofortification and phytoremediation. *Curr. Opin. Plant Biol.* 12:373–80.

Zhao, K., et al. 2007. An *Arabidopsis* example of association mapping in structured samples. *PLoS Genet.* 3(1):e4.

16 Insecticides with Novel Modes of Action
Mechanism and
Resistance Management

Murad Ghanim and Isaac Ishaaya

CONTENTS

16.1 INTRODUCTION

Progress has been made during the past three decades to develop novel compounds affecting developmental processes in insects such as chitin synthesis inhibitors, juvenile hormone mimics, and ecdysone agonists. In addition, efforts have been made to develop compounds acting selectively on some insect groups by inhibiting or enhancing the activity of biochemical sites such as respiration (diafenthiuron) and activating the acetylcholine receptor (neonicotinoids) or the GABA receptor (avermectins) (Ishaaya and Horowitz 1998; Ishaaya 2001; Horowitz and Ishaaya 2004). Selective compounds with minimum effect on natural enemies and the environment can be classified as biorational insecticides (Ishaaya et al. 2005; Horowitz et al. 2009).

Reducing the risks associated with insect pest management tactics by using selective insecticides along with their resistance management to maintain their activity during a prolonged period is of utmost agricultural importance (Horowitz and Ishaaya 1995).

This chapter aims to present an overview of novel groups of insecticides with selective properties, with emphasis on their modes of action, selectivity, and importance to serve as components in integrated pest management (IPM) programs. Special attention will be given to their modes of resistance and resistance management in order to prolong their life span for the benefit of man and agriculture.

16.2 INSECT GROWTH REGULATORS

During the twentieth century, significant progress in the synthesis of new chemicals has resulted in the synthesis of effective insecticides such as organophosphates, carbamates, and pyrethroids. Unfortunately, many of these chemicals are harmful to man and beneficial organisms, and in many cases they cause ecological disturbances. One of the early approaches that captured worldwide attention is the development of novel compounds affecting developmental processes in insects, such as the chitin synthesis inhibitors, the juvenile hormone mimics, and ecdysone agonists. Several recent reviews have been published regarding their activities and agricultural importance (Retnakaran et al. 1985; Ishaaya 1990; Ishaaya and Horowitz 1998; Smagghe and Degheele 1998; Oberlander and Smagghe 2001; Ishaaya 2001). We will try to review and collate the widespread literature in a concise and consolidated form.

16.2.1 Chitin Synthesis Inhibitors

This group of insecticides consists of various compounds acting on insects of different orders by inhibiting chitin formation, thereby causing abnormal endocuticular deposition and abortive molting (Mulder and Gijswijt 1973; Ishaaya and Casida 1974; Post et al. 1974). Among the inhibitors of chitin synthesis are the benzoylphenyl ureas (BPUs) and buprofezin (Figure 16.1). BPUs are compounds with selective properties, affecting the larval stage by inhibiting the molting process (Mulder and Gijswijt 1973; Ishaaya and Casida 1974). They act mainly by ingestion, but in some species they suppress fecundity and exhibit ovicidal and contact toxicity (Ishaaya and Horowitz 1998). Studies with diflubenzuron [1-(4-chlorophenyl)-3-(2,6-diflubenzuroyl)urea], the first commercial compound and the most investigated one, revealed that the compound alters cuticle composition—especially that of chitin—thereby affecting the elasticity and firmness of the endocuticle (Grosscurt 1978; Grosscurt and Anderson 1980). The reduced level of chitin in the cuticle seems to result from inhibition of biochemical processes leading to chitin formation (Post et al. 1974; Ishaaya and Casida 1974; Hajjar and Casida 1979; Van Eck 1979). Chitin synthetase is not the primary biochemical site for the reduced level of chitin since, in some studies, benzoylphenyl ureas do not inhibit its activity in cell-free systems (Cohen and Casida 1980; Mayer et al. 1981; Cohen 1985). Some of the reports indicate the possibility that benzoylphenyl ureas might affect the insect hormonal site, thereby resulting in physiological disturbances such as inhibition of DNA synthesis (Mitlin et al. 1977; DeLoach et al. 1981; Soltani et al. 1984), alter carbohydrase and phenoloxidase activities (Ishaaya and Casida 1974; Ishaaya and Asher 1977), or suppress microsomal oxidase activity (Van Eck 1979). Recent

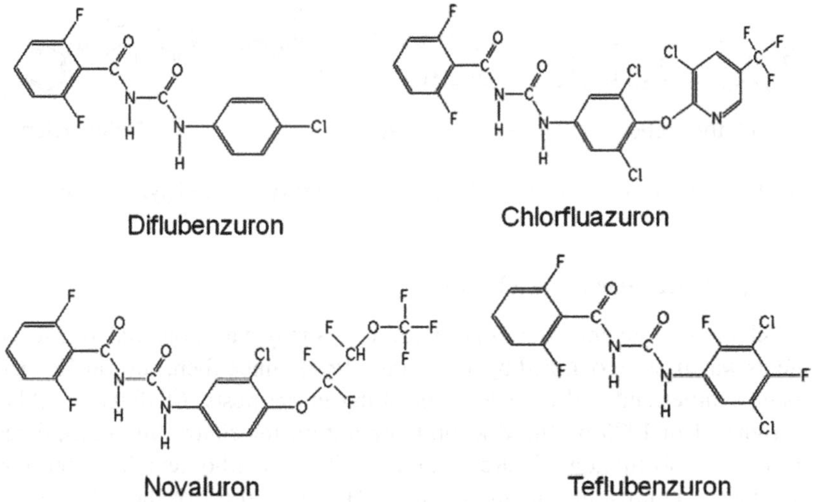

Diflubenzuron   Chlorfluazuron

Novaluron   Teflubenzuron

FIGURE 16.1 Molecular structures of diflubenzuron, chlorfluazuron, novaluron, and teflubenzuron.

studies, using imaginal discs and cell-free systems, indicate that benzoylphenyl ureas inhibit 20E-dependent GlcNAc incorporation into chitin (Mikolajczyk et al. 1994; Oberlander and Silhacek 1998). These findings suggest that benzoylphenyl ureas affect ecdysone-dependent biochemical sites, which leads to chitin inhibition. Several effective BPUs that are far more potent than diflubenzuron have been developed, such as chlorfluazuron, teflubenzuron, hexaflumuron, and novaluron (Figure 16.1). They are very effective in controlling insect pests of cotton and vegetable crops, such as *Spodoptera* and *Heliothis* species (Horowitz and Ishaaya 2004). In general, BPUs act by ingestion, but novaluron has some contact and trans-laminar activity, thereby affecting, in addition to lepidopteran and coleopteran larvae (by ingestion), whiteflies (by contact) and leafminers (translaminar) (Ishaaya et al. 1996, 2002b). BPUs affect generally the larval stages of insects, which synthesize chitin during the molting process. Hence, the adults of predators and parasitoids are not affected. BPUs are therefore considered important components in IPM programs.

Buprofezin is a thiadiazine-like compound that acts as a chitin synthesis inhibitor. It has both contact and vapor phase activity. It acts on the nymph stages of leafhoppers, planthoppers, scales, and whiteflies (Ishaaya et al. 1988; De Cock and Degheele, 1998). Its mode of action resembles that of BPUs, although its structure is not analogous. It inhibits incorporation of ^{3}H-glucosamin into chitin (Izawa et al. 1985; Uchida et al. 1985). As a result of chitin deficiency, the procuticle of the whitefly nymphs loses its elasticity and the insect is unable to complete the molting process (De Cock and Degheele 1988). Similarly to BPUs, buprofezin does not affect the adult stage. The compound has a mild effect on natural enemies (De Cock and Degheele, 1998) and is considered an important component in IPM programs for controlling whiteflies in various agricultural systems.

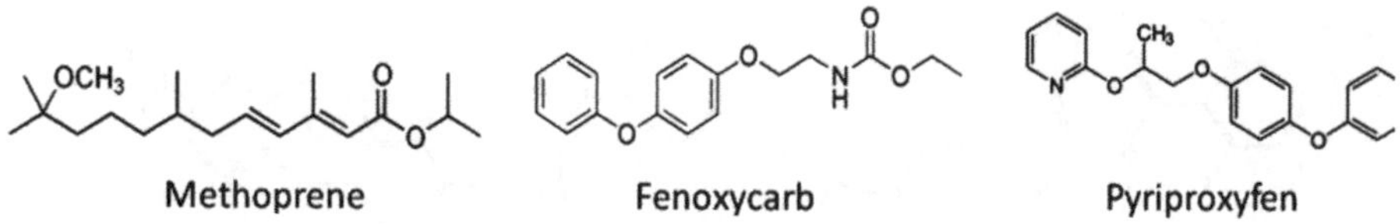

Methoprene   Fenoxycarb   Pyriproxyfen

FIGURE 16.2   Molecular structures of methoprene, fenoxycarb, and pyriproxyfen.

16.2.2   JUVENILE HORMONE MIMICS

The possibility that juvenile hormone (JH) analogs may have potential as insect control agents was first recognized by Williams (1967). Since then, several JH analogs have been synthesized and used for controlling insect pests. Methoprene (Altosid) was introduced in 1975 by the Zoecon Corporation for controlling mosquitoes by preventing adult formation (Figure 16.2). In addition, methoprene has been formulated for indoor control of dog and cat fleas. However, most of the JH analog synthesized at the early stages, such as farnesol derivatives, juvabione, methoprene, and kinoprene, were not sufficiently stable and could not be used for controlling agricultural insect pests.

Among the JH mimics, fenoxycarb and pyriproxyfen (Figure 16.2) exhibit reasonable field stability and high potency on agricultural insect pests. Fenoxycarb, ethyl [2-(4-phenoxyphenoxy)ethyl] carbamate, was the first commercial compound to be marketed for the control of agricultural pests (Dorn et al. 1981; Masner et al. 1987; Peleg 1988). Pyriproxyfen, 2-[1-methyl-2-(4-phenoxyphenoxy)ethoxy]pyridine, is a phenoxycarb derivative in which a part of the aliphatic chain has been replaced by pyridyl oxyethelene. It is a potent JH mimic affecting the hormonal balance in insects, resulting in strong suppression of embryogenesis, metamorphosis, and adult formation (Itaya 1987; Kawada 1988; Langley 1990; Koehler and Peterson 1991). Pyriproxyfen is considered a leading compound for controlling whiteflies (Ishaaya and Horowitz 1992, 1995; Ishaaya et al. 1994) and scale insects (Peleg 1988), and it is one of the most important components in insecticide resistance management programs in cotton fields worldwide (Horowitz and Ishaaya 1994; Dennehy and Williams 1997; Horowitz 1999). Extensive studies are in progress by several research institutions and chemical companies aiming at synthesizing new potent and selective JH mimics. Hence, this group of insecticides is considered a potential component to be used in IPM programs for controlling a diversity of insect pests.

16.2.3   ECDYSONE AGONISTS

During the past three decades, many investigations have been directed toward elucidating the possible use of the ecdysteroid receptor as a target site for developing novel insecticides (Bergamasco and Horn 1980; Horn et al. 1981; Riddiford 1985; Dhadialla et al. 1998). Several substituted dibenzoylhydrazines that act as ecdysone agonists have been synthesized by Rohm and Haas Co. (Spring House, Pennsylvania). Two compounds, tebufenozide and methoxyfenozide, have been commercialized and

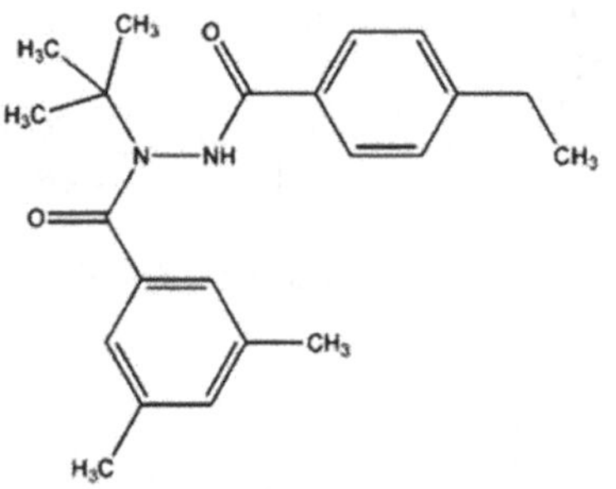

Tebufenozide 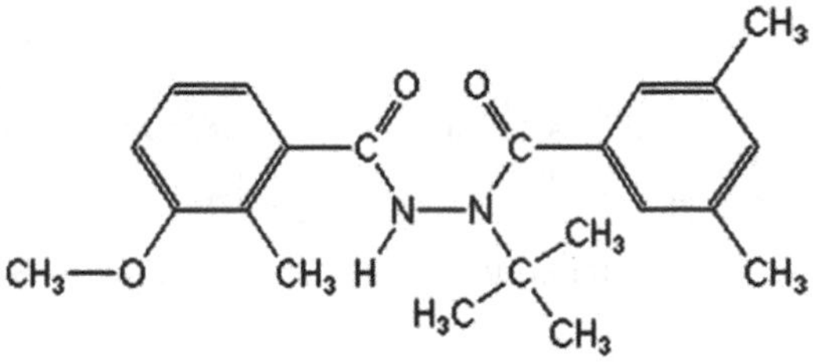   Methoxyfenozide

FIGURE 16.3 Molecular structure of tebufenozide and methoxyfenozide.

used for controlling lepidopteran pests (Ishaaya et al. 1995; Dhadialla et al. 1998; Smagghe and Degheele 1998). These compounds bind to the ecdysone receptors, thereby accelerating the molting process (Wing 1988; Wing et al. 1988; Retnakaran et al. 1995; Palli et al. 1996). Ecdysteroids are powerful toxicants acting specifically on lepidopteran pests such as *Manduca sexta* (Wing et al. 1988), *Plodia interpunctella* (Silhacek et al. 1990), *Spodoptera frugiperda* (Monthéan and Potter 1992), and *S. littoralis* (Smagghe and Degheele 1992; Ishaaya et al. 1995). Two of these compounds, tebufenozide and methoxyfenozide, have been commercialized for controlling agricultural insect pests, especially lepidopterans such as the codling moth *Carpocapsa pomonella* in apple (Heller et al. 1992), *S. frugiperda* in maize, the bollworm *Heliothis zea* (Chandler et al. 1992), and the armyworms *S. exigua*, *S. littoralis*, and *S. exempta* in cotton, field crops, vegetables, and ornamentals (Chandler et al. 1992; Smagghe and Degheele 1994; Ishaaya et al. 1995, Dhadialla et al. 1998; Smagghe and Degheele 1998) (Figure 16.3).

These compounds bind to the ecdysteroid receptors, accelerating the molting process and thereby disrupting the insect hormonal balance (Wing 1988; Palli et al. 1996). Methoxyfenozide is five- to tenfold more potent than tebufenozide on *S. littoralis* (Ishaaya et al. 1995) and exhibits more activity against budworm/bollworm and diamondback moth (Le et al. 1996; Dhadialla et al. 1998; Smagghe and Degheele 1998). These diacylhydrazines are considered highly selective with no harm to natural enemies (Dhadialla et al. 1998; Smagghe and Degheele 1998), and as such, they fit well into insect pest management programs.

16.3 NEONICOTINOIDS

Efforts have been made to develop nicotinyl insecticides with high affinity to the insect nicotinic acetylcholine receptor (nAChR), resulting in the development of a new group of insecticides (Abbink 1991; Tomizawa and Yamamoto 1992; Liu and Casida 1993; Elbert et al. 1998). Neonicotinoids interact with nicotinic acetylcholine receptors at the central and peripheral nervous system, resulting in excitation and paralysis, followed by death. Neonicotinoids of potential use in agriculture are imidacloprid, acetamiprid, and thiamethoxam (Figure 16.4). These compounds interact

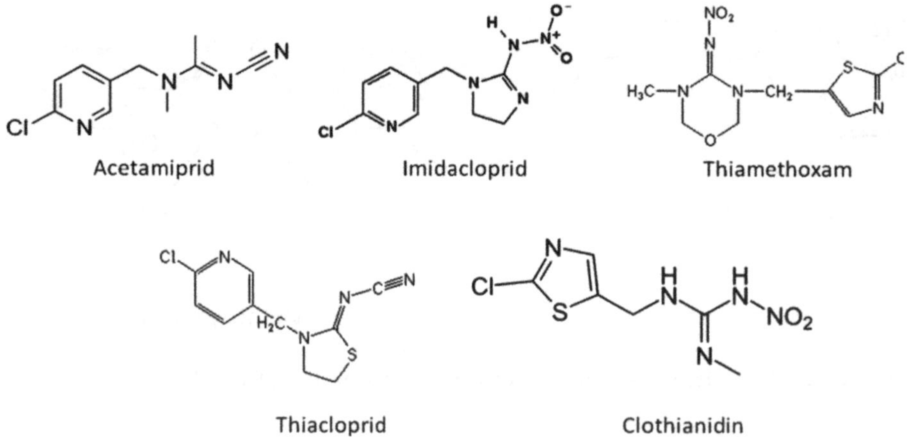

FIGURE 16.4   Molecular structures of acetamiprid, imidacloprid, thiamethoxam, thiacloprid, and clothianidin.

with nAChR in a structure-activity relationship (Tomizawa et al. 1995a,b), resulting in excitation and paralysis followed by death. Their selectivity results from a higher affinity to the insect nAChR than to that of vertebrates, in contrast to the original nicotine (Tomizawa et al. 1995b). Hence, it has been suggested that these new compounds are called neonicotinoids (Yamamoto et al. 1995).

Comparative toxicity was carried out, under laboratory conditions, of two neonicotinoids, acetamiprid and imidacloprid, against the whitefly *Bemisia tabaci* (Gennadius), using foliar and systemic applications on cotton seedlings (Horowitz et al. 1998). The ovicidal and nymphicidal activities of foliar application of acetamiprid on cotton seedlings were, according to LC_{50} and LC_{90} values, ten- and eighteen-fold more potent than imidacloprid, However, when applied to soil, imidacloprid was more potent than acetamiprid (Horowitz et al. 1998). In other assays, it was found that the comparative toxicity of the two neonicotinoids could be different when assayed on different host plants. The translaminar activity of imidacloprid on cabbage leaves against *Mysus persicae* was superior to that of acetamiprid, while against *Aphis gossypii* on cotton, the activity of imidacloprid was inferior to that of acetamiprid (Buchholz and Nauen 2002).

Several neonicotinoids have been commercialized for controlling insect pests: imidacloprid (Bayer CropScience), acetamiprid (Nipon Soda), nitanpyram (SumiTake), thiacloprid (Bayer CropScience), thiamethoxam (Syngenta), clothianidin (Sumitake/Bayer CropScience), and dinotefuran (Mitsui Chemicals) (Figure 16.4).

The neonicotinoids act specifically on sucking pests and have a mild or no effect on parasitoids and predators, and as such, they fit well in various IPM programs. Hence nAChR proved to be an important site for the development of a new group of insecticides able to control homopteran pests and others in various agricultural systems (Ishaaya 2001).

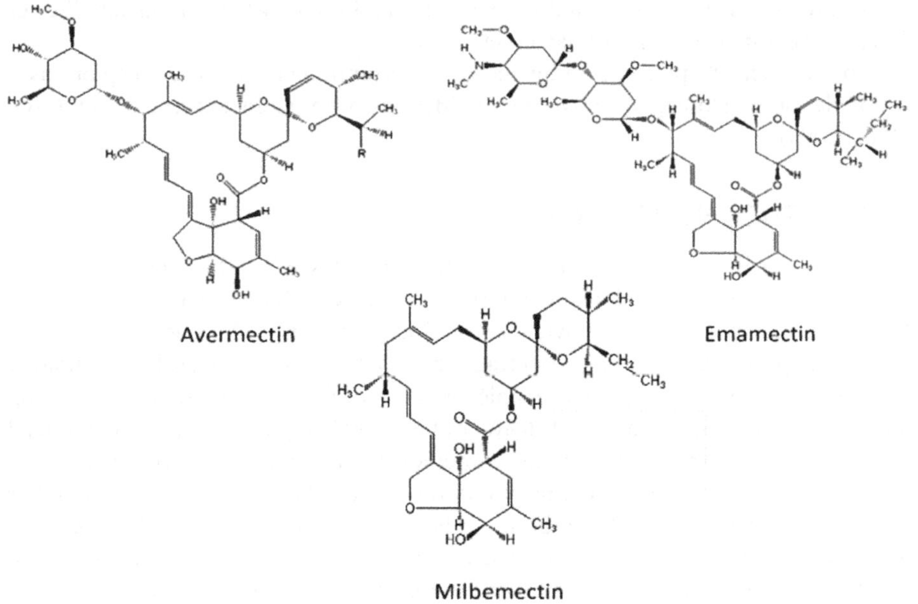

FIGURE 16.5 Molecular structures of avermectin, emamectin, and milbemectin.

16.4 AVERMECTINS

The avermectins are a group of macrocyclic lactones isolated from fermentation of the soil microorganism *Streptomyces avermitilis*. They bind with high affinity to sites in the head and muscle neuronal membranes of various insect species (Deng and Casida 1992; Roher et al. 1995), thereby acting as agonists for GABA-gated chloride channels (Mellin et al. 1983; Albrecht and Sherman 1987). They affect the nervous system of arthropods by increasing chloride ion flux at the neuromuscular junction, resulting in cessation of feeding and irreversible paralysis (Ishaaya 2001; Horowitz and Ishaaya 2004). Abamectin developed for agricultural use is a mixture of about 80% avermectin B_{1a} and 20% avermectin B_{1b} (Fisher and Mrozik 1989). It acts specifically on phytophagous mites, but also on agromizid leafminers, ants, cockroaches, and selected lepidopteran pests (Dybas 1989; Lasota and Dybas 1991). It is considerably less toxic to beneficial insects such as honeybees, parasitoids, and predators (Hoy and Cave 1985; Dybas 1989; Zhang and Sanderson 1990), and as such, it is an important addition to IPM programs.

The search for new avermectin derivatives resulted in the development of emamectin benzoate (Proclaim) by Merck (Whitehouse Station, New Jersey) and milbemectin (Milbeknock) by Sankyo (Tokyo, Japan) (Figure 16.5). Emamectin acts specifically on lepidopteran pests (Horowitz and Ishaaya 2004). Studies carried out in our laboratory indicate that emamectin is a powerful toxicant against *Spodoptera littoralis, Helicoverpa armigera, Frankliniella occidentalis*, and the whitefly *Bemisia tabaci* (Ishaaya et al. 2002a). Another avermectin derivative is milbemectin, which acts on a wide range of mites such as the two-spotted spider mite *Tetranychus urticae, T. cinnabarinus*, and the citrus red mite *Panonychus citri* (Sankyo 1997). It is a mixture of milbemycin A3 and milbemycin A4, both of which are metabolites of *Streptomyces hygroscopicus* (Barrett et al. 1985).

An excellent chapter regarding the use of γ-aminobutyric acid receptors as a rationale for developing selective insecticides has been published recently by Ozoe et al. (2009).

16.5 OTHER NOVEL GROUPS

Several novel insecticides acting on biochemical sites other than those described above have been developed during the past two decades. Pymetrozine, 4,5-dihydro-6-methyl-4-[(3-pyridinyl methylene)amino]-1,2,4-triazine-3(2H)one, is a novel azomethin pyridine insecticide affecting the nerve that controls the salivary pump of some sucking pests, causing irreversible cessation of feeding, followed by starvation and death (Schwinger et al. 1994; Kayser et al. 1994) (Figure 16.6). The compound is a powerful toxicant against aphids, whiteflies, and planthoppers (Fuog et al. 1998). It exhibits systemic and translaminar activities and can be used as drench or in foliar application (Flückiger et al. 1992). It is also effective in lessening aphid-transmitted diseases caused by persistent viruses (Fuog et al. 1998); in some cases, host plants have a major effect on the potency of pymetrozine due to its rate of penetration and translocation in various plants. It is a more powerful toxicant to whiteflies when applied on Bulgarian beans than when applied on cotton plants (Ishaaya and Horowitz 1998). Pymetrozine has no cross-resistance to other groups of insecticides, with no appreciable effect on natural enemies or the environment; hence, it can be

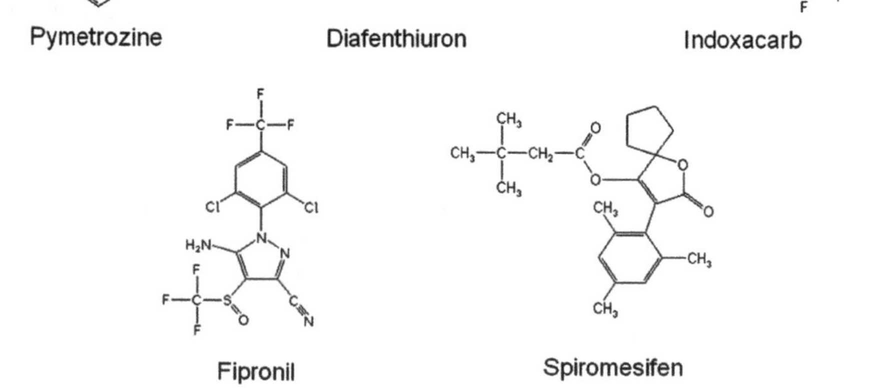

| Pymetrozine | Diafenthiuron | Indoxacarb |

| Fipronil | Spiromesifen |

FIGURE 16.6 Molecular structures of pymetrozine, diafenthiuron, indoxacarb, fipronil, and spiromesifen.

considered a potential component in IPM programs for controlling aphids and white-flies and for suppressing virus transmission.

Diafenthiuron, 3-(2,6-diisopropyl-4-phenoxyphenyl)-1-*tert*-butylthiourea, is a new type of thiourea developed by Syngenta that acts specifically on sucking pests such as mites, whiteflies, and aphids (Streibert et al. 1988; Kadir and Knowles 1991; Ishaaya et al. 1993) (Figure 16.6). It is photochemically converted in sunlight to its carbodiimide derivative, which is a more powerful toxicant than diafenthiuron (Steinmann et al. 1990). The carbodiimide acts as an adenosine triphosphatase (ATPase) inhibitor following metabolic activation to the corresponding carbodiim-ide (Petroske and Casida 1995). The carbodiimide metabolite inhibits mitochon-drial respiration by selective and covalent binding to the proteolipid ATPase (Ruder et al. 1992; Kayser and Eilinger 2001). Diafenthiuron inhibits ATPase activities in assays carried out with bulb mites *Rhizoglyphus echinopus*, the two-spotted spider mite *Tetranychus urticae,* and a freshwater fish, the blue gill *Lepomis macrochirus* (Kadir and Knowles 1991). Biological tests revealed a decreased toxicity of diafenthi-uron in the presence of piperonyl butoxide, indicating the importance of cytochrome P450 in metabolizing difenthiuron (Ruder et al. 1992). Diafenthiuron has a favorable mammalian toxicity coupled with a relatively low toxicity to beneficial insects and predatory mites (Streibert et al. 1988). As such, it is considered an important addi-tion in pest management programs in various agricultural systems for controlling the diversity of sucking pests.

Indoxacarb acts by inhibiting sodium ion entry into nerve cells, resulting in the paralysis and death of target insect pests (Lapied et al. 2001) (Figure 16.6). It is activated by the insect esterases following ingestion (Wing et al. 2000). It sup-presses the peak sodium currents in a time-dependent manner, and its potency on the insect neuron is at least one hundred times higher than that in the rat ganglian neuron (Yuji and Yoshiaki 2003). Various sucking insect pests can bioactivate indox-acarb more than lepidopterans (Wing et al. 2000). Indoxacarb acts on diversity of lepidopteran pests as well as on certain homopterans and coleopterans in various agricultural systems (Harder et al. 1996; Wing et al. 2000). Its efficacy has been demonstrated against important insect pests such as *Heliothis* and *Helicoverpa* spe-cies, *Spodoptera* sp., *Tricoplusia* sp., *Lygus* sp., *Empoaska* sp., and the Colorado potato beetle, *Leptinotarsa decemlineata* (Harder et al. 1997; Hammes et al. 1998; Michaelides and Irving 1998; Sullivan et al. 1999; Wing et al. 2000). Indoxacarb can play a useful role in resistance management programs. It has no cross-resistance with the current insecticides, such as pyrethroids, organophospates, and carbamates (Lapied et al. 2001). It has no appreciable effect on some important natural enemies, such as *Orius* and *Phytoseiulus* species (Bostanian and Akalach 2006). Hence, it can be considered an important addition to insect pest management programs.

Fipronil is a phenylpyrazole insecticide developed by Rhône Poulene in 1987 and placed on the market in 1993 (Figure 16.6). It exhibits both herbicidal and insecti-cidal activities (Yanase and Andoh 1989; Klis et al. 1991). It acts on γ-aminobutyric acid (GABA), thereby blocking chloride channels (Grant et al. 1990; Hainzl et al. 1998; Scharf and Siegfried 1999; Bloomquist 2001). Competitive binding has demonstrated that this compound has greater affinity to the target site in insects than in mammals, resulting in a high selectivity toward insects (Grant et al. 1990;

Cole et al. 1993; Hainzl and Casida 1996). Fipronil acts on a diversity of insect species, such as the diamondback moth (*Plutella xylostella*), *Spodoptera* and *Heliothis* species, and Colorado potato beetle, and against household pests, such as termites and cockroaches (Grant et al. 1990; Hamon et al. 1996). Fipronil is considered an important addition in IPM programs.

Spiromesifen is an insecticide from the new class of spirocyclic tetronic acids that effectively acts against whiteflies and mites (Bretschneider et al. 2003; Nauen et al. 2003, 2005; Nauen and Konanz 2005) (Figure 16.6). Spiromesifen acts as an inhibitor of acetyl-CoA-carboxylase, a lipid metabolism enzyme. Treatment with spiromesifen causes a significant decrease in total lipids. This compound has been introduced into several countries during the last few years and is becoming a major compound in whitefly and mite resistance management programs in conjunction with other effective insecticides, such as neonicotonoids and diafenthiuron. Several studies have shown the effectiveness of spiromesifen against whiteflies and mites, while no cross-resistance with classical (organophosphates, carbamates, pyrethroids, and endosulfan) and new biorational insecticides, such as neonicotonoids and pyriproxyfen, was observed (Kontsedalov et al. 2008). Spiromesifen is active in foliar application and mainly against juvenile stages, and exhibits transovarial activity against mites and whitefly adults in a dose-dependent manner. It has been reported that spiromesifen is moderately active against whitefly adults and has an ovo-larvicidal activity. At relatively high concentrations under laboratory conditions, it has contact and translaminar activity (Kontsedalov et al. 2008).

Botanical and microbial insecticides such as the azadirachtin extracted from the neem tree (*Azadirachta indica*) and toxins obtained from *Bacillus thuringiensis* are generally compatible with organic agriculture and, in general, have low toxicity to human and natural enemies. However, some such as rotenone, pyrethrin, and ryania may have a toxic effect on nontarget organisms, such as bees, fish, and natural enemies (Plimmer 1993). Biological insecticides such as viruses, bacteria, fungi, and nematodes are additional components in controlling insect pests; however, the main drawbacks of these compounds are their instability under conditions of heat and UV radiation (Horowitz and Ishaaya 2004).

16.6 RESISTANCE MODE OF ACTION

Resistance to insecticides among many insects is known to occur behaviorally or physiologically. Behavioral resistance mechanisms, in which insects avoid toxic plant substances, have been much less studied. This is a way by which insects select their food and survive a sublethal dose. Genetic or physiological resistance has been divided into two major classes: altering target site by which the chemical compound binds or increasing metabolic detoxification (Hemingway et al. 2004). Increased excretion, which facilitates transport of insecticides through the digestive system, is also a possible resistance mechanism. A combination of target site alteration and enhanced metabolic detoxification is a likely mechanism and usually provides more resistance against one or several chemical compounds.

Genetic mechanisms of insecticide resistance among many insect pests have been extensively studied, especially in the case of conventional insecticides such as organophosphates and carbamates, which have been widely used since the 1950s, and later pyrethroids, since 1970. Although novel insecticides became more important in many IPM programs, the extensive use of conventional insecticides led to many resistance problems over the years. Many insect pests have developed resistance to all major conventional insecticidal groups and, in recent years, also to major novel insecticides, most importantly neonicotinoids and insect growth regulators (IGRs), known as biorational insecticides. The need for a greater diversity of compounds acting effectively against many insect pests is being met by the introduction of several insecticides with new modes of action, which are less affected or are unaffected by existing resistance mechanisms (Ishaaya and Horowitz 1998; Nauen and Denholm 2005).

Biorational insecticides today include the neonicotinoid insecticides, such as imidacloprid and acetamiprid, that target nicotinic acetylcholine receptors in the insect nervous system; the IGRs, including inhibitors of chitin synthesis—buprofezin and benzoylphenyl ureas such as novaluron; and the juvenile hormone mimic pyriproxyfen. Other new biorational insecticides inhibit the electron chain flow in the mitochondria (diafenthiuron) or affect feeding behavior in sucking pests (pymetrozine). Abamectin, emamectin, and milbemectin are fermentation products of *Streptomyces avermitilis* and have been reported to be effective against many insect pests. All biorational insecticides are considered to be relatively safe to natural enemies, and are being incorporated into many pest control programs around the world, for better control and for delaying resistance problems. Although these biorational materials bring many advantages for controlling many agricultural pests, resistance problems against them have been arising in recent years. Resistance problems were reported among three major novel insecticide classes, buprofezin, pyriproxyfen, and neonicotinids; the genetic basis of resistance to neonicotinoids has been largely investigated, and some evidence exists to suggest a genetic basis for the resistance.

For buprofezin, which inhibits chitin synthesis in several hemipteran pests (Ishaaya et al. 1988) and for which the resistance mode of action is not fully understood, a decrease in its susceptibility occurred three years after its introduction on Israeli cotton in 1989 (Horowitz and Ishaaya 1992). Significant decreases in susceptibility to buprofezin were detected in the sweetpotato whitefly *Bemisia tabaci* populations collected from cotton fields in Israel (Horowitz 1999), UK, the Netherlands, and Spain (Cahill et al. 1996; Elbert and Nauen 2000). Despite the decreased susceptibility to buprofezin in many countries, many *B. tabaci* populations collected from several regions in California and Arizona in 1998 and 1999 showed an increase in susceptibility to buprofezin (Toscano et al. 2001). Although the genetic basis for the decreased susceptibility to buprofezin was not elucidated, the decreased, then increased susceptibilities in many countries suggest that metabolic enzymes are responsible for manipulating this compound once it has entered the body of the insect.

Pyriproxyfen, which inhibits egg hatching, directly or transovarially, and affects nymph and pupal mortality (Ishaaya and Horowitz 1992, 1995), has been one of the main agents for controlling many insect pests since its introduction in the early 1990s (Dennehy and Williams 1997; Horowitz 1999). A significant decrease in pyriproxyfen susceptibility that reached five-hundred-fold was observed when whitefly *B. tabaci* populations were treated with three successive applications in rose greenhouses (Horowitz and Ishaaya 1994). Resistance to pyriproxyfen in cotton was delayed because its use was limited to one treatment per season, but eventual resistance that brought about field failure was observed. To date, all confirmed cases of strong resistance to pyriproxyfen have been associated with the Q rather than the B biotype of *B. tabaci* (Horowitz et al. 2005). It is therefore possible that the present distribution of genes for pyriproxyfen resistance reflects the current gene flow associated with Q-type populations.

The mechanisms of resistance to pyriproxyfen in *B. tabaci*, as well as the mode of action of the compound itself, are not fully understood. Inheritance of resistance to this compound was found to be partially dominant, and is conferred primarily by a mutant allele at a single locus (Horowitz et al. 2003a,b).

The introduction of the neonicotinoids has resulted in reduced use of pyriproxyfen, even in locations with less severe resistance to this insecticide; thus, the levels of resistance in several locations have declined. In the United States, pyriproxyfen and buprofezin were first used as rotational alternatives in cotton resistance management programs beginning in Arizona in 1996 and in California in 1997. Initial monitoring of *B. tabaci* collected from cotton in Arizona from 1996 to 1998 showed no reductions in susceptibility to pyriproxyfen (Dennehy et al. 1999). However, a significant decrease in susceptibility was observed in populations collected from some Arizona cotton growing regions in 1999 and 2000 (Li et al. 2001). Monitoring of *B. tabaci* populations in southern California and southwestern Arizona revealed that regional differences in pyriproxyfen toxicity were minimal, and similarly to buprofezin, susceptibility to pyriproxyfen was maintained after three years of use (Toscano et al. 2001). To date, both buprofezin and pyriproxyfen remain highly effective and continue to provide economic control of *B. tabaci* in California and Arizona cotton fields (Palumbo et al. 2001).

Recent laboratory experiments that have employed *B. tabaci* microarrays have shown that the mechanism to pyriproxyfen resistance is likely to be an increased metabolic activity of P450 monooxygenase enzymes, as well as other stress-related enzymes (Ghanim and Kontsedalov 2007). Among the P450 genes that were significantly induced following exposure to pyriproxyfen were CYP9F2, which was induced with a 3.85-fold change, and many mitochondrial genes related to elevated stress response and detoxification by monooxegenases. Among these mitochondrial genes are the NADH ubiquinone oxidoreductase of complexes I and III, which were induced by more than fourfold after exposure to pyriproxyfen. Although the changes in the expression of these genes may hint about the resistance mode of action, the specific target site for pyriproxyfen and the specific action that disrupts the insect development are still not fully understood (Ghanim and Kontsedalov 2007). Other genes that were induced following pyriproxyfen treatment belong to general stress, detoxification, protein synthesis, metabolism, and lipid and carbohydrate metabolism. All these changes were reported to occur following insecticide use and exposure to toxic materials. Further evidence that supports the metabolic detoxification of pyriproxyfen is the decline in resistance to this compound in unselected laboratory strains that were kept for thirteen generations without pyriproxyfen exposure. Reversal of the resistance was accompanied by increased biotic fitness. The seasonal cost for resistance following this work was estimated to be 25% (Wilson et al. 2007).

Neonicotinoids, which became one of the most widely used insecticide groups against many sap-sucking pests, comprise one of the most studied groups at the molecular level. The first commercial compound was imidacloprid. Among others that have been introduced are acetamiprid, nitenpyram, thiamethoxam, and thiacloprid, all acting at the nicotinic acetylcholine receptor level. Since this group's mode of action is similar for all compounds, cross-resistance threatens the effectiveness of the group as a whole (Li et al. 2001). A few cases of resistance to neonicotinoids have been reported (Nauen and Denholm 2005), and thus it became important to develop resistance management programs for this group (Elbert et al. 1996). Resistance to imidacloprid was first reported in *B. tabaci* from greenhouses in southern Spain (Cahill et al. 1996). From the late 1990s, resistance to neonicotinoids increased and field strains exhibited more than a hundred-fold resistance to this group (Nauen and Denholm 2005). Most of these resistance cases were associated with the Q biotype of *B. tabaci* (Nauen et al. 2002; Horowitz et al. 2004; Dennehy et al. 2005). The involvement of target site modification and metabolic enzymes in neonicotinoid resistance was studied (Rauch and Nauen 2003). Biochemical analyses of metabolizing enzymes such as esterases, glutathione S-transferases (GSTs), and cytochrome P450-dependent monooxygenases showed that only the monooxygenase activity was correlated with imidacloprid, thiamethoxam, and acetamiprid resistance. The involvement of P450-dependent monooxygenase activity in resistance to neonicotinoids was also supported by molecular data. One P450 gene from the CYP6 class, termed CYP6CM1, was shown to be overexpressed only in resistant Q- and B-type strains (Karunker et al. 2008).

16.7 RESISTANCE MANAGEMENT

One of the main strategies to delay the development of resistance among pests is the use of pesticides with different modes of actions. Such an approach can be effected by alternating or subsequent use. Newer modes of actions and novel insecticides are less affected by existing resistance incidences, and lead to better control and significant delay of the resistance. Using pesticides with the same mode of action leads to rapid field failures in controlling the target pests, within fewer generations (Figure 16.7).

Resistance outbreaks are observed usually after uncontrolled use of insecticides in the absence of planned management. Tactics for avoiding or preventing the development of resistance have been mostly based on monitoring pest populations, and insecticide sprays combined with an understanding of biology of the pest. This information has been combined into models that provide probabilities for resistance emergence among pest populations (Denholm and Rowland 1992; Georghiou 1994). Many ways were proposed for managing pest populations, keeping in mind that repeated and uncontrolled use leads to resistance problems. The tactics mentioned below were proposed and sometimes used as individual or combined methods for controlling pests and delaying resistance (Georghiou 1983).

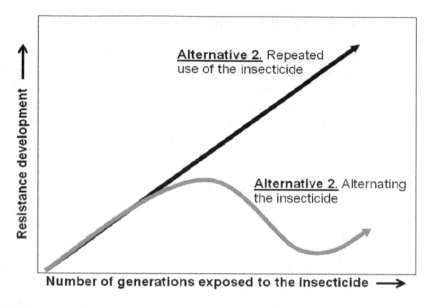

FIGURE 16.7 A general model for resistance development under different management regimes.

The first tactic to reduce selection for resistance and preserve susceptible insects in the population is the use of low application doses and less frequent applications. This approach is easy to apply and does not involve high risk. However, using a low concentration of any chemical in controlling pests involves some selection that could lead to resistance. Low concentrations may be sometimes sublethal and can cause artificial selection of resistant populations in the field (Roush 1989).

The second tactic involves using a high concentration of the insecticide to overcome any resistant populations in the field. This concentration should also be sufficient to overcome the action of any detoxification enzymes that might lead to resistance. This can be empowered by using several synergists.

The third tactic aims at using two or more insecticides with different modes of action. This way ensures better control and reduction in selecting resistant individuals. The insecticides can be applied simultaneously as mixtures, or in alternation. It is hypothesized that mixtures provide a greater benefit and are more effective than rotating the sprays (Roush 1989). However, mixtures rely on the assumption that there is no cross-resistance between the insecticides mixed.

A combination of the methods mentioned above was a strategy adopted by many countries and provided success in controlling many important pests for a certain period of time. This strategy involves the use of "windows," in which insecticides with different modes of action are applied at time intervals. This strategy was adopted to manage pyrethroid resistance in the bollworm, *Helicoverpa armigera*, on Australian cotton (Forrester et al. 1993) in the early 1980s. Although these methods were successful, they failed to prevent a gradual increase in the frequency of pyrethroid resistance in *H. armigera*.

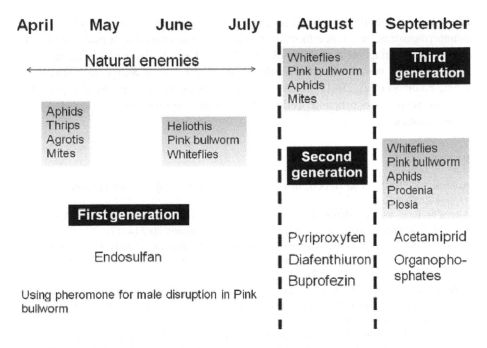

FIGURE 16.8 Resistance management program in Israeli cotton.

A similar resistance management program was introduced in Israel in the late 1980s to control *Bemisia tabaci* and coexisting cotton pests. This program relied heavily on the windows strategy, on restricting the use of key compounds to one application per season, and on rotating insecticides in a sequence that delays resistance (Horowitz et al. 1994) (Figure 16.8). This program resulted in decreasing the number of sprays over the season since different insecticides, which were intended to control certain pests, also controlled other, nontarget pests, thus reducing the overall number of sprays through the season. However, this strategy did not completely prevent resistance. As seen in Figure 16.8, endosulfan was used as a first compound that controlled many pests, such as aphids, thrips, argotis, and whiteflies. None of these is a major pest at the beginning of the cotton season, and endosulfan sprays are usually enough to ensure fields without these pests. However, during the mid-season, in August, the pest populations, especially whiteflies and the pink bullworm, require the use of novel chemicals such as pyriproxyfen, diafenthiuron, and buprofezin. During the late season, the remains of these populations are controlled with neonicotinoids and organophosphates. The main strategy noted here is the nonuse of compounds from the same group more than once during the cotton season. Some insecticides are multifunctional and can treat more than one pest when the populations are in low numbers, for example, endosulfan at the beginning of the season. During this time, the pest populations are usually not yet exposed to insecticides and are expected to be more susceptible than the next generations. Fewer compounds used at the beginning of the season ensure a minor effect on natural enemies that are found and becoming established for the rest of the season. Late use of novel compounds that do not affect natural enemies ensures their existence through the season. Despite all these countermeasures, resistance to novel insecticides was observed in many regions in Israel in the management program of whiteflies and pink bullworm. Similar management strategies were undertaken in the southwestern United States (Dennehy et al. 1996; Palumbo et al. 2001) for controlling B. *tabaci*, and the results also provided an opportunity to significantly reduce the number of sprays through the season and delay resistance.

REFERENCES

Abbink, J. 1991. The biochemistry of imidacloprid. *Pflanzenschutz-Nachur Bayer* 44:183–94.

Albrecht, C. P., and M. Sherman. 1987. Lethal and sublethal effect of avermectin B_1 on three fruit fly species (Diptera: Tephritidae). *J. Econ. Entomol.* 80:344–47.

Barrett, A. G. M., et al. 1985. The application of novel carbanion chemistry in milbemycin-avermectin synthesis. In *Recent advances in the chemistry of insect control,* ed. N. F. James, 257–71. London: Royal Society of Chemistry.

Bergamasco, R., and D. H. S. Horn. 1980. The biological activities of ecdysteroids and ecdysteroid analogues. In *Progress in ecdysone research,* ed. J. A. Hoffman, 299–324. Amsterdam: Elsevier.

Bloomquist, J. R. 2001. GABA and glutamate receptors as biochemical sites for insecticide action. In *Biochemical sites of insecticide action and resistance,* ed. I. Ishaaya, 17–41. Berlin: Springer.

Bostanian, N. J., and M. Akalach. 2006. The effect of indoxacarb and five other insecticides on *Phytoseiulus persimilis* (Acari: Phytosiidae), *Amblyseius fallacis* and nymphs of *Orius insidiosus* (Hemiptera: Anthocoridae). *Pest. Manage. Sci.* 62:334–39.

Bretschneider, T., et al. 2003. Spirodiclofen and spiromesifen—Novel acaricidal and insecticidal tetronic acid derivatives with a new mode of action. *Chimia* 57:697–701.

Buchholz, A., and R. Nauen. 2002. Translocation and translaminar bioavailability of two neonicotinoid insecticides after foliar application to cabbage and cotton. *Pest. Manage. Sci.* 58:10–16.

Cahill, M., et al. 1996. Insecticide resistance in *Bemisia tabaci*—Current status and implications for management. In *Proceedings of the Brighton Crop Protection Conference—Pests and Diseases*, Brighton, November 18–21, 1996, pp. 75–80. Farnham, UK: British Crop Protection Council.

Chandler, L. D., S. D. Pair, and W. E. Harrison. 1992. RH-5992, a new insect growth regulator active against corn earworm and fall armyworm (*Lepidoptera: Noctuidae*). *J. Econ. Entomol.* 85:1099–103.

Cohen, E., 1985. Chitin synthetase activity and inhibition in different insect microsomal preparations. *Experientia* 41:470–72.

Cohen, E., and J. E. Casida. 1980. Inhibition of *Tribolium* gut synthetase. *Pest. Biochem. Physiol.* 13:129–36.

Cole, L., R. Nicholson, and J. E. Casida. 1993. Action of phenylpyrazole insecticides at the GABA-gated chloride channels. *Pest. Biochem. Physiol.* 46:47–54.

De Cock, A., and D. Degheele. 1998. Buprofezin: A novel chitin synthesis inhibitor affecting specifically planthoppers, whiteflies and scale insects. In *Insecticides with novel modes of action: Mechanism and application,* ed. I. Ishaaya and D. Degheele, 74–91. Berlin: Springer.

DeLoach, J. R., et al. 1981. Inhibition of DNA synthesis by diflubenzuron in pupae of the stable fly *Stomoxys calcitrans* (L.). *Pest. Biochem. Physiol.* 15:172–80.

Deng, Y., and J. E. Casida. 1992. Housefly head GABA-gated chloride channel: Toxicological relevant binding site for avermectins coupled to site for ethynyl-bicycloortho benzoate. *Pest. Biochem. Physiol.* 43:166–22.

Denholm, I., and M. W. Rowland. 1992. Tactics for managing pesticide resistance in arthropods: Theory and practice. *Annu. Rev. Entomol.* 37:91–112.

Dennehy, T. J., and L. Williams. 1997. Management of resistance in *Bemisia* in Arizona cotton. *Pest. Sci.* 51:398–406.

Dennehy, T. J., et al. 1996. Monitoring and management of whitefly resistance to insecticides in Arizona. In *Proceedings of the Beltwide Cotton Production Research Conference*, Nashville, TN, January 9–12, 1996, ed. P. Dugger and D. A. Richter, 135–40. Memphis, TN: National Cotton Council of America.

Dennehy, T. J., et al. 1999. Arizona whitefly susceptibility to insect growth regulators and chloronicotinyl insecticides: 1998 season summary. In *Cotton, a college of agriculture report*, ed. J. C. Silvertooth, 376–91. Series P-116. Tucson, AZ: University of Arizona, College of Agriculture.

Dennehy, T. J., et al. 2005. New challenges to management of whitefly resistance to insecticides in Arizona. In *Vegetable report*, ed. D. N. Byrne and P. Baciewicz, 1–31. Series P-144. College of Agriculture and Life Sciences, University of Arizona. http://cals.arizona.edu/pubs/crops/az1382/az1382_2.pdf.

Dhadialla, T. S., G. R. Carlson, and D. P. Le. 1998. New insecticides with ecdysteroidal and juvenile hormone activity. *Annu. Rev. Entomol.* 45:545–69.

Dorn, J., et al. 1981. A novel non-neurotoxic insecticide with a broad activity. *Z. Pflanzenkr Pflanzenschutz* 88:269–75.

Dybas, R. A. 1989. Abamectin use in crop protection. In *Ivermectin and abamectin,* W. C. Campbell, 287–310. Berlin: Springer.

Elbert, A., and R. Nauen. 2000. Resistance in *Bemisia tabaci* (Homoptera: Aleyrodidae) to insecticide in southern Spain with special reference to neonicotinoids. *Pest Manage. Sci.* 56:60–64.

Elbert, A., R. Nauen, and W. Leicht. 1998. Imidacloprid, a novel chloronicotinyl insecticide: Biological activity and agricultural importance. In *Insecticides with novel modes of action: Mechanism and application,* ed. I. Ishaaya and D. Degheele, 50–73. Berlin: Springer.

Elbert, A., et al. 1996. Resistance management with chloronicotinyl insecticides using imidacloprid as an example. *Pflanzen-Nachrich Bayer* 49:5–53.

Fischer, M. H., and H. Mrozik. 1989. Chemistry. In *Ivermectin and abamectin,* ed. W. G. Campbell, 1–23. Berlin: Springer.

Flückiger, C. R., et al. 1992. CGA 215'944—A novel agent to control aphids and whiteflies. *Brighton Crop Prot. Conf.* (Pests and Diseases) 1:43–50.

Forrester, N. W., et al. 1993. Management of pyrethroid and endosulfan resistance in *Helicoverpa armigera* (Lepidoptera: Noctuidae) in Australia. *Bull. Entomol. Res.* (Suppl. 1), 1–132.

Fuog, D., S. J. Fergusson, and C. Flückiger. 1998. Pymetrozine: A novel insecticide affecting aphids and whiteflies. In *Insecticides with novel modes of action: Mechanism and application,* ed. I. Ishaaya and D. Degheele, 40–49. Berlin: Springer.

Georghiou, G. P. 1983. Management of resistance in arthropods. In *Pest resistance to pesticides,* ed. G. P. Georghiou and T. Saito, 769–92. New York: Plenum Press.

Georghiou, G. P. 1994. Principles of insecticide resistance management. *Phytoprotection* 75(Suppl.):51–59.

Ghanim, M., and S. Kontsedalov. 2007. Gene expression in pyriproxyfen-resistant *Bemisia tabaci* Q biotype. *Pest Manage. Sci.* 63:776–83.

Grant, D. B., et al. 1990. A comparison of mammalian and insect GABA receptor chloride channels. *Pest. Sci.* 30:355–56.

Grosscurt, A. C. 1978. Effect of diflubenzuron on mechanical penetrability, chitin formation, and structure of the elytra of *Leptinotarsa decemlineata*. *J. Ins. Physiol.* 24:827–31.

Grosscurt, A. C., and S. O. Anderson. 1980. Effect of diflubenzuron on some chemical and mechanical properties of the elytra of *Leptinotarsa decemlineata*. *Proc. K. Ned. Akad. Wet.* 83C:143–50.

Hainzl, D., and J. E. Casida. 1996. Fipronil insecticide: Novel photochemical desulfinylation with retention of neurotoxicity. *Proc. Natl. Acad. Sci. USA* 93:12764–67.

Hainzl, D., L. M. Cole, and J. E. Casida. 1998. Mechanism for selective toxicity of fipronil insecticide and its sulfone metabolite and desulfinyl photoproduct. *Chem. Res. Toxicol.* 11:1529–35.

Hajjar, N. P., and J. E. Casida. 1979. Structure activity relationship of benzoylphenyl ureas as toxicants and chitin synthesis inhibitors in *Oncopeltus faciatus*. *Pest. Biochem. Physiol.* 11:33–45.

Hammes, G. G., D. Sherrod, and W. H. Mitchell. 1998. A five year summary of beneficial arthropod studies with "Steward" (DPX MP 062). In *Proceedings of the Beltwide Cotton Conference,* ed. C. P. Drugger and D. A. Richter, 940. Memphis, TN: National Cotton Council.

Hamon, N., R. Shaw, and H. Yang. 1996. Worldwide development of fipronil insecticide. In *Proceedings of the Beltwide Cotton Conference, National Cotton Production Conference,* ed. C. P. Dugger and D. A. Richter, 759–65. Memphis, TN: National Cotton Council of America.

Harder, H. H., et al. 1996. DPX-MPO62; a novel broad-spectrum, environmentally soft, insect control compound. In *Proceedings of the 1996 Brighton Crop Protection Conference— Pests and Diseases,* Vol. 2, pp. 449–54.

Harder, H. H., et al. 1997. A novel broad-spectrum, environmentally soft, insect control compound. In *Proceedings of the 1997 Beltwide Cotton Production Conference, National Cotton Production Conference,* ed. C. P. Dragger and D. A. Richter, 48–50. Memphis, TN: National Cotton Council of America.

Heller, J. J., et al. 1992. Field evaluation of RH-5992 on lepidopteran pests in Europe. *Brighton Crop Prot. Conf.* (Pests and Diseases) 1:59–66.

Hemingway, J., et al. 2004. The molecular basis of insecticide resistance in mosquitoes. *Insect Biochem. Mol. Biol.* 34:653–65.

Horn, D. H. S., et al. 1981. Moulting hormones LIII. The synthesis and biological activity of some ecdysone analogues. *Aust. J. Chem.* 34:2607–18.

Horowitz, A. R. 1999. Insecticide resistance in whiteflies: Current status and implications for management. In *Proceedings of an ENMARIA Symposium on Combating Insecticide Resistance,* ed. I. Denholm and P. Ioannidis, 86–98. Thessaloniki, Greece.

Horowitz, A. R., P. C. Ellsworth, and I. Ishaaya. 2009. Biorational pest control—An overview. In *Biorational control of arthropod pests: Application and resistance management,* ed. I. Ishaaya and A. R. Horowitz, 1–20. Dordrecht: Springer.

Horowitz, A. R., G. Forer, and I. Ishaaya. 1994. Managing resistance in *Bemisia tabaci* in Israel with emphasis on cotton. *Pest. Sci.* 42:113–22.

Horowitz, A. R., and I. Ishaaya. 1992. Susceptibility of the sweetpotato whitefly (Homoptera: Aleyrodidae) to buprofezin during the cotton season. *J. Econ. Entomol.* 85:318–24.

Horowitz, A. R., and I. Ishaaya. 1994. Monitoring resistance to IGRs in the sweetpotato white-fly (Homoptera: Aleyrodidae). *J. Econ. Entomol.* 87:866–71.

Horowitz, A. R., and I. Ishaaya. 1995. Chemical control of *Bemisia*—Management and application. In *Bemisia 1995: Taxonomy, biology, damage, control and management,* ed. D. Gerling and R. T. Mayer, 537–56. Andover, Hants, UK, Intercept.

Horowitz, A. R., and I. Ishaaya. 2004. Biorational insecticides—Mechanism, selectivity and importance in pest management. In *Insect pest management; field and protected crops,* ed. A. R. Horowitz and I. Ishaaya, 1–28. Berlin: Springer.

Horowitz, A. R., S. Kontsedalov, and I. Ishaaya. 2004. Dynamics of resistance to the neonico-tinoids acetamiprid and thiamethoxam in *Bemisia tabaci* (Homoptera: Aleyrodidae). *J. Econ. Entomol.* 97:2051–56.

Horowitz, A. R., et al. 1998. Comparative toxicity of foliar and systemic application of acetamiprid and imidacloprid against the cotton whitefly, *Bemisia tabaci* (*Hemiptera: Aleyrodidae*). *Bull. Entomol. Res.* 88:437–42.

Horowitz, A. R., et al. 1999. Managing resistance to the insect growth regulator pyriproxyfen in *Bemisia tabaci*. *Pest. Sci.* 55:272–76.

Horowitz, A. R., et al. 2003a. Inheritance of pyriproxyfen resistance in the whitefly, *Bemisia tabaci* (Q biotype). *Arch. Insect Biochem. Physiol.* 54:177–86.

Horowitz, A. R., et al. 2003b. Biotype Q of *Bemisia tabaci* identified in Israel. *Phytoparasitica* 31:94–98.

Horowitz, A. R., et al. 2005. Biotypes B and Q of *Bemisia tabaci* and their relevance to neonicotinoid and pyriproxyfen resistance. *Arch. Insect Biochem. Physiol.* 58:216–25.

Hoy, M. A., and F. E. Cave. 1985. Laboratory evaluation of avermectin as a selective acaricide for use with *Metasciulus occidentalis* (Nesbitt) (Acarina: Phytoseiidae). *Exp. Appl. Acarol.* 1:139–52.

Ishaaya, I. 1990. Benzoylphenyl ureas and other selective control agents—Mechanism and application. In *Pesticides and alternatives,* ed. J. E. Casida, 365–76. Amsterdam: Elsevier.

Ishaaya, I. 2001. Biochemical processes related to insecticide action: An overview. In *Biochemical sites of insecticides action and resistance*, ed. I. Ishaaya, 1–16. Berlin: Springer.

Ishaaya, I., and K. R. S. Asher. 1977. Effect of diflubenzuron on growth and carbohydrate hydrolases of *Tribolium castaneum. Phytoparasitica* 5:149–58.

Ishaaya, I., and J. E. Casida. 1974. Dietary TH-6040 alters cuticle composition and enzyme activity of housefly larval cuticle. *Pest. Biochem. Physiol.* 4:484–90.

Ishaaya, I., A. De Cock, and D. Degheele. 1994. Pyriproxyfen, a potent suppressor of egg hatch and adult formation of the greenhouse whitefly (Homoptera: Aleyrodidae). *J. Econ. Entomol.* 87:1185–89.

Ishaaya, I., and A. R. Horowitz. 1992. Novel phenoxy juvenile hormone analog (pyriproxyfen) suppresses embryogenesis and adult emergence of the sweet potato whitefly (Homoptera: Aleyrodidae). *J. Econ. Entomol.* 85:2113–17.

Ishaaya, I., and A. R. Horowitz. 1995. Pyriproxyfen, a novel insect growth regulator for controlling whiteflies—Mechanism and resistance management. *Pest. Sci.* 43:227–32.

Ishaaya, I., and A. R. Horowitz. 1998. Insecticides with novel modes of action: An overview. In *Insecticides with novel modes of action: Mechanism and application,* ed. I. Ishaaya and D. Degheele, 1–24. Berlin: Springer.

Ishaaya, I., S. Kontsedalov, and A. R. Horowitz. 2002a. Emamectin, a novel insecticide for controlling field crop pests. *Pest Manage. Sci.* 58:1091–95.

Ishaaya, I., S. Kontsedalov, and A. R. Horowitz. 2005. Biorational insecticides: Mechanism and cross-resistance. *Arch. Insect Biochem. Physiol.* 58:192–99.

Ishaaya, I., Z. Mendelson, and A. R. Horowitz. 1993. Toxicity and growth suppression exerted by diafenthiuron in the sweetpotato whitefly *Bemizia tabaci. Phytoparasitica* 21:199–204.

Ishaaya, I., Z. Mendelson, and V. Melamed-Madjar. 1988. Effect of buprofezin on embryogenesis and progeny formation of sweetpotato whitefly (Homoptera: Aleyrodidae). *J. Econ. Entomol.* 81:781–84.

Ishaaya, I., S. Yablonski, and A. R. Horowitz. 1995. Comparative toxicity of two ecdysteroid agonists, RH-2485 and RH-5992, on susceptible and pyriproxyfen-resistant strains of the Egyptian cotton leafworm, *Spodoptera littoralis. Phytoparasitica* 23:139–45.

Ishaaya, I., et al. 1996. Novaluron (MCW-275), a novel benzoylphenyl urea, suppressing developing stages of lepidopteran, whitefly and leafminer pests. In *Proceedings of the 1996 Brighton Crop Protection Conference—Pests and Diseases,* pp. 1013–20. BCPC: London.

Ishaaya, I., et al. 2002b. Novaluron (Rimon), a novel IGR—Mechanism, selectivity and importance in IPM programs. *Med. Landbouww Rijksuniv. Gent.* 67:617–26.

Itaya, W. 1987. Insect juvenile hormone analogue as an insect growth regulator. *Sumitomo Pyrethroid World* 8:2–4.

Izawa, Y., et al. 1985. Inhibition of chitin biosynthesis by buprofezin analogs in relation to their activity controlling *Nilaparvata lugens* Stål. *Pest. Biochem. Physiol.* 24:343–47.

Kadir, H. A., and C. O. Knowles. 1991. Toxicological studies of the thiourea diafenthiuron in diamondback moth (Lepidoptera: Yponomeutidae), two-spotted spider mite (Acari: Tetranychidae), and bulb mite (Acari: Acaridae). *J. Econ. Entomol.* 84:780–84.

Karunker, I., et al. 2008. Over-expression of cytochrome P450 CYP6CM1 is associated with high resistance to imidacloprid in the B and Q biotypes of *Bemisia tabaci* (Hemiptera: Aleyrodidae). *Insect Biochem. Mol. Biol.* 38:634–44.

Kawada, H. 1988. An insect growth regulator against cockroach. *Sumitomo Pyrethroid World* 11:2–4.

Kayser, H., and F. Eilinger. 2001. Metabolism of diafenthiuron by microsomal oxidation: Procide activation and inactivation as mechanism contributing to selectivity. *Pest Manage. Sci.* 57:975–80.

Kayser, H., L. Kaufmann, and F. Schürmann. 1994. Pymetrozine (CGA 215'944); a novel compound for aphid and whitefly control. An overview of its mode of action. In *Proceedings of the 1994 Brighton Crop Protection Conference—Pests and Diseases,* Vol.2, pp. 737–42. BCPC: London.

Klis, S. F. L., H. P. M. Vijverberg, and J. Van den Berken. 1991. Phenylpyrazoles, a new class of pesticides: Electrophysiological investigation into basis effects. *Pest. Biochem. Physiol.* 39:210–18.

Koehler, P. G., and R. J. Patterson. 1991. Incorporation of pyriproxyfen in German cockroach (Dictyoptera: Blattellidae) management program. *J. Econ. Entomol.* 84:917–21.

Kontsedalov, S., et al. 2008. Toxicity of spiromesifen on the developmental stages of *Bemisia tabaci* biotype B. *Pest Manage. Sci.* 65:5–13.

Langley, P. 1990. Control of tsetse fly using a juvenile hormone mimic, pyriproxyfen. *Sumitomo Pyrethroid World* 15:2–5.

Lapied, B., F. Grolleau, and D. B. Sattelle. 2001. Indoxacarb, an oxadiazine, blocks insect neuronal sodium channels. *Br. J. Pharmacol.* 132:587–95.

Lasota, J. A., and R. A. Dybas. 1991. Avermectin, a novel class of compounds: Implications for use in arthropod pest control. *Annu. Rev. Entomol.* 36:96–117.

Le, D. P., et al. 1996. RH-2485: A new selective insecticide for caterpillar control. In *Proceedings of the 1996 Brighton Crop Protection Conference—Pests and Diseases,* Vol. 2, pp. 481–86. BCPC: London.

Li, A. Y., et al. 2001. Sustaining Arizona's fragile success in whitefly resistance management. In *Proceedings of the Beltwide Cotton Production Research Conference,* Anaheim, CA, January 9–13, 2001, ed. P. Dugger and D. A. Richter, 1108–14. Memphis, TN: National Cotton Council of America.

Liu, M.-Y., and J. E. Casida. 1993. High affinity binding of [^{3}H]-imidacloprid in the insect acetylcholine receptor. *Pest. Biochem. Physiol.* 46:40–46.

Masner, P., M. Angst, and S. Dorn. 1987. Fenoxycarb, an insect growth regulator with juvenile hormone activity: A candidate for *Heliothis virescens* (F.) on cotton. *Pest. Sci.* 18:89–94.

Mayer, R. T., A. C. Chen, and J. R. DeLoach. 1981. Chitin synthesis inhibiting insect growth regulators do not inhibit chitin synthase. *Experientia* 37:337–38.

Mellin, T. N., R. D. Busch, and C. C. Wang. 1983. Postsynaptic inhibitions of invertebrate neuromuscular transmission by avermectin B1a. *Neuropharmacology* 22:89–96.

Michaelides, P., and S. N. Irving. 1998. Cotton insect pest control with indoxacarb: A novel insecticide. In *Proceedings of the World Cotton Research Conference 2,* Athens, Greece, pp. 773–76.

Mikolajczyk, P., et al. 1994. Chitin synthesis in *Spodoptera frugiperda* wing imaginal discs. I. Chlorfluazuron, difluazuron and teflubenzuron inhibit incorporation but not uptake of [^{14}C]-*N*-acetyl-D-glucosamine. *Arch. Insect Biochem. Physiol.* 25:245–58.

Mitlin, N., G. Wiygul, and J. W. Haynes. 1977. Inhibition of DNA synthesis in boll weevils (*Anthonomus grandis* Boheman) sterilized by Dimilin. *Pest. Biochem. Physiol.* 7:559–63.

Monthéan, C., and D. E. Potter. 1992. Effect of RH-5849, a novel insect growth regulator, on Japanese beetle (Coleoptera: Scarabaeidae) and fall armyworm (Lepidoptera: Noctuidae) in turfgrass. *J. Econ. Entomol.* 85:507–13.

Mulder, R., and M. T. Gijswijt. 1973. The laboratory evaluation of two promising new insecticides which interfere with cuticle deposition. *Pest. Sci.* 4:737–45.

Nauen, R., T. Bretschneider, A. Elbert, R. Fischer, and R. Tieman. 2003. Spirodiclofen and spiromesifen. *Peticide Outlook* 14:243–46.

Nauen, R., and I. Denholm. 2005. Resistance of insect pests to neonicotinoid insecticides: Current status and future prospects. *Arch. Insect Biochem. Physiol.* 58:200–15.

Nauen, R., and Konanz, S. 2005. Spiromesifen as a new chemical option for resistance management in whiteflies and spider mites. *Pflanzenschutz-Nachrichten Bayer* 58:485–502.

Nauen, R., N. Stumpf, and A. Elbert. 2002. Toxicological and mechanistic studies on neonicotinoid cross resistance in Q-type *Bemisia tabaci* (Hemiptera: Aleyrodidae). *Pest Manage. Sci.* 58:868–74.

Nauen R., H. J. Schnorbach, and A. Elbert. 2005. The biological profile of spiromesifen (Oberon)—A new tetronic acid insecticide/acaricide. *Pflanzenschutz-Nachrichten Bayer* 58:417–40.

Oberlander, H., and D. L. Silhacek. 1998. New perspectives on the mode of action of benzoylphenyl urea insecticides. In *Insecticides with novel modes of action: Mechanism and application*, ed. I. Ishaaya, 92–105. Berlin: Springer.

Oberlander, H., and G. Smagghe. 2001. Imaginal discs and tissue cultures as targets for insecticide action. In *Biochemical sites of insecticide action and resistance,* ed. I. Ishaaya, 133–50. Berlin: Springer.

Ozoe, Y., M. Takeda, and M. Kazuhiko, M. 2009. γ-Aminobutyric acid receptors: A rationale for developing selective insect pest control chemicals. In *Biorational control of arthropod pests: Application and resistance management,* ed. I. Ishaaya and A. R. Horowitz, 131–62. Heidelberg: Springer.

Palli, S. R., et al. 1996. Cloning and developmental expression of *Choristoneura* hormone receptor 3, an ecdysone-inducible gene and a member of the steroid hormone receptor superfamily. *Insect Biochem. Mol. Biol.* 26:485–99.

Palumbo, J. C., A. R. Horowitz, and N. Prabhaker. 2001. Insecticidal control and resistance management for *Bemisia tabaci*. *Crop Protect.* 20:739–65.

Peleg, B. A. 1988. Effect of a new phenoxy juvenile hormone analog on California red scale (Homoptera: Coccidae) and the ectoparasite *Aphytis holoxanthus* DeBache (Hymenoptera: Aphelinidae). *J. Econ. Entomol.* 81:88–92.

Petroske, E., and J. E. Casida. 1995. Diafenthiuron action: Carbodiimide formation and ATPase inhibition. *Pest. Biochem. Physiol.* 53:60–74.

Plimmer, J. R. 1993. Regulatory problems associated with natural products and biopesticides. *Pest. Sci.* 39:103–8.

Post, L. C., B. J. de Jong, and W. R. Vincent. 1974. 1-(2,6-Disubstituted benzoyl)-3-phenylurea insecticides: Inhibitors of chitin synthesis. *Pest. Biochem. Physiol.* 4:473–83.

Rauch, N., and R. Nauen. 2003. Identification of biochemical markers linked to neonicotinoid cross-resistance in *Bemisia tabaci* (Hemiptera: Aleyrodidae). *Arch. Insect Biochem. Physiol.* 54:165–76.

Retnakaran, A., J. Granett, and T. Ennis. 1985. Insect growth regulators. In *Comprehensive insect physiology, biochemistry and pharmacology,* ed. G. A. Kerkut and L. I. Gilbert, 529–601. Vol. 12. Pergamon Press: Oxford.

Retnakaran, A., et al. 1995. Molecular analysis of the mode of action of RH-5992, a lepidopteran-specific, non-steroidal ecdysteroid agonist. *Insect Biochem. Mol. Biol.* 25:109–17.

Riddiford, L. M. 1985. Hormone action at the cellular level. In *Comprehensive insect physiology, biochemistry and pharmacology,* ed. G. A. Kerkut and L. I. Gilbert, 37–84. Vol. 18. Oxford, Pergamon Press.

Roush, R. T. 1989. Designing resistance management programs: How can you choose? *Pest. Sci.* 26:423–41.

Ruder, F. J., J. A. Benson, and H. Kayser. 1992. The mode of action of the insecticide/acaricide diafenthiuron. In *Insecticides: Mechanism of action and resistance,* ed. D. Otto and B. Weber, 263–76. Andover, UK: Intercept.

Sankyo Co. 1997. *Technical data sheet on milbemectin.* Tokyo: Sankyo Co.

Scharf, M. E., and D. B. Siegfried. 1999. Toxicity of neurophysiological effects of fipronil and fipronil sulfone on the western corn rootworm (Coleoptera: Chrysomelidae). *Arch. Insect Biochem. Physiol.* 40:150–56.

Schwinger, M., P. Harrewiju, and H. Katser. 1994. Effect of pymetrozine (CGA 215′944), a novel aphicide, on feeding behavior of aphids. In *Proceedings of the 8th IUPAC International Congress on Pesticide Chemicals,* Washington, DC, Vol. 1, p. 230.

Silhacek, D. L., H. Oberlander, and P. Procheron. 1990. Action of RH-5849, a nonsteroidal ecdysteroid mimic, on *Plodia interpunctella* (Hübner) *in vivo* and *in vitro. Arch. Insect Biochem. Physiol.* 15:201–12.

Smagghe, G., and D. Degheele. 1992. Effect of RH-5849, the first non-steroidal ecdysteroid agonist, on larvae of *Spodoptera littoralis* (Boisd.) (*Lepidoptera: Noctuidae*). *Arch. Insect Biochem. Physiol.* 21:119–28.

Smagghe, G., and D. Degheele. 1994. Action of a novel nonsteroidal ecdysteroid mimic, tebofenozide (RH-5992), on insects of different orders. *Pest. Sci.* 42:85–92.

Smagghe, G., and D. Degheele. 1998. Ecdysone agonists: Mechanism and biological activity. In *Insecticides with novel modes of action: Mechanism and application,* ed. I. Ishaaya and D. Degheele, 25–39. Berlin: Springer.

Soltani, N., M. T. Bosson, and J. Delachambre. 1984. Effect of diflubenzuron on the pupal-adult development of *Tenebrio molitor* L. (Coleoptera: Tenebrionidae): Growth and development, cuticle secretion, epidermal cell density and DNA synthesis. *Pest. Biochem. Physiol.* 21:256–64.

Steinmann, A., E. Stamm, and B. Frei. 1990. Chemodynamics in research and development of new plant protection agents. *Pest. Outlook* 1(3):3–7.

Streibert, H. P., J. Drabek, and A. Rindisbacher. 1988. CGA 106630—A new type of acaricide/insecticide for the control of the sucking pest complex in cotton and other crops. *Brighton Crop Prot. Conf.* (Pests and Diseases) 1:25–33.

Sullivan, M. J., S. G. Turnipseed, and D. Robinson. 1999. Insecticidal efficacy against a complex of fall and beet armyworms and soybean looper in South Carolina Cotton. In *Proceedings of the 1999 Beltwide Cotton Production Conference,* ed. C. P. Dugger and D. A. Richter, 1034–36. Memphis, TN: National Council of America.

Tomizawa, M., and I. Yamamoto. 1992. Binding of nicotinoids and the related compounds to the insect nicotinic acetylcholine receptor. *J. Pest. Sci.* 17:231.

Tomizawa, M., et al. 1995a. Pharmacological characteristics of insect nicotinic acetylcholine receptor with its ion channel and the comparison of the effect of nicotinoids and neonicotinoids. *J. Pest. Sci.* 20:57–64.

Tomizawa, M., et al. 1995b. Pharmacological effects of imidacloprid and related compounds on the nicotinic acetylcholine receptor with its ion channel from the *Torpedo* electric organ. *J. Pest. Sci.* 20:49–56.

Toscano, N. C., et al. 2001. Inter-regional differences in baseline toxicity of *Bemisia argentifolii* (Homoptera: Aleyrodidae) to the two insect growth regulators, buprofezin and pyriproxyfen. *J. Econ. Entomol.* 94:1538–46.

Uchida, M., T. Asai, and T. Sugimoto. 1985. Inhibition of cuticle deposition and chitin biosynthesis by a new insect growth regulator buprofezin in *Nilaparvata lugens* Stål. *Agric. Biol. Chem.* 49:1233–34.

Van Eck, W. G. 1979. Mode of action of two benzoylphenyl ureas as inhibitors of chitin synthesis in insects. *Insect Biochem.* 9:295–300.

Williams, C. M. 1967. Third generation pesticides. *Sci. Am.* 217:13–17.

Wilson, M., et al. 2007. Reversal of resistance to pyriproxyfen in the Q biotype of *Bemisia tabaci* (Hemiptera: Aleyrodidae). *Pest Manage. Sci.* 63:761–68.

Wing, K. D. 1988. RH-5849, a non steroidal ecdysone agonist: Effect on *Drosophila* cell line. *Science* 241:467–69.

Wing, K. D., R. A. Slawecki, and G. R. Carlson. 1988. RH-5849, a non steroidal ecdysone agonists: Effect on larval Lepidoptera. *Science* 241:470–72.

Wing, K. D., et al. 2000. Bioactivation and mode of action of the oxadiazine indoxacarb in insects. *Crop Protect.* 19:537–45.

Yamamoto, I., et al. 1995. Molecular mechanism of selective toxicity of nicotinoids and neonicotinoids. *J. Pest. Sci.* 20:33–40.

Yanase, D., and A. Andoh. 1989. Porphyrin synthesis involvement in dephenyl ether-like mode of action of TNPP-ethyl, a novel phenylpyrazole herbicide. *Pest. Biochem. Physiol.* 35:70–79.

Yuji, T., and K. Yoshiaki. 2003. Effects of the oxadiazine insecticide, indoxacarb, on voltage-gated sodium channels in dorsal unpaired median neurons isolated from the American cockroach, *Periplaneta americana* (Linnaeus) (Orthoptera: Blattidae). *Jpn. J. Appl. Entomol. Zool.* 47:29–32.

Zhang, Z., and J. P. Sanderson. 1990. Relative toxicity of abamectin to the predatory mite *Phytoseiulus persimilis* (Acari: Phytoseiidae) and the two-spotted spider mite (Acari: Tetranychidae). *J. Econ. Entomol.* 83:1783–90.

17 Conclusions

Claude Amiard-Triquet and Michèle Roméo

CONTENTS

This book allows us to have a complete view of the state of the art in the field of tolerance to environmental contaminants, considering (1) the frequency of occurrence of this phenomenon in different taxa exposed to many classes of chemicals, (2) the mechanisms of defence and the acquisition of tolerance to chemical stress, and (3) the ecological and ecophysiological consequences of being tolerant. In this concluding chapter, the large body of knowledge presented earlier can be brought together to respond to a number of very important questions:

- Does tolerance occur in all taxa exposed to any chemical contaminant in their environment under chronic or subchronic exposure?
- How does tolerance work?
- What are the main mechanisms involved in tolerance?
- What are the efficiency limits of tolerance?

• What are the consequences of tolerance for organisms in their environment, and for the assessment of environmental health and environmental quality using ecotoxicological tools?

17.1   IS TOLERANCE A WIDESPREAD PHENOMENON?

However tolerance is acquired in organisms chronically exposed to chemical pollution (through acclimation or a genetic process), it appears to be a widespread phenomenon, although counterexamples do exist and have been reported in the literature.

As outlined in Chapter 2, tolerance is particularly well documented for metals, but a number of studies have also shown that many organisms are able to cope with organic contaminants. Data that are highlighted in Chapters 2 and 13 are briefly summarised in Table 17.1. Data were acquired from field studies or preexposing organisms in the laboratory. It is very striking that species belonging to most of the major taxonomic groups and organisms living in different media (terrestrial, freshwater, estuarine, and marine organisms) have developed tolerance to chemicals.

Tolerance in bacteria covers a large range of different chemicals. This is particularly well documented in the case of human health, with bacteria becoming resistant to antibiotics. The presence of antibacterial resistant bacteria has also been demonstrated near fish farms using medicated fish food (Herwig et al. 1997). Some bacteria possess high tolerance to various metals and have been considered potential candidates for removal from wastes (Malik 2004). Tributyltin-resistant bacteria have been observed in estuarine and freshwater sediments (Wuertz et al. 1991). Adaptation to alkanes has been demonstrated in marine bacteria, allowing them to contribute to the biodegradation of hydrocarbon compounds in polluted environments (Ballihaut et al. 2004). Organic solvent-tolerant bacteria have been reviewed by Sardessai and Bhosle (2002). As regards microbial communities, pollutant-induced community tolerance (PICT) is a measure of biological community (especially microbial) sensitivity to a contaminant, combined with taxonomic analysis. This concept demonstrates that the acquisition of tolerance is also frequent, considering either soils or freshwater ecosystems, and that it depends not only on the community itself, but also on its interaction with the toxicant (Chapters 4 and 14).

Tolerance to metals has been widely reported. When associated metal accumulation has been studied, it appears that mechanisms that limit uptake or enhance elimination are not as important as the storage of accumulated metal in nontoxic physicochemical form, such as in metal concretions and by metal binding with metallothioneins (Chapter 13). Enhanced detoxified storage allows the survival of such tolerant forms to metal exposure in the organisms' environment, including exposure via the diet of animals. Metal hyperaccumulation has been reported in higher plants, and has evolved through adaptations of metal homeostasis processes, including metal uptake, chelation, trafficking, and storage. The accumulated concentration of a metal in an organism does not depend mainly on the trophic level occupied, but on the metal accumulation mechanism shown by an organism irrespective of its position in a food chain. Biomagnification of metals in the marine food chain is therefore not a common phenomenon. The most striking exception is the biomagnification of methyl mercury

TABLE 17.1
Contaminant Tolerance of Species Belonging to Different Taxonomic Groups (for details, see Chapters 2 and 13)

	Taxonomic Group	Medium
Inorganic Contaminants		
Metals	Ciliates	Soil, FW
	Microalgae	FW
	Macroalgae	SW
	Annelids	Soil, FW, SW
	Bryozoa	SW
	Gastropods	FW
	Bivalves	SW
	Crustaceans	FW, SW
	Insects	FW
	Fish	FW, SW
	Nematodes (communities)	SW
Fluoride	Microalgae	SW
Nitrates	Amphibians	Terrestrial, FW
Residual chlorine	Fish	FW
Organic Contaminants		
PCBs	Microalgae	SW
	Fish	SW
PAHs	Annelids	SW
	Fish	SW
	Nematodes	SW
Diesel	Copepods, ostracods	SW
Herbicides	Microalgae	Soil, FW, SW
	Crustaceans	FW
Biocide (antifouling)	Nanoplankton, microplankton, phytoplankton communities	SW
OP insecticide	Microalgae	FW
	Crustaceans	FW
Pesticides	Cyanophyceae	FW
	Crustaceans	FW
	Amphibians (toad embryos)	FW
Pharmaceuticals	Crustaceans	FW

Note: FW = freshwater, SW = estuarine or marine waters.

along the food chain, and major outbreaks of severe methyl mercury poisoning via the food chain to human consumers have occurred in Japan (Chapter 13).

Tolerance to different classes of organic contaminants has been demonstrated in several taxa, including microorganisms, invertebrates, and vertebrates. Even if counterexamples exist, tolerance to organic contaminants is clearly a frequent phenomenon. Phase I enzymes allow the biotransformation of PAHs, PCBs, and TCDD-like compounds, leading to tolerance mainly in vertebrates (Chapter 8). The efficiency of mechanisms of biotransformation and elimination of PAHs has also been recognised in certain invertebrates, such as some marine polychaetes. Low biotransformation favours food chain biomagnification, whereas high metabolisation can lead to the excretion of an important proportion of the contaminants that have entered the organism. However, in certain cases, the molecules resulting from biotransformation are hydrophobic, nonmetabolisable compounds (Chapter 13), and these have a high bioaccumulation and toxicity potential (particularly for hydroxylated and methyl-sulfonated metabolites of PCBs). In counteraction, ABC transporters play a significant role in xenobiotic bioavailability and constitute an effective toxicological process that could confer resistance generally to a wide range of organic compounds (Chapter 10).

Klerks and Weis (1987) have identified a potential bias in the reporting of field examples of tolerance, since it is questionable whether negative results have an equal chance of being published in the relevant scientific literature. Johnston (Chapter 2) listed focal species and evidence for differential tolerance in the papers that she reviewed, and she pointed out that occasionally a study is published in which organisms are chronically exposed to very high contaminant loads but do not develop enhanced tolerance. Kovatch et al. (2000) found no difference (measured as whole-sediment reproduction bioassays) in the population sensitivity of meiobenthic copepods (*Microarthridion littorale*) exposed to a highly contaminated sediment mixture. Acute or chronic exposures to uranium evoked little difference in the response of populations of the tropical cladoceran species (*Moinodaphnia macleayi*) compared to reference populations of this water flea (Semaan et al. 2001). The flatworm *Polycelis tubifex* exposed to cadmium did not acquire tolerance and mortality occurred (Indeherberg et al. 1999). In the case of the fish *Fundulus heteroclitus,* Nacci et al. (2002) used benzo[a]pyrene (B[a]P) to test the tolerance of two populations that differed in their tolerance to a dioxin-like compound. No difference in survival between fish from the two populations was found, but B[a]P exposure elicited elevated ethoxyresorufin *O*-deethylase (EROD) activity compared with fish from the polychlorinated biphenyl (PBC)-contaminated harbor.

17.2 HOW DOES TOLERANCE WORK?

17.2.1 TOLERANCE AT DIFFERENT LEVELS OF BIOLOGICAL ORGANISATION

The importance of the variability of sensitivity to stress has been documented in numerous taxa. In invertebrates, it is generally accepted that interspecific variability in tolerance is linked to very important variability in adaptive strategies (limiting the influx of contaminants, storage in detoxified physicochemical forms, increased elimination). Phylogenetic differences have been reported (Rainbow 1998), and even within relatively restricted zoological groups with similar modes of feeding, striking interspecific differences may be observed, such as those reported in filter-feeding bivalves exposed to Ag (Berthet et al. 1992). In addition to the tolerance characteristics of different species, many studies have reported that, within the same species, populations chronically exposed to chemicals in their environment have been able to cope more efficiently than "naïve" individuals. The best known examples are the resistance of bacteria to antibiotics and other chemicals, of terrestrial plants to metals (Chapter 15), and of insects to pesticides (Chapter 16). Many examples of inter- and intraspecific variability of tolerance are described in this book, particularly in Chapter 3.

Interspecific variability of tolerance is the basis of the pollution-induced community tolerance (PICT) concept proposed by Blanck et al. (1988). A biological community is composed of different species, the inherent sensitivity of which toward a given toxicant is highly variable. Thus, in a contaminated environment, the most sensitive organisms are lost as a consequence of pollutant pressure, whereas tolerant organisms are maintained. Consequently, the new community as a whole is more tolerant to the toxicant responsible for selection than another community, initially identical, but which has never been exposed to this toxicant. Such a PICT has been demonstrated in many studies on freshwater microbial communities (Chapter 4), and soil bacterial communities (Chapter 14). Similarly, monitoring of macrofaunal communities for the assessment of the ecological health status in our environment takes root in the fact that environmental stressors will be responsible for the loss of the more sensitive species/taxa, whereas tolerant species/taxa will be favoured (Chapter 3).

17.2.2 GENETIC ADAPTATION VERSUS PHYSIOLOGICAL ACCLIMATION

Depending on its nature physiological acclimation or genetic adaptation—the persistence of tolerance in the environment will be limited to a shorter or longer period in the life of an individual, or the tolerance will last much longer, being transmitted from one generation to the next. Thus, distinguishing between these possibilities is important. However, nearly two decades apart, the reviews by Klerks and Weis (1987) and Wirgin and Waldman (2004) both stress that only a few studies have really established the genetic origin of tolerance, at least in macroinvertebrates and vertebrates. All but one of the twenty-three aquatic invertebrate studies captured by the literature search in Chapter 2 examined tolerance to metals. This likely reflects the persistent and ubiquitous nature of metal contamination in aquatic systems, as well as the relative focus of contaminant expertise in these systems. In Chapter 6, Rainbow and Luoma concentrated on examples of genetically based adaptation to metal exposure in invertebrates and fish. Recent studies have focussed on the adaptation of fish to contaminant mixtures, such as those encountered in North American estuaries (Chapter 8) or on the Atlantic coast of France (Marchand et al. 2004).

Exposure to chemicals can exert a selection pressure leading to the presence of resistant genotypes in fish living in impacted areas. In these areas, an increased frequency of resistant genotypes was reported, allowing the maintenance of DNA integrity associated with the duplication of specific genes (Figure 17.1). However, negative consequences of being resistant may be observed, such as decreased fitness (Chapter 12) and decreased adaptability to new environments or stressors, thus increasing the probability of local extinction. On the other hand, in nonresistant

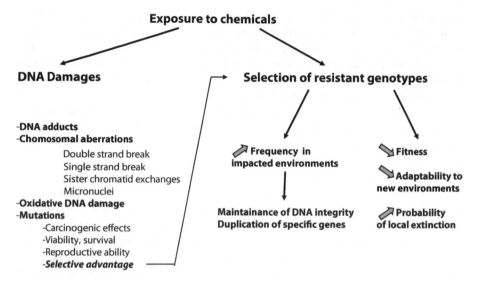

FIGURE 17.1   DNA response to chemical exposure and genetic adaptation.

organisms, exposure to chemicals can lead to DNA damage (Figure 17.1), the consequences of which may be limited by DNA repair (Peterson and Côté 2004). Mutations frequently have noxious effects, including carcinogenesis, and when affecting germinal tissues, they are heritable and can also affect future generations, provided that the offspring are viable and able to survive and reproduce. In some case, mutations can confer a selective advantage, leading to the selection of resistant genotypes.

Many more studies have been devoted to resistance in microorganisms, initially as a consequence of bacterial resistance to antibiotics in human medicine. In fact, bacterial resistance is a much more general phenomenon, as exemplified for soil bacteria in Chapter 14. Similarly, pesticide resistance in insects became a topic of concern when it was demonstrated that mosquitoes—pests very important as a vector of human diseases and a nuisance for the recreational use of coastal areas—became able to cope with DDT and other successive compounds created by the chemical industry to combat such acquired resistance (Chapter 16). In metalliferous sites where the total metal exposure can be up to ten- to one-thousand-fold higher than that of uncontaminated sites, only a small number of plant species have evolved tolerance to such concentrations, providing the focus of many evolutionary studies, in which it has been argued that metal tolerance could evolve rapidly following exposure to metal stress. The occurrence of both metallicolous and nonmetallicolous populations of rockcress *Arabidopsis halleri*, its high level of metal tolerance and accumulation, and its phylogenetic proximity with *A. thaliana*, providing numerous molecular tools for genetic studies, make *A. halleri* an excellent model species for investigating the evolutionary origin, intraspecific variability, and genetic architecture of metal tolerance and hyperaccumulation (Chapter 15).

17.3 WHAT ARE THE MAIN MECHANISMS INVOLVED IN TOLERANCE?

It is generally accepted that tolerance due to either physiological acclimation or genetic adaptation can result from limited uptake of contaminants, increased storage in nontoxic forms, or increased elimination favoured by biotransformation processes. In the case of metals, biodynamic modelling as developed by Rainbow and Luoma (Chapter 6) can identify which (if any) of the physiological parameters—uptake rate (from solution or diet), efflux rate (after uptake from solution or diet), or (by subtraction) detoxification rate (e.g., Croteau and Luoma 2009)—is significantly altered in a given metal-tolerant population, thereby identifying the mechanistic basis of the selection for metal tolerance in a particular population.

Tolerance seems generally to result from biological mechanisms involved in the normal metabolism of endogenous compounds. Concerning xenobiotics, these mechanisms are responding initially to normal constituents of the environment, at least at low levels, such as metals, petroleum hydrocarbons, or plant toxins. In polluted environments, detoxification enzymes and cellular ligands are strongly inducible, thus contributing to the organism's defence against contaminants. Genome sequencing programmes have revealed that metallothioneins (MTs), cytochrome P450s (CYPs), catalase, superoxide dismutases, heat shock proteins (HSPs), and ABC transporters constitute superfamilies of numerous genes that are highly conserved throughout their evolution.

The highly conservative character of MTs, together with their ubiquity (from yeast to higher mammals), leads to the conclusion that these proteins play an essential role in vital processes. At the moment, most published studies only report total MT concentrations, whereas the many isoforms of MT that coexist within organisms have been found to play different physiological roles. Molecular biological advances using MT gene amplification or duplication have confirmed that the functions of different isoforms are different, some of them being involved in metal homeostasis and others in nonessential metal detoxification. Expression of MT genes shows interspecific diversity. MT gene organisation and regulation sequences (metal-responsive elements (MREs)) need to be explored more deeply in each species to understand more fully the roles of the MTs. Exploring the role of untranslated regions as well as the MRE sequences is also needed to understand regulatory elements of MT synthesis. Novel genes coding for MTs are still being characterised (Amiard et al. 2006). Regulation of MT gene transcription is due to the presence of MREs, the core region of which is well conserved across species (Imbert et al. 1990). Multiple copies of MRE have been shown to activate MT gene transcription in parallel. Apart from studies of the expression of MT genes in the presence of particular toxic metals, new nuclear genetic markers have been developed, and extensive studies have now been carried out on the polymorphism of selected genes in relation to environmental stress factors such as pollution (Tanguy et al. 2002).

The number of ABC transporters represents a significant proportion of the genes in the species for which the complete genome has been sequenced, and an increasing number of ABC members are being discovered in numerous species. This indicates the important role of these transporters for living organisms (Chapter 10).

CYPs are ubiquist proteins. Their presence has been demonstrated in all organisms: bacteria (even archeobacteria studied by McLean et al. 1998), fungi and yeasts, protists, plants, and animals. Cytochrome P450 enzymes are involved in both synthetic and catabolic reactions. CYP enzymes metabolise endogenous substrates, including fatty acids, prostaglandins, leukotrienes, steroids, and vitamins, which can affect cellular functions (Nebert 1990). The activity of some P450s may be induced by numerous chemicals after their absorption by organisms. This pecularity may be considered an adaptive phenomenon provoked by the presence of these exogenous substances. Some of these inducer molecules enhance the level of the P450 enzymes, which then metabolise them. The resulting enhanced enzymatic activity may come from various regulation mechanisms at the mRNA level, as described in Chapter 8 (its transcription, persistence in the cell, and translation efficiency), and at the protein level (including its stabilisation, posttranslation modification, or degradation).

As a result of their elevated conservation and wide distribution across evolutionary scales, catalase and superoxide dismutases are considered to be components of an essential system evolved to allow organisms to cope with aerobic environments (Chapter 7; see also Fridowich 1975). All organisms, from bacteria to mammals, exposed to different environmental stressors, respond by synthesising a small number of highly conserved proteins called heat shock proteins (HSPs) that represent a general response of cells to various environmental stress proteotoxic situations (Chapter 9).

17.4 WHAT ARE THE EFFICIENCY LIMITS OF TOLERANCE?

17.4.1 INTERFERENCES BETWEEN CHEMICAL STRESS AND OTHER STRESSORS

Field and laboratory experiments have demonstrated that accumulated contaminants reduce the tolerance of mussels to aerial exposure, probably caused by a failure to reduce their scope for metabolic depression and inhibited ATP generation during anaerobiosis. Using the resistance of mussels to aerial exposure, Eertman and de Zwaan (1994) have proposed a test superimposing exposure to air, a natural stressor, over the effects of pollutant stressors. Viarengo et al. (1995) called the test "stress on stress response" and showed that mussels exposed to copper, dimethyl benzo anthracene (DMBA), and Aroclor 1254 have their capacity to survive in air significantly reduced. Mussel watch studies carried out in contaminated areas (Prince William Sound after the *Exxon Valdez* oil spill, Halifax Harbour [Canada], canals of Venice) show reduced survivability in air, which reveals a reduced status of health of the animals (Thomas et al. 1999; Hellou and Law 2003; Pampanin et al. 2005).

The reverse has also been clearly established (Chapter 5); animals that are submitted to natural stress in their environment are generally more sensitive to any additional chemical stress. In this chapter, the authors have focussed on organisms that are already living under extreme or fast changing natural conditions as occur in estuaries or at the limit of distribution of a species. Under such environmental conditions, it might be expected that, because of strongly developed (instantaneous) acclimation and (seasonal) acclimatisation capabilities, these species might better tolerate the additional stress of chemical disturbances. In fact, environmental stress factors like salinity, temperature, or food condition are able to enhance the effects of chemical stress in a number of marine bivalves, aquatic crustaceans, and fish. In the context of global climate change, special attention must be directed toward the effects of higher temperatures together with associated indirect changes (for instance, oxygen depletion, enhanced physiological activity) on the toxic effects of pollutants, since interaction between toxic and nontoxic stresses may accelerate the local extinction of species. Because coping with chemical stress induces an increased energy demand (Chapter 12), it is not surprising that food availability is among the natural factors able to modulate the tolerance of organisms to environmental contaminants.

17.4.2 COTOLERANCE

Cotolerance may occur when organisms or communities that have been exposed to one toxicant, but not to another one, become tolerant to both toxicants. It is most probable that cotolerance occurs for those compounds that have similar chemical structures and activities and induce common tolerance mechanisms. Depending on the kind of targeted activities (e.g., photosynthesis or substrate-induced respiration), cotolerance assessment can be variable, and the combination of different endpoints may provide a better approach (Chapter 4). In addition, cotolerance may arise due to the fact that genes for resistance to, or transformation of, different contaminants are found on the same mobile genetic element, such as a plasmid or a transposon, thus eliciting cotolerance to contaminants that are *a priori* unrelated structurally or functionally (Top and Springael 2003, and Wright et al. 2008 quoted in Chapter 14). Baker-Austin et al. (2006) propose that several mechanisms underlie the coselection process, including coresistance (different resistance determinants are present on the same genetic element) and cross-resistance (the same genetic determinant is responsible for resistance to antibiotics and metals).

The existence of cotolerance has been demonstrated in a number of cases, and it is particularly well documented for microbial resistance to metals and antibiotics. In the estuarine environment, Oger et al. (2003) reported eleven distinct *Staphylococcus* species (including the pathogenic *S. aureus*), and also bacteria belonging to the genera *Micrococcus* and *Halobacillus* carrying the cadA gene. This cadA determinant was mostly plasmid-borne in the *Staphylococcus* genus, and IS257 sequences, which are known to participate in antibiotic resistance gene dissemination in *S. aureus*, were found to be located near the cadA gene in sixteen of thirty-one cadmium-resistant *Staphylococcus* strains and one *Micrococcus* strain. Thus, IS257 could contribute to the dissemination of resistant genes among microbial communities. Stepanauskas et al. (2006) provided the first experimental evidence that the exposure of freshwater environments to individual metals and antibiotics selects for multiresistant microorganisms, including opportunistic human pathogens. However, in the experiments of Schmitt et al. (2006) with soil microcosms, cotolerance occurred only between antibiotics of the same group (oxytetracycline and tetracycline). Cotolerance between oxytetracycline and tylosin in soil microcosms exposed to oxytetracycline only was low, as was the reverse. The authors concluded that tolerance depended on the actual selection pressure instead of reflecting a more general pattern of multiresistance.

In an extensive study of metal tolerance in freshwater benthic algae, Takamura et al. (1989) showed that Cu-tolerant diatoms tended also to be Zn tolerant, whereas

Cd-tolerant Chlorophyceae tended also to be Zn resistant. A Cd-resistant strain of *Euglena gracilis* developed by Bariaud and Mestre (1984, quoted by Cosper et al. 1987) was cross-resistant to Co and Zn. Resistance to PCB and the concomitant development of cross-resistance to DDT was induced in a laboratory culture of a clone of the marine diatom *Ditylum brightwellii*; it was maintained for fifty generations in the absence of toxicant stress, indicating that the traits responsible for tolerance were genetically transmitted (Cosper et al. 1987). Cross-resistance to pesticides by insects is a serious point to be considered when aiming at pest control (Chapter 16).

Cotolerance has also been described for metals in fish (Xie and Klerks 2004). It would be too optimistic to consider that a species that is tolerant to a given environmental contaminant—or to natural stress—will also be tolerant to others. On the other hand, Klerks (1999, quoted by Wirgin and Waldman 2004) concluded that an increase in the number of contaminants present at a location may reduce the selection intensity for an individual contaminant.

17.4.3 SATURATION OF DEFENCE MECHANISMS

Tolerance is not universally guaranteed. It is well known that several biochemical mechanisms involved in the defence of organisms may be overwhelmed. In the laboratory, it is well documented that the relationship between doses and effects often fits well a bell-shaped curve, as exemplified for metallothionein (MT) and ethoxyresorufin *O*-deethylase (EROD) (see Figure 1.3).

Saturation has been also exemplified in the case of metallothionein induction in oysters, *Crassostrea gigas*. Experimental (Géret 2000) and field (Geffard 2001) results have been compared in oysters *C. gigas* analysed using the same laboratory procedures. Oysters from a clean site (Bay of Bourgneuf, France) were exposed to dissolved Cd (200 µg.L^{-1}), while other specimens were translocated for six months to a metal-rich estuary (Gironde, France). In the laboratory, Cd was accumulated strongly in the soft tissues. as shown for the gills, the concentration being about one order of magnitude higher than in the field (Figure 17.2). Despite this much more important enhancement of metal bioaccumulation under laboratory conditions, MT concentrations showed no greater changes than those observed in the oysters translocated into the metal-rich Gironde estuary.

Various polycyclic aromatic hydrocarbons are inducers of CYP1A1 in rat hepatocyte cultures, but the highest doses of dibenzo[a,h]anthracene, benzo[a]pyrene, benzo[k]fluoranthene, chrysene, and acenaphthylene evoked a saturation of the biotransformation mechanism with decreasing EROD activities. Further analysis revealed that the low efficacy of acenaphthylene as an inducer of CYP1A1 protein and EROD activity is due to its marked cytotoxicity (Till et al. 1999). According to the studies carried out in fish species from North American estuaries, it seems generally considered that resistance is associated with reduced inducibility of CYP1A. Enzymes of the cytochrome P4501A, particularly EROD, have long been considered as contributing to organic compound detoxification (Newman and Unger 2003). Since deleterious effects may result from biotransformation of these compounds, it is necessary to understand what other processes are involved in resistance (for more details and interpretation, see Chapter 8).

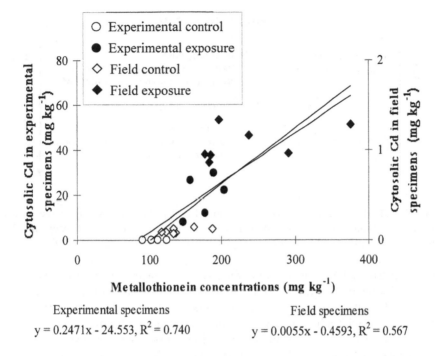

FIGURE 17.2 Relationship between cadmium and metallothionein concentrations in gills of oysters *Crassostrea gigas* contaminated with chronic doses of cadmium in the field or high doses in the laboratory.

In the case of HSPs, the existence of a plateau at increasing doses is reported for clams *Ruditapes decussatus* exposed to tributyltin (Solé et al. 2000). Other mechanisms of defence, such as the induction of catalase (CAT), glutathione S-transferase (GST), superoxide dismutase (SOD), and multixenobiotic resistance (MXR), show the same pattern with increasing concentrations/activities at moderate contaminant levels, followed by a plateau at intermediate levels, and finally a decrease at highly toxic levels (Allen and Moore 2004; Dagnino et al. 2007).

17.5 CONSEQUENCES OF TOLERANCE FOR ORGANISMS AND THEIR ENVIRONMENT

17.5.1 CONSEQUENCES AT THE INDIVIDUAL LEVEL

Allen and Moore (2004) and Dagnino et al. (2007) have proposed that different response profiles of biochemical defence mechanisms may be related to the development of a stress syndrome, assessed by using lysosomal stability, protein carbonyls, and cell death. In healthy organisms, levels of multidrug resistance and activity of superoxide dismutase are stable or slightly increasing. In stressed organisms, these responses reach their maximum, then decrease in parallel with progression of disease to a curable or incurable health status. This general pattern

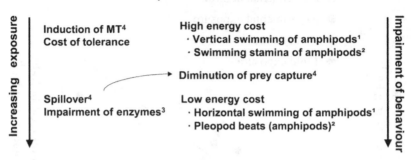

FIGURE 17.3 Relationship between behavioural and metabolic disruption: the case of aquatic invertebrates exposed to increasing levels of metals. ([1]Wallace and Estephan 2004; [2]Lawrence and Poulter 1998; [3]Neuberger-Cywiak et al. 2003; [4]Wallace et al. 2000).

may be described with more detail in the case of crustaceans submitted to increasing metal exposure (Figure 17.3). Coping with toxicants has an associated metabolic cost (Chapter 12). At low exposure, crustaceans are able to detoxify metals, but behaviours with a high energy cost are significantly impaired. With increasing exposure, the detoxification capacity of metallothioneins is overwhelmed, a process that has been termed spillover (Brown and Parsons 1978), and this results in increasing metal concentrations bound to high molecular weight proteins, i.e., enzymes. With the impairment of enzymes, toxic effects may appear, such as the diminution of prey capture observed in the grass shrimp *Palaemonetes pugio*. At this stage, even behaviours with lower energy cost may be disturbed.

In the case of stress proteins (HSPs), the relationship between their concentrations and stress indices (scope for growth, condition index) is not so clearly established, as discussed by Solé et al. (2000).

More generally, the level at which any saturation of defence mechanisms occurs is far from being precisely determined. Other ways of defence, such as avoidance behaviour, have only a limited protective role; it is for this reason that we have not designated a chapter to avoidance in this book. Tolerance, due to either acclimation or adaptation, can induce different kinds of costs (physiological costs in terms of energy metabolism and fitness, side effects of reactive metabolites, lesser ability to cope with new stress). Thus, Wirgin and Waldman (2004) ask provocatively if an adapted population is necessarily an unhealthy one?

Along the Atlantic coast of North America, populations of the fish *Microgadus tomcod* exhibit phenotypes that are resistant to aromatic hydrocarbon (AH) contaminants, including polychlorinated biphenyls (PCBs), polychlorinated dibenzo-*p*-dioxins (PCDDs), and polycyclic aromatic hydrocarbons (PAHs). However, reduced abundance, cancer, and truncated age structure were observed in Hudson River tomcod (Wirgin and Waldman 2004). In an invertebrate, the worm *Nereis diversicolor*, a decrease of abundance and fitness was observed in the multipolluted Seine estuary (France) despite the tolerance of this species to at least one class of contaminants (metals) and the inducibility of GST (involved in the biotransformation of organics) (Amiard-Triquet and Rainbow 2009). In such cases, the critical abundance—the specific density below which the durability of the population is at risk (Maltby et al. 2001)—may be reached within a short time period.

17.5.2 CONSEQUENCES FOR BIODIVERSITY AND ECOSYSTEM FUNCTIONING

The tolerance discussed by most papers in this volume is a result of populations in contaminated environments being subjected to selection pressure resulting in them becoming more tolerant to those contaminants. This evolutionary response can enable species to persist, and perhaps even thrive, and thus contribute to the diversity of species and a normal (or subnormal) functioning of ecosystems, particularly for biogeochemical cycles of both nutrients and contaminants and transfer of matter and energy in food webs. The inherent or induced tolerance of species enables them to persist, and therefore contributes to increasing the overall diversity of the communities in which they live. However, tolerance comes with costs, and it has frequently been noted that populations adapted to one set of environmental stressors have increased susceptibility to other stressors. Pollutants can act as powerful selective forces by altering genetic variability. If populations in uncontaminated areas have a wide variation in tolerance, and those inhabiting polluted areas are selected for tolerance, it would be expected that the genetic diversity in the polluted sites would be reduced. On the other hand, if contaminants are mutagenic, one might predict that polluted populations would have greater genetic diversity (Figure 17.1).

The responses to stress associated with tolerance (or not) are highly variable, depending on the species and, within the same species, the type of stressor. The role of tolerance in the conservation of communities in their initial stress-free status is limited to a relatively narrow gradient of contamination, since increasing levels of stress result in contrasting responses of susceptible versus resistant species/populations. In some cases, tolerance may not be beneficial for environmental conservation since it seems that invasive species are often more able to cope with stress than autochthonous ones (Chapter 11).

Because of the high variability of tolerance, it is important to examine in depth the responses of these species/taxa/functional groups, the role of which is crucial in order to avoid ecological disorders, for instance, bacterial communities (Chapter 14), primary producers such as microalgae in aquatic biofilms (Chapter 4) or metal-tolerant terrestrial plants (Chapter 15), bioturbating organisms such as earthworms in soils (Chapter 6) or annelids in mudflats (Mouneyrac et al. 2009; Denis et al. 2009), shredders responsible for the first steps of plant material degradation (Chapter 6), and organisms constituting important links in food webs (Chapter 13).

17.5.3 HUMAN HEALTH AND WELL-BEING

It is not good news that species that are threats to human health and well-being (bacterial pathogens, mosquitoes, agricultural pests, highly contaminated food species) have evolved tolerance. On the other hand, certain aspects of tolerance can have interesting applications, such in bioremediation or biofortification.

Some plants exhibit adaptation to very contaminated soils, and can accumulate large quantities of metals in a nontoxic form in their aboveground biomass. These are called hyperaccumulators. Members of the *Brassicaceae*, particularly *Thlaspi caerulescens* and *Arabidopsis halleri,* are well-known Zn and Cd hyperaccumulators and have been extensively studied in this respect (Chapter 15). Work has been done to identify the genes that influence metal tolerance and hyperaccumulation; in particular, a gene *HMA4* encoding a protein from the metal-transporting ATPase family and responsible for the root-to-shoot translocation of metals has been found in *Arabidopsis halleri* (Dräger et al. 2004). Unfortunately, such model species (*A. halleri* but also *A. thaliana*) are not relevant for direct applied use due to their low biomass production, which prevents their use in the phytoremediation of metal-polluted environments. They also cannot serve as crops, and therefore cannot directly increase the content of bioavailable nutrients in food.

Hyperaccumulators may provide biological tools for phytoremediation, which consists of the use of green plants either to remove the pollutants from soils through phytoextraction or to render them harmless through phytostabilisation. This simple and cost-effective concept fits with green technology, which corresponds more and more to citizens' demands. A review by Khan et al. (2009) presents the recent developments in the utilisation of soil phytoremediation, and metal remediation of water using plants is also well documented (Liao and Chang 2004; Krishnani and Ayyapan 2006).

Another investigation of the ability of plants to hyperaccumulate chemical (in particular mineral) elements from soils could lead to biofortification. This aims to increase micronutrient concentrations in the edible parts of plants through breeding. It is considered to be a cost-effective way to alleviate micronutrient malnutrition in rural populations in developing countries where the problem is most prevalent. Enhancement of trace element bioavailability in the rhizosphere, translocation from roots to shoots, and redistribution toward grain tissues are the obvious targets of biofortification (Zhao and McGrath 2009). In fact, significant progress can only be made through a better understanding of the underlying mechanisms of ion uptake, transport, and homeostasis in plants. Furthermore, any potential technologies should be evaluated under real conditions. This means assessment of trace element bioavailability to humans in any biofortified products and of potential downsides, such as enhanced accumulation of toxic metals (e.g., Cd), which may share the same transporters as essential elements (e.g., Fe, Zn, and Ca), and assessment of the efficacy and environmental risks of phytoextraction methods in contaminated soils under field conditions.

The joint FAO/WHO Food Standards Programme is aware of the problem of biofortified food. It has established a codex committee on nutrition and foods for special dietary use whose purpose is to provide guidance for the maintenance or improvement of the overall nutritional quality of foods through addition of essential nutrients for the purpose of fortification, restoration, and nutritional equivalence. Human beings, as top predators of both aquatic and terrestrial food chains, are particularly at risk of eating biofortified foods, whether hyperaccumulator plants, breast milk, or fish. For instance, it is necessary to consider the benefit-risk balance of the consumption of seafood products, which are sources of both toxic methyl mercury and omega-3 fatty acids that improve cardiovascular health, as underlined in Chapter 13. Fish are also known potentially to accumulate lipophilic persistent organic pollutants (dioxins and PCBs) in fat. It is of concern to estimate the optimal amount of fish fat to be consumed in order to achieve the nutritional requirement without exceeding any toxicological thresholds. Taking into account the consumption of fish from the western coastal areas of France, Verger et al. (2008) showed that the nutritional recommendation for long-chain n-3 polyunsaturated fatty acids (PUFAs n-3, 500 mg d^{-1}) is generally compatible with the toxicological threshold of total dioxins, whereas higher intakes of PUFA are likely to be associated with contaminant doses exceeding the threshold (Figure 17.4).

It is not the aim of this book to develop the question of bacterial resistance to antibiotics in medicine, but interactions between antibiotics and contaminants in the environment are evoked in terms of cotolerance (Section 17.4.2). Moreover, in the field of human health, the tolerance induced by pesticides in insects that are vectors of serious and even lethal illnesses is of great importance, such as exemplified in the emergence of mosquito species resistant to insecticides widely used in malaria and dengue fever control. The potential for resistance to develop in vectors has been apparent since the 1950s, but the scale of the problem has been poorly documented. Few new public health insecticides have been developed for control of disease vectors for the past three decades, and without good stewardship, these insecticides will cease to be effective for vector control. The management of insecticide resistance in field populations of insects is crucial to the long-term sustainability

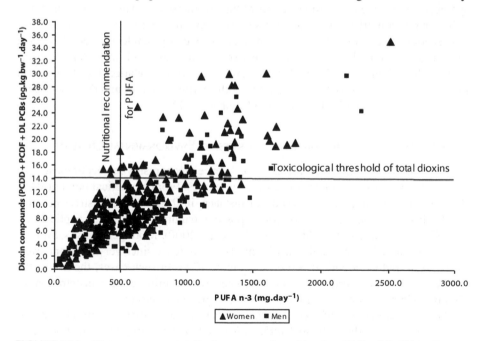

FIGURE 17.4   Dietary exposure to dioxin compounds and intake of LC n-3 PUFA by French fish consumers. (Adapted from Verger, P., et al., *Food Addit. Contam.*, 25, 765–71, 2008.)

of insecticide-based disease control campaigns (Clark et al. 2009), and the same situation has arisen for agricultural pests (Chapter 16).

17.6 OPERATIONAL CONSEQUENCES OF TOLERANCE

To improve environmental quality assessment as well as risk assessment, it is necessary to put more "eco" into ecotoxicological tools (biomarkers, bioassays) by using keystone species for the area being assessed (Chapman 2002). Because organisms chronically exposed to contaminants in their environment may have developed tolerance (either genetically or physiologically based), the origin of the specimens involved in a laboratory experiment can consistently influence the response to chemical stress. In addition, the so-called ecotoxicological tests are performed on organisms that are easily cultured in the laboratory and genetically stable. If animals from the real world are used to improve the realism of toxicological assessment, the implications of the occurrence of tolerance must be carefully taken into account in order to avoid false negatives or positives in our conclusions.

Depending on the tolerance of species to different kinds of contaminants and to different levels of exposure, it is indispensable to take into account a sufficient number of biological models. Because the determination of many different biomarkers/toxicological parameters in different species presents great resource implications, it is necessary to find relevant criteria for the selection of models. Thus, the choice of species to be studied must consider (1) their representativeness of their environment and (2) their role in the structure and functioning of ecosystems. By focussing on key species, it is assumed that impairments of their responses in terms of biochemical or behavioural biomarkers, reproductive success, growth, or survival will reveal a risk of cascading deleterious effects right through to the ecosystem level. Disturbances at the supra-individual levels affecting these organisms may result in ecological disorders, such as changes in biogeochemical cycles and trophic relationships that govern nutrient/food availability (with respect to both energy resources and the risk of contaminant transfer) (Amiard-Triquet and Rainbow 2009).

17.6.1 THE ROLE OF BIOMARKERS IN ASSESSING ENVIRONMENTAL QUALITY

We have discussed above that saturation of defence mechanisms can occur for several core biomarkers used in biomonitoring programmes. From a practical point of view, the existence of a plateau, observed, for instance, in the response of stress proteins (HSPs) in clams *Ruditapes decussatus* exposed to tributyltin, limits the applicability of HSPs as biomarkers in field studies (Solé et al. 2000). More generally, saturation poses a problem for the use of many biomarkers of defence since very different levels of exposure induce identical responses (Figure 17.5). In such cases, the use of the methodology of biomarkers must be reinforced by determining the concentrations of chemicals known as potential inducers of each given biomarker.

Another possible bias of toxicological studies is revealed after the careful comparison of responses of the killifish *Fundulus heteroclitus* belonging to populations from North American estuaries, differing by the degree and the nature of the major contaminants (for details, see Chapter 8). In a number of cases, when the expression of genes

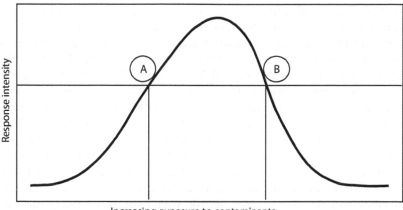

Increasing exposure to contaminants

FIGURE 17.5   Response intensity of biomarkers (MT, EROD, HSPs, CAT, GST, SOD, MXR) along a pollution gradient. In case of saturation of the defence mechanisms, the response intensity B may be identical to A, obtained for a much lower exposure.

or the induction of detoxificatory systems is at a high level in the medium of origin, experimental stress will hardly be able to provoke any additional significant response.

EROD and, more recently, following the development of molecular biology, CYP genes are frequently used as core biomarkers of exposure to organic xenobiotics. However, the example of Atlantic tomcod *Microgadus tomcod* and killifish *Fundulus heteroclitus* (Chapters 3 and 8) underlines the fact that, depending on species and sites, responses may be contrasted with the possibility of false negatives or positives when considered for a biomonitoring programme. Also, observations on hepatic carcinoma reveal interspecific differences in sites polluted by the same major class of contaminants (PCBs in Hudson River or New Bedford Harbor). Again, population effects (abundance) are marked in the Atlantic tomcod and the killifish. In terms of environmental quality assessment, the use of one species or the other can result in contrasted interpretations, leading to either over- or underevaluations. Similarly, when considering taxonomically closely related species sharing similar feeding habits (oysters and mussels) or similar niches (ragworm *Nereis diversicolor* and clam *Scrobicularia plana*), different interspecific responses must be taken into account for the interpretation of biomarkers (Chapter 3).

17.6.2   Ecotoxicity Tests

Many toxicologists aiming to find the details of mechanisms of ecotoxicity at the molecular, cellular, or physiological level often use high experimental doses of exposure, expecting clearer results. This procedure is clearly irrelevant since the use of such acute doses immediately overwhelms any detoxificatory processes (for instance, the induction of biodegradation or antioxidant enzymes, biomineralisation of toxic metals, or the induction of intracellular ligands allowing the sequestration of the contaminant in nontoxic form) and then does not allow organisms to cope with pollutants as they might do in the field. At sublethal, more realistic doses, detoxificatory processes are functional, and ecotoxicity studies carried out under these conditions are more representative of effects of chronic exposure in the field.

Both acute and sublethal tests have been conducted in different countries, acute tests allowing a first screening of the toxic concentrations. In the United States, the American Society for Testing and Materials (ASTM) has published numerous reports describing standard practices for conducting acute toxicity tests with fish, macroinvertebrates, and amphibians, since the 1980s (e.g., ASTM 1992, 1995, 2004). Bioassays follow standardised international protocols (e.g., OECD, ISO, USEPA, etc.). The European Community has launched the REACH programme concerning the toxicity of chemicals. REACH considers registration, evaluation, authorisation, and restriction of chemicals. The goal is to assess nearly all chemical compounds at the European or national level, and then to take into account local conditions and assess nonspecific risks from a site, representing European or national environment conditions. Toxicity tests include the freshwater compartment (including sediment), microorganisms of wastewater treatment plants, the terrestrial compartment (and subterrestrial waters), the atmosphere, top predators, and the marine environment. In all cases, the main point is that toxicity tests involve bioassays performed in the laboratory (with a single biological species, bioassays follow a European norm). These assays have to deal with different trophic levels. For instance, if aquatic toxicity is assessed, the bioassay must encompass algae, small crustaceans, and fish. Acute toxicity is studied by a battery of tests, such as growth inhibition of algae, acute toxicity to invertebrates, inhibition of microbial activity (leading to the evaluation of the median effect concentration (EC50), the lethal concentration (LC50), and the inhibition concentration (IC50)), whereas sublethal effects are observed by prolonged toxicity study with the cladoceran *Daphnia magna* (21 days) and further toxicity studies with fish and growth tests on higher plants. Sublethal tests lead to the evaluation of a no-observed-effect concentration (NOEC) and a lowest-observed-effect concentration (LOEC). Tests are widely used; however, some limitations can be evoked, such as sensitivity changes in strains kept over the long term in the laboratory or the concurrent use of organisms reared/cultivated in the laboratory or taken from the natural environment. Semaan et al. (2001) compared tolerance (or nontolerance) to uranium in the cladoceran *Moinodaphnia macleayi* from three populations: one maintained for ten years in the laboratory, a reference population, and an impacted population located on a uranium mine site in Australia. Although the species was more sensitive to uranium than any other cladoceran tested, no significant differences in tolerance were found between the populations. In this case, the use of laboratory stock populations in further regulatory testing is justified. Other studies, however, question the relevance of toxicity data obtained with organisms that have adapted or acclimated to laboratory conditions containing unnaturally low levels of trace elements. Muyssen and Janssen (2001) tested subsequent generations of two field-collected natural clonal lineages versus a clone that had been raised for many generations in the absence of natural levels of zinc. The field-collected *D. magna* clones were initially more tolerant to acute zinc toxicity (up to a factor of 4), but gradually lost their Zn tolerance and became more similar to the laboratory clone. Another study by the same team showed significant copper acclimation of laboratory *D. magna* cultures over three generations. These results clearly indicate that the background bioavailable copper concentrations present in culture media have to be considered in the evaluation of the results of toxicity tests (Bossuyt and Janssen 2003).

If organisms used in ecotoxicity tests are collected in the wild (for instance, with the aim of having species representative of regional conditions in order to put more "eco" into ecotoxicology (Chapman 2002)), the interpretation of the results must be soundly informed of any possible tolerance of the biological model. The fact that reference toxicants are commonly used in standardised tests may help to avoid this potential bias, but only in the case of species for which the sensitivity is well documented, with background data on their response to the reference toxicant. For more examples, refer to Chapter 3.

17.6.3 SAFETY FACTORS

The use of results of bioassays to assess the concentration of a chemical below which no deleterious effects will occur for many species in the field need to use uncertainty or safety factors. In the case of aquatic environments in Europe, the calculation of predicted no-effect concentrations (PNECs) is described in detail in a Technical Guidance Document (TGD 2003). Similar procedures exist in all developed countries. Considering the presence of tolerance in certain tests organisms, do safety factors currently used in agreement with official recommendations ensure the effectiveness of such procedures? Johnston (Chapter 2) considers that, in general, the development of differential tolerance may result in populations that are more than twice as tolerant to contaminant exposure than unimpacted populations. If differences in tolerance ratios are relatively small, then they may be encompassed by conventional safety factors used to establish protective guidelines. This is consistent with the results of Medina et al. (2007), who argued that existing safety margins in regulations are likely to be protective even with the complicating effect of tolerant populations. However, given the large range in tolerance ratios reported here (Nacci et al. 2009; Chapters 2 and 3), such safety margins may, in fact, not be adequate for all contaminant classes, and it is crucial that we continue to investigate the extent to which tolerance can modify an organism's response.

17.6.4 CONCURRENT USE OF SENSITIVE AND TOLERANT SPECIES

Interspecific differences of tolerance provide opportunities for the use of various species according to the different tasks that contribute to a sound evaluation of the health status of environments at risk. Monitoring of the chemical quality of environment is often based on chemical analysis of numerous contaminants in bioaccumulators such as mussels and related species in the coastal environment or lichens in continental areas. In highly contaminated sites, sensitive species may have become extinct. Consequently, for coastal research, Funes et al. (2006) recommend oysters as model sentinels for adaptation studies, while mussels, more sensitive to pollution, can be more useful as indicator organisms in communities. In addition, the tolerance of oysters provides a biological matrix, even in highly contaminated environments, very useful for the determination of contaminant concentrations as well

TABLE 17.2

Contrasted Responses of Two Endobenthic Invertebrates (the Clam Scrobicularia plana and the Ragworm *Nereis diversicolor*) to Environmental Conditions in French Estuaries, the Seine (Highly Polluted) and the Loire (Moderately Polluted) Estuaries

	Scrobicularia plana		*Nereis diversicolor*	
	Loire	Seine	Loire	Seine
Biochemical Markers				
Metallothionein-like proteins	↗	↗	⊘	⊘
Glutathione S-transferase	↗	↗	NS	↗
Acetylcholine esterase	↘	↘	↘	↘
Digestive enzymes	NS	↘	↘	↘
Physiological Markers				
Energy reserves	NS	NS	NS	↘
Condition index	NS	NS	↘	↘
Behavioural Markers				
Burrowing rate	↘	↘	NS	↘
Prey capture	⊘	⊘	↘	⊘

Note: The response is increased ↗ or decreased ↘ by comparison with a reference site as clean as possible.

NS: no significant reference with the reference.

⊘: not determined.

as biomarkers, provided that interspecific differences are well established (RNO 2006).

The comparison of paired species for their responses to environmental contamination does not always reveal situations as contrasting, as has been shown for oysters and mussels. Results recently obtained in our research group on *Scrobicularia plana* (Fossi Tankoua, personal communication), aggregated with previous work on *Nereis diversicolor* (Kalman et al. 2009; Mouneyrac et al. 2009), are presented in Table 17.2. In the less contaminated Loire estuary, *N. diversicolor* is less efficient than *S. plana* in revealing the presence of inducers of the phase II enzyme glutathione S-transferase (Chapter 8), and it is also less sensitive when burrowing behaviour is tested. On the other hand, they are more sensitive than *S. plana* when considering digestive enzyme activities and condition indices. As a whole, the fate of local populations up to potential extinction depends both on the nature of the pollutants involved (mainly metals or contaminant mixtures) and the levels of exposure (see Chapter 3 for details). In terms of representativeness of their environment (estuarine and coastal mudflats) and of their role in the structure and functioning of these ecosystems, both species are relevant models. In practice, it is probably impossible to use such nearly redundant biological models because of great resource implications. But again, keep in mind that apparently small differences may have serious consequences.

Last but not least, interspecific variability of tolerance is at the basis of a number of methodologies widely used in assessing environmental quality: pollution-induced community tolerance, which is very relevant for microbial ecology (Chapters 4 and 14), and the monitoring of macrofaunal communities with bioindicators and biotic indices (Chapter 3).

REFERENCES

Allen, J. J., and M. N. Moore. 2004. Environmental prognostics: Is the current use of biomarkers appropriate for environmental risk evaluation? *Mar. Environ. Res.* 58:227–32.

Amiard, J. C., et al. 2006. Metallothioneins in aquatic invertebrates: Their role in metal detoxification and their use as biomarkers. *Aquat. Toxicol.* 76:160–202.

Amiard-Triquet, C., and P. S. Rainbow, 2009. *Environmental assessment of estuarine ecosystems. A case study.* Boca Raton, FL: CRC Press.

ASTM. 1992. *Standard practice for conducting an interlaboratory study to determine the precision of a test method.* Philadelphia: American Society for Testing and Materials.

ASTM. 1995. *Standard test methods for measuring the toxicity of sediment-associated contaminants by benthic invertebrates.* EI706–95, Technical Report. Philadelphia: American Society for Testing and Materials.

ASTM. 2004. *Standard guide for conducting static acute toxicity tests starting with embryos of four species of saltwater bivalve molluscs.* E724–1998/2004. Philadelphia: American Society for Testing and Materials.

Baker-Austin, C., et al. 2006. Co-selection of antibiotic and metal resistance. *Trends Microbiol.* 14:176–82.

Ballihaut, G., et al. 2004. Analysis of the adaptation to alkanes of the marine bacterium *Marinobacter hydrocarbonoclasticus* sp17 by two dimensional gel electrophoresis. *Aquat. Living Resour.* 17:269–72.

Berthet, B., J. C. Amiard, C. Amiard-Triquet, M. Martoja, and A. Y. Jeantet. 1992. Bioaccumulation, toxicity and physico-chemical speciation of silver in bivalve molluscs: Ecotoxicological and health consequences. *Sci. Total Environ.* 125:97–122.

Blanck, H., S. A. Wängberg, and S. Molander. 1988. Pollution-induced community tolerance— A new tool. In *Function testing of aquatic biota for estimating hazards of chemicals,* ed. J. J. Cairns and J. R. Pratt, 219–30. STP 998. Philadelphia: American Society for Testing and Materials.

Bossuyt, B. T. A., and C. R. Janssen. 2003. Acclimation of *Daphnia magna* to environmentally realistic copper concentrations. *Comp. Biochem. Physiol.* 136C:253–64.

Brown, D. A., and T. R. Parsons. 1978. Relationship between cytoplasmic distribution of mercury and toxic effects to zooplankton and chum salmon (*Oncorhynchus keta*) exposed to mercury in a controlled ecosystem. *J. Fish. Res. Bd. Can.* 35:880–84.

Chapman, P. M. 2002. Integrating toxicology and ecology: Putting the "eco" into ecotoxicology. *Mar. Pollut. Bull.* 44:7–15.

Clark, J. M., J. R. Bloomquist, and H. Kawada. 2009. *Advances in human vector control.* Vol. 1014, American Chemical Society Symposium Series. Washington, DC: American Chemical Society.

Cosper, E. M., et al. 1987. Induced resistance to polychlorinated biphenyls confers cross-resistance and altered environmental fitness in a marine diatom. *Mar. Environ. Res.* 23:207–22.

Croteau, M. N., and S. N. Luoma. 2009. Predicting dietborne metal toxicity from metal influxes. *Environ. Sci. Technol.* 43:4915–21.

Dagnino, A., et al. 2007. Development of an expert system for the integration of biomarker responses in mussels into an animal health index. *Biomarkers* 12:155–72.

Denis, L., et al. 2009. Dynamic diagenetic modelling and impacts of biota. In *Environmental assessment of estuarine ecosystems—A case study*, ed. C. Amiard-Triquet and P. S. Rainbow, 303–22. Boca Raton, FL: CRC Press.

Eertman, R. H. M., and A. de Zwaan. 1994. Survival of the fittest: Resistance of mussels to aerial exposure. In *Biomonitoring of coastal waters and estuaries*, ed. K. J. M. Kramer, 269–82. Boca Raton, FL, CRC Press.

Fridovich, I. 1975. Superoxide dismutases. *Annu. Rev. Biochem.* 44:147–59.

Funes, V., et al. 2006. Ecotoxicological effects of metal pollution in two mollusc species from the Spanish South Atlantic littoral. *Environ. Pollut.* 139:214–23.

Geffard, A. 2001. Réponses du biota à la contamination polymétallique d'un milieu estuarien, la Gironde, Fr: exposition, imprégnation, induction d'une protéine de détoxication, la métallothionéine, impact au niveau individuel et populationnel. PhD thesis, University of Nantes.

Géret, F. 2000. Synthèse de métallothionéines chez deux bivalves (l'huître et la moule) en réponse à une contamination métallique par voie directe et par voie trophique. PhD thesis, University of Nantes.

Hellou, J., and R. J. Law. 2003. Stress on stress response of wild mussels, *Mytilus edulis* and *Mytilus trossulus*, as an indicator of ecosystem health. *Environ. Pollut.* 126:407–16.

Herwig, R. P., J. P. Gray, and D. P. Weston. 1997. Antibacterial resistant bacteria in surficial sediments near salmon net-cage farms in Puget Sound, Washington. *Aquaculture* 149:263–83.

Imbert, J., et al. 1990. Regulation of metallothionein gene by metals. *Adv. Inorg. Biochem.* 8:139–64.

Indeherberg, M. B. M., N. M. Van Straalen, and E. R. Schockaert. 1999. Combining life-history and toxicokinetic parameters to interpret differences in sensitivity to cadmium between populations of *Polycelis tenuis* (Platyhelminthes). *Ecotoxicol. Environ. Saf.* 44:1–11.

Kalman, J., et al. 2009. Assessment of the health status of populations of the ragworm *Nereis diversicolor* using biomarkers at different levels of biological organisation. *Mar. Ecol. Pgor. Ser.* 393:55–67.

Khan, M. S., et al. 2009. Role of plant growth promoting rhizobacteria in the remediation of metal contaminated soils. *Environ. Chem. Lett.* 7:1–19.

Klerks, P. L. 1999. The influence of contamination complexity on adaptation to environmental contaminants. In *Genetics and ecotoxicology*, ed. V. E. Forbes, 103–21. Philadelphia: Taylor & Francis.

Klerks, P. L., and J. S. Weis. 1987. Genetic adaptation to heavy metals in aquatic organisms: A review. *Environ. Pollut.* 45:173–205.

Kovatch, C. E., et al. 2000. Tolerance and genetic relatedness of three meiobenthic copepod populations exposed to sediment-associated contaminant mixtures: Role of environmental history. *Environ. Toxicol. Chem.* 19:912–19.

Krishnani, K. K., and S. Ayyappan. 2006. Heavy metals remediation of water using plants and lignocellulosic agrowastes. *Res. Environ. Contam. Toxicol.* 188:59–84.

Lawrence, A. J., and C. Poulter. 1998. Development of a sub-lethal pollution bioassay using the estuarine amphipod *Gammarus duebeni*. *Water Res.* 32:569–78.

Liao, S. W., and W. L. Chang. 2004. Heavy metal phytoremediation by water hyacinth at constructed wetlands in Taiwan. *J. Aquat. Plant Manage.* 42:60–68.

Malik, A. 2004. Metal bioremediation through growing cells. *Environ. Int.* 30:261–78.

Maltby, L., et al. 2001. Linking individual-level responses and population-level consequences. In *Ecological variability: Separating natural from anthropogenic causes of ecosystem impairment*, ed. D. J. Baird and G. A. J. Burton, 27–82. Pensacola, FL: Society of Environmental Toxicology and Chemistry (SETAC).

Marchand, J., et al. 2004. Physiological cost of tolerance to toxicants in the European flounder *Platichthys flesus*, along the French Atlantic Coast. *Aquat. Toxicol.* 70:327–43.

McLean, M. A., et al. 1998. Characterization of a cytochrome P450 from the acidothermophilic archea *Sulfolobus solfataricus*. *Biochem. Biophys. Res. Commun.* 252:166–72.

Medina, M. H., J. Correa, and C. Barata. 2007. Micro-evolution due to pollution: Possible consequences for ecosystem responses to toxic stress. *Chemosphere* 67:2105–14.

Mouneyrac, C., et al. 2009. Linking energy metabolism, reproduction, abundance and structure of *Nereis diversicolor* populations. In *Environmental assessment of estuarine ecosystems—A case study*, ed. C. Amiard-Triquet and P. S. Rainbow, 159–81. Boca Raton, FL: CRC Press.

Muyssen, B. T. A., and C. R. Janssen. 2001. Multigeneration zinc acclimation and tolerance in *Daphnia magna*: Implications for water-quality guidelines and ecological risk assessment. *Environ. Toxicol. Chem.* 20:2053–60.

Nacci, D. E., et al. 2002. Effects of benzo[a]pyrene exposure on a fish population resistant to the toxic effects of dioxin-like compounds. *Aquat. Toxicol.* 57:203–15.

Nacci, D. E., et al. 2009. Evolution of tolerance to PCBs and susceptibility to a bacterial pathogen (*Vibrio harveyi*) in Atlantic killifish (*Fundulus heteroclitus*) from New Bedford Harbor (MA, USA) harbor. *Environ. Pollut.* 157:857–64.

Nebert, D. W. 1990. Drug metabolism: Growth signal pathways. *Nature* (London) 347:709–10.

Neuberger-Cywiak, L., Y. Achituv, and E. M. Garcia. 2003. Effects of zinc and cadmium on the burrowing behavior, LC_{50} and LT_{50} on *Donax trunculus* Linnaeus (*Bivalvia-Donacidae*). *Bull. Environ. Contam. Toxicol.* 70:713–22.

Newman, M. C., and M. A. Unger. 2003. *Fundamentals of ecotoxicology*. Boca Raton, FL: Lewis Publishers.

Oger, C., J. Mahillon, and F. Petit. 2003. Distribution and diversity of a cadmium resistance determinant (*cadA*) and occurrence of IS257 insertion sequences in staphylococcal bacteria isolated from a contaminated estuary (Seine, France). *FEMS Microbiol. Ecol.* 43:173–83.

Pampanin, D. M., et al. 2005. Physiological measurements from native and transplanted mussel (*Mytilus galloprovincialis*) in the canals of Venice. Survival in air and condition index. *Comp. Biochem. Physiol.* 140A:41–52.

Peterson, C. L., and J. Côté. 2004. Cellular machineries for chromosomal DNA repair. *Genes Dev.* 18:602–16.

Rainbow, P. S. 1998. Phylogeny of trace metal accumulation in crustaceans. In *Metal metabolism in aquatic environments*, ed. M. J. Bebianno and W. J. Langston, 285–319. London: Chapman & Hall.

RNO. 2006. *Surveillance de la qualité du milieu marin*. Paris: Ministère de l'Environnement and Institut Français de Recherche pour l'Exploitation de la Mer (Ifremer).

Sardessai, Y., and S. Bhosle. 2002. Tolerance of bacteria to organic solvents. *Res. Microbiol.* 153:263–68.

Schmitt, H., et al. 2006. On the limits of toxicant-induced tolerance testing: Cotolerance and response variation of antibiotic effects. *Environ. Toxicol. Chem.* 25:1961–68.

Semaan, M., D. A. Holdway, and R. A. Van Dam. 2001. Comparative sensitivity of three populations of the cladoceran *Moinodaphnia macleayi* to acute and chronic uranium exposure. *Environ. Toxicol.* 16:365–76.

Solé, M., Y. Morcillo, and C. Porte. 2000. Stress-protein response in tributyltin-exposed clams. *Bull. Environ. Contam. Toxicol.* 64:852–58.

Stepanauskas, R., et al. 2006. Coselection for microbial resistance to metals and antibiotics in freshwater mesocosms. *Environ. Microbiol.* 8:1510–14.

Takamura, N., F. Kasai, and M. M. Watanabe. 1989. Effects of Cu, Cd and Zn on photosynthesis of freshwater benthic algae. *J. Appl. Phycol.* 1:39–52.

Tanguy, A., et al. 2002. Polymorphism of metallothionein genes in the pacific oyster *Crassostrea gigas* as a biomarker of response to metal exposure. *Biomarkers* 7:439–50.

TGD. 2003. Technical Guidance Document on Risk Assessment in support of Commission Directive 93/67/EEC on Risk Assessment for new notified substances, Commission Regulation (EC) n° 1488/94 on Risk Assessment for existing substances and Directive 98/8/EC of the European Parliament and of the Council concerning the placing of biocidal products on the market. EUR 20418 EN/2. European Commission, Joint Research Centre.

Till, M., et al. 1999. Potency of various polycyclic aromatic hydrocarbons as inducers of CYP1A1 in rat hepatocyte cultures. *Chem.-Biol. Interact.* 117:135–50.

Thomas, R. E., P. M. Patricia, and S. D. Rice. 1999. Survival in air of *Mytilus trossulus* following long-term exposure to spilled *Exxon Valdez* crude oil in Prince William Sound. *Comp. Biochem. Physiol.* 122C:147–52.

Verger, P., et al. 2008. Balancing the risk of dioxins and polychlorinated biphenyls (PCBs) and the benefit of long-chain polyunsaturated fatty acids on the n-3 variety of French fish consumers in western coastal areas. *Food Addit. Contam.* 25:765–71.

Viarengo, A., et al. 1995. Stress on stress response: A simple monitoring tool in the assessment of a general stress syndrome in mussels. *Mar. Environ. Res.* 39:245–48.

Wallace, W. G., and A. Estephan. 2004. Differential susceptibility of horizontal and vertical swimming activity to cadmium exposure in a gammaridean amphipod (*Gammarus lawrencianus*). *Aquat. Toxicol.* 69:289–97.

Wallace, W. G., T. M. Hoexum Broewer, and M. Brouwer. 2000. Alterations in prey capture and induction of metallothioneins in grass shrimp fed cadmium-contaminated prey. *Environ. Toxicol. Chem.* 19:962–71.

Wirgin, I., and J. R. Waldman. 2004. Resistance to contaminants in North American fish populations. *Mutat. Res.-Fund. Mol. M.* 552:73–100.

Wuertz, S., et al. 1991. Tributyltin-resistant bacteria from estuarine and freshwater sediments. *Appl. Environ. Microbiol.* 57:2783–89.

Xie, L., and P. L. Klerks. 2004. Fitness cost of resistance to cadmium in the least killifish (*Heterandria formosa*). *Environ. Toxicol. Chem.* 23:1499–503.

Zhao, F., and S. McGrath. 2009. Biofortification and phytoremediation. *Curr. Opin. Plant Biol.* 12:373–80.

Index